Lecture Notes in Networks and Systems 1592

The series "Lecture Notes in Networks and Systems" publishes the latest developments in Networks and Systems—quickly, informally and with high quality. Original research reported in proceedings and post-proceedings represents the core of LNNS.

Volumes published in LNNS embrace all aspects and subfields of, as well as new challenges in, Networks and Systems.

The series contains proceedings and edited volumes in systems and networks, spanning the areas of Cyber-Physical Systems, Autonomous Systems, Sensor Networks, Control Systems, Energy Systems, Automotive Systems, Biological Systems, Vehicular Networking and Connected Vehicles, Aerospace Systems, Automation, Manufacturing, Smart Grids, Nonlinear Systems, Power Systems, Robotics, Social Systems, Economic Systems and other. Of particular value to both the contributors and the readership are the short publication timeframe and the worldwide distribution and exposure which enable both a wide and rapid dissemination of research output.

The series covers the theory, applications, and perspectives on the state of the art and future developments relevant to systems and networks, decision making, control, complex processes and related areas, as embedded in the fields of interdisciplinary and applied sciences, engineering, computer science, physics, economics, social, and life sciences, as well as the paradigms and methodologies behind them.

Indexed by SCOPUS, EI Compendex, INSPEC, WTI Frankfurt eG, zbMATH, SCImago.

All books published in the series are submitted for consideration in Web of Science.

For proposals from Asia please contact Aninda Bose (aninda.bose@springer.com).

Daniela Doina Cioboată · José Machado
Editors

International Conference on Reliable Systems Engineering (ICoRSE) - 2025

Springer

Editors
Daniela Doina Cioboată
Bucharest, Romania

José Machado
School of Engineering
University of Minho
Guimarães, Portugal

ISSN 2367-3370 ISSN 2367-3389 (electronic)
Lecture Notes in Networks and Systems
ISBN 978-3-032-02507-4 ISBN 978-3-032-02508-1 (eBook)
https://doi.org/10.1007/978-3-032-02508-1

This Springer imprint is published by the registered company Springer Nature Switzerland AG
The registered company address is: Gewerbestrasse 11, 6330 Cham, Switzerland

Organization

Scientific Committee

Karolina Macuchova	HiLASE Centre, Institute of Physics CAS, Czech Republic
Umidjon Mardonov	Tashkent State Technical University, Uzbekistan
Mihai Gabriel Matache	National Institute of R&D for Machines and Installations Designed for Agriculture and Food Industry-INMA, Romania
Paulo E. Miyagi	University of Sao Paulo, Brazil
Vlad Mureşan	Technical University of Cluj-Napoca, Romania
Dumitru Mnerie	«Politehnica» University of Timişoara, Romania
Mihai Mărgăritescu	INCDMTM, Romania
Constantin Niţu	«Politehnica» University of Bucharest, Romania
Adrian Olaru	«Politehnica» University of Bucharest, Romania
Vasile Păiş	Research Institute for Artificial Intelligence "Mihai Drăgănescu" of the Romanian Academy, Romania
Andrei Păun	Research Institute for Artificial Intelligence "Mihai Drăgănescu" of the Romanian Academy and Faculty of Mathematics and Computer Science of the University of Bucharest, Romania
Lukas Prasol	Ansbach University of Applied Sciences, Germany
Pierluigi Rea	University of Cagliari (CA), Italy
Ciprian Rizescu	«Politehnica» University of Bucharest, Romania
Michal Rogalewicz	«Politecknica» University of Poznan, Poland
Adriano A. Santos	Polytechnic of Porto, Portugal
Vladimir Serbezov	Technical University of Sofia, Bulgaria
Laurenţiu Slătineanu	"Gheorghe Asachi" Technical University of Iasi, Romania
Natalia Smetankina	National Academy of Sciences of Ukraine, Ukraine
Dănuţ Stanciu	INCDMTM, Romania
Ion Stiharu	«Concordia» University, Montreal, Canada
Mihail Tîţu	«Lucian Blaga» University of Sibiu, Romania
Michael Todorov	Technical University of Sofia, Bulgaria
Pedro Torres	Polytechnic Institute of Castelo Branco, Portugal
Justyna Trojanowska	Poznan University of Technology
Yevgen Tsegelnyk	«O.M. Beketov» National University of Urban Economy in Kharkiv, Ukraine
Viktoriia Serhiivna Vakal	Research Institute of Mineral Fertilizers and Pigments of Sumy State University, Ukraine
Alexander Skvorchevsky	Research Center of Robotics, Mechatronics and Production Informatization, Ukraine

Contents

Analysis of Thermocouple Measurement Errors in Various Contact Methods During Turning Operations

Umidjon Mardonov[1](✉), Ye Yang[2], Xiaojun Wang[2], Otabek Khasanov[1], Sarvar Tursunbaev[1], and Shakhrizoda Ozodova[1]

[1] Tashkent State Technical University, Tashkent, Uzbekistan
u.mardonov@tdtu.uz

[2] North China University of Technology, Beijing, China

Abstract. This manuscript studied several schemes of the installation of natural thermocouples on the metal cutting process in measuring the thermo EMF generated during metal cutting. Absolute and relative measurement errors of thermocouples were analyzed in three diverse installation methods. The contact end of the natural thermocouples on the machined workpiece was installed using a needle, wheel, and brush which are the most popular methods of measuring temperature in machining experiments. Analytical analysis of the experimental results shows that the most accurate method of installment is a brush-contacted thermocouple with 0.18 mV and 2% absolute and relative errors respectively. Indicating the highest absolute and relative errors 1.62 mV and 165% respectively, the wheel contact method showed the most inaccurate results. Moreover, the influence of the cutting speed on the variability of the data in the measurement of thermo EMF in the cutting zone was analyzed. Experimental studies were conducted on the three different cutting speeds while feed rate and cutting depth remained constant in turning cylindrical part. It is concluded that the increase in the cutting speed can cause the improvement of the thermo EMF in the dry turning process.

Keywords: Absolute error · Relative error · Machining · Measurement · Thermocouple · Turning

1 Introduction

Many scientists have worked on predicting the temperature at the chip-tool interface during machining. In the studies, the metal cutting parameters were cutting speed, feed rate, depth of cut, and tool nose radius as the main factors affecting the cutting temperature. It is clear that cutting speed, feed rate, depth of cut, and then tool nose radius are the parameters that significantly influence the chip-tool interface temperature. The results show that increasing cutting speed, feed rate, and depth of cut increases the cutting temperature while increasing nose radius decreases the cutting temperature [1, 2].

The temperature at the chip-tool interface is an important parameter in the analysis and control of the machining process. Due to high shear and friction, energy is dissipated during the machining process, and the temperature in the primary and secondary

D. D. Cioboată and J. Machado (Eds.): ICoRSE 2025, LNNS 1592, pp. 1–14, 2025.
https://doi.org/10.1007/978-3-032-02508-1_1

cutting zones is usually very high, thus affecting the chip deformation and tool wear. The temperature affects the mechanical properties of the workpiece material and affects the wear of the tool on the surfaces. Therefore, it is very important to determine the temperatures of the tool and chip interface for process analysis or control [3–5].

Many studies have been conducted on the measurement of temperature generated during cutting [6–8]. Thermocouple methods are based on the principle of a thermocouple, in which two contacting materials generate an electromotive force (EMF) due to the temperature difference between the cold and hot junctions [9–11]. The tool-work thermocouple has become a popular method for measuring temperature during metal cutting [12–14]. This method is very useful for showing the effects of cutting speed, feed, depth of cut, and tool parameters on temperature. In a tool-work thermocouple, the chip-tool interface forms a hot junction, and the tooltip forms a cold junction [15–17].

Sushil D. Ghodam worked on measuring the temperature of the cutting tool during turning using a work thermocouple. He said that the temperature at the cutting point of the tool is a crucial parameter in controlling the machining process. With the development of machining processes, special attention has been paid to the tool life in his research [18, 19].

D. O'Sullivan et al. studied the measurement of cutting temperature in a turning operation. In the research, they focused on the infrared thermal camera and embedded thermocouple methods to control the cutting temperature in turning [20].

N.A. Abukhshim et al. analyzed the heat generation model and temperature measurement in machining. They briefly reviewed the temperature prediction techniques applied in high-speed machining processes. A thermal imaging camera was used to predict the cutting temperature in their experiments. They emphasized some advantages of the method they used in high-speed machining. However, the thermal imaging camera has some limitations in using a metal cutting process with cutting fluid conditions [21].

Several methods have been developed and tested to measure tool, chip, and workpiece temperatures. However, none are perfect, and some are better in one situation but fail in another. The appropriate technique for a given problem depends on the situation under consideration, such as ease of use, situational dynamics, and the required accuracy [22].

Abhijeet Amritkar and his team studied and investigated a calibration devise to research a relationship between obtained thermo emf during metal cutting process and the cutting temperature. Moreover, they developed a tool-part thermocouple setup to measure the machining temperature in the cutting process, they emphasized that it is the simplest and most economical technique of temperature measurement in the metal cutting process [23].

Considering the foregoing, it is clear that the measurement of cutting temperature is one of the important experiences in metal cutting process in both industrial and research applications. Natural thermocouple is popular, and accurate contact measuring method in machining operation because of having a high measuring rate which is usually occurred in metal cutting operations. On the other hand, measurement errors occur while contacting a continuous rotating workpiece during the machining process. This negative impacts on the accuracy and reliability of obtained results in machining operations. Therefore, improving the measurement accuracy of the thermocouple, and proposing reliable method to apply in machining operations is one of the actual problems in both industrial and research fields. In thins research, three types of contact ways for thermocouple on rotating workpiece in turning operation is analytically and experimentally studied. Consequently, the most accurate, easy to set method of installing natural thermocouple in turning process is proposed.

2 Methods

Experimental studies were carried out under the same conditions. In this case, the same values were chosen for the angles of the cutting tool and all the experiments were performed at the selected values of the angles of the cutting tool. Table 1 below shows the laboratory conditions in which the laboratory experiments were carried out.

Table 1. Experimental laboratory conditions.

1	Atmospheric pressure	751 mmHg
2	Height above sea level	440 m
3	Cutting duration	75 s
4	Tool material	T15K6
5	Workpiece material	AISI 1045
6	Machine tool (lathe) model	1M63MF101

During the experiments, three different states of metal cutting environment were created. First, the cylindrical workpiece was machined in a dry condition at 3 different cutting speeds ($V_{C1} = 55$ m/min, $V_{C2} = 45$ m/min, $V_{C3} = 35$ m/min), where the cutting depth was set to 0.5 mm and the feed was set to 2 mm/rev. A natural thermocouple method was studied to measure and analyze the heat generated during the cutting process. A Hantek DSO2C10 model oscillograph (Fig. 1) was used to determine Thermal Electromotive Force (thermo EMF) indicators. An oscilloscope was connected to the desktop monitor for convenient analysis of thermal temperature diagrams and obtaining results.

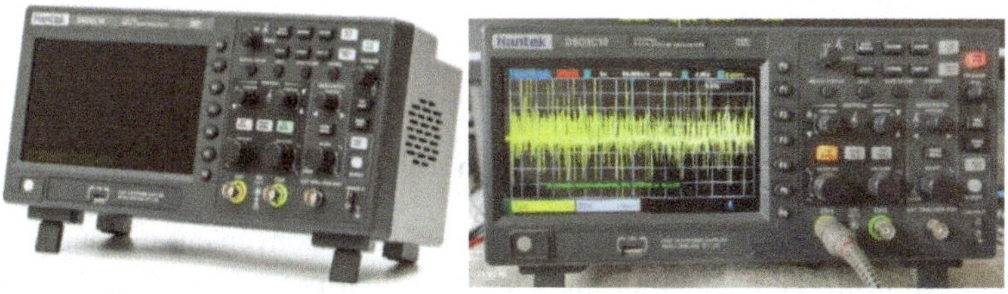

Fig. 1. The Hantek DSO2C10 model oscilloscope used in the experimental study.

The natural thermocouple in this experiment was installed in three different ways to analyze the errors of measurement. This difference in installment is where the contact method of conductor cable on the workpiece. Thermocouple cables are contacted on the workpiece with a) a needle, b) a wheel, and c) a brush. Each method under investigation was studied at three chosen cutting speeds.

2.1 Natural Thermocouple Contacted with Needle on Workpiece

To conduct the experiment in the 1st method (needle contact), a copper wire (as a conductor) and a copper needle were used. Assembling this scheme presents some difficulties. We need to install the copper needle on the back side of the spindle using a wooden plug and connect one side of the copper wire (conducting wires) to the needle, and the other side is connected to the oscilloscope. Then the wire allows us to transmit the generated thermo emf in the cutting process to the oscilloscope and measure it (Fig. 2).

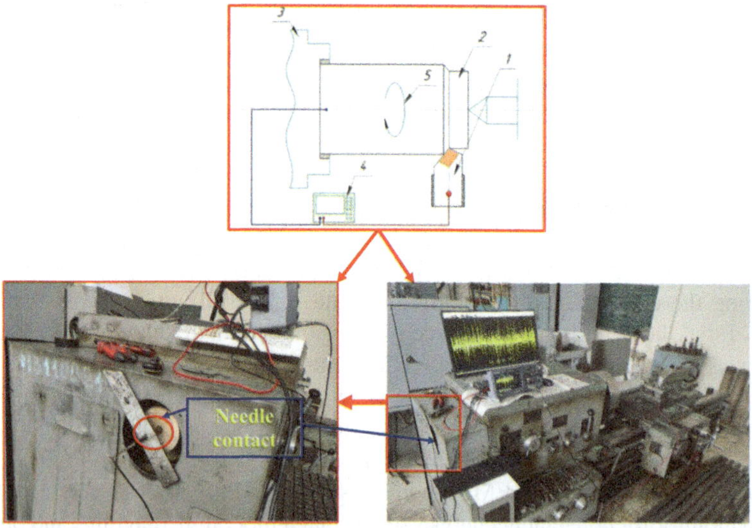

Fig. 2. The installation scheme of a natural thermocouple in the turning process using the needle contact method. 1 – cutting tool, 2 – workpiece, 3 – work holder, 4 – oscillograph, 5 – rotation direction of the workpiece.

2.2 Natural Thermocouple Contacted with a Wheel on the Workpiece

The copper wheel is fastened to the special tripod to be convenient to contact it on the workpiece. To keep the balance of the wheel, we first put a spring on the end of the tripod. The spring is fixed with nonconductor material to the tripod not to dissipate the EMF generated in cutting zone (Fig. 3).

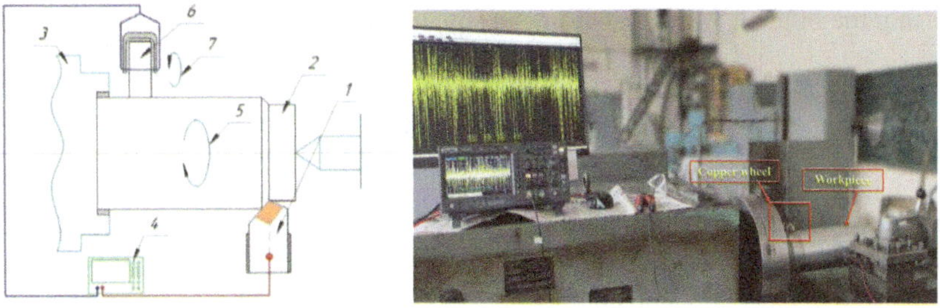

Fig. 3. The installation scheme of a natural thermocouple in the turning process using the wheel contact method. 1 – cutting tool, 2 – workpiece, 3 – work holder, 4 – oscillograph, 5 – rotation direction of the workpiece, 6 – contact wheel, 7 – rotation direction of the wheel.

2.3 Natural Thermocouple Contacted with a Brush on the Workpiece

In the 3rd method, a copper brush was used to contact the conductors on the workpiece (Fig. 4). The brush was fixed on a tripod and put the end in contact with the workpiece. The copper wire is connected to the brush to the switch and connects the other end to the oscilloscope. This helps us to accurately measure the thermo EMF.

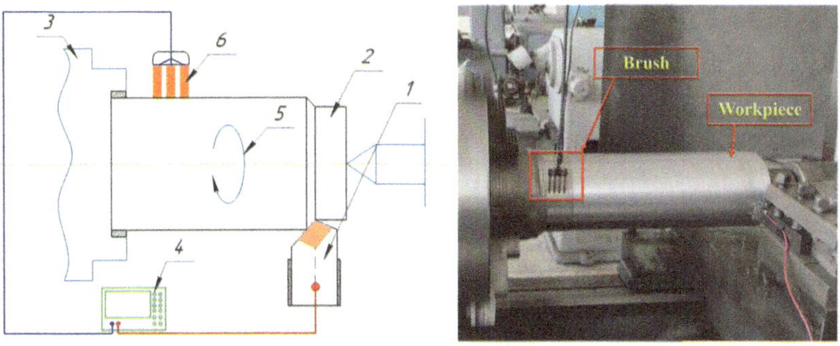

Fig. 4. The installation scheme of a natural thermocouple in the turning process using the brush contact method. 1 – cutting tool, 2 – workpiece, 3 – work holder, 4 – oscillograph, 5 – rotation direction of the workpiece, 6 – brush.

2.4 Calculation of the Errors of Direct Measurement Results

In order to calculate the error in the results of direct measurement of the amount of the measured thermo EMF U, it is necessary to carry out several measurements of this value. The calculation is continued according to the following sequence [24].

1. Using the Eq. 1, we calculate the arithmetic mean value of U,

$$U_{mean} = \frac{U_1 + U_2 + \ldots U_n}{n} = \frac{1}{n}\sum_{i=1}^{n} U_i \tag{1}$$

where, n - number of measurement results

2. Using the FOLLOWING Eq. 2, mean square error of U is calculated,

$$\sigma_x = \sqrt{\frac{1}{n(n-1)} \sum_{i=1}^{n} (\Delta U_i)^2} \tag{2}$$

where, $\Delta U_i = U_i - U_{mean}$ – absolute error of each measured result of n.
3. The values of the squares of these absolute errors must contain twice as many numbers as the values of the measurement results, including zeros, and the values of the mean square error have the same number of digits as the values of the measurement results. Then, the initial absolute error of the measured value by multiplying its mean square error by the Student's coefficient $t_\alpha(n)$ is calculated [25].

$$\Delta U = \sigma \cdot t_\alpha(n) \tag{3}$$

The value of this error should contain the same number of digits as the values of the measurement results. For a given number n and confidence probability $\alpha = 95\%$, the value of Student's coefficient $t_\alpha(n)$ is obtained from a special table [25].
4. The average value of the measured numbers is determined by comparing the value of the absolute error. The number showing the average value of the measured results should be rounded, leaving all the digits up to the last digit in the final recording of the absolute error. We write the measurement result as the sum of the rounded average value and the absolute error as follows:

$$U = U_{mean} \pm \Delta U \tag{4}$$

5. The relative error of the measured value was also calculated using the following equation,

$$\varepsilon_U = \frac{\Delta U}{U_{mean}} \tag{5}$$

The relative error is usually expressed as a percentage. The value of this error should be rounded, leaving only two significant figures. So, the value of the relative error is determined by the specific conditions of the measurement process, but this value itself is not known. The method described above allows us to estimate the magnitude of this error. In addition, as the number of measurements increases, the result of the error estimation is closer to the true value of the relative error. However, increasing the number of measurements will never reduce the error itself.

3 Results

3.1 Needle Contact Method

Figure 5 shows the indicators obtained from the oscillograph of a natural thermocouple connected to the cutting process by the needle contact method. In the graph, the change diagram of thermo EMF during $t = 80$s is depicted in the experiments performed at three different cutting speeds. It can be seen from the figure that the decrease in the cutting speed leads to a decrease in the thermo EMF generated in the process. From this, we can once again investigate that the temperature generated during the cutting process is directly proportional to the cutting speed.

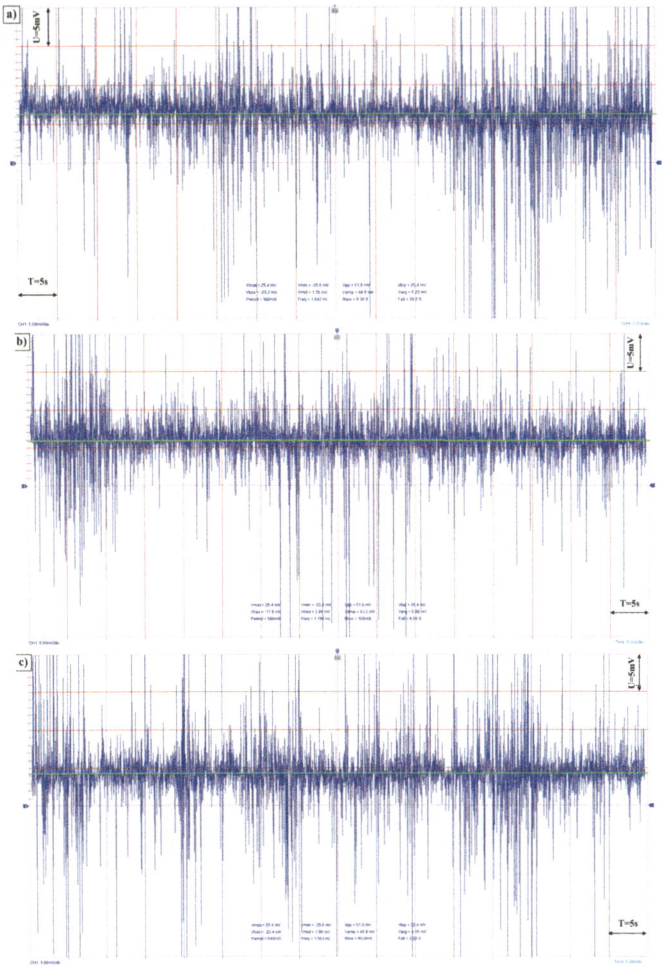

Fig. 5. Time dependence diagram of thermo EMF generated during machining in experiment 1 (needle contact method): cutting depth $a_p = 0.5$ mm, feed $f = 2$ mm/rev; a) $V_{C1} = 55$m/min, b) $V_{C2} = 45$ m/min, c) $V_{C3} = 35$ m/min.

Also, in all three cutting speeds, we can observe sharp changes in the frequency of Thermo EMF in the first 10 s of the cutting process. We can see that these changes lasted for a shorter time at higher cutting speeds. A similar sharp change was observed in the process of the experiment conducted at the highest cutting speed. The main values of the results obtained during the experiment are given in Table 2.

3.2 Wheel Contact Method

Figure 6 shows the parameters obtained during experiments using a natural thermocouple, one end of which is connected to the workpiece through a special wheel. We can see that the obtained results differ sharply from the results of the first method. That is,

Table 2. The main values of thermo EMF (mV) obtained as a result of the research carried out by the needle contact method in thermocouple.

	U_{max}	U_{min}	U_{avg}	U_{mid}	U_{amp}
V_{C1}	25.4	25.6	6.22	1.2	48.6
V_{C2}	25.4	25.6	5.98	3.80	43.2
V_{C3}	25.4	25.6	4.16	1	48.8

in the process of cutting performed under the same conditions, the difference between the values of thermo EMF was very large.

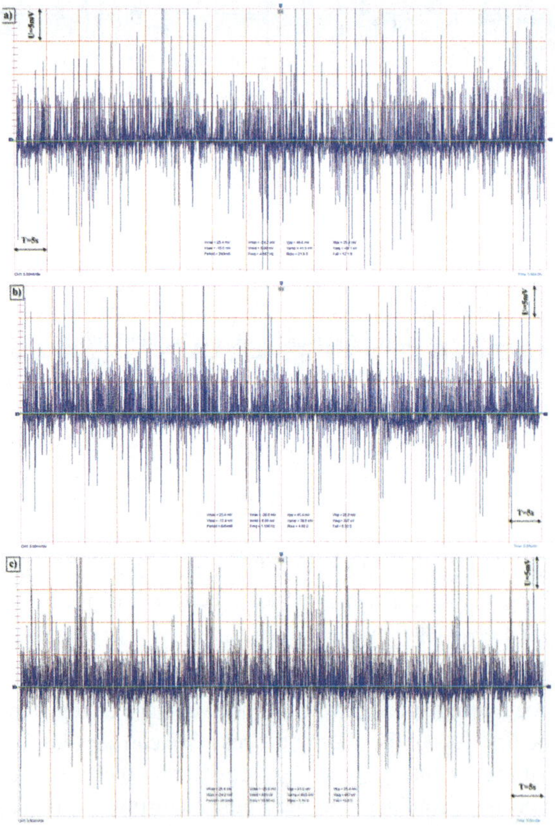

Fig. 6. Time dependence diagram of thermo EMF generated during machining in experiment 2 (wheel contact method): cutting depth $a_p = 0.5$ mm, feed $f = 2$ mm/rev; a) $V_{C1} = 55$ m/min, b) $V_{C2} = 45$ m/min, c) $V_{C3} = 35$ m/min.

Using the wheel contact method, based on the results obtained from the experiment (Table 3), we can estimate that the error is much more noticeable in this method.

Table 3. The main values of thermo EMF (mV) obtained as a result of the research carried out by the wheel contact method in thermocouple.

№	U_{max}	U_{min}	U_{avg}	U_{mid}	U_{amp}
V_{C1}	25.4	24.2	0.049.1	5	41
V_{C2}	25.4	20	0.397	6	38.6
V_{C3}	25.4	25.6	0.487	0.6	49.6

3.3 Brush Contact Method

The results of thermo EMF can be seen in Fig. 7 below in the experiments carried out by installing a natural thermocouple in the process of cutting with a brush contact method. We can see that the results are very similar to those of the needle contact method. Table 4 shows the significant values of the experimental results.

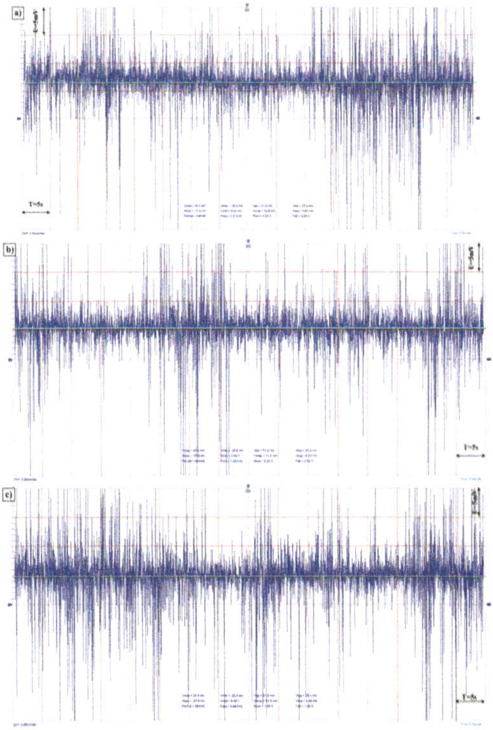

Fig. 7. Time dependence diagram of thermo EMF generated during machining in experiment 3 (brush contact method): cutting depth $a_p = 0.5$ mm, feed $f = 2$ mm/rev; a) $V_{C1} = 55$ m/min, b) $V_{C2} = 45$ m/min, c) $V_{C3} = 35$ m/min.

Table 4. The main values of thermo EMF (mV) obtained as a result of the research carried out by the brush contact method in thermocouple.

№	U_{max}	U_{min}	U_{avg}	U_{mid}	U_{amp}
V_{C1}	25.4	25.6	6.64	4	42.8
V_{C2}	25.4	25.6	5.53	0	51
V_{C3}	25.4	25.6	4.98	0	51

4 Discussions

To calculate the absolute and relative errors in the results of direct measurement of the value of thermo EMF U_n measured by the above three methods, the values of this amount at $n = 7$ chosen points (t = 10 s, 20 s, 30 s, 40 s, 50 s, 60 s, 70 s) were marked. Based on the measured values, the absolute and relative errors of the thermo EMF values obtained from the natural thermocouples installed in the needle contact, wheel contact, and brush contact methods, the results obtained as a result of calculations using Eqs. (1), (2), (3), (4), (5) are given in Table 5 below.

Table 5. Absolute and relative errors in experiments using the technology of installation of natural thermocouples using the needle, wheel, and brush contact methods.

	U_{V1}, mV	U_{V2}, mV	U_{V3}, mV
Needle contact method			
U_{avg}	6.86	5.54	4.42
$\Delta U, [mV]$	1.2139	0.7698	0.7546
$\varepsilon_{Un'}, [\%]$	18	14	17
Wheel contact method			
U_{avg}	0.98	0.91	0.84
$\Delta U, [mV]$	1.6250	1.1750	1.0309
$\varepsilon_{U'}, [\%]$	165	128	122
Brush contact method			
U_{avg}	7.1	5.95	5.42
$\Delta U, [mV]$	0.1877	0.4801	0.7411
$\varepsilon_{U'}, [\%]$	2	8	3

Based on the results obtained in the table above, we can say that the relative and absolute errors of measuring the thermo EMF generated during the metal cutting process by the natural thermocouple method, which is a common method compared to others, by installing a needle contact through a spindle to the workpiece, are slightly higher. That is, in the experiments carried out at the cutting speed of V_3 = 55 m/min, the amount of relative error was 18%, while at the value of $V_1 = 45$ m/min was 14%. The average absolute error of measurement in this method was 0.76 mV.

However, in the results presented in Table 5, we can see that the relative and absolute errors of measuring the thermo EMF generated in the process of turning by the natural thermocouple method, by installing a special wheel on the lathe, have increased greatly. That is, we can see that the absolute error exceeds the measured value in the experiments conducted at the highest cutting speed, and the relative error is the largest in this method, up to 165%.

The results obtained in the table show that the relative and absolute errors of measurement in the experiment carried out by the brush contact installation method of natural thermocouple in the measurement of thermo EMF in machining showed the best results compared to the previous two installation methods. That is, while the amount of relative error was 13% in the experiments conducted at the lowest cutting speed, it was surprising that the amount of relative and absolute error decreased as the cutting speed increased. In the experiments conducted at the highest cutting speed, the relative error was 2%, while the value of the absolute error was ~ 0.18 mV. Below are comparative graphs of absolute and relative error values determined based on experiments conducted at three different cutting speeds by installing a natural thermocouple in three different methods (Fig. 8).

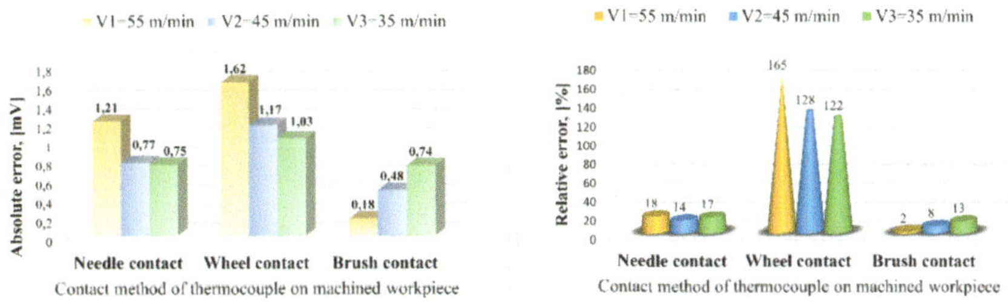

Fig. 8. Comparison graphs of absolute (a) and relative (b) error values determined based on experiments conducted at three different cutting speeds by installing a natural thermocouple in three different ways.

In these experimental studies, using the natural thermocouple method of measuring the thermo EMF generated in the metal cutting process, several different schemes of its installation in the cutting process were analyzed, as well as the data for measuring the thermo EMF generated in the cutting zone. The effect of cutting speed on the variability was also analyzed. Based on the analysis, it was concluded that the most accurate and easiest installation of the natural thermocouple on the machined part to measure the thermo EMF in the machining process is the brush contact method.

5 Conclusions

This research studies a diverse approach of installations of natural thermocouples in the machining process to measure the thermo EMF. According to experimental data, and mathematical calculations, the following conclusions can be drawn:

The installation scheme of the thermocouple with needle contact method is more difficult, and time taking those other experimented methods. However, its accuracy is better than the wheel contact method. On the other hand, needle contacted installation method is recommended to use in measuring thermo EMF in machining with a lubricating-cooling external environment. Unlike this method, in the other two methods, flooding cutting fluid on the workpiece surface can affect the accuracy of the obtained results.

Showing the highest relative and absolute errors $\varepsilon_U = 162\%$ and $\Delta U = 1.62$ mV in measuring the thermo EMF in the metal cutting process by natural thermocouple, the wheel-contacted installation method of thermocouple is concluded as the most undesirable method.

Based on the experimental analysis, it was concluded that the most accurate method of measuring the thermo EMF in the metal cutting process using the natural thermocouple method is the brush contacted installation method. Moreover, its installation scheme is complex and easy to experiment with. However, while machining in a cutting fluid environment, cutting fluid between the workpiece and the contact brush on it can interfere with taking accurate EMF in the process. Thus, the brush contact installation method of thermocouple is highly recommended in dry machining processes.

References

1. Lazoglu, I., Altintas, Y.: Prediction of tool and chip temperature in continuous and interrupted machining. Int. J. Mach. Tools Manuf **42**(9), 1011–1022 (2002). https://doi.org/10.1016/S0890-6955(02)00039-1
2. Kamruzzaman, M., Rahman, S.S., Ashraf, M.Z.I., Dhar, N.R.: Modeling of chip–tool interface temperature using response surface methodology and artificial neural network in HPC-assisted turning and tool life investigation. Int. J. Adv. Manuf. Technol. **90**, 1547–1568 (2017). https://doi.org/10.1007/s00170-016-9467-6
3. Lata, S., Rana, R.: Investigation of chip-tool interface temperature: effect of machining parameters and tool material on ferrous and non-ferrous metal. Mater. Today: Proc. **5**(2), 4250–4257 (2018). https://doi.org/10.1016/j.matpr.2017.11.689
4. Khasanov, S., Mardonov, U., Ongboyev, A.: Influence of a magnetic field on HSS tool wear/life and the intensity of external machining environments in turning. Int. J. Mechatron. Appl. Mech. (16), 163–170 (2024). https://doi.org/10.17683/ijomam/issue16.19
5. Ogedengbe, T.S., et al.: The effects of heat generation on cutting tool and machined workpiece. In Journal of Physics: Conference Series, vol. 1378, no. 2, p. 022012. IOP Publishing (2019). https://doi.org/10.1088/1742-6596/1378/2/022012
6. Longbottom, J.M., Lanham, J.D.: Cutting temperature measurement while machining–a review. Aircr. Eng. Aerosp. Technol. **77**(2), 122–130 (2005). https://doi.org/10.1108/00022660510585956
7. Dubey, V., Sharma, A.K., Singh, R.K.: A technological review on temperature measurement techniques in various machining processes. In: Advances in Metrology and Measurement of

Engineering Surfaces: Select Proceedings of ICFMMP 2019, pp. 55–67 (2020). https://doi.org/10.1007/978-981-15-5151-2_6

8. Akhil, C.S., Ananthavishnu, M.H., Akhil, C.K., Afeez, P.M., Akhilesh, R., Rahul, R.: Measurement of cutting temperature during machining. IOSR J. Mech. Civil Eng. **13**(2), 108–122 (2016). https://doi.org/10.9790/1684-130201108122
9. Seker, U., Korkut, İ., Turgut, Y., Boy, M.: The measurement of temperature during machining. In International conference power transmissions (2003)
10. Mardonov, U., Tuyboyov, O., Abdirakhmonov, K., Tursunbaev, S.: Mathematical approach to the flank wear of high-speed steel turning tool in diverse external cutting environments. Int. J. Mechatron. Appl. Mech. (14), 19–26 (2023). https://doi.org/10.17683/ijomam/issue14.3
11. Kaminise, A.K., Guimarães, G., da Silva, M.B.: Development of a tool–work thermocouple calibration system with physical compensation to study the influence of tool-holder material on cutting temperature in machining. Int. J. Adv. Manuf. Technol. **73**, 735–747 (2014). https://doi.org/10.1007/s00170-014-5898-0
12. Masek, P., Zeman, P., Kolar, P.: Cutting temperature measurement in turning of thermoplastic composites using a tool-work thermocouple. Int. J. Adv. Manuf. Technol. **116**, 3163–3178 (2021). https://doi.org/10.1007/s00170-021-07588-0
13. Stephenson, D.A.: Tool-work thermocouple temperature measurements—theory and implementation issues. ASME. J. Eng. Ind. **115**(4), 432–437 (1993). https://doi.org/10.1115/1.2901786
14. Leshock, C.E., Shin, Y.C.: Investigation on cutting temperature in turning by a tool-work thermocouple technique. ASME. J. Manuf. Sci. Eng. **119**(4A), 502–508 (1997). https://doi.org/10.1115/1.2831180
15. Dhar, N., Islam, S., Kamruzzaman, M., Ahmed, T.: The calibration of tool-work thermocouple in turning steels. In: Proceedings of National Conference on Industrial Problems on Machines and Mechanisms (IPROMM-2005), pp. 459–466 (2005)
16. Mardonov, U., Khasanov, S., Jeltukhin, A., Ozodova, S.: Influence of using cutting fluid under the effect of static magnetic field on chip formation in metal cutting with HSS tools (turning operation). Manuf. Technol. **23**(1), 73–80 (2023). https://doi.org/10.21062/mft.2023.006
17. Sarvar, T., Nodir, T., Mardonov, U., Saydumarov, B., Kulmuradov, D., Boltaeva, M.: Effects of germanium (ge) on hardness and microstructure of al-mg, al-cu, al-mn system alloys. Int. J. Mechatron. Appl. Mech. (16), 179–184 (2024). https://doi.org/10.17683/ijomam/issue16.21
18. Ghodam, S.D., Student, M.: Temperature measurement of a cutting tool in turning process by using tool work thermocouple. Int. J. Res. Eng. Technol **3**, 831–835 (2014)
19. Mardonov, U., Khasanov, O., Ismatov, A., Baydullayev, A.: Studies concerning water-based coolants under magnetic field during a metal-cutting process (turning). In International Conference on Reliable Systems Engineering, pp. 308–317. Springer, Cham (2023). https://doi.org/10.1007/978-3-031-40628-7_26
20. O'Sullivan, D., Cotterell, M.: Temperature measurement in single point turning. J. Mater. Process. Technol. **118**(1–3), 301–308 (2001). https://doi.org/10.1016/S0924-0136(01)00853-6
21. Abukhshim, N.A., Mativenga, P.T., Sheikh, M.A.: Heat generation and temperature prediction in metal cutting: a review and implications for high-speed machining. Int. J. Mach. Tools Manuf **46**(7–8), 782–800 (2006). https://doi.org/10.1016/j.ijmachtools.2005.07.024
22. Saez-de-Buruaga, M., Soler, D., Aristimuño, P.X., Esnaola, J.A., Arrazola, P.J.: Determining tool/chip temperatures from thermography measurements in metal cutting. Appl. Therm. Eng. **145**, 305–314 (2018). https://doi.org/10.1016/j.applthermaleng.2018.09.051
23. Amritkar, A., Prakash, C., Kulkarni, A.: Development of temperature measurement setup for machining. World J. Sci. Technol. **2**, 15–19 (2012)

24. Beketov VG.: Metodicheskie rekomendatsii po obrabotke rezultatov raschetov i izmereniy pri reshenii zadach i vipolnenii laboratornix rabot po fizike [Methodological recommendations for processing the results of calculations and measurements when solving problems and performing laboratory work in physics]. Moskva (2009)
25. Zaxarov, I.P., Klimova, Ye.A.: Raschet znacheniy koeffitsienta Styudenta dlya drobnogo chisla stepeney svobodi [Calculation of Student coefficient values for a fractional number of degrees of freedom]. Sistemi obrobki informasoiï (4), 43–47 (2010)

The Influence of High-Chromium Powder Coatings on Wear Resistance of Working Parts

Lazizbek Raufov[1(✉)], Darob Berdiev[1], Mahmuda Umarova[1], Ravshan Tashmatov[2], Zarip Sadullayev[2], Dustmurat Fazilov[2], and Shokhida Khojibekova[2]

[1] Tashkent State Technical University Named after Islam Karimov, Tashkent, Uzbekistan
raufov.lazizbek@tdtu.uz

[2] Almalyk Branch of Tashkent State Technical University Named after Islam Karimov, Almalyk, Uzbekistan

Abstract. The article proposes a method for increasing the workability of a part by creating friction coatings from high-chromium metal powder on the working surface of the working parts. In order to increase the wear resistance of the working parts, tests were carried out in which iron-based powder with a high chromium content was applied using the plasma coating method, the result of which was an increase in its wear resistance by 3–4 times. A technology for the efficiency of high-chromium iron-based powder from local raw materials has been developed, as a result it allows increasing the efficiency of the powder by 85%. It was determined that when forming friction coatings from a high-chromium powder mixture using a gas flame, the porosity of the coating decreased to 7%, and the hardness reached 54…56 HRC. Also, the service life of a share made of steel 40X (alloy carbon steel) increased by 3.5–4 times compared to a share made of steel L53 (ploughshare steel), subject to heat treatment.

Keywords: Wear resistance · Coatings · Hardness · Porosity · Working body · Soil · Ploughshare

1 Introduction

One of the most energy-intensive processes in the agricultural industry is soil cultivation, which consumes 30–40% of the energy consumed in agriculture [1–3]. A number of working parts of agricultural machine parts (shares, cultivator knives, segments of cutting machines, etc.) often fail prematurely, as a result of which a large number of spare parts have to be prepared for such working parts [4]. For example, for the production of plow shares, carbon steels L53, L65, sometimes 65G are often used, and their mechanical properties are increased by heat treatment, as a result, the hardness of these working parts of the shares is in the range HB 321 - 541 (HRC 33 - 52) [5, 6].

Many studies have recommended applying various coatings to improve the performance of parts operating in various wear and corrosion environments. A technology has been developed for applying a coating to the surface of AISI 316L stainless steel

D. D. Cioboată and J. Machado (Eds.): ICoRSE 2025, LNNS 1592, pp. 15–26, 2025.
https://doi.org/10.1007/978-3-032-02508-1_2

using laser current, which provides wear resistance 1.4–1.5 times higher than that of the base. [7], corrosion resistance increases by 20–22% due to impregnation of the surface of carbon steels with CrMoV9 metals. [8], the effect of time and temperature of heat treatment on Ni-20 wt. % Al coatings applied by flame spraying was studied. The objective is to find the optimal sample in terms of microstructure and compare its oxidation resistance with the coating in the after-sprayed state. The coatings were applied by flame spraying onto a low-carbon steel substrate (St 37). The samples were subjected to heat treatment in vacuum at 850 °C and 950 °C for 1 h and at 1050 °C for 1, 3 and 6 h, as a result of which the corrosion and abrasive resistance of the coating increased by 30–35%. [9], various composite coatings [Cr-Fe_4 Co_4 Ni_4]Cr_3 reinforced with nano-TiC have been developed using the cluster-plus-glue-atom model and manufactured on 904L stainless steel by laser cladding; compared to 904L stainless steel, the corrosion current density of the [Cr-Fe_4Co_4 Ni_4]Cr_3 -15vol%TiC composite coating decreased by almost an order of magnitude, and the impedance increased by 3.5 times [10], and it has been established that the effective use of titanium carbides TiC, silicon SiC and tungsten WC in the powder composition increases the corrosion and wear resistance of the working parts of agricultural tillage machinery by 35–40%. [11], the influence of various coating application methods on increasing the wear resistance of copper and copper alloys was studied, coated with a Ni- Cr - Si powder base, increasing their corrosion and antifriction properties by 1.2–1.3 times [12], the wear resistance of working surfaces was increased by 2–3 times due to the coating of Fe- Ti -Ni, Fe- Cu - Mo, Fe- Ti -Al materials with a composition prepared from a powder mixture using a gas arc [13].

Analysis shows that the wear resistance of soil-cultivating working bodies is formed due to the high level of hardening of the cutting part. Currently, several methods of high-quality hardening of the cutting part of soil-cultivating working bodies are known. However, previous studies have shown that all existing methods do not provide wear resistance when working with the cutting part of soil-cultivating tools under conditions of increased abrasive wear and stress.

When using a ploughshare, large counteracting forces are applied to the working surfaces. The main indicator influencing the change in the geometric dimensions and external shape of the working elements is the composition of the soil being processed [14].

Analysis has shown that a plowshare can be taken out of service for two reasons: firstly, when the working element becomes unsuitable for use, and secondly, as an exception, when the technical and economic indicators exceed the required standard.

The methods of technology widely used for finishing the working surfaces of earth-processing organs in production include thermal and chemical-thermal treatment, application of wear-resistant coatings and combined coating, etc. Currently, there are many technologies designed to increase the service life of working organs, but the cost of these technologies is becoming more and more expensive.

In developed countries, the surfaces of the working parts of soil-cultivating organs are coated with metal powders using a gas flame, forming a corrosion-resistant and wear-resistant coating layer, and this method increases the service life of the working organs. Adhesion resistance is determined mainly by the nature of the formation of coatings from fine-grained materials in coatings from metal powder particles.

In works [15–18] it is shown that the quality of the coating or modified layer on the surface of a part largely depends on the physical and mechanical state of the material of the surface layer, which, in turn, is determined by its energy state.

In the process of applying a functional coating to the surface of a part or creating a modified layer on it, the initial value of the specific accumulated deformation energy changes, and this change depends not only on the process of forming a coating or layer on the surface of the part, but also on its initial state: structure, technological residual stresses, thickness of the layer with modified physical and mechanical properties, distribution of dislocation density and other indicators (characteristics). These indicators of the state of the material of the surface layer of the part depend on the physical and mechanical characteristics of the material being processed and the method of processing the part. Considering the fact that the working parts of agricultural machines operate in an abrasive and highly dynamic environment, in developed countries, working surfaces are improved by applying various coatings to the surface that provide resistance to strong external influences.

The aim of the study is to develop a technology for applying a wear-resistant coating based on high-chromium iron-containing alloy powder in order to improve the performance of the working parts of soil-cultivating machines.

Research objectives:

- to study existing methods for increasing the wear resistance of working parts of soil-cultivating machines;
- conduct an analysis of the characteristic features of wear of cutting elements of working bodies under operating loads;
- develop a technology for applying a wear-resistant coating to the cutting part of the working element using high-chromium iron-based metal powder obtained from local raw materials, using flame spraying.

Object of study – ploughshare made of grade 40X steel (Table 1), the working parts of which are coated with powder of high-chromium iron-containing alloy obtained from local raw materials.

2 Materials and Research Methods

In this work, an attempt was made to increase the machinability of the part by creating friction coatings from high-chromium metal powder on the working surface of plow shares using a gas flame.

A ploughshare made of 40X steel was chosen as the object of study. The studied metal powder with a high chromium content contained 26% Cr, 4% C and the rest Fe.

The granulometric composition of the powder particles was determined according to GOST 6613–86, weight 200 g. Powder particles are passed through a special sieve, granular particles up to 60 μm in size.

Determination of powder moisture content 0.0001 g. Precision analytical electronic scales 220 V–50 Hz, a drying oven SNOL 67/350 was used for drying the powder (heating temperature 100 °C for 60 min), powder particle size is 40–60 μm. Being granular in size it ensured good flowability.

To create a flame during the formation of the coating, a GAL type burner was used from compressed gas and oxygen, as well as acetylene to create pressure.

The composition of metals and powders using the EXPLORER 5000 PORTABLE ANALYZER, metallographic analysis was carried out on a Metrohm 850 Professional IC ion chromatograph with a magnification of 100 to 1000 times, and a Zeiss EVO MA 10/Aztec Energy Advanced X-Act scanning electron microscope was also used in the study, and macroanalysis was carried out on an MBS-3 microscope at magnification from 4 to 100 times. The coating hardness was determined using PMT-3M microhardness testers [17, 19].

Table 1. Chemical composition of steel samples for research.

№	Steel grades	The content of the element, % by weight					
		C	Mn	Si	S	P	Cr
1	40X	0.41	0.74	0.35	0.025	0.022	0.92
2	L53	0.47–0.57	0.5–0.8	0.15–0.40	0.05	0.045	–

3 Results

Ensuring high mechanical properties of the coating during the formation of the coating also depends on the composition of the metal powder. In our studies, the analysis of the characteristics of metal powders is carried out according to three properties: chemical properties, physical properties and technological properties.

The analysis revealed that powders of the same chemical composition can have different physical properties and differ sharply in technological properties, which leads to a significant change in the conditions for converting the powder into a finished product and affects their mechanical properties.

According to our research, on average 74–75% of the metals used in coatings are elemental iron powder. Iron-based powder is finely dispersed particles of dark gray state and is used as the main powder element in the formation of coatings to improve the workability of the working surfaces of the ploughshare. According to the analysis, the iron powder contains carbon - 0.05%, silicon - 0.08%, manganese - 0.20%, sulfur - 0.02%, phosphorus - 0.02% and it was found that oxygen is 0.5%.

25–26% of chromium element powder was added to iron powder to create a coating with high properties [2, 3, 21–23]. One of the advantages of iron-chromium element powder coating is that it meets the requirements of high pressure, abrasion resistance, and does not require heat treatment. When the powder particles are spherical or granular, they have good fluidity.

To apply coatings from high-chromium iron-based powder, two main technological processes were implemented: coating layer formation technology and powder particle spraying processes.

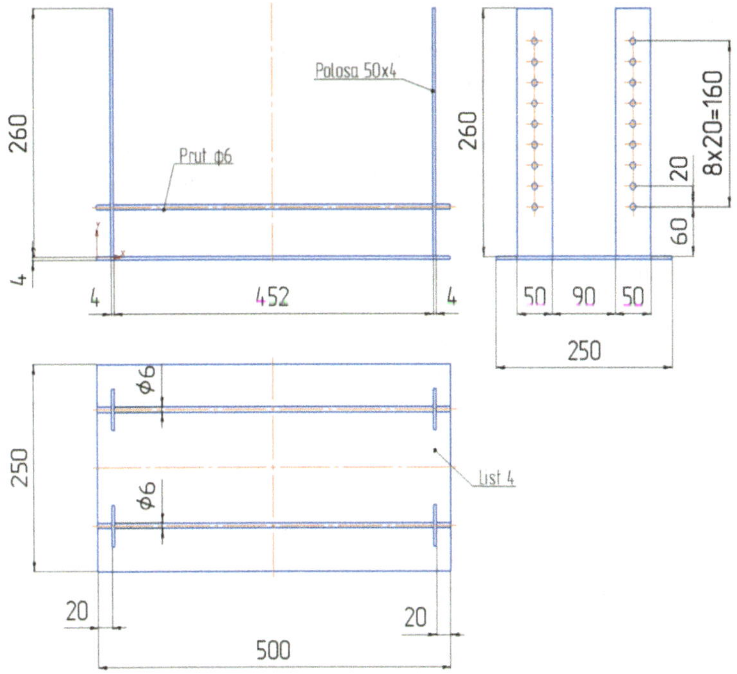

Fig. 1. The process of determining the useful use of a powder material for a specially designed device

The research used the method of spraying powder particles heated in a gas flame to form a coating. As a result of heating the material, feeding it to the cleaned surface under high pressure and acting on it with strong pressure, the powder particles form a deformation element and serve as a basis for forming a durable layer on the working surface. The coating material layer is formed on the surface of the cleaned surface and receives a high level of thermal and kinetic energy under the influence of the gas flame flow. When using acetylene (C_2H_2) in oxy-fuel gas, the temperature of the combustible gas flame jet increases to 3000° C, the speed of the powder particles leaving the burner tip is 150–160 m/s. Powder particles, entering the high-temperature gas flame jet, pass into a state of melting or high plasticity and fly out of the burner hole at a speed of 20–80 to 1000 m/s. It follows from the analysis that obtaining an effective coating when forming a powder coating in a gas flame depends on a number of parameters, namely, on the design of the device used (how high the ratio of combustible gas and oxygen in the mixture is), the distance between the end of the burner nozzle and the sprayed area of the product, the number of particles of the sprayed powder bonded to the sprayed surface of the product, the amount of adhesion, its density, and the granularity of the powder.

In our research, a special device was developed to determine the exact distance between the tip of the burning torch and the area to be coated with spraying (Fig. 1), this is a special 100 mm by 180 mm. Meanwhile, special burner installers were adapted and experimental tests were carried out [23–25].

When developing a method for forming layers of coating from iron-based powder, it is advisable to study the efficiency coefficient (EC) during adhesion of powder particles, which is one of the main indicators for determining the efficiency coefficient during adhesion of powder particles to the metal surface.

Based on experimental tests, a process mode was developed and a result was obtained based on the value of the coefficient of use of the material of high-chromium iron-based powder, which ensures the production of coating layers' resistant to aggressive environments and wear [18]. The working area of the working element of the earth-processing ploughshare, a high coefficient of use of the material for the formation of a coating using high-chromium iron-based powder, the maximum distance between the area of the spray coating with the tip of the burning torch is 140 mm. It was found that the efficiency is up to 85%, compared with other intermediate distances.

The coating layer serves to provide the desired properties of the product surface. The base metal of the product provides the necessary strength of the structure [17, 19–21]. The coating provides wear resistance, heat resistance, thermal resistance, corrosion resistance and other properties required by the metal surface.

The essence of the coating application process can be reduced to heating the material added to the metal to a state close to melting, and ensuring the adhesion of this material to the base of the part. As a result of research, it has been established that when creating a coating, it is extremely important to meet a number of technological requirements:

It is required that the resulting coating layer has a minimal effect on the base metal, i.e. does not dissolve the surface of the metal. Therefore, when applying the coating, it is necessary to form a minimal melting process on the surface of the base metal, otherwise, when forming the coating layer, the amount of the base metal will increase. As a result, the property [26] is formed on the surface of the coating metal;

When coating a metal surface, it is necessary to minimize the heat-affected zone and create minimal stresses and deformations. It has been established that this requirement can be met by controlling the minimum surface melting and other parameters of the base metal [25].

Our studies have shown that with an increase in the distance between the end of the burner nozzle generating the gas flame and the working surface of the product, the hardness value of the coating layer first increased and then decreased (Table 2).

Table 2. Values of the distance between the gas spray tip and the working surface of the product

T/r	Distance between the gas spray tip and the working surface, mm	Thickness of the resulting coating, mm	Coating hardness, HRC	Porosity level, %
1	100	1.6	54…56	up to 6%
2	120	1.6	54…56	up to 7%
3	140	2	54…56	up to 7%
4	160	2	50…51	up to 9%
5	180	2.3	48…49	up to 12%

In Table 1, when coating the working surface of the working element of the ploughshares with high-chromium iron-based powder, the maximum hardness (54…56 HRC, average thickness 2 mm) in powder mixtures, the distance between the tip of the flame sprayer and the working surface is 140 mm. It is evident that the coating thickness in other dimensions is on average 2 mm, but the hardness value is different.

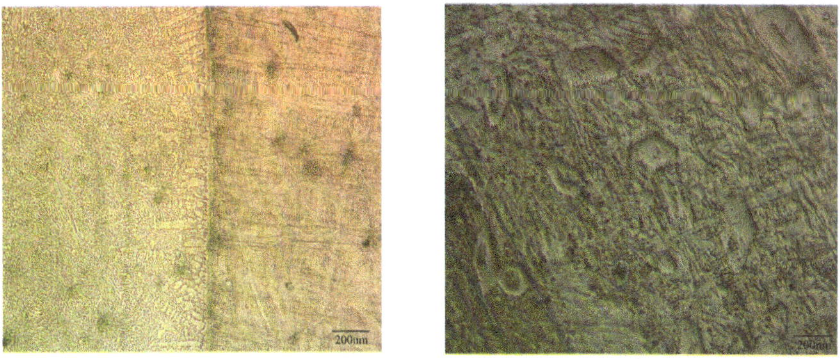

a b

Fig. 2. Macrostructure of high-chromium powder coating: *a* - appearance of a high-quality connection with the base metal ($\times 100$); *b* - appearance coatings ($\times 100$)

From the macro analysis of the coating with the highest measured hardness, it is evident that the coating is strongly bonded to the metal substrate (Fig. 2a). The porosity level observed in this coating reaches up to 7% (Fig. 2b), and the hardness of the coating was in the range of 54–56 HRC. The macrostructure images shown in Fig. 2 were obtained under laboratory conditions using a high-resolution digital optical microscope IE500M brand. To identify the microstructure of the spraying layer, a 3% solution of nitric acid in ethyl alcohol was used. Prior to imaging, the coated specimens were sectioned, ground, polished, and etched with a 4% nitric acid solution in alcohol to reveal the coating-substrate interface. The flame spraying process was carried out at a nozzle-to-surface distance of 140 mm when applying the coating to the working surface of the 40X steel ploughshare, which is the subject of this study (Fig. 3).

Fig. 3. Model of a ploughshare made of 40X steel, the working surface of the working element of the ploughshare, coated with iron-based powder with a high chromium content

For the cultivation of soils in Uzbekistan, steel grade L53 (Table 1) made of sheet steel is used in the main ploughshare. Steel grade L53, selected for the manufacture of ploughshares, is insufficient to ensure the required level of soil cultivation [5–8]. In our study, it is recommended to replace this grade with steel grade 40X. The purpose of this is that the service life of the produced ploughshare is ensured mainly by the coating covering the working part, which is the basis for ensuring economic efficiency.

Table 3. Results of field tests of ploughshares in the farm "Ibodullaev Khaidaraly" Piskent district of Tashkent region

Soil composition	Number of tested ploughshares, pcs	Number of tested plowshares, pcs		C_2/C_1
		For steel L53, C_1, *ha*	Steel 40X, C_2, *ha*	
Clay soil, moisture 12.2 – 15.7%, hardness 2.47 – 5.8 MPa	10	8–9	36–37	4.1
Sand and soil, humidity 10.4 – 13.5%, hardness 2.02 – 3.7 MPa	10	10.5–11	34–35	3.4

Note: C_1 - serviceability of plowshares made of serially produced steel L53;
C_2 - serviceability of plowshares, the working parts of which are covered with grade 40X steel with a spray coating.
ha – hectare ($1\text{ha} = 10{\cdot}10^3\ \text{m}^2$).

For the tests, a ploughshare made of commercially available L53 steel and heat-treated at the Bakhtselmash joint venture and a ploughshare made of 40X steel with a flame-resistant coating on the cutting surface made of high-chromium iron-based powder were selected. Field test analyses of the research results were conducted during soil cultivation under various conditions at the "Ibodullaev farm. Khaidaraly" Piskent district of Tashkent region (Table 3) and in the farm "Bekzod Sherzod Buston" city of Gijduvon district, Bukhara region. For processing, we used a CLASS ARION 630C tractor equipped with a four-body rotary plow Lemken, and a New Holland T 7060 tractor equipped with a three-body rotary plow Lemken. The speed of movement of the units was 7–9 km/h. The serial and test shares were successively installed in the plow. Sequential installation of the shares allowed us to obtain results of maximum accuracy. The share installed at the end has a smaller effect on the share body than the share installed first. The composition of the processed soil was determined according to GOST 20915-2011. The composition of the processed soil is medium sandy, processed to a depth of 30 cm.

According to the results of field tests, the ploughshare made of L53 series steel on medium sandy soil was 8–9, and during subsequent treatments the ploughshare parameters changed and a large wear process was observed (Table 4).

Preliminary conclusions can be made based on the condition of the tested ploughshare, mainly based on the analysis of the wear of the cutting end of the working element. The ploughshare made of 40X steel with a coating on the working surface,

Table 4. Change in ploughshare mass during testing

Hardening technology	Initial weight, g	Reduction in mass during processing of 1 *ha*/mass, g	Mass reduction during processing of 3 *ha*/mass, g	Mass reduction during processing of 5 *ha*/mass, g	Reduction in mass during processing of 9 *ha*/mass, g
Hardening	4212	4050.5/161.5	3974.5/237.5	3588/624	3297/915
Powder spraying	4251	4187.5/63.5	4069/182	3935.5/315.5	3727/524

when cultivating 9 hectares of land (ha), determined a change in mass of 524 g, the average weight of one ploughshare made of L53 series steel in this cultivation, when measuring, it was found that the average mass of one ploughshare was reduced by 915 g. The parameters of the tested ploughshare before its complete unsuitability for work were 14.5 ha.

The reduction in the mass of the plow (ΔM, %) due to scratching and friction with substances contained in the soil also allows us to assess the level of wear resistance (Fig. 4).

When analyzing the results of processing (8–10 ha) of soil, the plow we studied determined that the mass of the working element of the ploughshare made of grade 40X steel decreased by 9–10%, in contrast to the mass of the base of the ploughshare made of L53 steel, which decreased by 20–22% for the same amount of work.

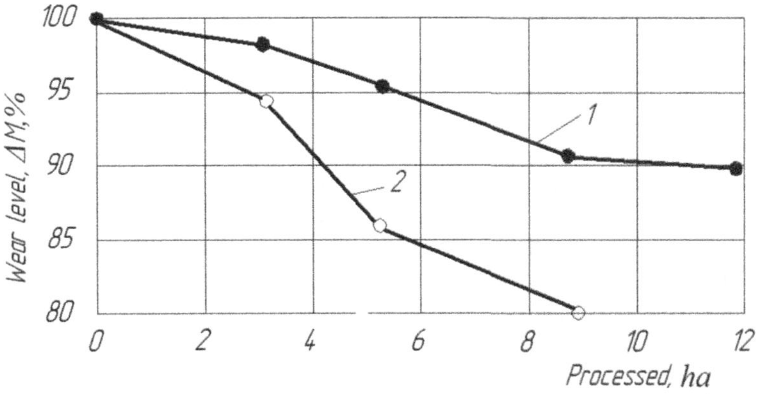

Fig. 4. Reducing the weight of the plow share: dynamics of bending of the share with a blade made of coated steel grade 40X; 2- dynamics of bending of the cutting part of the blade made of steel L53.

Based on the analysis of the obtained results, it was determined that the production of coatings from high-chromium iron-based powder made from grade 40X steel will increase the service life of earthwork parts by 1.6–1.7 times.

When analyzing the corrosion characteristics of the tested ploughshares, it was found that the length of the cutting edge of the ploughshare heated to a standard temperature changed by 4.34 mm/ha, and the length of the cutting edge of the coated ploughshare by 1.89 mm/ha. In this case, the blade of the ploughshare with standard temperature firing has a width of 0.64 mm/ha, and the blade of the ploughshare with a coating has a width of 0.28 mm/ha [27].

4 Discussion of Results

As a result of the analysis of existing works and the studied data, it became clear that no studies were conducted on obtaining wear-resistant coatings from local raw materials using gas flame from high-chromium iron-based powder mixture and studying the effect of only iron and chromium elements on the coating properties.

The use of proven methods for increasing the wear resistance of metal surfaces requires high costs, which leads to an increase in the cost of the part. The method proposed in our study is to obtain wear-resistant coatings using a gas flame from a high-chromium iron-based powder alloy from local raw materials, which is inexpensive and provides high wear resistance.

In the scientific and technological center of JSC "Uzbek Technological Metallurgical Plant" using local raw materials, the developed technology of powder coating of working surfaces of soil-cultivating plowshares was introduced, as a result, the service life of the plowshares increased by 1.7–1.8 times, the cost of capital expenditures for the production of spare parts decreased by 30–35%, wear resistance increased by 50–60%.

5 Conclusions

As a result of the conducted research, a high-chromium powder based on iron from local raw materials was developed, ensuring the formation of a high-quality coating with a hardness of more than 54–56 HRC and reduced porosity to 7%. A technology for applying a powder coating to the working surfaces of a plowshare was created, the use of which made it possible to increase the service life of the plowshare by 1.7–1.8 times.

It has been established that spraying the working surface of a ploughshare made of 40X steel with iron-containing high-chromium powder using the gas flame method makes it possible to significantly reduce the need for expensive L53 steel, thereby reducing the cost of production.

References

1. Betenya, G.F.: Increasing the durability of soil-cutting elements of agricultural machinery by surfacing and freezing. In: BelNIINTI, Minsk, p. 44 (1986)
2. Panteleenko, F.I.: Restoration of Machine Parts, p. 672. Mashinostroenie (2003)
3. Konstantinov, V.M.: Increasing wear resistance when hardening plow shares with diffusion-alloyed cast iron shavings. Repair Restorat. Modernizat. (5), 17–20 (2003)
4. Al-Sayed S. R., Elshazli A. M., Hussein A. H. A. Laser surface hardening of Ni-hard white cast iron // 2020 Metals. 2020. V. 10(6). N. 795. P. 1–19

5. Zhaburenok, S.N.: Increasing the durability of plow shares by surfacing diffusion-alloyed alloys from cast iron shavings and subsequent heat treatment: Author's abstract. dis. Ph.D. tech. Sciences, p. 20. Polotsk State. univ., Novopolotsk (2004)
6. Garkunov, D.N.: Tribotechnics. In: MSKhA, p. 356 (2005)
7. Chikarakara, E., Naher, S., Brabazon, D.: Process mapping of laser surface modification of AISI 316Lstainless steel for biomedical applications. Appl. Phys. A **101**, 367–371 (2010). https://doi.org/10.1007/s00339-010-5843-5
8. Krastev, D.: Improvement of corrosion resistance of steels by surface modification. Corros. Resist. 295–316 (2012). https://doi.org/10.5772/33144
9. Shabani, A., Nahvi, S.M., Raeissi, K.: Effect of heat treatment on structure and oxidation resistance of flame-sprayed Ni-20 wt.% Al on carbon steel. J. Thermal Spray Technol. **30**(3), 739–753 (2021). https://doi.org/10.1007/s11666-021-01158-2
10. Shang, X., et al.: Nano- TiC reinforced [Cr-Fe $_4$ Co $_4$ Ni $_4$]Cr $_3$ high-entropy-alloy composite coating fabricated by laser cladding. J. Mater. Res. Technol. **21**, 2076–2088 (2022). https://doi.org/10.1016/j.jmrt.2022.10.048
11. Günen, A., Soylu, B., Karakaş, Ö.: Titanium carbide coating to improve surface characteristic, wear and corrosion resistance of spheroidal graphite cast irons. Surf. Coat. Technol. **437**, 128280 (2022). https://doi.org/10.1016/j.surfcoat.2022.128280
12. Meng, F., et al.: Development and research status of wear-resistant coatings on copper and its alloys: review. Crystals **15**(3), 204 (2025). https://doi.org/10.3390/cryst15030204
13. Novikov, A.E., Motorin, V.A., Lamskova, M.I., Filimonov, M.I.: Composition and tribological properties of hardened cutting blades of tillage machines under abrasive deterioration. J. Frict. Wear **39**(2), 158–163 (2018). https://doi.org/10.3103/S1068366618020137
14. Mikhalchenkov, A.M.: Technological methods for increasing the service life of ploughshares. In: Rural Machine Operator, no. 2, pp. 39–41 (2008)
15. Tursunbaev, S., Turakhodjaev, N., Odilov, F., Mardanokulov, S., Zokirov, R.: Change in wear resistance of alloy when alloying aluminium alloy with germanium oxide. In: E3S Web of Conferences, vol. 401 (2023). https://doi.org/10.1051/e3sconf/202340105001
16. Novikov, A.E., Motorin, V.A., Lamskova, M.I., Filimonov, M.I.: Composition and tribological properties of hardened cutting blades of tillage machines under abrasive deterioration. J. Frict. Wear **39**(2), 158–163 (2018)
17. Machado, H.D., Aristizabal-Sierra, R., Garcia-Mateo, C., Toda-Caraballo, I.: Effect of the starting microstructure in the formation of austenite at the intercritical range in ductile iron alloyed with nickel and copper. Int. J. Metalcasting **14**(3), 836–845 (2020)
18. Yu, D., Zhou, T., Zhou, H., Bo, H., Lu, H.: Non-single bionic coupling model for thermal fatigue and wear resistance of gray cast iron drum brake. Optics Laser Technol. **111**, 781–788 (2019)
19. Bataev, V.A., Bataev, A.A., Alkhimov, A.P.: Methods of structural analysis of materials and quality control of parts, p. 224. Nauka (2007)
20. Sarvar, T., Nodir, T., Mardonov, U., Saydumarov, B., Kulmuradov, D., Boltaeva, M.: Effects of germanium (Ge) on hardness and microstructure of Al-Mg, Al-Cu, Al-Mn system alloys. Int. J. Mechatron. Appl. Mech. (16), 179–184 (2024). https://doi.org/10.17683/ijomam/issue16.21
21. Panneerselvam, S., Martis, C.J., Putatunda, S.K., Boileau, J.M.: An investigation on the stability of austenite in Austempered Ductile Cast Iron (ADI). Mater. Sci. Eng. A **626**, 237–246 (2015)
22. Tursunbayev, S., Turakhodjayev, N., Mardanokulov, S., Zokirov, R., Odilov, F.: The effect of lithium on the mechanical properties of alloys in the Al-Li system. E3S Web Conf. **390**, 05046 (2023). https://doi.org/10.1051/e3sconf/202339005046
23. Berdiev, D.M., Tashmatov, R.K.: Umarova MA increasing die durability in cold stamping by quenching with intermediate tempering. Russ. Eng. Res. **42**(10), 1011–1013 (2022)

24. Berdiev, D.M., Yusupov, A.A., Umarova, M.A.: Kamilova GM improving the wear resistance of high-precision dies. Russ. Eng. Res. **44**, 1553–1555 (2024)
25. Berdiev, D.M., Yusupov, A.A., Umarova, M.A.: Decreasing gear wear by improved thermocyclic treatment. Russ. Eng. Res. **43**(1), 52–54 (2023)
26. Berdiev, D.M., Konstantinov, V.M., Umarova, M.A.: Increasing the wear resistance of engineering parts using thermal cycling treatment. Adv. Transdisc. Eng. **58**, 243–249 (2024)
27. Wang, S., et al.: Exploring the hardness-independent wear behavior of typical wear-resistant materials under dynamic and static conditions. J. Mark. Res. **33**, 6798–6809 (2024)

Highly-Compatible Geometric Calibration Method for Quadrupole Electromagnets Used in Particle Accelerators

Marius Popa[1], Victor Eduard Albei[1,2], Dragos Ovezea[1], Cristinel Ilie[1], Nicolae Tanase[1](✉), Daniel Lipcinski[1], Dan Enache[1], Florentina Badea[3], and Sorin Ionuț Badea[3]

[1] National Institute for Research and Development in Electrical Engineering ICPE-CA, Bucharest, Romania
{marius.popa,eduard.albei,dragos.ovezea,cristinel.ilie, nicolae.tanase,daniel.lipcinski,dan.enache}@icpe-ca.ro

[2] Center for Research and Training in Innovative Techniques of Applied Mathematics in Engineering "Traian Lalescu" (CiTi UPB), University Politehnica of Bucharest, Bucharest, Romania

[3] National Institute of Research and Development in Mechatronics and Measurement Technique (INCDMTM), Bucharest, Romania
{florentina.badea,sorin.badea}@incdmtm.ro

Abstract. This paper aims to demonstrate the method by which a single experimental setup may be used in order to determine the geometric calibration constants and the adherence to manufacturing tolerances for a large number of particle accelerator electromagnet models, regardless of their size or function within the linear or storage ring accelerator assemblies. Our method is meant to serve as a more flexible alternative to the widely-used rotating search coil magnetometer method, which may be intractable or impossible for harmonic content measurements involving electromagnet apertures which are either oversized or, respectively, undersized relative to the rotating coil assembly. The hereby proposed method is heavily reliant on the usage of a triaxial Gaussmeter Hall probe together with certain signal processing techniques and an application-specific mathematical interpretation of the magnetostatic context within the core of an accelerator electromagnet. As part of this paper, we build upon both fundamental concepts of electromagnetic field theory and recent advancements in order to provide the proofs and mathematical insight on why this method is justified and how it relates to the rotating coil method, which it attempts to substitute for. Our method's application is showcased on a quadrupole beam-focuser electromagnet model for the linear accelerator (LINAC) part of Variable Energy Gamma (VEGA) system from ELI-NP. We then compare the results of the experimental activity with FEM simulations of the ideally-occurring field inside the electromagnet apertures and conclude our paper by attempting to explain the inconsistencies between the two sets of results.

Keywords: Electromagnet · Harmonic content · Magnetostatics · Quality assurance · Calibration · Electron beam · Quadrupole · Hall probe

D. D. Cioboată and J. Machado (Eds.): ICoRSE 2025, LNNS 1592, pp. 27–40, 2025.
https://doi.org/10.1007/978-3-032-02508-1_3

1 Introduction

This paper aims to disseminate our findings regarding the tasks of calibration and quality assurance in the manufacturing of particle accelerator electromagnets. The 3D hall probe method presented below is supported by fundamental concepts of electromagnetic field theory and is proven to be more flexible than the rotating coil method [1–5]. These tasks were originally attributed to INCDIE ICPE-CA's Electromechanical Systems and Technologies Department (STE Dept.) as part of our collaboration with the Extreme Light Infrastructure for Nuclear Physics (ELI-NP, Romania) research institute. Our joint working committee was interested in developing the technological means for manufacturing research-grade particle accelerator electromagnets locally, with the purpose of aiding in the construction of the Variable Energy Gamma (VEGA) System at ELI-NP. This extensive piece of research infrastructure is a gamma-ray source based on a storage ring, designed to provide users with gamma-ray beams covering a variable energy range from 1 MeV to 19.5 MeV. The VEGA system relies on the Inverse Compton Scattering (ICS) phenomenon, where gamma photons are generated by the inverse scattering of laser photons off a beam of relativistic electrons. It utilizes an electron storage ring and a high-gain optical cavity. The electron beams are produced in an RF electron source, accelerated in a linear accelerator (LINAC), and then transported for injection into the storage ring, which operates in a stable regime with a quasi-constant current, with losses being compensated by regular electron top-ups [6–10].

Building upon ICPE-CA's partnership [11] with the Facility for Antiproton and Ion Research (FAIR, Germany), our department has proposed the implementation of a general-purpose experimental setup capable of providing quality assurance for a wide array of electromagnet models. Our work is therefore focused[1] on acquiring and processing electromagnetic harmonic content data. We are aware of the fact that the theoretical basis on which our work unfolds has been thoroughly documented by others [12]; the novelty factor of our work pertains entirely to the technological implementation of a highly-optimized setup capable of contributing to series production processes. Following our success in accomplishing these quality assurance tasks, our department was restructured to also provide services in the field of validating and designing solutions for electromagnetic devices, henceforth becoming the STEE Department.

1.1 Motivation for Developing the Method

The main motivation behind implementing a method that leverages the small-form factor of the Gaussmeter 3D Hall probe is the mismatch between the mechanical dimensions of our on-site rotating coil measurement device and the much smaller aperture of our devices under test (DUTs), see Fig. 2. Simply put, our more conventional device doesn't fit inside the quadrupole magnet's aperture, as shown in Fig. 1.

Figure 2 presents the 3D model of a quadrupole beam-focuser electromagnet for the linear accelerator (LINAC), that we were tasked with characterizing and calibrating.

[1] For the sake of completeness, we insist on including mathematical proofs of even the simpler concepts that we have utilized, it is our belief that papers such as this one should be student-accessible and stand-alone readable, even if they reference other works.

a b

Fig. 1. The on-site rotating coil measurement device (a) compared to the entire in-line-four electromagnet array (b), on its mounting frame—for reference, the micrometer screw in (a), indicated by the red bounding box, is roughly the diameter of the electromagnet's aperture.

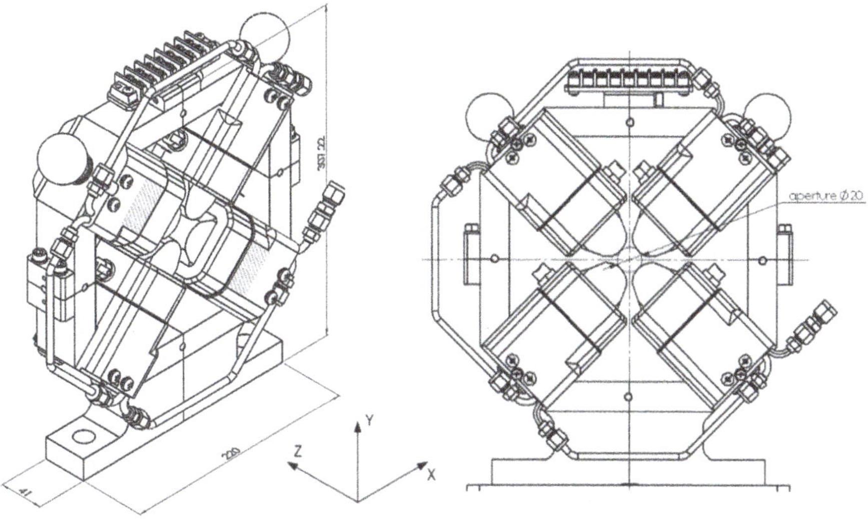

Fig. 2. The 3D model and front view of a quadrupole beam-focuser electromagnet for which our experimental setup was validated, reference dimensions are in millimeters [6–9].

1.2 Physical and Mathematical Constraints

Our test stand was designed to enforce a minimum number of constraints on the electromagnetic DUTs, allowing for increased portability across multiple research and manufacturing processes. Among the physical constraints, the most significant is the DUT's mechanical extent across its ideal magnetic axis (here, the Z-axis, see Fig. 2), potential violation of this constraint normally only occurs when approaching multi-part electromagnet assemblies, but does not represent an issue for small-size assemblies such as the in-line-four focusing array shown in this paper. In such unfavorable cases, our solution allows for the individual testing of each assembly component individually, following a computer-aided compilation of the testing data across multiple components.

Regarding the mathematical constraints that arise from the method's means of optimizing the number of measurements, we may list the following:

1. Our work unfolds under the assumption of a magnetostatic context, meaning no B-field time variation and no electrical currents existing within the space associated to the measurements;

$$\nabla \times \boldsymbol{B} = \mu_0\big(\boldsymbol{J} + \varepsilon_0 \tfrac{\partial E}{\partial t}\big) = 0 \text{ since } \boldsymbol{J} = 0;\ \tfrac{\partial E}{\partial t} = 0 \tag{1}$$

2. Measurements are only valid when taken as close as possible to the electromagnet's Z-plane of symmetry, near the coils, where the variation of the B-field across the Z axis is negligible and variations on the X and Y axes reach their peaks, given the mechanical footprint of the electromagnet and the constraints regarding gaussmeter probe positioning;

$$\frac{\partial B_z}{\partial z} = 0 \tag{2}$$

2 Experimental Instrumentation and Methods

Our quality assurance tasks included: the validation of coil positioning around the electromagnet's aperture, functional coil matching, determination of the deviation of the magnetic axis relative to the mechanical Z-axis and the local coordinates of the intersection of the magnetic axis with the mechanical Z-plane of symmetry.

All of the characteristics listed above are figures of merit (FoM's) for the electromagnet and they may be efficiently determined by means of processing harmonic content[2] data.

Therefore, we are working with a spatial Fourier decomposition, not the time-to-frequency transform that our readers may be accustomed to from classical signal processing. It will be shown that obtaining a functional representation of the B-field vector for any point along the circular contour also allows for the estimation of the B-field vector in any point within the same circular contour. It will further be shown that this functional approximation of the B-field is also valid for points outside of the contour, but only under certain conditions (8).

Incorrect coil positioning or substantial coil mismatching introduce harmonic distortions. Ideally, an N-pole electromagnet should generate a harmonic content that is purely concentrated in its $N/2$ th harmonic.

[2] The harmonic content of an electromagnet is the Complex Fourier Series decomposition of the X and Y components of the B-field vector as function of the angle along a circular contour of a given reference radius, defined within a plane that is parallel to the Z-plane of symmetry of the electromagnet.

The misalignment of the magnetic axis may be fully characterized by performing a three-dimensional linear fit across the coordinates of the B-field minima determined from multiple aperture sections[3] using the B-field's harmonic representation[4].

2.1 Harmonic Content Computation

We begin by stating that the Z component of the B-field vector is mostly irrelevant to our study. This is because, under normal circumstances, the Z component is parallel to the electron beam, thus exerting no Lorentz force on the charged particles.

As such, we may begin our work by assuming that we're only interested in a function that describes the dependency of the X and Y components of the B-field in relation to the spatial coordinates X and Y within an aperture section:

$$B : \mathbb{R}^2 \to \mathbb{R}^2 \tag{3}$$

Or, since we can represent any element $(x, y) \in \mathbb{R}^2$ as an element $z = x + jy \in \mathbb{C}$:

$$B : \mathbb{C} \to \mathbb{C} \tag{4}$$

In order for us to approximate the B-field along a circular contour by its Complex Fourier Series decomposition, we must first be able to approximate it by a power series [4]:

$$B(z) = -\frac{\mu_0 I}{2\pi z_0} \sum_{n=0}^{\infty} \left(\frac{z}{z_0}\right)^n = -\frac{\mu_0 I}{2\pi} \sum_{n=0}^{\infty} \frac{z^n}{z_0^{n+1}} \tag{5}$$

Consider the polar coordinates representation ($z = \rho e^{j\theta}$) of $B(z)$, the Complex Fourier Series decomposition is now self-evident:

$$B(z) = \sum_{n=0}^{\infty} \left(-\frac{\mu_0 I}{2\pi} \cdot \frac{\rho^n}{\rho_0^{n+1}} \cdot e^{-j(n+1)\theta_0} \right) e^{jn\theta} = \sum_{n=0}^{\infty} \mathrm{CFS}_n e^{jn\theta} \tag{6}$$

Therefore, given a circular contour of known radius ρ_{ref}, centered in the origin of the aperture section's local coordinate system, we may sample the B-field in 2 N equidistant points on the contour in order to determine the first N harmonic components $\mathrm{CFS}_{n,\text{ref}}$ $n \in \overline{0, N-1}$ of the B-field on that contour, according to the Nyquist-Shannon sampling theorem.

Then, the B-field in all other points within the circular contour may be determined using the normalized coefficients, c_n:

$$c_n = \frac{CFS_{n,\text{ref}}}{\rho_{\text{ref}}^n} = -\frac{\mu_0 I}{2\pi} \cdot \frac{1}{\rho_0^{n+1}} \cdot e^{-j(n+1)\theta_0} \Rightarrow B(z) = \sum_{n=0}^{\infty} c_n z^n \tag{7}$$

[3] I.e.: planes parallel to the Z-plane of symmetry, but on different Z-coordinates.

[4] In a more naive approach, the coordinates of the B-field minima may be determined by performing a form of stochastic gradient descent within the chosen aperture section, however this method is far too slow for series production tasks and its resolution is limited by the minimum step of the machine used for positioning the Gaussmeter 3D Hall probe.

Remark: Equations (6) and (7) are only valid under:

$$\left|\frac{z}{z_0}\right| < 1 \tag{8}$$

2.2 Magnetostatic Potential Computation

The particular magnetostatic context (1) within our aperture section allows for the definition of a scalar potential function[5] ϕ_m, such that $\boldsymbol{B} = -\nabla\phi_m$. Then, ϕ_m uniquely defines the vector field $\boldsymbol{B}$ and it may be computed as:

$$\phi_{m,P_0}(P_1) = \int_{P_0}^{P_1} \boldsymbol{B} \cdot \boldsymbol{dl} \tag{9}$$

where P_0 is the point for which the reference potential $\phi_m = 0$ is considered.

By substituting (7) into (9), we may express $\phi_m(z)$ in terms of $B(z)$'s harmonic content, as an integral over a real interval[6]:

$$\phi_m(z_1) = \mathrm{Re}\left\{\sum_{n=0}^{\infty} -j\frac{c_n}{n+1}z_1^{n+1}\right\} \tag{10}$$

Since:

$$\boldsymbol{A} \cdot \boldsymbol{B} \equiv \mathrm{Re}\{\overline{z_A} \cdot z_B\};\ \sum_{n=0}^{\infty} j \cdot \overline{c_n\left(\frac{z_1}{|z_1|}s\right)^n} \equiv \boldsymbol{B};\ \frac{z_1}{|z_1|}ds \equiv \boldsymbol{dl} \tag{11}$$

Here, point P_0 is the origin of the local coordinate system of the aperture section, hence why integration starts in 0.

Finally, by once again using the polar coordinates representation:

$$\phi_m(z) = \sum_{n=0}^{\infty} \frac{\rho_{c,n}}{n+1}\rho_z^{n+1}\sin\left[(n+1)\theta_z + \theta_{c,n}\right] \tag{12}$$

From this point onward, the process of determining quantities such as the quadrupole field strength B' [5] on the basis of the $\phi_m(z)$ scalar potential function coincides with the more classical approach of the rotating coil method [1]. This becomes evident when comparing (10) with expression (13) of the magnetic flux $\phi_{m,\mathrm{coil}}(z)$ through the perimeter described by a rotating coil of length L and N_t turns:

$$\phi_{m,\mathrm{coil}}(z_1) = -N_t L\mathrm{Re}\left\{\int_0^{z_1} B(z)dz\right\} \propto \mathrm{Re}\left\{\sum_{n=0}^{\infty} \frac{c_n}{n+1}z_1^{n+1}\right\} \tag{13}$$

[5] Our interest in the ϕ_m scalar function is justified by the significance of its harmonic series coefficients in changing the properties of the charged particle beam that is passed through the electromagnet's aperture, as it is shown in [5].

[6] We find that such integrals over real intervals are simpler to describe to interpreters such as MATLAB's.

Hence our choice of denoting the scalar potential function as $\phi_m(z)$, the two are closely related.

2.3 Data Acquisition and Preprocessing

The physical labor required for the measurement tasks was carried out by means of an application-specific assembly consisting of:

1. LS Mecapion custom 3-DOF Cartesian Robotic Platform (Fig. 3b);
2. Wenzel XOrbit 87 Coordinate Measuring Machine (Fig. 3a);
3. Lake Shore Model 460 3-Axis Gaussmeter (Fig. 1b);
4. Lake Shore MMZ-2536-UH 3-Axis Gaussmeter Hall Probe (Fig. 3b);
5. FAST-PS-2020-400 and FAST-PS-M-3030-900 DC current source assembly (Fig. 1b);
6. Locally CNC machined high-precision 3-DOF positioning stand (Fig. 3a).

All of our data acquisition and processing tasks were carried out using the National Instruments LabWindows/CVI data acquisition framework, the Wenzel Metrosoft QUARTIS measuring software and the MathWorks MATLAB computing software. Our results have been checked against models obtained using the COMSOL Multiphysics 5.0 simulation software.

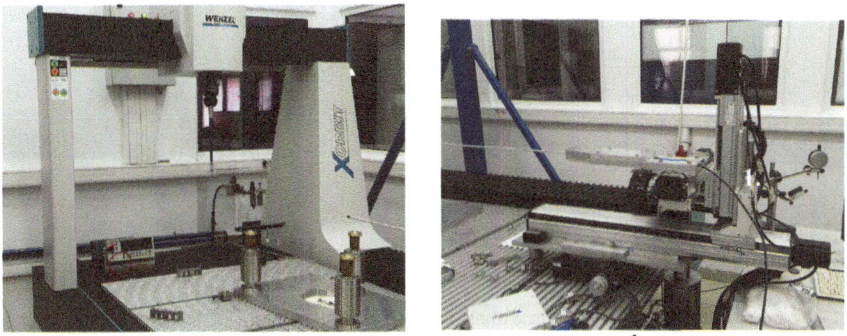

a b

Fig. 3. The main components of our experimental setup—the coordinate measuring machine (left), the DUT positioning stand (a, lower-right corner) and the Cartesian robotic platform (right).

Before engaging in any measurements of the B-field, the controlled positioning of the Gaussmeter 3D Hall probe tip was characterized by means of the coordinate measuring machine. This allowed us to compensate for the probe's slight bending due to manufacturing tolerances and its own weight, when positioned horizontally. Further deformation of the probe under stationary conditions was reduced by means of a temperature-controlled environment, kept at a constant 21 °C. Furthermore, we accounted for the displacement between the probe's tip[7] and the center of each Hall Effect active area of the probe,

[7] Our work defines "the probe's tip" as the center of the square surface perpendicular to the Z-axis, on the measuring end of our probe model.

one for each B-field axis that is measured, see Fig. 4. The mounting frame of the electromagnet assembly itself was similarly positioned using the same coordinate machine, together with the positioning stand's adjustable columns.

While carrying out measurements of the B-field, the electromagnets were powered using the high-precision laboratory DC current supply and were temperature-controlled by means of a proprietary hydraulic cooling system. The Gaussmeter 3D Hall probe's exponentially-decaying mechanical oscillations in response to the Cartesian robotic platform's acceleration forces were studied, and corresponding time delays were added between each repositioning instruction and the subsequent B-field measurement, giving the exponential decay enough time to manifest itself in a useful manner. To further compensate for any mechanical perturbation and any noise present in the electronic system, B-field measurements were carried out using the gaussmeter's time-averaging function. This is proven to increase the SNR, given our signal's assumed ergodicity.

In order to compensate for systematic errors related to the mechanical backlash of the Cartesian robotic platform's actuators, we parse the list of measurement coordinates spatially, from the bottom-up and from left-to-right, with a deterministic number of direction changes, taking place in deterministic positions.

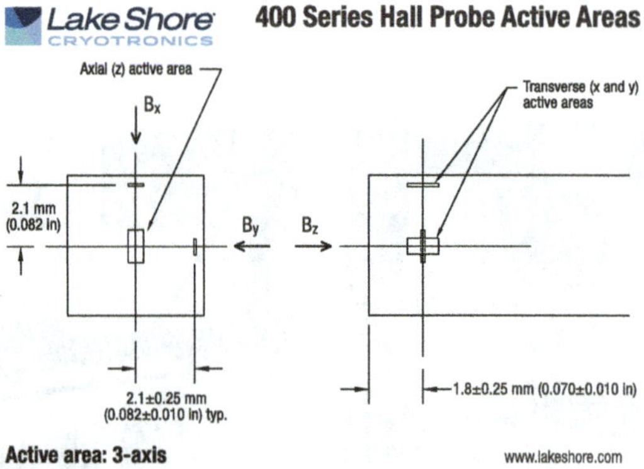

Fig. 4. Manufacturer-provided approximation for the displacement of the Gaussmeter 3D Hall probe active areas. Graphics courtesy of Lake Shore Cryotronics, Inc.

In order to validate the alignment between the magnetic axis and the mechanical Z-axis of the electromagnet model, our method consists of utilizing the c_n harmonic coefficients together with MATLAB's function minimizing algorithms in order to determine the spatial coordinates of B-field minima across multiple slices of the electromagnet model along the mechanical Z-axis. The path defined by all points of B-field cancellation is the axis on which the focusing effect of the quadrupole field takes place in an ideal manner.

If time allows it, after the B-field minima linear fit is performed and the geometric calibration constants are determined, the electromagnet's positioning may be subsequently adjusted such that another round of measurements is carried out, having the previously identified magnetic axis coincide with the mechanical Z-axis of the Cartesian robotic platform[8]. This second round of measurements gives a confirmation of the geometric calibration and may be additionally used in the highly-accurate determination of the electromagnet's intended main harmonic strength. Our work showcases the effect of poor geometric calibration upon the perceived harmonic content of the electromagnet, see Fig. 5 and Fig. 6.

As previously stated, computing the first N harmonics of the B-field requires a theoretical minimum number of 2 N samples taken along the ρ_{ref} radius circular contour. Despite this, displacement step-size limitations of the Cartesian robotic platform impose the requirement for linear interpolation of multiple measurements. Since displacements on the Z-axis do not require an increased level of precision, we only perform interpolation between four neighboring points within the respective Z-plane. Moreover, we account for the displacement of the Hall Effect active areas by performing each measurement with three different compensation offsets. These precautions increase our minimum number of measurements up to 24 N samples in the worst case scenario under which no measurement coordinates happen to usefully overlap.

3 Results

The measurement data obtained using our experimental setup is visualized in this part of our work. Based on multiple B-field planes determined as in Fig. 5 and Fig. 6, we determined the defining parameters of our electromagnet's magnetic axis, in terms of the mechanical coordinate system of the in-line-four assembly's mounting frame, see Fig. 7. These parameters effectively define the geometric calibration of the electromagnet[9].

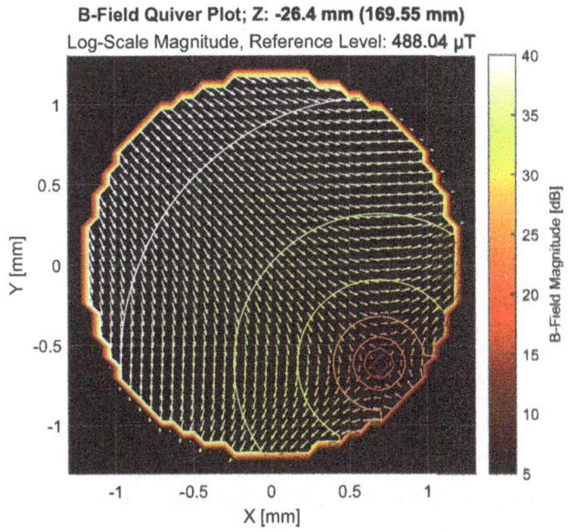

Fig. 5. Quiver plot of the B-field within the electromagnet's aperture; note the offset of the magnetic axis.

[8] I.e.: Having the second round of measurements be performed on the geometrically calibrated electromagnet, hence confirming the validity of the calibration constants.

[9] By knowing these parameters, micrometer adjustment screws on the mounting frame may be used to properly align all of the magnetic axes of the electromagnets with the main axis required by the VEGA System.

FEM simulations of the electromagnet's aperture offer highly similar results. The area represented in Fig. 5 corresponds to the center of Fig. 10, with an extra offset whose accurate determination is effectively our main task, i.e.: geometric calibration.

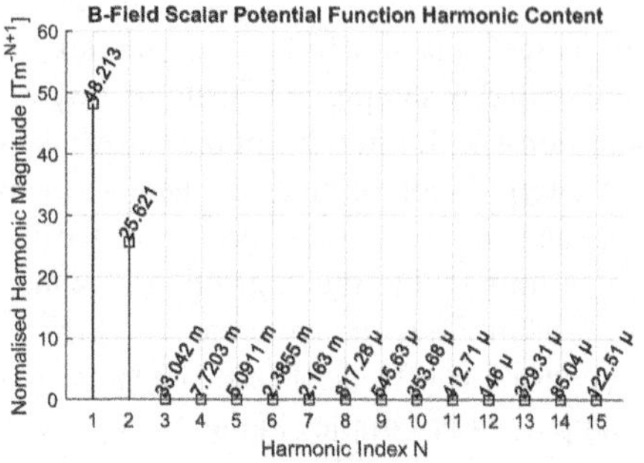

Fig. 6. Harmonic content of the electromagnet's potential function ϕ_m; note the 1st harmonic component's magnitude, it is off the charts due to the electromagnet's magnetic axis not being properly aligned with the assumed mechanical Z-axis, an offset in the Z plane introduces this type of 1st harmonic unbalance.

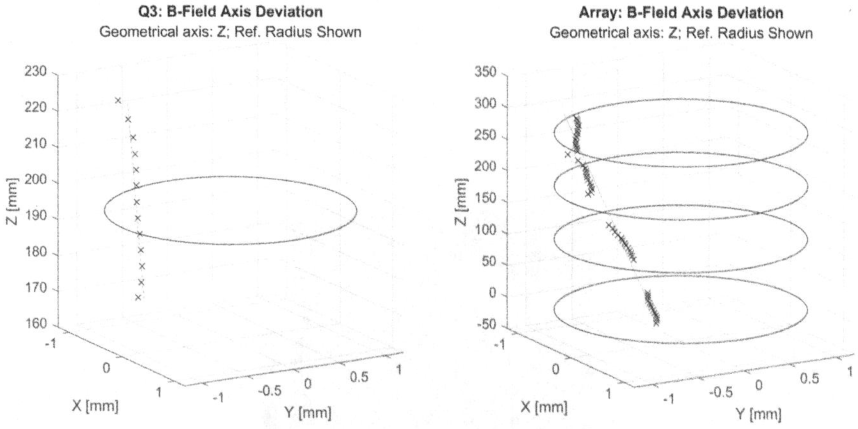

a b

Fig. 7. Visualisation of the magnetic axis determined based on the measurement data—measurements for a single electromagnet (a) and the entire in-line-four focusing array (b), the reference radius ρ_{ref} (7) circles are shown, both plots use the mounting frame's coordinate system.

The COMSOL Multiphysics 5.0 simulation software was used to assess the accuracy of our measurements. A stationary magnetic field model with a parametric sweep of the current $I_{\mathrm{coil}} \in [1,15]$ [A], was used for the numerical modeling of the quadrupole electromagnet, as follows:

The coil subdomains of our model were defined by relation (14) while the iron yoke subdomains followed (15). Lastly, the B-H constitutive relation (16) was used by the magnetization model.

$$J_e = \frac{NI_{\text{coil}}}{A} \tag{14}$$

where J_e [A/m^2] is the external current density, N is the number of turns for each coil, I_{coil} [A] is the supply current, A [m^2] is the cross–sectional area of the conductor in the coil.

$$\nabla \times \boldsymbol{H} = \boldsymbol{J} \tag{15}$$

where **H** [A/m] is the magnetic field strength, **J** [A/m^2] is the current density.

$$\boldsymbol{B} = f(\|\boldsymbol{H}\|)\frac{\boldsymbol{H}}{\|\boldsymbol{H}\|}\nabla \times \boldsymbol{H} = \boldsymbol{J} \tag{16}$$

A magnetic insulation boundary condition $\mathbf{n} \times \mathbf{A} = 0$ encloses the model, where A [T/m] is the magnetic vector potential, **n** is the normal outward from the plane.

The geometric model consists of four subdomains representing the iron yoke, eight subdomains representing four coils, while the rest of the geometry is defined as a set of air subdomains, as shown in Fig. 8 and Fig. 9.

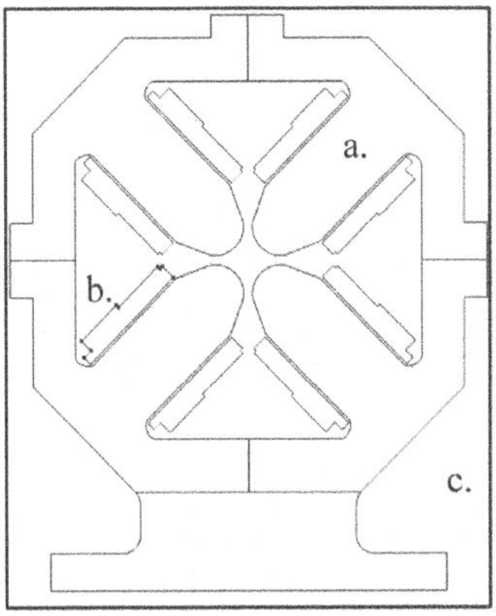

Fig. 8. Computational domain for the electromagnet—a. Four-part iron yoke (soft iron, lossless); b. Eight coils (copper); c. Air (all other geometry).

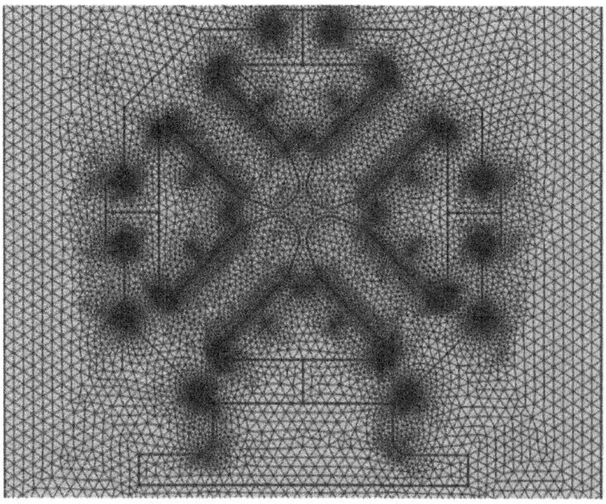

Fig. 9. The mesh used for the task of computing the FEM simulation of the quadrupole electromagnet (consisting of 71429 free triangular elements); the magnetic insulation boundary condition may be seen on the lateral sides of the figure.

Figure 10 displays the B-field generated by the quadrupole. The maximum reached value is 2.37 T, an acceptable value which does not saturate the iron yoke.

Figure 11a depicts the excitation curve of the electromagnet, simulated for $I_{\text{coil}} \in [1,15]$ [A]. A slow saturation occurs starting at a current of $\approx$12 A. We observe a high degree of similarity between our measurements in Fig. 11b and the FEM simulation.

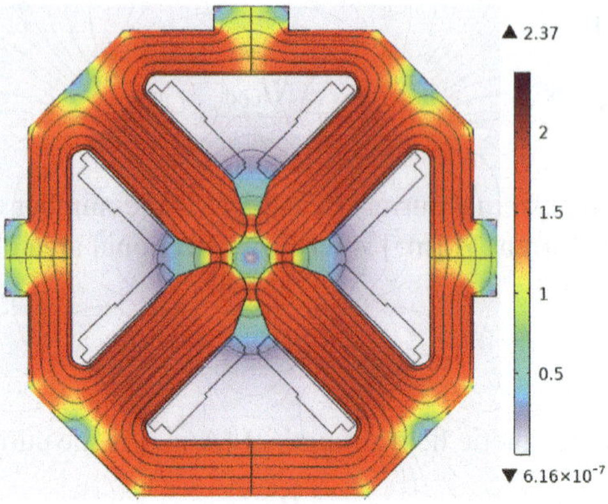

Fig. 10. Simulated magnetic flux density (for $I_{\text{coil}} = 15$ A)—color map and streamline.

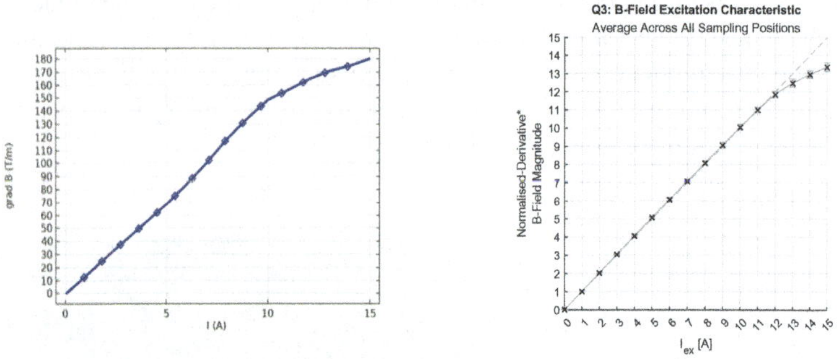

a b

Fig. 11. FEM-simulated excitation characteristic (a), compared to an averaged[10] excitation characteristic (b) along multiple points in space within the electromagnet's aperture.

4 Conclusions

To conclude our work in developing a highly compatible experimental setup by means of which particle accelerator electromagnets may be calibrated and checked for compliance, we may summarize our findings.

By using the Wenzel XOrbit 87 Coordinate Measuring Machine inside a temperature-controlled environment with the purpose of ensuring a minimum number of unknown process parameters, we have demonstrated that our Gaussmeter 3D Hall probe method offers reasonably accurate results, when compared to the rotating coil method—the commonly accepted method for such measurements. This claim is indirectly supported in

[10] It is apparent that the two plots are not to scale and do not use the same units, this is because the characteristic in Fig. 11b is obtained by measuring the excitation characteristic in multiple points in space within the electromagnet's aperture, then by normalizing all measured characteristics such that their derivative in $I_{\text{coil}} = 0$ is unitary and finally performing an average on all normalized characteristics. This process is justified since we are strictly interested in obtaining a general description of the non-linearity of the behavior.

our work, not by physically comparing the two methods, but by comparing our measurement results with FEM simulation data. Further, we may argue that due to the reduced electromechanical complexity of the Gaussmeter 3D Hall probe and its respective handling system, our method is less prone to user error due to poor configuration. In contrast, other works [1] have shown that the rotating coil method often involves sophisticated coil assemblies that must be thoroughly characterized before being involved in a measurement setup.

Regarding the more prominent differences between our measurements and the FEM simulation data, we observe that they are strictly due to the mechanical tolerances of the electromagnet's mounting frame, such differences are expected and, in fact, our main task is their accurate description by means of the geometric calibration constants. It must be noted that the main drawback of our method is its time-consuming data acquisition stage. Considering the need for time-averaging the Gaussmeter 3D Hall probe signals and the high number of 24 N measurements required for determining the first N terms of the electromagnet's harmonic content, a highly-accurate measurement run may take up to eight hours. The data acquisition stage may be accelerated by using a Gaussmeter 3D Hall probe which implements a more expensive triaxial MEMS device, instead of the current configuration, consisting of three monoaxial devices closely encased together. This would remove the need for the three separate compensation offsets, effectively reducing the number of measurements to 8 N. Other improvements include a more rigid probe structure that would be less subjected to mechanical oscillation and thus would require less settling time; a time-averaging scheme that is purposefully defined to cancel out the most harmful noise harmonics detected in the Gaussmeter 3D Hall probe's signal under null B-field conditions (realistically, the only such harmonic would be the 50 Hz power grid noise) and lastly, a method for confirming geometric calibration data by intentionally introducing known mechanical offsets between two measurement runs and characterizing the differences between the two resulting sets of calibration constants.

References

1. Köster, O.: Combining rotating-coil measurements of large-aperture accelerator magnets, (Doctoral dissertation). Darmstadt Technical University, Germany (2016)
2. Jain, A.K.: Basic theory of magnets (1998)
3. Mess, K.H., Wolff, S.: Superconducting Accelerator Magnets, pp.65–79. World Scientific, Singapore (1996)
4. Todesco, E., Prestemon, S., Ferracin, P.: Unit 5—field harmonics. In: USPAS course "Superconducting Accelerator Magnets". Lawrence Berkeley National Lab, US (2015).https://indico.cern.ch/event/440690/contributions/1089751/attachments/1143438/1638603/U5rr.pdf. Accessed 16 Apr 2025
5. Spentzouris, L., Barletta, W., Harms, E.: Magnetic Elements Lecture. In: USPAS course "Fundamentals of Accelerator Physics and Technology with Simulations and Measurements Lab (undergraduate level)". Michigan State University, US (2012). https://uspas.fnal.gov/materials/12MSU/magnet_elements.pdf. Accessed 16 Apr 2025
6. *** Extreme Light Infrastructure - Nuclear Physics. www.eli-np.ro
7. *** Lyncean Technologies, Inc. Technical Design Report Variable Energy Gamma (VEGA) System, unpublished (2020)

8. Cîrdei, C.: Variable Energy Gamma (VEGA) System – overview. https://www.i-tech.si/wp-content/uploads/2023/05/Cirdei.C_compressed.pdf. Accessed 19 May 2025
9. Cîrdei, C.: Overview of RF generation, distribution and LLRF systems for ELI/NP-VEGA project. https://www.i-tech.si/wp-content/uploads/2024/04/C.C.pdf. Accessed 19 May 2025
10. Chen, G.-L., Tian, S.-Q., Schubert, H., Udup, E., Matei, C.: A tool for beam optics design and its application on the VEGA transport line at ELI-NP, JINST 19 T04008 (2024). https://doi.org/10.1088/1748-0221/19/04/T04008
11. Daniel, D., Chiriță, I., Tănase, N., Mălureanu, S.: Analysis, characterization and optimization of a steerer magnet prototype. In: The 9th International Symposium on ATEE. Bucharest, Romania (2015)
12. Campmany, J., Marcos, J., Massana, V.: Determination of magnetic multipoles using a Hall probe. In: IPAC14, vol. 4025 (2014)

Development of an Automated Control System for Laboratory Hydroponic Cultivation

Pierluigi Rea[1](✉), Giovanni E. Castellet y Ballarà[1], Maurizio Ruggiu[1], and Erika Ottaviano[2]

[1] University of Cagliari, Cagliari, Italy
pierluigi.rea@unica.it
[2] University of Cassino and Southern Lazio, Cassino, Italy

Abstract. In recent years, the increasing demand for sustainable and efficient agricultural solutions has driven the widespread adoption of innovative techniques. Hydroponics, a soil-free cultivation method that uses aqueous nutrient solutions, offers several advantages, such as more efficient water usage, higher planting density, and the ability to grow in controlled environments. However, the success of hydroponic cultivation largely depends on maintaining optimal growing conditions, which requires precise monitoring and control of variables like temperature, humidity, pH, and nutrient concentration. In this context, automation plays a critical role. Integrating automated control systems can significantly enhance operational efficiency and crop quality while reducing human intervention and minimizing the potential for errors. This work aims to design and develop an automated control system for hydroponic cultivation that incorporates various sensors for real-time data collection and utilizes control algorithms to regulate environmental conditions automatically. The system can continuously monitor key growth parameters, ensuring an ideal environment for plant growth. The research focuses on developing a laboratory prototype to demonstrate its effectiveness and scalability.

Keywords: Mechatronics · PLC Programming · Laboratory set-up · Hydroponics

1 Introduction

Hydroponics is a plant cultivation technique that uses aqueous nutrient solutions instead of soil. In a hydroponic system, plant roots are suspended in channels through which a nutrient solution flows, providing all the essential nutrients for growth, including nitrogen, phosphorus, potassium, calcium, magnesium, and trace elements [1, 2]. This solution is continuously monitored and adjusted to maintain optimal pH levels and nutrient concentrations [3]. Research and business development in this field have notably surged in recent years. Studies indicate that the number of hydroponics-related publications has grown significantly, with a 381% increase between 2010 and 2019. 2019 saw the highest number of publications, marked by a peak interest in the Nutrient Film Technique (NFT)

D. D. Cioboată and J. Machado (Eds.): ICoRSE 2025, LNNS 1592, pp. 41–50, 2025.
https://doi.org/10.1007/978-3-032-02508-1_4

[4]. Brazil emerged with the most hydroponics-related publications, followed by Italy, the United States, and Germany. Together, these four countries accounted for 40.8% of the total published studies in 2019. These trends underscore the growing recognition of hydroponics, especially the NFT method, as a sustainable agricultural practice, with varying levels of research and development activity across different nations [5]. There are several types of hydroponic systems, each offering distinct characteristics and benefits [6]:

- **Nutrient Film Technique (NFT)**: Plant roots are suspended in this system's thin, continuously flowing film of nutrient solutions. It is highly efficient in terms of both water and nutrient usage.
- **Drip System**: The nutrient solution slowly drips onto the plant roots. This versatile system can be used for a wide variety of crops.
- **Deep Water Culture (DWC)**: Plant roots are submerged directly in an oxygenated nutrient solution. This straightforward method is effective for promoting rapid growth in certain crops.
- **Ebb and Flow System**: The nutrient solution periodically floods the grow bed before draining away, providing the roots with oxygen between cycles.
- **Aeroponics**: Plant roots are suspended in the air and misted with a fine spray of nutrient solution. This system promotes excellent oxygenation of the roots and encourages fast plant growth.

Monitoring systems for hydroponic cultivation can vary from basic setups, which control only the circulation pump and nutrient dosing, to more advanced and comprehensive systems that monitor a wide range of parameters, including water quality and the growing environment. These systems may also integrate IoT technology for intelligent, remote management, enabling sensor data to be stored in the cloud for easy access and manipulation [7]. The most advanced systems typically include lighting, refrigeration, ventilation, hydraulic pumps, and other innovative technologies to regulate plant growth, often leveraging cameras and predictive analysis algorithms to reduce human intervention.

Numerous studies have explored specific challenges in hydroponic cultivation, aiming to develop machine learning models that address key aspects such as nutrient management, pH levels, water oxygenation, light intensity, and determining the optimal light spectrum for photosynthesis, root development, and product quality. The application of IoT solutions provides a wealth of opportunities to enhance these processes.

A fully integrated hydroponic management system generally consists of [8]:

- **Water Control System**: Monitoring nutrient solution parameters, tanks, and acidity regulators.
- **Irrigation System**: Management of hydraulic pumps, channels, and water reserve levels.
- **Microclimate Monitoring System**: Measurement of temperature, relative humidity, and CO2 levels.
- **Ventilation and Heating System**: Control the microclimate to maintain optimal conditions.
- **Lighting System**: Full control in closed environments or integration with natural sunlight.

- **Supervision System**: Centralized software to control and manage all the aforementioned subsystems.

All these control and implementation systems are essential for meeting the various requirements necessary for plant development, ensuring continuous monitoring and maintenance. The water control system plays a crucial role in managing the nutrient and chemical conditions of the solution [9]. Plants require both macronutrients, such as nitrogen, phosphorus, and potassium, and micronutrients, such as zinc, copper, manganese, boron, and iron, as well as secondary nutrients, such as calcium, magnesium, and sulfur. Electrical conductivity measurements are directly linked to the concentration of conductive ions in the solution, although some limitations in this area are still under study. The acidity or alkalinity of the solution also influences plant growth and root health, as pH levels affect nutrient availability and the roots' ability to absorb them. In hydroponic cultivation, the pH is typically between 5.5 and 6.5 for optimal plant growth.

The temperature of the nutrient solution is closely tied to the amount of oxygen absorbed by the roots, with a negative correlation to the amount of oxygen dissolved in the solution. The temperature of the root zone significantly impacts both plant growth and the chemical composition of the nutrient solution. Additionally, air temperature is crucial, as it affects how plants convert carbon dioxide, light, and water into nutrients through photosynthesis. Incorrect temperatures can hinder photosynthesis and leaf respiration, reducing plant growth and productivity. The optimal temperature range for hydroponic cultivation generally falls between 18 °C and 26 °C to support healthy plant development. However, more specific temperature ranges may be necessary for different types of crops (for instance, tomatoes require higher temperatures than spinach).

Humidity management is crucial as it influences leaf respiration and the development of harmful ecosystems, such as fungi, molds, and diseases. Stomatal conductance generally decreases as the relative humidity deficit in the air increases. In conditions of low relative humidity (high vapor pressure deficit), the rate of evaporation from the leaves rises, leading to excessive water loss. As a result, stomata close, preventing CO2 absorption, and photosynthesis halts.

Light intensity is key in various physiological processes related to plant growth and the photochemical reactions that convert CO2 into carbohydrates. It is considered a critical factor in regulating plant biosynthesis. Light intensity directly impacts the movement of CO2 and H2O through stomata during photosynthesis and transpiration. Optimal light intensity can enhance the photosynthetic rate and boost dry mass production. Numerous studies have demonstrated that light intensity is the primary factor influencing plant productivity and light use efficiency, which is defined as the ratio of the accumulated dry-weight chemical energy in the plant shoot to the total light energy received by the plant canopy. Appropriate fans are essential for managing air velocity, an important environmental factor affecting plant growth in controlled environments. Fans help regulate temperature, relative humidity, and CO2 concentration within the plant canopy, promoting gas and water exchange between plants and the surrounding air. Proper air velocity increases photosynthesis and transpiration rates, thereby fostering plant growth. Additionally, CO2 use efficiency is closely linked to air velocity. Air velocity has direct effects—through the transfer of energy and mass between leaves and surrounding air—and indirect effects by influencing metabolic processes. Optimal airflow ensures uniform environmental conditions within the canopy. One study found that air velocities between 0.3 and 0.7 m/s are ideal for plant growth, although the optimal range may vary depending on species, structure, and canopy depth.

CO2 concentration positively affects photosynthesis, but when it exceeds the saturation point, photosynthetic capacity plateaus and may even be inhibited. CO2 influences photosynthesis and transpiration rates by directly impacting stomata, which tend to close at high CO2 concentrations. Additionally, high light intensity and low CO2 levels cause stomata to open, whereas high CO2 concentrations, even under high light intensity, lead to stomatal closure. CO2 enrichment can reduce plant growth at low air temperatures. Effectively managing these parameters is essential for ensuring healthy growth in hydroponic systems. However, manually monitoring these factors would be highly labor-intensive and costly. Hydroponics offers a promising solution for sustainable agriculture, addressing the challenges of resource scarcity and food security through automation and IoT technologies [10, 11]. By implementing automated and intelligent systems, the full potential of hydroponics can be realized, optimizing production and contributing to a more efficient and sustainable agricultural future, taking advantage of the advances in mechatronics solutions and applications both in industrial [12] and non-conventional applications [13].

2 Description of the System

This work aims to develop an automated control system for hydroponic cultivation that can actively monitor and regulate the process. The system uses low-cost sensors and actuators, all managed through a PLC with an HMI interface. The system must meet key requirements such as reliability, ease of use, and versatility. It is designed to read the desired control values for process variables and provide real-time feedback to the operator. This includes displaying the values of process variables and the status of actuators on the HMI screen and storing this data in a file for future reference [14]. The chosen cultivation method for this system is the Nutrient Film Technique (NFT), where plants are nourished by a continuous film of nutrient-rich water flowing by gravity through tubes that house the plants. A circulation pump pushes the water into the tubes, where it sprays the roots, and the water then returns to the tank by gravity, recirculating within the system without waste.

A general scheme of the hydroponic system is shown in Fig. 1, which includes:

- Sensors for temperature, pH, TDS, and water level; sensors for air temperature, humidity, and carbon dioxide.
- Equipment including water heat exchanger, motorized 2-way and 3-way valves, propeller mixer, peristaltic pumps, circulation pump, fan, lights, radiator, motorized windows, CO2 release valve, and humidifier nozzle valve.

A Siemens S7-1200 PLC is used for process control and is equipped with an HMI. Due to the high cost of sensors and the need to create a more cost-effective miniature system, the Arduino UNO R4 Wifi microcontroller was chosen. This microcontroller will connect to low-cost sensors and outputs, allowing for more efficient distribution of the process controls. The PLC will communicate with the Arduino by reading sensor values and exchanging necessary data via Snap7, as shown in Fig. 2. The PLC is connected to the HMI through an Ethernet switch, as depicted in the "Topology View" screen in the TIA Portal. The Arduino UNO R4 Wi-Fi board has an internal ESP32 Wi-Fi module, enabling wireless communication. The PLC supports Ethernet communication with the HMI, so the physical system requires a Wireless Access Point to facilitate communication between the Arduino and the PLC.

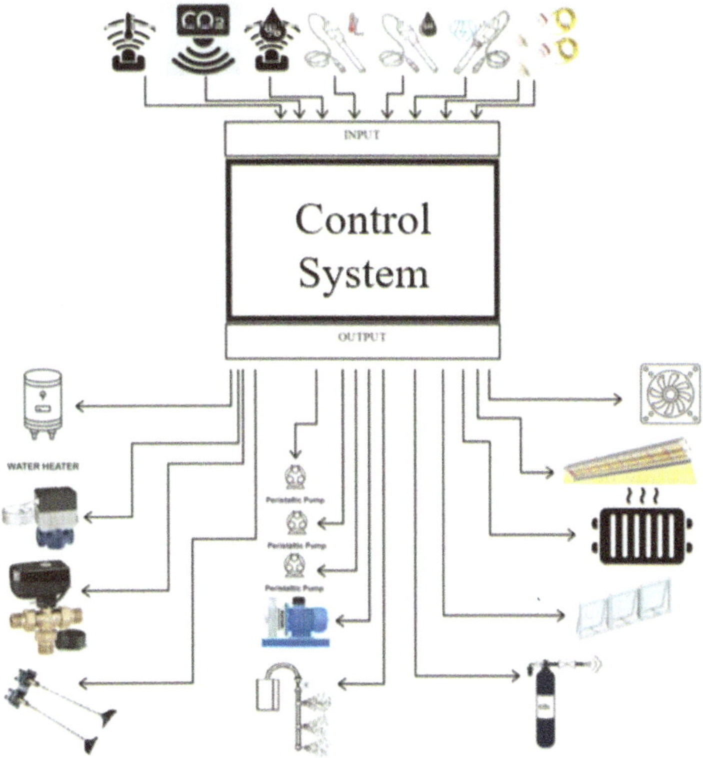

Fig. 1. A general control scheme for the hydroponic system.

The complete system is shown in Figs. 3 and 4, with a Wi-Fi repeater configured as an Access Point, complete with its IP and MAC addresses. The S7-1200 CPU features 14 digital inputs (24 V), 10 digital outputs, 2 analog inputs, and 2 analog outputs [15], while the Arduino has 14 digital I/O pins (including 6 PWM pins) and 6 analog input channels. An Arduino system connects the sensors and controls the peristaltic pumps, the mixer, and the feed pump (Fig. 3). This decision was made since the sensors require four analog inputs, but the PLC only has two. Additionally, the PLC provides only ten digital outputs, whereas we need to manage fourteen outputs. The Arduino transmits sensor data to the PLC via the TL-WA850RE Access Point and also reads the output statuses from the PLC for control purposes. The dosing tank contains the complete input/output system to

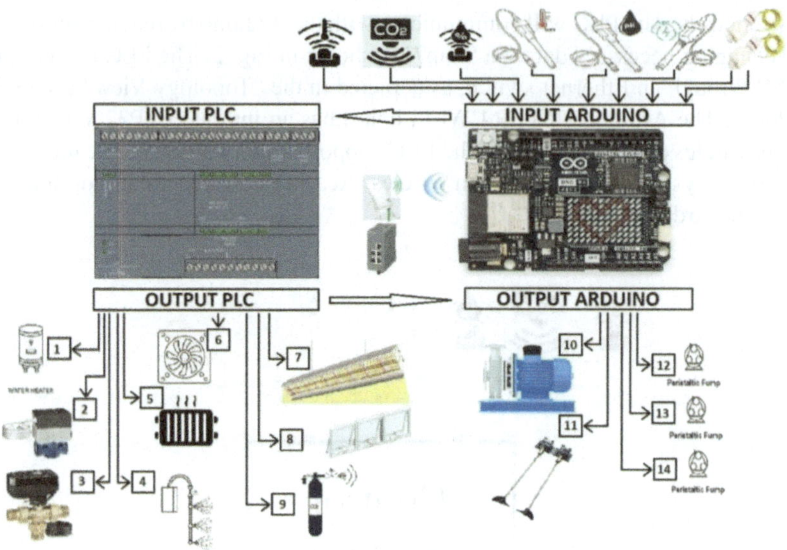

Fig. 2. A general control scheme for the hydroponic system.

monitor and control the nutrient solution. Specifically, it consists of a plastic tank and a plywood framework that supports the pH, TDS, temperature, and level sensors, as well as the three peristaltic pumps and the propeller mixer [16–19]. In direct contact with the water in the tank, the level sensors are mounted on a plastic interlocking framework that attaches to the main wooden structure (Fig. 3). This framework also supports Arduino, the breadboard for connecting the sensors and the relays, along with their associated control and power circuits for managing the pumps and mixer. The structure shown in Fig. 3 has approximate dimensions of 300 mm in length, 250 mm in width, and 300 mm in height. The container has a volume of 4 L, making it suitable for laboratory purposes only.

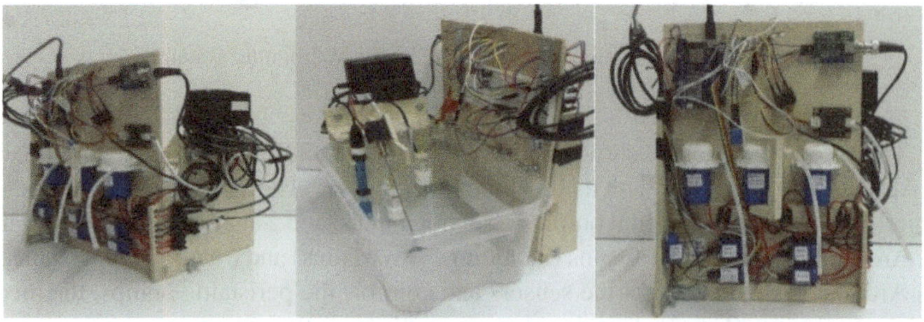

Fig. 3. Laboratory setup of the hydroponic tank prototype.

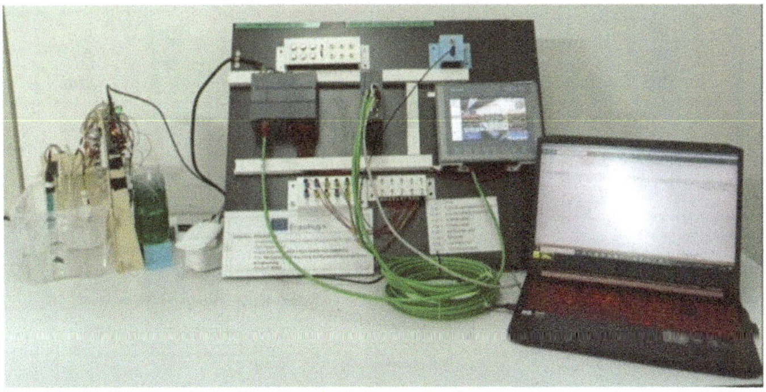

Fig. 4. A Laboratory setup of the control and operation of the hydroponic system.

3 Description of Operations

The HMI allows for managing both the program and the processes (Figs. 5 to 7). Using the touchscreen, it is possible to view sensor readings, monitor values calculated by the program, observe the real-time status of the actuators through animations, modify and access the properties required for program operation, and control the state of the CPU [20, 21]. Figure 6 represents the dosing tank management, and Fig. 7 the environmental monitoring.

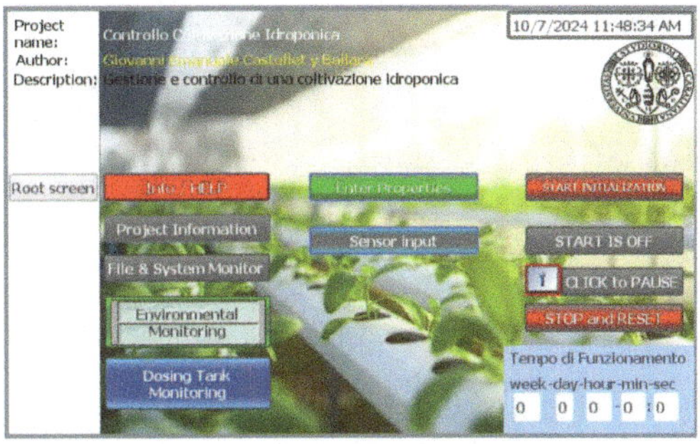

Fig. 5. Main page.

Interfaces are created for dosing the water and supplements, as well as the heating control. HMI allows monitoring the process or assigning/programming the nutrients and environmental conditions. The S7-1200 PLC manages the sensors through an Arduino interface using a Wi-Fi connection in a bidirectional way. This operational mode allows extending the analog I/O from 2 to 6 I/O. Computational operations are then developed by the PLC S7-1200, and Arduino is used as an interface between sensors and actuators only. This choice allows cost reduction while not sacrificing the operational features.

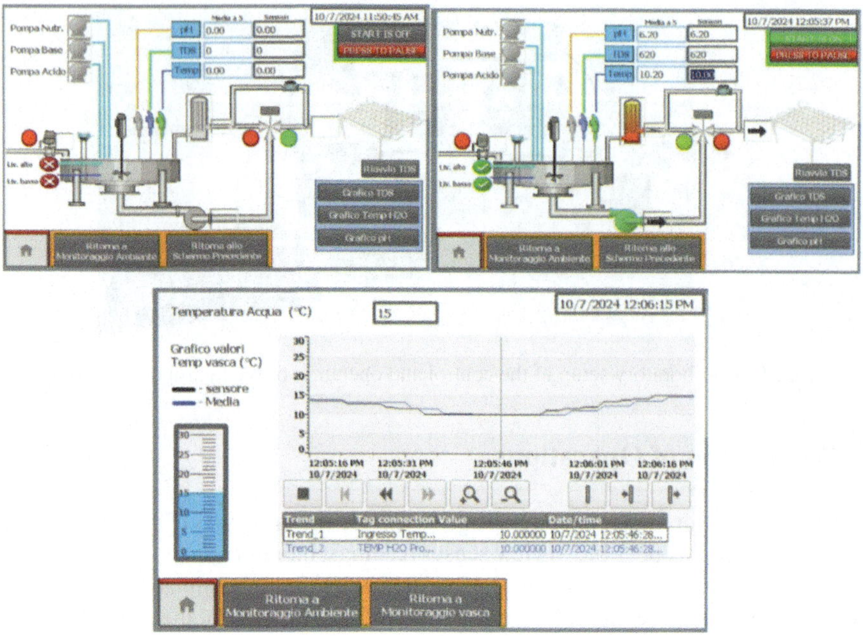

Fig. 6. Dosing tank: supply, heating.

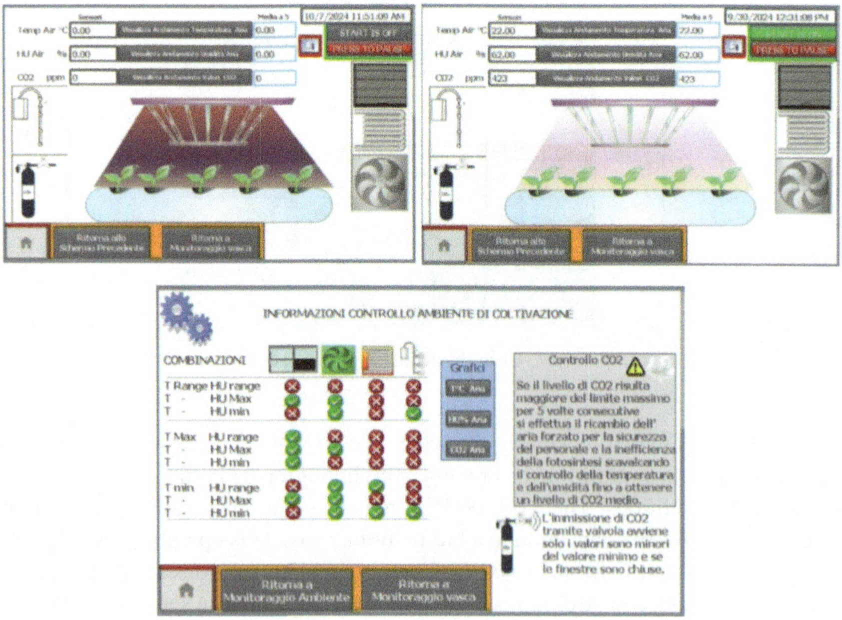

Fig. 7. Environmental monitoring.

4 Conclusions

This project aimed to design and develop an automated control system for hydroponic cultivation, integrating sensors for real-time data collection and using control algorithms to regulate environmental conditions automatically. The system continuously monitors key growth parameters, ensuring an optimal environment for plant development. The research focuses on a laboratory prototype to demonstrate its effectiveness and scalability.

Acknowledgments. Part of this work is related to ERASMUS++ 2022-1-ES01-KA220-HED-000089155 MISCE - Mechatronics for Improving and Standardizing Competencies in Engineering Description, 2022–2025 MISCE.

References

1. Nikolov, N.V., Atanasov, A.Z., Evstatiev, B.I., Vladut, V.N., Biris, S.-S.: Design of a small-scale hydroponic system for indoor farming of leafy vegetables. Agriculture **13**(6), 1191 (2023). https://doi.org/10.3390/agriculture13061191
2. Agrawal, P., Bhagwate, S., Singhaniya, D., Tamrakar, A., Agrawal, P.S.: Automated hydroponics system. Int. J. Res. Appl. Sci. Eng. Technol. (IJRA) **11**(III), 1718–1726 (2023). https://doi.org/10.22214/ijraset.2023.49606
3. Luna-Maldonado, A., Márquez-Reyes, J., Breceda, H., Rodriguez, H., et al.: Automation and Robotics Used in Hydroponic System (2019). https://doi.org/10.5772/intechopen.90438
4. de Castro Silva, M.G., Hüther, C.M., Ramos, B.B., da Silva Araújo, P., da Silva Hamacher, L., Pereira, C.R.: AA global overview of hydroponics: nutrient film technique. Revista Engenharia na Agricultura-REVENG **29**(Contínua), 138–145 (2021). https://doi.org/10.13083/reveng.v29i1.11679
5. Hesham, A.A., Tong, Y.-X., Yang, Q.-C.: Optimal control of environmental conditions affecting lettuce plant growth in a controlled environment with artificial lighting: a review. South Afr. J. Botany **130**, 75–89 (2020). https://doi.org/10.1016/j.sajb.2019.12.018
6. Driesen, E., Van den Ende, W., De Proft, M., Saeys, W.: Influence of environmental factors light, CO2, temperature, and relative humidity on stomatal opening and development: a review. Agronomy **10**(12), 2020 (1975). https://doi.org/10.3390/agronomy10121975
7. Bazel, M., Mohammed, F., Baarimah, A.O., et al.: The era of Industry 5.0: an overview of technologies, applications, and challenges (2024). https://doi.org/10.1007/978-3-031-59707-7_24
8. Valette, E., El-Haouzi, H.B., Demesure, G.: Industry 5.0 and its technologies: a systematic literature review upon the human place into IoT- and CPS-based industrial systems. Comput. Ind. Eng. **184**, 10942 (2023). https://doi.org/10.1016/j.cie.2023.109426
9. Haji Razali, M., Syazili, R., Ssomad, A., et al.: Design and development of mechatronic application in agricultural irrigation device. World J. Eng. Techno. **04**, 450–459 (2016)
10. Figliolini, G., Rea, P.: Mechatronic design and experimental validation of a novel robotic hand. Ind. Robot **41**(1), 98–108, 17103865 (2014)
11. Rea, P., Ottaviano, E.: Analysis and mechanical design solutions for sit-to-stand assisting devices. Am. J. Eng. Appl. Sci. **9**(4), 1134–1143 (2016)
12. Rezwan, S., Ahmed, F., Nusrat, R., Rumana, T., Md, E., Sumaiya, S.: Emerging Applications of Mechatronics (2024). https://doi.org/10.1007/978-981-97-7117-2_7

13. Ceccarelli, M. Ottaviano, E., Galvagno, M.: A 3-DOF parallel manipulator as earthquake motion simulator. In: Proceedings of 7th International Conference on Control, Automation, Robotics and Vision, pp. 944–949 (2002)
14. Rijwani, T., et al.: Industry 5.0: a review of emerging trends and transformative technologies in the next industrial revolution. Int. J. Interact. Des. Manuf. (IJIDeM) **19**(2), 667–679 (2024). https://doi.org/10.1007/s12008-024-01943-7
15. Siemens User Manual and product Overview Simatic S7. https://www.siemens.com/global/en/products/automation/systems/industrial/plc/s7-1200.html
16. DHT11 Humidity & Temperature Senso.: https://www.alldatasheet.com/datasheet-pdf/pdf/1132088/ETC2/DHT11.html. Accessed 15 May 2025
17. Temperature Sensor DS18B20. https://www.alldatasheet.com/datasheet-pdf/pdf/58557/DALLAS/DS18B20.html. Accessed 15 May 2025
18. TDS sensor Datasheet by Keystudio. https://wiki.keyestudio.com/KS0429_keyestudio_TDS_Meter_V1.0. Accessed 15 May 2025
19. CCS811. https://learn.adafruit.com/adafruit-ccs811-air-quality-sensor. Accessed 15 May 2025
20. Basics on Creating HTMLs for SIMATIC CPUs SIMATIC STEP 7. http://support.automation.siemens.com/WW/view/de/68011496. Accessed 15 May 2025
21. AWP commands S7-1200. https://support.industry.siemens.com/. Accessed 15 May 2025

Modeling of Marine Gas Turbine Rotor Impeller Stress-Strain State Caused by Forced Vibration and Creep Load

Natalia Smetankina[1](✉), Serhii Morhun[2], Liu Weixing[3], and Zhang Zhiyang[3]

[1] Anatolii Pidhornyi Institute of Power Machines and Systems of the National Academy of Sciences of Ukraine, 2/10, Komunalnykiv Street, Kharkiv 61046, Ukraine
nsmet@ipmach.kharkov.ua

[2] Admiral Makarov National University of Shipbuilding, 9 Heroiv Ukrainy Avenue, Mykolaiv 54001, Ukraine

[3] Jiangsu Ocean University, 59 Cangwu Road, Lianyungang 222005, Jiangsu, China

Abstract. The problem of the gas turbine rotor blade stress-strain state caused by creep and forced vibration is considered. In this paper, a refined finite element mathematical model describing creep processes in the rotor of a marine gas turbine engine is developed. Such mathematical model uses the specially developed finite elements, that gives us an opportunity to describe geometry of the rotor elements. On the base of such highly précised mathematical model the turbine rotor most highly loaded first impeller displacements and equivalent stresses fields were found for several most dangerous forced vibration modes. The calculated data of the first impeller equivalent stresses and durability has been compared with experimental results and numerical data for the case, that doesn't take creep into consideration. The divergence between numerical and experimental data is less than 10%. Obtained results will be used for the next stage of studies, concerning the rotor durability and reliability problems.

Keywords: Gas turbine rotor · Finite Element Method · Creep · Equivalent stresses · Forced Vibration

1 Introduction

In the modern world the problems of energy efficiency for the offshore regions are the cutting edge of gas turbine engines building. The main way for this problem solution is the incensement of the engine's capacity and durability by means of computational methods without usage of extremely expensive experimental techniques. To achieve this goal, a number of problems need to be solved in various fields such as fluid flow dynamics, heat transfer, solid mechanics, etc. Among them of the special importance is the complex influence of the forced vibration and creep processes on the gas turbine rotor. Similar problems arise when modeling welding, gas cutting, technological processes in metallurgy, etc. For example, the problem of determining the deformation of aircraft

D. D. Cioboată and J. Machado (Eds.): ICoRSE 2025, LNNS 1592, pp. 51–61, 2025.
https://doi.org/10.1007/978-3-032-02508-1_5

structures during supersonic heating is very important [1]. Another method of analyzing accident conditions in nuclear reactors is applied [2].

Thus, forced vibrations and creep of materials can significantly limit the long-term properties of structural elements of gas turbine engines, especially the most heated ones, such as blades of the first rotor impeller [3, 4]. In most cases, significant creep deformations occur at elevated temperatures when the working flow temperature exceeds 0.3–0.5 of the melting point of the blade material [5].

2 Literature Review

To take into account the dependence of the creep deformation rate function on temperature, various dependences are used, one of the most common and physically justified of which is exponential [6]. To obtain the temperature distribution in the structure, solutions of stationary or non-stationary thermal conductivity problems are used. At present, most works (see, for example, [7, 8]) perform calculations at constant temperatures, or when they change in limited ranges in time or according to the coordinates of blade cross-sections. A special case is the cyclic change of temperature in time, which, in particular, is possible under the conditions given above. To model this process, an effective method of averaging over the period of temperature change over time is used and averaged equations of state are obtained. In them the values of the temperature that are constant over time are used [9, 10].

Considering these problems, it is possible to conclude that in some of them, the determination of temperature fields precedes the calculations of the reliability, in which the previously found temperature distributions play the role of initial conditions. Incidentally, in practice, processes are characterized by a constant increase in temperature at some points of a solid deformable body due to its constant heating. In this case, such bodies are under the influence of significant mechanical loads. To adequately assess the change in the stress state in a loaded solid body during its constant heating, it is necessary to apply the method of creep deformation rate from temperature in the entire range of its changes.

In this work, an attempt is made to provide such a statement, method and algorithms, which are the basis for solving problems of gas turbine rotor blade creep that work under the influence of turbulent gas flow, also causing the rotor forced vibration. This study would be held taking into account the refined mathematical model on the base of the finite elements method (FEM).

Since the advent of FEM, one of the areas of its application has been strength calculations. The properties of finite element models are determined by the finite elements used and, depending on the blades under study, existing or new finite elements were developed [4]. The authors of [5] used triangular finite elements of variable thickness with a flat middle surface to study the vibrations of gas turbine engine blades in non-stationary flow. The finite elements, which are used to obtain the frequencies and shapes of free oscillations in the field of centrifugal forces, contain three nodes at the corners of the middle surface with seven degrees of freedom at each node (three displacements, two angles of rotation of the normal to the middle surface relative to the coordinate axes of the element, and two components of shear strain).

The application of FEM to some classes of problems is complicated by the presence of complex-configured boundaries of the studied regions. In these cases, it is necessary to reduce the size of the elements and increase their number, which complicates the problem. Therefore, the authors of [6–11] developed a triangular bending element of a plate, which differs from the previously considered triangular elements by the presence of one curvilinear boundary. The authors of [12] used FEM to study the vibration behavior of pre-twisted working blades with cracks, presented in the form of flat cantilever plates. They modeled a pre-twisted plate by a set of triangular elements of constant thickness with nine degrees of freedom.

The authors of [13–19] developed a curved shell-type finite element for calculating the vibrations of turbomachine blades. The use of this type of element made it possible to model objects of complex geometry and take into account the effect of transverse deformation. A similar element was developed in [20] taking into account the effects of rotations, excluding the influence of Coriolis acceleration.

After the literature analysis, dealing with the problems of gas turbine rotors creep and forced vibration problems, it can be concluded that there are several unsolved problems which are of great importance. One of them problem is the determination of rotor creep and forced vibration parameters determination. This problem can be solved by means of the developed refined mathematical model.

3 Research Object and Aim

The research object is the turbine rotor mostly heated blades equivalent stress parameters, taking thermal forced vibration and creep problems into account. Turbine rotor impellers can be considered as a cyclically symmetric system. In the rotor impellers the turbulent gas flow causes their forced vibration [11, 12] and the heat exchange between the flow and impellers. Such heat exchange causes the start of creep processes in the blades.

So, the researchers need to develop a refined mathematical model that describes complex forced vibration, heat exchange and creep processes in the turbine rotor. The thermoelastic problems of the modern marine gas turbine rotors have been studied and experimentally verified in papers [13, 14]. So the main attention should be paid to the creep problem.

In order to achieve this goal, the following tasks need to be accomplished:

- Develop a refined mathematical model describing the creep processes in rotor mostly heated blades on the basis of FEM;
- Determine the field of the impeller displacements taking into account the forced vibration and creep processes;
- Determine the field of the blade's equivalent stresses.

4 Problem Formulation

Let's take into consideration the mostly heated first impeller of the rotor. As the stationary coordinate system, the Cartesian right-handed xyz coordinate system with the center at point O is taken (Fig. 1). This coordinate system rotates together with the turbine rotor at a constant angular velocity Ω equal to the angular velocity of rotor.

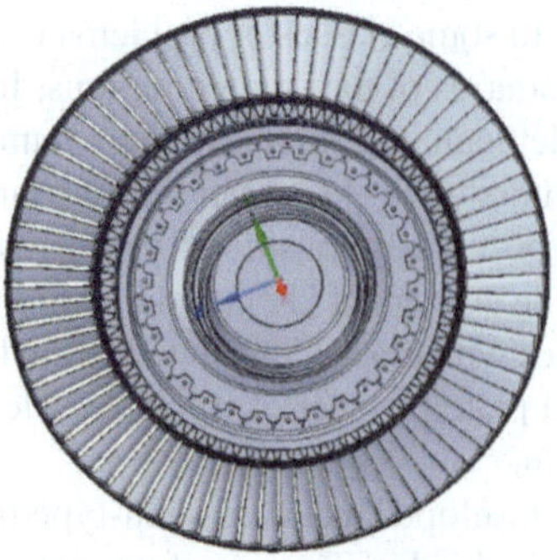

Fig. 1. The 3D model of the gas turbine rotor mostly heated first impeller.

Let's solve the three-dimensional thermal creep problem in a closed region of solid continuum, which is bounded by the surface S. This surface is given in the form of a Navier equilibrium system (1), written in vector form [6]

$$\frac{d\{\sigma\}}{d\{\delta\}} = \{F\}, \tag{1}$$

where $\{\sigma\}$ is the stress vector; $\{\delta\}$ is the generalized displacement vector; $\{F\}$ is the vector of gas dynamic forces.

According to the generalized Hooke's law the solid body stress-strain state can be described by the next matrix equation [6–8]

$$\{\sigma\} = [D]\{\varepsilon\}, \tag{2}$$

where $[D]$ is the material elasticity matrix; $\{\varepsilon\}$ is the strain vector.

To study the deformation of a solid body at each of its points, we consider the total strain tensor, which by virtue to the hypothesis of the additivity of strains is represented by the sum of the elastic strain vector $\{\varepsilon_e\}$, the vector of thermal strains $\{\varepsilon_T\}$, and the vector of creep strains $\{\varepsilon_c\}$, respectively:

$$\{\varepsilon\} = \{\varepsilon_e\} + \{\varepsilon_T\} + \{\varepsilon_c\} \tag{3}$$

The elastic strain vector has the following components [11, 13]

$$\begin{gathered} \{\varepsilon_e\} = \left\{\varepsilon_x, \varepsilon_y, \varepsilon_z, \gamma_{xy}, \gamma_{xz}, \gamma_{yz}\right\}^T = \\ = \left\{\frac{\partial\delta_x}{\partial x}, \frac{\partial\delta_y}{\partial y}, \frac{\partial\delta_z}{\partial z}, \frac{\partial\delta_x}{\partial y} + \frac{\partial\delta_y}{\partial x}, \frac{\partial\delta_x}{\partial z} + \frac{\partial\delta_z}{\partial x}, \frac{\partial\delta_y}{\partial z} + \frac{\partial\delta_z}{\partial y}\right\}^T, \end{gathered} \tag{4}$$

where δ_x, δ_y and δ_z are components of the generalized displacement vector.

The vector of thermal strains [13, 15] has the form

$$\{\varepsilon_T\} = \left\{\varepsilon_{xT}, \varepsilon_{yT}, \varepsilon_{zT}, 0, 0, 0\right\}^T = \alpha\left\{T_x, T_y, T_z\right\}^T, \tag{5}$$

where α is the material thermal extension coefficient; T_x, T_y and T_z are temperature field projections.

In [16, 17], the problems of elastic and thermal strains of gas turbine rotor blades have been successfully solved. The components of the vector of creep strains can be found according to the following dependency

$$\{\varepsilon_c\} = \mu \frac{\partial c}{\partial\{\sigma\}}, \tag{6}$$

where $\dot{c}$ is the creep strain rate potential depending on the components of the stress tensor and its invariants; μ is the Poisson's ratio.

Depending on the type of the potential given, one or another creep hypothesis is distinguished, for example, flow or hardening.

The material elasticity matrix is formed according to [17].

To describe the process of hidden damage accumulation due to creep, we use the fracture mechanics approach. Thus, the kinetic equation for the damage parameter is taken as follows

$$\dot{\omega}_c = \frac{D_c(\sigma_E)^r}{(1-\omega_c)^l}, \tag{7}$$

where D_c, l and r are the coefficients determined from the blade material creep curves; σ_E are equivalent stresses.

Let's take into consideration that in the initial moment of the gas turbine engine work t_0 the damage parameter caused by creep ω_c and the creep strain vector ε_c are equal to zero

$$\omega_c(t_0) = 0, \quad \varepsilon_c(t_0) = 0 \tag{8}$$

Boundary conditions for the gas dynamic forces acting on the rotor blade surfaces can be set like this

$$\{\sigma\}\{n\} = \{F\}, \tag{9}$$

where $\{n\}$ is normal vector to the blade surfaces.

The system of matrix Eqs. (2)–(6) with appropriate initial and boundary conditions (7)–(9) gives an opportunity to determine the turbine rotor blade stress-strain state, taking into account the creep processes.

5 Problem Solution

To solve the problem, we use FEM. The algorithm of FEM approximation of turbine rotor blades based on a curvilinear eight-node finite element of hexagonal type and the shape function of the element are given in [11, 13, 14, 18, 19].

For each finite element we can write next variation functional.

$$\int_{V_e} \dot{\sigma}_e \dot{\varepsilon}_e dV_e = \int_{S_e} \delta_i^S \dot{p}_e dS_e + \sum\nolimits_{i=1}^{m} \delta_i^{n_e} \dot{F}^{ne}, i = 1, \ldots m, e = 1, \ldots, k, \tag{10}$$

where $\dot{\sigma}_e$ is the stress rate tensor; $\dot{\varepsilon}_e$ is the strain rate tensor; $\delta_i^{S_e}$ is the generalized displacement rate tensor caused by the influence of gas flow pressure; $\dot{\delta}_i^{n_e}$ is the generalized displacement rate tensor caused by the influence of nodal forces; $\dot{p}_e$ is the gas flow pressure rate tensor; $\dot{F}^{n_e}$ is the nodal forces rate tensor; V_e is finite element volume; S_e is finite element surface area; n is number of finite elements nodes; k is number of finite elements.

The problem solution is found in the form of generalized displacement rate. Thus, we can use next dependence.

$$\{\delta_i\} = \begin{Bmatrix} \sum_{i=1}^{n} \delta_{xi} N_i \\ \sum_{i=1}^{n} \delta_{yi} N_i \\ \sum_{i=1}^{n} \delta_{zi} N_i \end{Bmatrix}, i = 1, \ldots, m, \tag{11}$$

where δ_{xi}, δ_{yi} and δ_{zi} are rates of the finite element generalized displacements according to the *x*, *y* and *z* axes; N are the finite element shape functions [13].

Relations between the tensors of strain rate and generalized displacement rate, and the tensors of stress and strain rate are given

$$\dot{\varepsilon}_e = B^e \dot{\delta}_e, \quad \dot{\sigma}_e = D\dot{\varepsilon}_e, e = 1, \ldots, k \tag{12}$$

where B^e is the finite element gradient matrix [13, 14].

The right-hand parts of dependence (13) are the convolution of tensors with "deaf" indices. Substituting (12) into expression (10), we obtain the following functional.

$$\int_{V_e} (\dot{\varepsilon}_e)^T D\dot{\varepsilon}_e dV_e = \int_{S_e} \delta_i^{S_e} \dot{p}_e dS_e + \sum_{i=1}^{m} \dot{\delta}_i^{n_e} \dot{F}^{n_e}, i = 1, \ldots, m, e = 1, \ldots, k \tag{13}$$

Substituting into expression (13) the strain rate tensor expressed in terms of (2), we have

$$\begin{aligned} &\int_{V_e} \left(\dot{\delta}_{el}^e\right)^T (B^e)^T DB^e \dot{\delta}_{el}^e dV_e + \int_{V_e} \left(\dot{\delta}_T^e\right)^T (B^e)^T DB^e \dot{\delta}_T^e dV_e + \int_{V_e} \left(\dot{\delta}_c^e\right)^T (B^e)^T DB^e \dot{\delta}_c^e dV_e = \\ &= \int_{S_e} \delta_i^{S_e} \dot{p}_e dS_e + \sum_{i=1}^{m} \dot{\delta}_i^{n_e} \dot{F}^{n_e}, i = 1, \ldots, m, e = 1, \ldots, k \end{aligned} \tag{14}$$

where $\dot{\delta}_{el}^e$ is the finite element generalized displacement rate caused by elastic strain; $\dot{\delta}_T^e$ is the finite element generalized displacement rate caused by thermal strain; $\dot{\delta}_c^e$ is the finite element generalized displacement rate caused by creep strain.

Due to the fact that the components of the local generalized displacement vector of the finite element ($\delta_{el}^e, \delta_T^e, \delta_c^e, \delta_i^{V_e}, \delta_i^{S_e}, \delta_i^{n_e}$) are included in the global one, the variation in local components is reduced to the variation of the global velocity vector.

Finding the variation of expression (14) in components of generalized displacement vector we obtain the equation of the thermal creep problem in the following form

$$\begin{aligned} &\sum_{e=1}^{k} \left[\int_{V_e} (B^e)^T DB^e dV_e\right] \cdot \{\dot{\delta}_{el}^e\} = \sum_{e=1}^{k} \left[\int_{V_e} (B^e)^T D\dot{\varepsilon}_{el}^e dV_e\right] + \sum_{e=1}^{k} \left[\int_{V_e} (B^e)^T D\dot{\varepsilon}_T^e dV_e\right] \\ &+ \sum_{e=1}^{k} \left[\int_{V_e} (B^e)^T D\dot{\varepsilon}_c^e dV_e\right] \int_{S_e} \delta_i^{S_e} \dot{p}_e dS_e + \sum_{e=1}^{k} \sum_{i=1}^{m} \dot{\delta}_i^{n_e} \dot{\delta}_i^{n_e} \dot{F}^{n_e} \end{aligned} \tag{15}$$

Simplifying (15), we obtain the following matrix equation

$$[K^e]\{\dot{\delta}^e\} = \{\dot{F}_{el}^e\} + \{\dot{F}_T^e\} + \{\dot{F}_c^e\} + \{\dot{p}_e\} + \{\dot{F}^{n_e}\}, \tag{16}$$

where $[K^e] = \int_{V_e} (B^e)^T DB^e dV_e$, $\{\dot{F}^e_{el}\} = \int_{V_e} (B^e)^T D\dot{\varepsilon}^e_{el} dV_e$, $\{\dot{F}^e_T\} = \int_{V_e} (B^e)^T D\dot{\varepsilon}^e_T dV_e$, $\{\dot{F}^e_c\} = \int_{V_e} (B^e)^T D\dot{\varepsilon}^e_c dV_e$, $e = 1, \ldots, k$; $[K^e]$ is the finite element stiffness matrix.

For the whole gas turbine rotor blade model, the solving system of matrix equation is given below

$$[K]\{\dot{\delta}\} = \{\dot{F}_{el}\} + \{\dot{F}_T\} + \{\dot{F}_c\} + \{\dot{p}\} + \{\dot{F}^n\}. \tag{17}$$

The system of matrix Eqs. (17) is solved by the Cholesky method, since the matrix of the initial system is positively defined using the numerical Euler integration scheme at each time step.

After finding the generalized displacement vector, it is substituted to expressions (11) and on its basis we find the components of the stress and strain vectors from Eqs. (2)–(6).

6 Results and Discussion

The investigated blades are made from the special heat-resistant Ni-Ti alloy with the following mechanical characteristics: $\rho = 8100$ kg/m^3(blade material density), $E = 1.69 \cdot 10^5$ MPa (Young's modulus), $\mu = 0.3$ (Poisson's ratio), $\sigma_E = 600$ MPa. The working flow temperature coming to the first impeller blades is approximately 1038^{0}C. The temperature of cooling air coming to the first impeller blade curvature is 400^{0}C. The rotor angular velocity $\Omega = 11330$ rev/m.

The study was carried out for the first three modes of forced vibrations of the blades of the most loaded first impeller of the turbine rotor. The power of the gas turbine under study is 25 MW. The total pressure at the stage inlet is 0.7578 MPa.

As it is obvious from the Eqs. (2), (3), and (18), the level of equivalent dynamic stresses of the turbine rotor impeller is greatly influenced by magnitude of the rotor impeller deformations caused by its forced vibrations and creep. So, first of all, the fields of the first impeller displacements have been determined for the first three harmonic indices of forced vibrations. These modes of forced vibrations are the most dangerous, as they can cause a resonance condition with the rotor rotation frequency. Therefore, it is obligatory to obtain the distribution of displacements (mm) for them. The results of such calculations in graphic form are given below in Fig. 2. The solution of the problem of forced vibrations of rotor impellers of modern marine gas turbines is given in [14].

Analyzing data represented on the Fig. 2 it is obvious that the impeller displacement increases with the growth of its forced vibration frequency value. The main factors, provoking the biggest displacement of the blades are the working fluid flow and thermal extension. Thermal extension is the main reason for the creep problem and creep deformation of the impeller. Thus, for the third harmonic index forced vibration mode, impeller blades displacement is about 5–10 mm on the top of the blades peripheral section for the vibration frequency of 1015.3 Hz. All these negative factors provoke the start of crack processes in the blades and their sharp growth can be predicted.

But on the other hand, it should be predicted that the impeller's resistance to low cycle fatigue and creep failure will be more sufficient. To give the correct answer on this question we should calculate the impeller equivalent stresses rate, taking into account

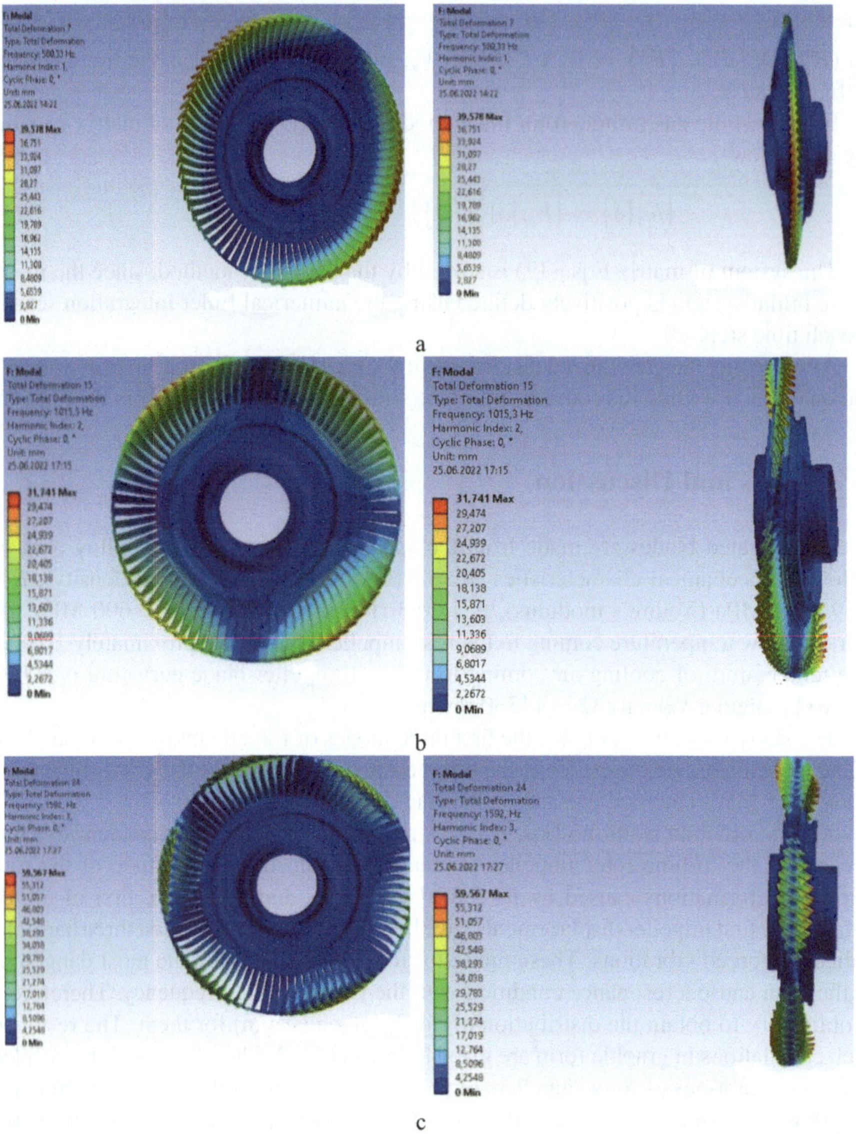

Fig. 2. Gas turbine rotor first impeller displacement fields: (*a*) harmonic index $k = 1$, vibration frequency $f = 536.6$ Hz; (*b*) harmonic index $k = 2$, vibration frequency $f = 1015.3$ Hz; (*c*) harmonic index $k = 3$, vibration frequency $f = 1592.8$ Hz.

the creep problem, and compare with the equivalent stresses, calculated without creep (Tables 1 and 2).

The obtained equivalent stresses have been compared with experimental data, received by the usage of methodology given in [18].

Table 1. The influence of creep on the first impeller equivalent dynamic stresses rate, σ_E, MPa.

Forced vibration modes index, k	Forced vibration frequency f, Hz	Maximum equivalent dynamic stresses rate σ_E, MPa		
		Numerical data	Experimental data	Divergence, %
		Without creep		
1	548.2	384.2	371.9	3.2
2	1023.9	497.9	470.2	5.6
3	1602.8	520.4	507.9	2.4
		Taking into account creep		
1	536.6	375.2	371.9	1.8
2	1015.3	479.5	470.2	2.9
3	1592.8	518.3	507.9	3.4

Table 2. The influence of creep on the first impeller number of cycles till impeller's failure, N.

Type of load	Number of cycles till failure		
	Numerical data	Experimental data	Divergence, %
Without creep	1.54·108	1.48·108	3.9
Taking account creep	1.32·108	1.28·108	2.8

Analysis of the data presented in Tables 1 and 2 shows that the growth of the blade surface temperature and forced vibration frequency causes the decrease of their durability. Such conclusion can be made because of the serious difference between the values of equivalent dynamic stresses and number of cycles till the blade failure when we take and don't take creep into consideration. The lowest number of cycles before failure of the blades have been observed in the case of creep taking into account. Obtained results show that for the surface temperature of nearly 1000^0C the maximum equivalent stresses are approximately equal to the blade material endurance limit. But we should mention that presence of the cooling channels in the studied marine gas turbine rotor first impeller blades plays the positive role, because the blade's feather is designed as a shell of negative Gauss curvature and the cooling channels walls can be taken into consideration as stringers, that increases the strength of whole impeller. In addition, the presence of cooling air in the curvature reduces the surface temperature on nearly 7% [13, 20] and thus slows down the creep processes.

It should also be noted that all numerical results, given in Tables 1 and 2 are in good agreement with the experimental data, the divergence is less than 10%.

7 Conclusions

The study of equivalent stresses and displacements of the rotor of a modern gas turbine engine caused by forced vibrations and creep is carried out. To increase the reliability of the obtained results a more precise refined mathematical model has been developed by means of FEM. Using the developed mathematical model, the problem of the modern single-shaft marine gas turbine engine rotor most highly loaded impeller creep has been studied for different forced vibration modes. Firstly, the values of the impeller displacements, caused by creep, have been calculated for the impeller's first three vibration modes. Then the values of maximum equivalent stresses have been determined for the two cases of taking and without taking creep problems into consideration. The obtained numerical data has been compared to experimental and good correlation has been found. It gives us an opportunity to conclude that not only the non-stationary gas flow causing the irregular pressure field and heat exchange on the rotor surfaces, but also creep are the main negative factors, causing such high equivalent dynamic stresses in the impeller blades.

This situation demands the new studies of the turbine rotor durability and the received data will be used as initial for such calculations.

References

1. Wang, S., Li, S., Luo, L., Zhao, Z., Du, W., Sundén, B.: A high temperature turbine blade heat transfer multilevel design platform. Numer. Heat Transfer Part A: Appl. **79**(2), 122–145 (2021). https://doi.org/10.1080/10407782.2020.1835104
2. Seok, S., Shahriar, A., Montoya, A.: A finite element approach for simplified 2D nonlinear dynamic contact impact analysis. Arch. Appl. Mech. **93**(9), 3511–3531 (2023). https://doi.org/10.1007/s00419-023-02451-y
3. Ziaei-Asl, A., Tayefe-Ramezanlou, M.: Thermo-mechanical behavior of gas turbine blade equipped with cooling ducts and protective coating with different thickness. Int. J. Mech. Sci. **150**, 656–664 (2019). https://doi.org/10.1016/i.jmecsci.2018.10.070
4. Kumar, R., Kumar, V.S., Butt, M.M., Sheikh, N.A., Khan, S.A., Afzal, A.: Thermo-mechanical analysis and estimation of turbine blade tip clearance of a small gas turbine engine under transient operating conditions. Appl. Therm. Eng. **179**, 115700 (2020). https://doi.org/10.1016/applithermaleng.2020.115700
5. Arnab, B., Prabhakar, S., Giacomo, C.: Natural frequency analysis of a functionally graded rotor-bearing system with a slant crack, subjected to thermal gradients. Int. J. Turbo Jet Eng. **40**(3), 243–255 (2023). https://doi.org/10.1515/tjj-2021-0002
6. Boychenko, N., Ishtyryakov, I.: Characterization of the stress-strain state in a gas turbine engine compressor disk taking into account damage accumulation. Frattura ed Integrita Struturalle. **13**(50), 54–67 (2019). https://doi.org/10.3221/IGF-ESIS.50.07
7. Bogdanov, V.L.: On one approach in fracture mechanics of composites with parallel cracks under the action of initial (residual) stresses. Mech. Compos. Mater. **59**(2), 239–262 (2023). https://doi.org/10.1007/s11029-023-10094-x
8. Miroshnikov, V., Denshchykov, O., Grebeniuk, I., Savin, O.: An analysis of the stress–strain state of a layer on two cylindrical bearings. Computation **12**(9), 182–187 (2024). https://doi.org/10.3390/computation12090182

9. Shakurova, L., Armenise, I., Kustova, E.: Effect of slip boundary conditions on nonequilibrium reacting air flows. J. Theor. Appl. Mech. **53**(3), 253–269 (2023). https://doi.org/10.55787/jtams.23.53.3.253
10. Skamniottis, C., Courtis, M., Cocks, C.F.A.: Multiscale analysis of thermomechanical stresses in double wall transpiration cooling systems for gas turbine blades. Int. J. Mech. Sci. **207**(7), 106657 (2021). https://doi.org/10.1016/j.ijmec.sci.2021.106657
11. Morhun, S., Vilkul, S.: Gas dynamic analysis of the modern single shaft gas turbine engine flow path. Int. J. Turbo Jet Eng. **40**(s1), s271–s278 (2023). https://doi.org/10.1515/tjeng-2022-019
12. Kauss, O., Tsybenko, H., Naumenko, K., Hutter, S., Kruger, M.: Structural analysis of gas turbine blades made of Mo-Si-B under transient thermo-mechanical loads. Compos. Mater. Sci. **165**, 129–136 (2019). https://doi.org/10.1016/j.commatsci.2019.04.23
13. Morhun, S.: Determination of modern single shaft gas turbine rotor thermal stresses. J. Theor. Appl. Mech. **54**, 46–50 (2024). https://doi.org/10.55787/jtams.24.54.1.046
14. Smetankina, N., Morhun, S.: The modern single shaft gas turbine rotor stress-strain state determination taking into account the contact thermoelasticity problem. Technicki vjesnik. **31**(2), 551–556 (2024). https://doi.org/10.17559/TV-20230628000770
15. Yershov, S.V., Yakovlev, V.A.: The influence of mesh resolution on 3D RANS flow simulations in turbomachinery flow path. J. Mech. Eng. **24**(1), 13–27 (2021). https://doi.org/10.15407/pmach2021.01.013
16. Narasimbachary, S.B., Smith, K., Schinde, R.S., Amann, C., Kadau, K., Iyer, S.: Life assessment of large turbine blades. In: ASME Turbo Expo 2019: Turbomachinery Technical Conference and Exposition Proceeding Paper. Phoenix, USA (2019). https://doi.org/10.1115/GT2019-91630
17. Duriagina, Z.A., Kulyk, V.V., Filimonov, O.S., Trostianchyn, A.M., Sokulska, N.B.: The role of stress-strain state of gas turbine engine metal parts in predicting their safe life. Prog. Phys. Metals **22**(4), 643–677 (2019). https://doi.org/10.15407/ufm.22.04.643
18. Smetankina, N., Morhun, S.: Determination of the modern marine single shaft gas turbine rotor blades fatigue strength parameters. Int. J. Turbo Jet-Eng. **42**(2), 271–276 (2025). https://doi.org/10.1515/tjj-2024-0057
19. Merculov, V., Kostin, M., Martynenko, G., Smetankina, N., Martynenko, V.: Force simulation of bird strike issues of aircraft turbojet engine fan blades. Lect. Notes Netw. Syst. **305**, 129–141 (2022). https://doi.org/10.1007/978-3-030-83368-8_13
20. Balitskii, A.I., et al.: Improvement of the mechanical characteristics, hydrogen crack resistance and durability of turbine rotor steels welded joints. Energies **15**(16), 6006–6015 (2022). https://doi.org/10.3390/en15166006

Mathematical Modeling of the Effect of the Lithium Element on the Fluidity of Al-Si and Al-Cu Alloys

Azizakhon Turakhujaeva[1](✉), Sarvar Tursunbaev[1], Sherzod Tashbulatov[1], Anvar Turaev[1], Sharofuddin Mardonakulov[1], Ravshanbek Murodqosimov[1], Furkat Odilov[2], Mokhinur Rakhmonova[1], and Sobirjon Murodov[1]

[1] Tashkent State Technical University, 100095 Tashkent, Uzbekistan
{toraxojayeva.azizaxon,tursunbayev.sarvar,
raxmonova.mohinur}@tdtu.uz, rmurodqosimov@bk.ru
[2] Andijan State Technical Institute, 170100 Andijan, Uzbekistan

Abstract. Mathematical modeling of applied research, based on the results of initial experiments, allows to determine further results without conducting experiments. In the article, the technology of increasing the liquefaction property by introducing lithium fluoride element into Al-Si and Al-Cu binary alloys from aluminum alloys is presented. The fluidity property was determined using a spiral sample. Based on the obtained results, a graph of the amount of lithium in the aluminum alloy, depending on its fluidity property, was developed. Using the developed graph as initial information, the effect of lithium element on the fluidity properties of aluminum alloys was mathematically modeled. In this case, a function was developed for the alloy that expresses the relationship between the fluidity and the amount of lithium fluoride included in the alloy. At the end of the article, the authors' conclusions about the conducted research are given.

Keywords: Lithium · Aluminum · Fluorine · Fluidity · Lagrange polynomials (Lagrange interpolation) · Mathematical modeling

1 Introduction

The operational properties of aluminum alloys are crucial for the production of cast parts. These properties can be improved by altering the composition of the alloy [1–4]. This article investigates the fluidity properties of aluminum alloys. The fluidity of aluminum alloys is the main factor that plays an important role in the production of castings of complex configurations and is responsible for product quality [5–8]. Castings with improved fluidity make it possible to obtain parts with fewer defects, such as under filling. Scientists from Russia, China, Turkey, and others have studied the effects of modification and alloying of aluminum alloys with metals or non-metals on their casting properties, including fluidity [9–12]. But not all of them have produced successful results.

Guodong Niu et al. found that addition of Ce to aluminum alloys gives a chance to increase the length of the fluidity from 93.2 to 130.2mm. The optimal amount of Ce was

D. D. Cioboată and J. Machado (Eds.): ICoRSE 2025, LNNS 1592, pp. 62–69, 2025.
https://doi.org/10.1007/978-3-032-02508-1_6

discovered to be 0.3 w in terms of both fluidity and grain refinement factors. Excess of the Ce content in aluminum alloys by more than 0.5 wt.% leads to the formation of primary dendrites and aggregation of intermetallic compounds, which affects the fluidity of the alloy. For the experiment as the component was used A356 alloy ingot with a weight of 1 kg and Ce content at 0.0, 0.1, 0.2, 0.7 wt. The results are shown in the diagram, which indicates that an increase of the content of Ce improves the quality of the fluidity of the alloys, yet after 0.3wt there can be seen a dramatic decrease. For the test H13 tool steel was employed to determine the main factors of the fluidity. The increase in Ce component not only increase the fluidity length, but also it gives a good refinement effect [13].

The same method focused on the modification of Al alloys was a research suggested by scientists from South China University of Technology. In the investigation, the content of Ni in the Al-Si alloys was analyzed and was found out that adding Ni can potentially increase the casting properties of the alloys. The best impact was recorded when the content of nickel was between 2wt% and 6wt%. On the contrary, when the Ni content reaches 8% by weight, the fluidity decreases notably. This is explained by the development of the aluminum nickel phase and the subsequent decrease in dendritic coherency temperature [14].

Another study, indicating an excellent improvement not only in the fluidity of aluminum alloys, but also in the enhancement of the quality of the surface roughness, prevention of hot cracking is the finding introduced by Guan-Cheng Chen and Xiaochun Li. An addition of a small amount of nanoparticles (TiC) to Al-X alloys offered an opportunity to improve both the mechanical and casting properties. The optimal percentage of the TiC was to be found 2vol%. According to the microstructural analysis and hot cracking tests, the best percentage of the TiC ranges from 1 to 2 vol%, since it completely prevents from cracking and also exhibits more equiaxed charasterstics in contrast with those of 0.5 vol%. Not only does a greater amount of TiC demonstrate a cracking prevention and better characteristics, but it also shows a smoother surface with an increased nanoparticle content. As a result of the experiment, the surface roughness was decreased by 58%, the hardness was improved from approx. 60.8 to 62.3, the dendritic arms were transformed into more spherical ones and hot cracking was completely mitigated [15].

But there are other studies showing the opposite result. The work presented by scientists National Research Technical University (Moscow) is focused on the study of the effect of alloying additives on the fluidity of the alloy. Additives such as strontium, titanium, and boron were used as modifiers, and alloy A356.2 was evaluated as an alloy. The authors used the vacuum suction method to assess fluidity in order to more accurately assess changes in the casting properties of the aluminum alloy. The AlSr10 and AlTi4 ligatures were employed to lead Sr and titanium into the alloy, respectively. The scientists concluded that the addition of AlSr10 and AlTi5B1 ligatures does not significantly change the fluidity of the alloy. But the AlTi 4 ligature with the same amount of titanium significantly impairs the liquid-flow properties of the alloy [16].

Similarly, the same opposite results were obtained when adding lithium to the Al alloys. The addition of lithium to aluminum alloys degrades casting properties despite significantly improved mechanical properties such as strength, hardness, density, good viscosity, etc., which makes it difficult to use Al-Li-X alloys in production where fluidity plays a key role. Scientists from Turkey studied the effect of adding lithium to aluminum alloys and came to the conclusion that with an increase in the percentage of lithium in the alloy, it had a bad effect on its fluidity. As noted in the article, the deterioration of fluidity could be due to the fact that when lithium is added to the alloy [17], the metal increases the rate of oxidation of the alloy, which in turn leads to increased slag formation and oxidation of lithium, thus deteriorating the fluidity of the alloy [18].

Due to the irreplaceable mechanical properties after alloying aluminum alloys with lithium and the excellent casting properties of fluorides, the effect of lithium fluorides on Al-Si and Al-Cu alloys has been studied [19, 20]. In this study, practical experiments were conducted on the mathematical modeling of the effect of lithium fluoride on the fluidity of aluminum alloys (Al-Si and Al-Cu), along with a comparative analysis using graphs [21, 22] aimed at improving the fluidity properties of the alloys.

2 Materials and Methods

Fluidity is an important casting property in the production of cast parts. In the research, to assess the fluidity of non-ferrous metals, spiral samples (Curie samples) were cast using sand-clay molds (Fig. 1). The molds were prepared in two flask sections:

- The bottom flask contained the spiral sample,
- The top flask contained the pouring channel.

As the research objects, binary aluminum alloys were selected:

- AK7 from the aluminum-silicon system,
- D16 from the aluminum-copper system.

To compare the samples, first of all, an aluminum alloy of the AK7 brand was cast, without the addition of a lithium-fluorine compound. The samples were marked on the scales before the charge was placed on the crucible so that they were of the same mass and an equal amount of charge was loaded into the crucible. The weight was taken at the rate of 130 g per sample. In the next step, a lithium fluoride compound was added to the aluminum alloy, packaging it in aluminum foil. In this case, a combination of lithium fluoride in the amount of 5% of the total weight was added to the first sample, a combination of lithium fluoride in the amount of 10% was added to the second sample, and 15% was added to the third sample.

3 Results and Discussions

After casting the samples into the molds, they were removed from the molds. below are examples of studies carried out on AK7 aluminum (Fig. 1).

The second stage of the experiment mentioned above was conducted. In the next stage, a lithium fluoride compound was introduced into the D16-grade aluminum alloy. The experiments were carried out in the same order as described above.

In this stage as well, experiments were conducted with mass fractions of 5%, 10%, and 15%. The cast samples were separated from the molds, and the spiral length of

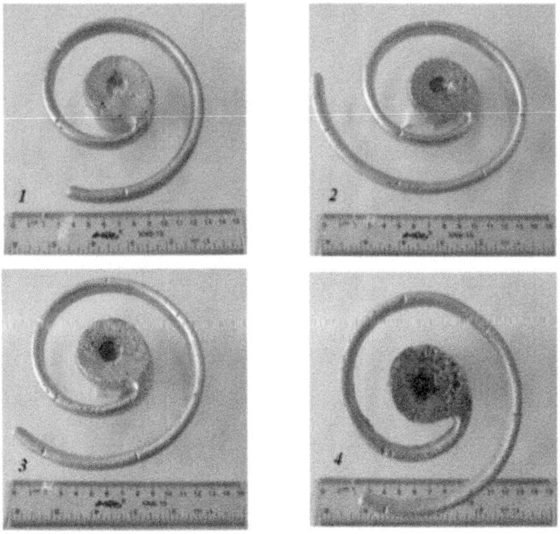

Fig. 1. Cast samples.1- AK7 without lithium fluoride; 2-sample with 5% addition of lithium fluoride; 3-sample with 10% addition of lithium fluoride; 4-lithium fluoride 15% added sample.

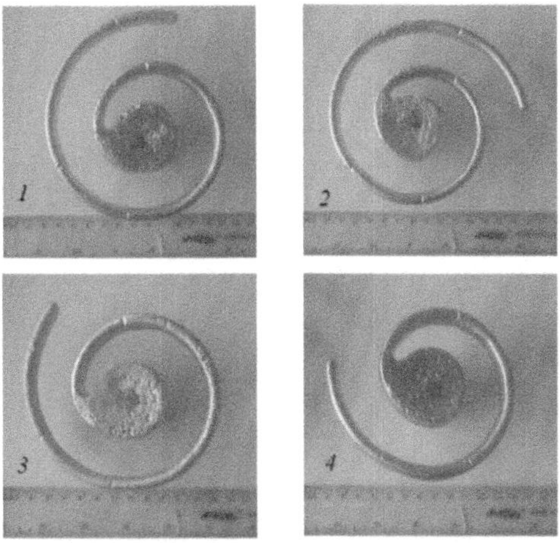

Fig. 2. Cast samples.1- D16 without lithium fluoride; 2-sample with 5% addition of lithium fluoride; 3-sample with 10% addition of lithium fluoride; 4-lithium fluoride 15% added sample.

the samples was determined. Below are the research samples conducted on D16-grade aluminum (Fig. 2). The results of measuring fluidity are given in Fig. 3.

Based on these measurement results, a graph illustrating the relationship between the lithium fluoride compound content and spiral fluidity was constructed (Fig. 4).

The graph in Fig. 4 was used as the basis for the mathematical modeling of the effect of the lithium element on the fluidity properties of aluminum alloys.

First, we will develop a mathematical model for the research conducted on the AK7 alloy.

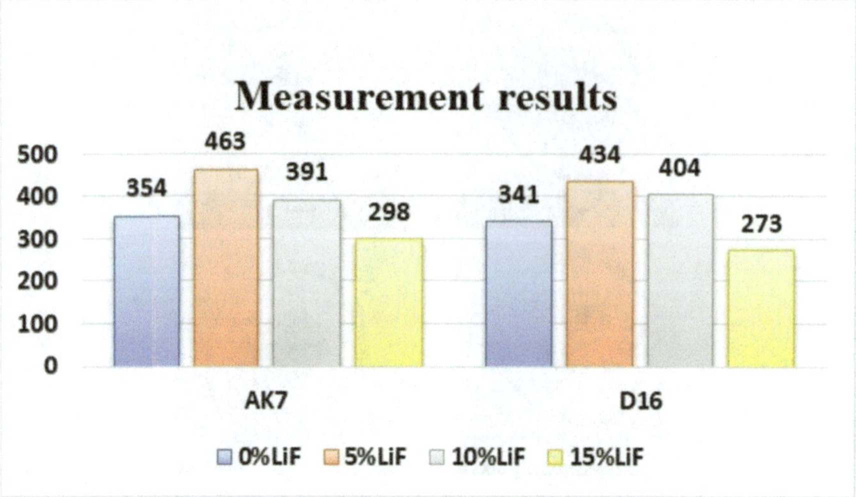

Fig. 3. The results of measuring fluidity.

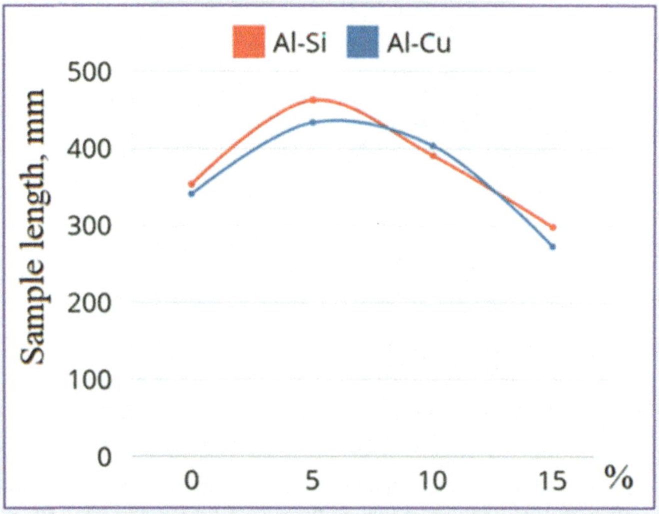

Fig. 4. Graph of the relationship between fluidity and the amount of lithium fluoride.

η = The amount of LiF in percentage;
y = fluidity (mm);
$f(\eta) = y$.

$$\begin{cases} 625\eta_1 + 25\eta_2 + 5\eta_3 = 440 \\ 1000\eta_1 + 1000\eta_2 + 10\eta_3 = 390 \\ 3375\eta_1 + 225\eta_2 + 15\eta_3 = 290 \end{cases} \quad (1)$$

First, we transform the system of equations into matrix form and use Cramer's rule to solve the system of complex nonlinear equations and determine the unknowns. Initially, we calculate the determinant Δ of the matrix formed by the system of equations.

$$\Delta = \begin{vmatrix} 625 & 25 & 5 \\ 1000 & 100 & 10 \\ 3375 & 225 & 15 \end{vmatrix} = -562500 \qquad (2)$$

2 Solving the Matrix, We Determine the Following Values of the Unknowns,

$$x_1 = \frac{\Delta_1}{\Delta} = \frac{-110000}{-562500} = \frac{44}{225}$$

$$x_2 = \frac{\Delta_2}{\Delta} = -\frac{4962500}{562500} = -\frac{397}{45}$$

$$x_3 = \frac{\Delta_3}{\Delta} = \frac{-60562500}{-562500} = \frac{323}{3}$$

After solving the above system and determining the unknowns, the function expressing the relationship between fluidity and the amount of lithium fluoride introduced into the alloy is as follows:

$$\varphi(\eta) = 0{,}195\eta^3 - 8{,}822\eta^2 + 107{,}67\eta \qquad (3)$$

Similarly, we will perform calculations for the D16 alloy in the same manner.

$$\begin{cases} 625\eta_1 + 25\eta_2 + 5\eta_3 = 450 \\ 1000\eta_1 + 1000\eta_2 + 10\eta_3 = 410 \\ 3375\eta_1 + 225\eta_2 + 15\eta_3 = 270 \end{cases} \qquad (4)$$

$$\Delta = \begin{vmatrix} 625 & 25 & 5 \\ 1000 & 100 & 10 \\ 3375 & 225 & 15 \end{vmatrix} = -562500 \qquad (5)$$

$$x_1 = \frac{\Delta_1}{\Delta} = \frac{-97500}{-562500} = \frac{13}{75}$$

$$x_2 = \frac{\Delta_2}{\Delta} = -\frac{5025000}{562500} = -\frac{134}{15}$$

$$x_3 = \frac{\Delta_3}{\Delta} = \frac{-63562500}{-562500} = 113$$

After solving the system in the manner described above and determining the unknowns, the function expressing the relationship between fluidity and the amount of lithium fluoride introduced into the D16-grade alloy is as follows:

$$\varphi(\eta) = 0{,}173\eta^3 - 8{,}933\eta^2 + 113\eta \qquad (6)$$

4 Conclusions

Based on the conducted experiments, it can be concluded that the addition of a lithium fluoride compound as an alloying element to AK7 and D16-grade aluminum alloys increases fluidity to a certain extent compared to samples without the additive. The most optimal addition amount is recommended to be 4–5% of the total mass. As seen from the graph above, the best results are achieved with a 4–5% addition. Lithium fluoride increases the fluidity property of the AK7 aluminum alloy by a maximum of 30–31% and that of the D16 aluminum alloy by a maximum of 26–28%.

Furthermore, using mathematical modeling, functions derived from Lagrange polynomials can be generated by selecting a sufficiently high polynomial degree based on the data from tables and graphs. The coefficient value before the highest-degree variable in these polynomials indicates that the limited dataset provides sufficient information. Additionally, these functions, based on initial experimental results, allow for predicting future outcomes without conducting additional experiments.

References

1. Lü, Y.Z., Wang, Q.D., Ding, W.J., Zeng, X.Q.: Fluidity of Mg-Al alloys and the effect of alloying elements. Zeitschrift Fuer Metallkunde/Mater. Res. Adv. Techn. **91**(6), 477–482 (2000)
2. Tursunbaev, S., Turakhodjaev, N., Zhang, L., Wang, Z., Mardonov, U., Saidova, M.: Mechanical properties and evolution of the microstructure of Al-Cu-Mg system alloys under the influence of alloying elements (GE and SI). Int. J. Mechatron. Appl. Mech. (18), 164–169 (2024). https://doi.org/10.17683/ijomam/issue18.19
3. Tashlykova-Bushkevich, I., Shikagawa, T., Suzuki, T., Shepelevich, V., Itoh, G.: Effect of Cr and Zr dopes on hydrogen behaviour in rapidly solidified aluminium foils. In: Materials Science Forum, vol. 638–642 (2010). https://doi.org/10.4028/www.scientific.net/MSF.638-642.465
4. Tursunbayev, S., Turakhodjayev, N., Mardanokulov, S., Zokirov, R., Odilov, F.: The effect of lithium on the mechanical properties of alloys in the Al-Li system. E3S Web Conf. **390**, 05046 (2023). https://doi.org/10.1051/e3sconf/202339005046
5. Tursunbaev, S., Turakhodjaev, N., Mardonakulov, S., Toshmatova, S.: Effect of germanium oxide on the properties of aluminum casting details in agricultural machinery. BIO Web Conf. **85**, 01024 (2024). https://doi.org/10.1051/bioconf/20248501024
6. Zhang, M., et al.: Effects of Ca, Al mass ratio on microstructure and properties of mg-Al-Ca-Mn Alloy l 犆犪犾质量比对犕犵犃犾犆犪犕狀合金组织和性能的影响. Tezhong Zhuzao Ji Youse Hejin/Special Cast. Nonferrous Alloys **43**(3), 296–299 (2023). https://doi.org/10.15980/j.tzzz.2023.03.002
7. Taghaddos, E., Hejazi, M.M., Taghiabadi, R., Shabestari, S.G.: Effect of iron-intermetallics on the fluidity of 413 aluminum alloy. J. Alloy. Compd. **468**(1–2), 539–545 (2009). https://doi.org/10.1016/j.jallcom.2008.01.079
8. Prukkanon, W., Srisukhumbowornchai, N., Limmaneevichitr, C.: Influence of Sc modification on the fluidity of an A356 aluminum alloy. J. Alloy. Compd. **487**(1–2), 453–457 (2009). https://doi.org/10.1016/j.jallcom.2009.07.169
9. Han, Z.-H., Wang, Z.-M., Sun, Z.-P., Zhang, B.-R., Rao, W.-F.: Influence of non-uniform ultrasonic vibration on casting fluidity of liquid aluminum alloy. China Foundry **19**(5), 380–386 (2022). https://doi.org/10.1007/s41230-022-2039-5

10. Tursunbaev, S., Turakhodjaev, N., Odilov, F., Mardanokulov, S., Zokirov, R.: Change in wear resistance of alloy when alloying aluminium alloy with germanium oxide. E3S Web Conf. **401**, 05001 (2023). https://doi.org/10.1051/e3sconf/202340105001
11. Šolc, M., et al.: Simulation Models in a Fluidity Test of the Al-Si Alloy. Metals **14**(4) (2024). https://doi.org/10.3390/met14040456
12. Turakhujaeva, S., Karimov, K., Turakhodjaev, N., Akhmedov, A.: Mathematical modeling of quantitative changes in hydrogen and oxide inclusions in aluminum alloy. E3S Web Conf. 365, 05016 (2023). https://doi.org/10.1051/e3sconf/202336505016
13. Niu, G., Mao, J., Wang, J.: Effect of Ce addition on fluidity of casting aluminum alloy A356. Metall. Mater. Trans. A **50**(12), 5935–5944 (2019). https://doi.org/10.1007/s11661-019-05458-9
14. Yang, L., Li, W., Du, J., Wang, K., Tang, P.: Effect of Si and Ni contents on the fluidity of Al-Ni-Si alloys evaluated by using thermal analysis. Thermochim. Acta **645**, 7–15 (2016). https://doi.org/10.1016/j.tca.2016.10.013
15. Chen, G.-C., Li, X.: Effect of TiC nano-treating on the fluidity and solidification behavior of aluminum alloy 6063. J. Mater. Process. Technol. **324**, 118241 (2024). https://doi.org/10.1016/j.jmatprotec.2023.118241
16. Баженов, В.Е., et al.: Изучение влияния добавок Ti, Sr и B на жидкотекучесть алюминиевого сплава A356. 2 (АК7пч). In: Bazhenov, V.E., et al.: Study of the effect of Ti, Sr, and B additives on the fluidity of aluminum alloy A356. 2 (AK7pch).) Izvestiya Vuzov. Tsvetnaya Metallurgiya **28**(4), 55–66 (2022)
17. Ozaydin, O., Kaya, Y., Dispinar, D.: Effect of Li additions and holding time on the mechanical properties of the AlSiM9mg alloys. Metallurgia Italiana **113**(11–12), 19–24 (2021)
18. Бокиев, Л.А., Ганиев, И.Н., Джайлоев, Д.Х.: ТЕПЛОЁМКОСТЬ АЛЮМИНИЕВОГО СПЛАВА AL5FE10SI С МАГНИЕМ. Вестник педагогического университета (Естественных наук), Bokiev, L.A., Ganiev, I.N., Dzhailoev, D.H.: The heat capacity of aluminum alloy AL5FE10SI with magnesium. Bull. Pedagogical Univ. (Nat. Sci.) **1–2**(9–10), 97–99 (2021)
19. Berdiev, D.M., Yusupov, A.A., Toshmatov, R.K., Abdullaev, A.K.: Increasing die durability in cold stamping by quenching with intermediate tempering. Russ. Eng. Res. **42**(10), 1011–1013 (2022). https://doi.org/10.3103/S1068798X22100057
20. Sarvar, T., Nodir, T., Mardonov, U., Saydumarov, B., Kulmuradov, D., Boltaeva, M.: Effects of germanium (Ge) on hardness and microstructure of Al-Mg, Al-Cu, Al-Mn system alloys. Int. J. Mechatron. Appl. Mech. (16), 179–184 (2024). https://doi.org/10.17683/ijomam/issue16.21
21. Nodir, T., Sarvar, T., Kamaldjan, K., Shirinkhon, T., Shavkat, A., Mukhammadali, A.: The effect of lithium content on the mass of the part when alloyed with lithium aluminum. Int. J. Mechatron. Appl. Mech. (11), 52–56 (2022). https://doi.org/10.17683/ijomam/issue11.7
22. Sarvar, T., Nodir, T., Shohista, S., Anvar, T., Mashhura, M.: Development of a mathematical model of the effect of lithium on the wear resistance properties of aluminum-lithium alloys. Am. J. Appl. Sci. Technol. **4**(01), 37–41 (2024)

Control of a Mechatronic Device for Measuring the Optometric Parameters of a Spectacle Wearer

George Baboianu and Cristian Gabriel Alionte(✉)

National University of Science and Technology POLITEHNICA Bucharest, Splaiul Independenței 313, 060042 Bucharest, Romania
cristian.alionte@upb.ro

Abstract. This article aims to describe software for a mechatronic system employed in the positioning of an optometric system for accurately determining the patient's anthropometric data at three distinct distances. Following the description of the hardware and software components that constitute the backbone of the proposed mechatronic platform, the focus shifts to the software for the robotic positioning system. The system provides more accurate measurements by allowing you to virtually follow the motions of the patient's eyes and the location of their pupils. To obtain clear vision and rapid adaption, the proper results of these measures help to the design of an appropriate progressive lens.

Keywords: Optometry · Positioning mechatronic systems · Autonomous systems

1 Introduction

A mechatronic system was engineered and produced to position a tablet that captures photos essential for accurately determining the patient's anthropometric data at three distinct distances. These parameters are crucial in the fabrication of customized progressive lenses for individuals with facial asymmetries. As is known, the asymmetries of the human face create problems when wearing progressive lenses. The eyes travel different distances and are not in the same horizontal or vertical plane. A person's convergence differs for each eye. The lens is attached to the eyeglass frame and rests on the nose and ears. The ears are not symmetrically positioned and a person may have a deviated septum. All these facial asymmetries must be measured and analyzed so that the progressive lens can be built taking into account the movements of each eye and the position of the lens in front of the eye. The patient's optical parameters must be measured for several reference distances in order to be able to build a correct progressive lens. The tablet has three degrees of freedom, comprising two translational motions and one rotating movement. There are people with major facial asymmetries for whom a standard progressive lens does not provide the desired visual comfort. For these people, these personalized lenses are built using special software programs.

D. D. Ciobotă and J. Machado (Eds.): ICoRSE 2025, LNNS 1592, pp. 70–82, 2025.
https://doi.org/10.1007/978-3-032-02508-1_7

The tablet's position can be adjusted via the device control to three predetermined locations. Consequently, we can quantify the optometric parameters for the reader (350 mm), laptop (450 mm), and display (650 mm) (Fig. 1).

Fig. 1. Mechatronic tablet positioning device.

2 Controlling and Powering the Mechatronic Device

The control of the mechatronic system will be carried out from two directions. A "remote control" is integrated within the Eye Fit software for positioning during nearest picture capture, alongside another "remote control" console comprising buttons on the optometric device. The tablet can be moved from both locations for enhanced accessibility by the optician. The ARDUINO board controls the tablet's position, using drivers which allow the stepper motors to accomplish precise positioning. The control software is developed in C++ and the power supply takes from a constant 220 V source, with the transformer included into the device delivering a supply current of 12 V at 3.4 A to the three motors and electronic boards. From an electrical perspective, the apparatus has the following components:

- 2 Arduino Uno
- 5 lightning settings buttons
- 1 Bluetooth Module
- 1 CNC Shield Expansion Plate
- 3 stepper Motor Drivers A4988
- 3 stepper motors
- 3 normally open end-of-stroke limiters

Figure 2 offers a visual representation of the wiring diagram for the electrical assembly.

When it comes to microcontroller platforms, Arduino is among the most user-friendly options. It is possible to conceive of it as a minicomputer, as it possesses the computational capability of a typical computer from fifteen years ago. It can gather information from the surrounding environment and respond to it. The Arduino is the center of an ecosystem of gadgets that has been developed to an extraordinarily high level. There is a very good possibility that you will be able to locate an Arduino gadget that can provide you with whatever it is that you require, whether it be information that you want to gather from the environment or connections with other systems.

Based on an ATmega328P microcontroller, the Arduino UNO provides a wide range of functions that may be utilized in a variety of applications. These characteristics can be utilized in anything from straightforward LED control circuits to intricate automation and robotics systems. It is the intention of this paper to investigate the characteristics, applications, and components of the Arduino UNO control board, as well as the common applications that it has in the process of developing electronic projects using Arduino.

There are several hardware components that are necessary for the Arduino UNO to function correctly. These components are what make up the Arduino UNO. In terms of its design, the board is both tiny and simple to operate, making it an excellent choice for both novices and seasoned professionals in the field of electronics. There are a number of hardware components that are most crucial:

- ATmega328P microcontroller: This component, which serves as the brain of the Arduino UNO board, is accountable for processing and executing the code that is responsible for controlling the circuit. The ATmega328P is an 8-bit microcontroller that has 32 KB of flash memory for the storing of programs, 2 KB of SRAM memory for the storage of temporary data, and 1 KB of EEPROM memory for the storage of permanent data. This microcontroller is fitted with a clock that operates at 16 megahertz, which is responsible for determining the rate at which instructions are carried out.
- Digital and analog input/output ports: The Arduino UNO features 14 digital I/O pins, which are input/output pins. Six of these pins can also be used to generate pulse width modulation (PWM) signals. Additionally, there are six analog pins on the board, which enable the reading of analog signals from many signal sources such as sensors, transducers, and other similar devices. It is possible to transform analog signals into digital values thanks to the presence of an internal analog-to-digital converter (ADC), which is attached to these analog pins.
- Power connector: The Arduino UNO board can be operated via a USB cable, that provides both the power required to operate the board and a method of recharging programs. As an alternative, it can be supplied by an external source, such as a 9 V–12 V conversion or a battery.
- A voltage regulator is a component that is included in the Arduino UNO. This component is responsible for converting the voltage from the external source into a stable voltage of 5 V. This voltage is then utilized to power the board as well as any external components.
- Using the USB connector, the user is able to load software into the ATmega328P microcontroller and observe the communications that take place between the Arduino and the personal computer. When the board is linked to a computer, this port is also utilized to supply power to the board.

3 Arduino IDE Programming Interface and Software

Users take advantage of the Arduino IDE, which stands for the Integrated Development Environment, in order to interface with the Arduino UNO board. An easy-to-use development environment that enables you to develop, compile, and load programs on the Arduino board is called the Arduino Integrated Development Environment (IDE). The user is responsible for writing the code that governs the behavior of the board and the circuit. The development process is done in a programming language that is comparable to C and C++. Listed below is an overview of some of the most significant characteristics of the Arduino IDE:

- A basic editor that allows users to develop code in the programming language used by Arduino is provided by the Code Editor. Furthermore, the integrated development environment (IDE) provides suggestions for libraries and functions that are accessible for many different types of hardware.
- Among the components of the Arduino Integrated Development Environment (IDE) is a compiler, which is accountable for converting code that is written into instructions in binary format that the microcontroller is able to interpret and carry out.
- Serial Console: This functionality enables communication between the Arduino board and the computer using the serial port. This communication is helpful for debugging and monitoring the data that is delivered to and from the Arduino board.
- Predefined libraries: The Arduino Integrated Development Environment (IDE) comes with a significant number of libraries that make it easier to build fundamental and sophisticated functionality. These libraries include motor control, sensor reading, LCD screen control, and many others. The fact that users do not have to build code from scratch is one of the ways in which these libraries save time and resources.
- The Arduino UNO board is used in a wide range of applications, thanks to its versatility and the large community that is constantly developing new projects and solutions. Here are some examples of applications in which Arduino UNO plays a key role:
- Educational projects: The Arduino UNO is widely used in specialized education because it allows students and beginners to understand the basic principles of electronics and programming. Simple projects, such as turning on an LED or reading a temperature sensor, are used to introduce fundamental concepts.
- Automations and equipment control: The Arduino UNO can be used to control various devices, such as motors, pumps, lights, or even complex automation systems, such as those used in smart homes. The board can be connected to external relays and modules to control devices that require more current than the board can provide.
- Robotics: Due to its ability to interact with motors, sensors, and other electronic components, the Arduino UNO is an ideal tool for developing robot prototypes. For example, it is used for motion control, obstacle detection or autonomous navigation.
- Sensor and IoT (Internet of Things) projects: The Arduino UNO can be used to create environmental monitoring systems, such as measuring temperature, humidity, pollution, or other parameters of interest. It is also often used in IoT applications, connecting to the internet through additional modules (such as Wi-Fi or Bluetooth) to send data or control devices remotely.

It is important to take into consideration the following benefits that the Arduino UNO board offers:

Ease of use: The Arduino UNO is incredibly simple to use, even for novices, as a result of its comprehensive documentation and extensive community of support.
It is inexpensive, which makes it an excellent choice for rapid prototyping and electronic experimentation because of its low cost.
Adaptability: Since the board is versatile, it may be utilized for a broad variety of applications, ranging from straightforward prototypes to intricate systems.
An active community: There is a sizable community of users and developers who offer projects, tutorials, and solutions to the platform.
Because of the aforementioned, if we are talking about the process of collecting data from the environment around us, the following are some instances of sensors: Fire sensors, LPG gas, carbon monoxide, accelerations of moving devices, current consumed by various household devices, pressure force, degree of rotation, RFID cards, distances, lighting level, the direction of North, human presence, sound, temperature, humidity, atmospheric pressure, or video are some examples of sensors that can be used to determine the amount of alcohol present in the air while breathing.

There are Ethernet network boards for the Arduino that are capable of communicating information over the Internet, devices that are capable of transmitting data through a radio connection, WIFI network boards, GSM devices for the Arduino (capable of sending and receiving SMS, initiating voice calls, or sending data through the 3G network), and Bluetooth connectors for connecting the Arduino with the mobile phone or laptop. If we are considering the possibility of connecting with other systems, then we are referring to it.

Fig. 2. Arduino UNO control board.

In the field of mechanical engineering, there are DC motors, which are utilized for robots, stepper motors, which are typically utilized in the industrial sector, and servo motors, which are regulated as precisely as possible. There are a variety of LCD screens available for the Arduino, ranging from the most basic (a text LCD with 16 characters) to graphic LCD panels, which are used to display the information that has been retrieved.

Components that define the Arduino board and the functions that they do are as follows:

1. Reset button – This will restart any code that is loaded on the Arduino board
2. AREF - Stands for "Analog Reference" and is used to set an external reference –voltage
3. Ground Pin - Ground Pins on Arduino
4. Digital Input/Output – Pins 0–13 can be used for digital input or output
5. PWM - Pins marked with the symbol (~) can simulate analog output
6. USB Connection – Used for powering Arduino and uploading sketches
7. TX/RX - Transmits and receives data and indicates by lighting LEDs
8. ATmega Microcontroller - This is the brain and is where the programs are stored
9. Power LED Indicator – This LED lights up whenever the board is connected to the power source
10. Voltage regulator – This controls the amount of voltage entering the Arduino board
11. Power Barrel Jack – This is used to power the Arduino with extra power
12. 3.3 V Pin – This pin delivers 3.3 V of power
13. 5 V Pin – This pin provides 5 V of power
14. Ground Pin - Ground Pins on Arduino
15. Analog pins – These pins can read the signal from an analog sensor and convert it to a digital signal

Fig. 3. Bluetooth module HC-06.

The HC-06 Bluetooth module is a very popular wireless communication module that is widely used for interfacing a variety of electrical devices to microcontrollers. This is especially common in instances when automation and robotics projects are being accomplished. In addition to being a straightforward and low-cost module that is easy to deploy, it is a member of the family of serial Bluetooth modules that are based on Bluetooth 2.0 technology. Its purpose is to facilitate the transmission of data between devices using Bluetooth technology. Applications that require a data transfer between a microcontroller (for example, Arduino) and a mobile device, such as a smartphone or computer, are supposed to be the primary target audience for the HC-06 module. It is possible to utilize it to establish wireless connections for a variety of applications, including control of robotics, monitoring of sensors, Internet of Things applications, and more (Fig. 3).

A Bluetooth chip and a set of pins that allow for simple connection to other circuits are both included in the HC-06 module, which happens to be a rather small device. In addition to a transmitter and receiver, it is equipped with an embedded CPU that is responsible for managing Bluetooth communications and indicators. The module has the capability of transmitting data at a speed of up to 3 Mbps, which is sufficient for the majority of applications requiring wireless control and communication.

Main features of the HC-06 module are:

- Bluetooth standard: The HC-06 uses Bluetooth 2.0 + EDR (Enhanced Data Rate), which means it can reach a transfer speed of 3 Mbps.
- Supply voltage: The module operates at a supply voltage of 3.3 V to 5 V, making it compatible with most development boards, including Arduino.
- Serial communication: The HC-06 uses the serial communication protocol (UART), having a two-line interface: RX (Receive) and TX (Transmit), through which data is exchanged between connected devices.
- Frequency range: The module operates on the 2.4 GHz frequency, which is typical for Bluetooth devices.
- Range: The maximum communication distance is about 10 m, but this may vary depending on the environment and signal interference.
- Operating modes: The HC-06 module can operate in "Master" mode (to initiate Bluetooth connections) and "Slave" mode (to accept connections initiated by other devices). The HC-06 is usually configured in "Slave" mode, and the connection is established by the device sending commands.
- LED indicators: The HC-06 module includes LEDs to signal the connection status. Usually, the LED is lit constantly when the connection is established.

Utilizing the UART interface, the HC-06 module establishes a connection with microcontrollers. To put this into perspective, the device will require two pins for transmitting and receiving data (also known as TX and RX), and it will also require two more pins for conducting power. When it comes to the suitable power supply for the HC-06 module, the voltage that is utilized is typically either 3.3V or 5V. This is determined by the particular model of the microcontroller that is being utilized.

This is the procedure that must be followed in order to use the HC-06 module with a microcontroller like an Arduino:

1. Connecting the modules to the microcontroller:
   a. Connecting the HC-06 module's TX pin to the Arduino's RX pin.
   b. Connect the HC-06 module's RX pin to the Arduino's TX pin.
   c. Connecting the VCC pin to a 3.3 V or 5 V power source.
   d. Connecting the GND pin to the circuit ground.
2. Arduino programming: After physical connection, the user can write a program on the Arduino to establish communication with the HC-06 module. In this case, the user will use the serial functions (Serial.begin(), Serial.print(), etc.) to send and receive data.

Through the use of AT instructions, the HC-06 module can be programmed to establish connections with a variety of devices. The user has the ability to change the settings of the module by sending these commands through the serial port. These commands enable us to change the device name, the connection password, or the communication speed and more.

- AT Mode: To enter AT mode, it is necessary to connect the HC-06 module to a serial terminal and send AT commands. If the module responds correctly to AT commands, it means that the module is configured correctly. For example, the AT command will send a response message, confirming that the module is working properly.
- Set the device name: Through the AT command, it is possible to change the name of the device. For example, the AT + NAME = MyDevice command will change the name of the device to "MyDevice".
- Password setting: One can also set a password for Bluetooth connections, which adds a layer of security to the login process.

The HC-06 module is used in a wide range of applications, due to its simplicity and low cost. Some of the most common applications include:

- Robot Control: HC-06 is often used to control robots via Bluetooth. A smartphone or PC can send commands to move the robot in different directions.
- Home automation: The HC-06 module can be used to control various devices in a smart home, such as lights, thermostats, or outlets, through a mobile app.
- Sensor monitoring: It can be used to send data from sensors (e.g., temperature or humidity sensors) to a mobile device for real-time monitoring.
- User Interfaces: The module can be used to create simple interfaces for controlling and monitoring electronic devices.

For the aim of converting the serial port into a Bluetooth port, the Bluetooth serial module is utilized. The master and slave modes are both available on this module. The Bluetooth operating module does not require the unit and is able to interact with the other Bluetooth device that has serial. However, in order for the two Bluetooth modules to communicate with one another, there must be at least two conditions met:

1. The master and the slave are the only ones who are permitted to communicate.
2. The password must be entered correctly. At a frequency of 2.4 GHz, communication is carried out wirelessly via serial RX and TX.

While holding down the green button, the stand will begin to operate and will be initialized. During the early stage, the system will progress to the beginning position, also known as the zero position, which is indicated by the limiters that are located at the conclusion of each stroke. Following the completion of the initialization process, the system can be controlled in two different ways: either over Bluetooth, from the tablet that is held on the stand, or manually, from the console that contains buttons.

In order to manually control the stand, there are five buttons available: the Start button (green), the Stop button (red), and three blue buttons that allow the stand to be positioned in each of the three reading zones. The buttons are linked to digital pins 4 through 9 in the circuit. It is necessary for the device that is connecting to the stand to transmit data that corresponds to the system locations that are contained in the program that is loaded into the Arduino Uno in order to enable Bluetooth control. If you are in the first, second, or third position, the orders are as follows: 1, 2, 3.

The Bluetooth module and the Arduino Uno microcontroller are able to communicate with one another through the use of a serial protocol (UART) that is connected to digital pins 2 and 3. The utilization of the Serial Software library, which has the capability to transform any two digital pins into RX/TX pins in order to accomplish serial connection between two devices, makes this particular endeavor feasible.

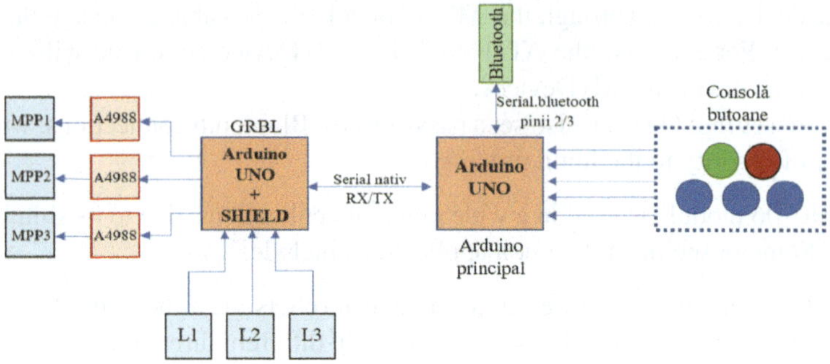

Fig. 4. Connection diagram.

Additionally, from the perspective of Bluetooth communication, a command can be sent to the tablet to take a snapshot of the patient who is being studied by pushing the green button on the console. This eliminates the need to touch the tablet or approach the patient, which could potentially cause discomfort.

The second Arduino UNO, which is mounted with the CNC shield (Fig. 4), is responsible for ensuring that the stand is positioned on the appropriate location. Electronic pins 0 and 1 are responsible for establishing the connection between the native serial port (hardware) and the first Arduino, which is the main Arduino. This allows it to communicate with the first Arduino. It gets a line of code of type GCODE each serial, and the line of code is determined by the intended location respectively. A machine code language that was developed specifically for CNC systems is being discussed here. The GRBL library, which is loaded onto the second Arduino UNO, is utilized in order to accomplish the task of interpreting this code. Additionally, this library is responsible for the startup process, which begins at the very beginning of the system (Fig. 5).

When the second Arduino UNO has finished receiving and interpreting the positioning signal, it will then send a series of electrical pulses to the A4988 drivers. These pulses will indicate the number of steps that the stepper motors need to perform in order to move the system to the intended position. It should be noted that there is an additional signal that indicates the direction in which the motor is rotating. Following the system's arrival at the intended position, the motor windings continue to receive power in order to ensure that the system continues to remain in the same position. In order to relocate the system to a different location, it is necessary to issue a new command, which can be received from the console or by Bluetooth.

Pressing and holding the stop button will cause the system to retract to the zero position and will cause all control methods to be interrupted. This will allow you to stop the system.

Fig. 5. The control system comprised of Arduino Uno, CNC Shield, and A4988 driver.

4 Working Procedure of the Mechatronic System

The system was developed to evaluate the optical parameters of a patient in three primary positions: the reading position, which is considered to be in the near field of vision; the position of the laptop, which is considered to be in the intermediate field of vision; and the position of the monitor screen, which is considered to be in the broad field of vision. The reading position is taken into consideration as the starting point of the system. In order to accomplish the transmission for the x translation axis, T2.5/6 belts and 16/T2.5 belt wheels are utilized. A Plusivo 23HS5628 stepper motor that is capable of providing a holding torque of 1.26 Nm is linked to the drive belt wheel. The M8X4 trapezoidal screw and a Plusivo 17HS8401S stepper motor with a holding torque of 0.59 Nm are the components that are utilized in the production of the z-axis transmission. A plastic tablet holder that is coupled to a belt wheel that is driven by a Plusivo 17HS8401S stepper motor is utilized by the system in order to tilt the tablet (Fig. 6).

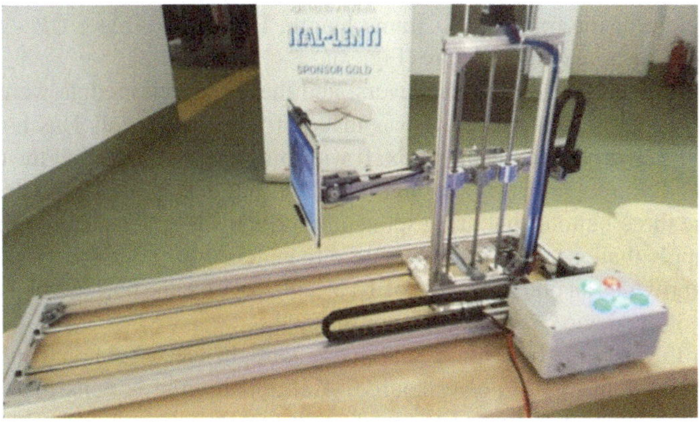

Fig. 6. Mechatronic tablet positioning device.

When it comes to the command and control techniques of the opto-mechatronic system, there are two alternatives available: wireless control through the EYE FIT software that is installed on the tablet, and local control through the use of the fixed control console that is situated next to the system.

It is necessary to push and hold the green (ON) button throughout the process of turning on and initializing the media. The initial location, also known as the origin of the system, will be reached by the system during the early stage. This position is constrained by the stroke limiters, which are L1, L2, and L3. On (the green button), off (the red button), and three blue buttons for setting the stand in the three test positions are the five buttons that are connected to the Arduino UNO, which is designated the Master Controller. These buttons are used for operating the stand manually. For the purpose of controlling the system over Bluetooth, the Master controller establishes software serial communication with an HC-06 Bluetooth module, which is responsible for reading data from the tablet. As soon as a positioning command has been established, the Master Controller will transmit the appropriate instructions to the second Arduino Uno, which is loaded with the GRLB library. The second Arduino Uno will then send the relevant orders to the A4988 motor drivers, which will then send them to the motors.

During the consultation, the image of the patient's eyes can be captured by either the local console, which requires the patient to press and hold the green button for two seconds, or by the visual and detecting subsystem, which requires the patient to hit the software button on the tablet. The system can be turned off by holding down the red button (which will bring it back to its original position) or it can receive another instruction to test a new patient after it has captured the photos that are required to measure the interpupillary distance in each of the three places.

According to the measurements with the tablet, the following optometric parameters for reference distances result where the following notation is used Dip - interpupillary distance; MpD - Interpupillary distance for right eye; MpS - interpupillary distance for the left eye; Vx - vertex distance (from cornea to lens); HD and HS - mounting height; BH - the height of the frame; BL - frame length; Br - nasal bridge; UP – pantograph angle; UC - frame angle of curvature (Fig. 7):

The Arduino boards, the motor drivers, and the Bluetooth module were all placed inside of a box that had a cover and was made of plastic material. This box might be considered the electrical connection box. It was joined to the support of the orthopathic system. In addition to mounting the on, off, and off buttons on the lid of this box, I also mounted the three buttons that are responsible for activating the established positions. A power supply that operates at 220 V is used to generate the power supply. This power supply also provides an output current of 12 V and 3.4 A for the purpose of the electrical installation of the device that has been constructed.

In order to verify the functionality of the system, the electrical boards of the device were powered by the 12 V power source. Additionally, the buttons that control the location of the tablet were tested to ensure that they functioned properly. The tablet is positioned in each of the three reference positions, and operation of the gadget is determined by commands.

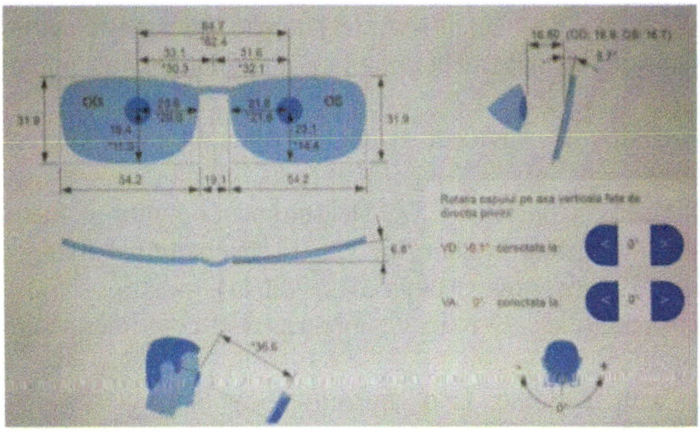

a)

						* Dimensiuni pentru vederea la aproape		
Dip:	64.7	mm	BH:	31.9	mm	*Dip:	62.4	mm
MpD:	33.1	mm	BL:	54.2	mm	*MpD:	30.3	mm
MpS:	31.6	mm	Br:	19.1	mm	*MpS:	32.1	mm
PpD:	23.6	mm	Vx:	16.8	mm	*PpD:	20.0	mm
PpS:	21.8	mm	VxOD:	16.9	mm	*PpS:	21.0	mm
HD:	19.4	mm	VxOS:	16.7	mm	*HD:	11.0	mm
HS:	23.1	mm	UP:	5.7	°	*HS:	14.4	mm
			UC:	6.8	°	Distanta la citire: *36.5 cm		

b)

Fig. 7. Optical parameters: a) software interface, b) some measured values.

5 Conclusions

There are three stepper motors that are responsible for rendering the movements of the tablet, and the skeleton of the mechatronic device is constructed out of aluminum profiles that are connected together using specific fasteners. The tablet undergoes two translational motions and one rotating movement as a consequence of these motors, which are driven by two electronic boards that are based on the Arduino Uno platform. The optomecatronic device is a prototype that, at this moment, is not particularly convenient to use in terms of its size and weight. It is possible to make improvements to it from this perspective; nevertheless, the most essential thing is that we acquire repeatability of measurements with a high level of precision. The examinations and tests that were carried out with this apparatus demonstrate that it has a high level of functionality. The measurements obtained from it are more accurate, being able to practically follow the motions of the patient's eyes and the location of their pupils. In order to achieve clear vision and rapid adaptability, the correct results of these measures contribute to the design of a progressive lens that is correct.

References

1. Arduino Hardware. https://www.allaboutcircuits.com/technical-articles/understanding-arduino-unohardware-design/. Accessed 03 Mar 2025

2. Arduino ports. https://i2.wp.com/www.technobyte.org/wp-content/uploads/2016/08/Arduino-Unomultiple-input-management-circuit-op-amp-comparator-min.jpg. Accessed 03 Mar 2025
3. Arduino Software. http://www.ntu.edu.sg/home/ehchua/programming/arduino/arduino.html. Accessed 03 Mar 2025
4. Baboianu, G., Niță, I.E., Comeaga, C.D.: Measurement of anthropometric parameters using opto-mechatronic positioning system. In: The 12 International Symposium on Advanced Topics in Electrical Engineering – March 25–27 (2021)
5. Arduino Microcontroller. https://components101.com/microcontrollers/arduino-uno. Accessed 06 Mar 2025
6. Getting started with Arduino Uno. https://www.youtube.com/watch?v=zusVAX7bgtA. Accessed 06 Mar 2025
7. https://electronics.stackexchange.com/questions/61537/what-is-the-cap-for-inarduino-reset-circuit. Accessed 07 Mar 2025
8. Baboianu, G., Nitu, C., Comeaga, C.D.: Processing of captured digital images for measuring the optometric parameters required in the construction of ultra-personalized special lenses. In: Proceedings of the International Conference of Mechatronics and Cyber-MixMech (2019)
9. Measuring planar objects. https://www.mathworks.com/help/vision/ug/measuring-planar-objects-with-a-calibrated-camera.html;jsessionid=c627065c7b49ba5320b6887b93f8. Accessed 12 Mar 2025
10. Arduino Board. https://learn.sparkfun.com/tutorials/what-is-an-arduino/whats-on-the-board. Accessed 06 Apr 2025
11. What Is USB. http://users.utcluj.ro/~baruch/sie/labor/Magistrala-USB.pdf. Accessed 03 Apr 2025
12. Arduino I/O ports. https://www.arduino.cc/en/Hacking/PinMapping168. Accessed 02 Apr 2025
13. Stengel-Rutkowski, S., Schimanek, P.: Chromosomal und nicht-chromosomale Dysmorphie syndrome. Enke Verlag, Stuttgardt (1985)
14. Ophtalmix Blog Homepage. https://oftalmix.wordpress.com. Accessed 02 Apr 2025
15. Image processing. https://www.dcode.fr/image-histogram. Accessed 06 Apr 2025
16. Concepcion-Grande, P., Chamorro, E., Cleva, J.M., Alonso, J., Gómez-Pedrero, J.A.: Evaluation of an eye-tracking-based method for assessing the visual performance with progressive lens designs. Appl. Sci. **13**(8), 5059 (2023). https://doi.org/10.3390/app13085059
17. Andersson, R., Nyström, M., Holmqvist, K.: Sampling frequency and eye-tracking measures: how speed affects durations, latencies, and more. J. Eye Mov. Res. **3**, 1–12 (2010). https://doi.org/10.16910/jemr.3.3.6
18. Constantin, V., Comeagă, C.D., Grămescu, B., Besnea, D., Moraru, E.: The design and testing of an additive manufacturing-obtained compliant mechanism for the complex personalisation of lenses in clinical optometry. Appl. Sci. **13**(24), 13010 (2023)
19. Liangliang, L., Liu, T., Yan, W., Fan, R.: Research on indoor real-time positioning technology of LED visible light communication based on arbitrary three axis rotation angle. J. Phys. Conf. Ser. **2492**(1) (2023). https://doi.org/10.1088/1742-6596/2492/1/012023

Dynamic Response of a Human Arm Model Using Lagrangian Multipliers Method Under Various Input Displacements

Zahraa Abd Al-Elah Kahaleel[1], Fawaz F. Al-Bakri[2](✉), and Muslim Ali[3]

[1] Mechanical Engineering Department, University of Kerbala, Kerbala, Iraq
zahraa.abdalelah@s.uokerbala.edu.iq

[2] Department of Biomedical Engineering, College of Engineering, University of Babylon, Babylon, Iraq
fawaz.al-bakri@uobabylon.edu.iq

[3] Prosthetics and Orthotics Engineering Department, College of Engineering, University of Kerbala, Kerbala, Iraq
muslim.m@uokerbala.edu.iq

Abstract. Prosthetics play broadly a significant role to enhance mobility and independence for disabilities people. These mechanical devices are utilized to provide functional movements of a human arm in diverse applications such as rehabilitation, sports, and innovation. Therefore, it becomes essential important to assess the dynamic response of human prosthetics. This study focuses on studying the dynamic response of a five-degree-of-freedom prosthetic human arm. The human system was modeled by using Lagrange-Euler method to obtain a five-second-order ordinary differential equations. Then it was linearized to facilitate the optimization process required to gain local optimal system parameters. A comparative analysis was then conducted between the linear and nonlinear systems responses to ensure that the linearized system can successfully able to compensate for the real system. The arm parameters (masses, damper coefficients, and stiffness) were selected relying on a Lagrangian multipliers method while achieving short settling time and low overshoot Eventually, the system performance of the proposed system was investigated while involving a wide range of initial excitations. The results showed that all parts of the human arm (elbow, forearm, and hand) oscillated, providing well-suited responses with a short settling time and low overshoot, despite varying initial conditions.

Keywords: Modeling · Linearization · Optimization · Lagrangian Multipliers Method

1 Introduction

The development of artificial human arms significantly improves the mobility and independence of individuals who have lost their arms, enabling them to perform daily activities with greater ease. As a result, the modeling and enhancement of prosthetic arms has

D. D. Cioboată and J. Machado (Eds.): ICoRSE 2025, LNNS 1592, pp. 83–98, 2025.
https://doi.org/10.1007/978-3-032-02508-1_8

become an increasingly important area of research, particularly in light of recent advancements in robotics and computational methods. [1] explained the use of artificial neural networks for analyzing the kinematics of a robot with two degrees of freedom, addressing both forward and inverse kinematics through neural network-based solutions. In [2], human arm movement was modeled and simulated using a two-dimensional, three-part coupled pendulum system, focusing on both forward and inverse dynamic approaches to arm motion. In [3], a mathematical model for direct forward kinematic analysis of a 5-degree-of-freedom articulated robot arm was developed, employing the Denavit-Hartenberg (D-H) method to calculate joint coordinates using MATLAB. Furthermore, [4] developed a comprehensive kinematic model for a robot arm, incorporating complete kinematic analyses and deriving the necessary equations through the D-H method.

On the other hand, the work [5], developed a kinematic model of the human hand with 16 degrees of freedom, aimed at determining joint torques. Moreover, [6] analyzed a human arm movement during dancing by modeling the arm as an inverted triangular pendulum. The three segments of the human arm (upper arm, forearm, and hand) were analyzed using the Lagrange-Euler equation and the results simulated using both Maple and MATLAB programs. In more recent work, [7] developed a robotic arm capable of sorting objects based on color sensing, modeling the arm with three degrees of freedom and using MATLAB GUI for visualization purposes. According to [8], the authors designed a heavy-duty robotic arm for harvesting pumpkins and watermelons, incorporating accelerometers as sensors to collect data on arm movements. The D-H method and MATLAB were used to solve the kinematic equations and program the robotic arm, respectively. In the same regard, [9] presented a kinematic model of a three-degree-of-freedom articulated robot arm, calculating forward kinematics using the D-H method and simulating the results using a MATLAB GU while the design of the robotic arm was done using SolidWorks. Likewise, [10] used electric cylinder to convert the rotating mechanical force generated by the electric motor to linear motion. The influence of the external forces, the force of gravity and inertia were taken into account while neglecting friction forces. Additionally, in [11], the Newton-Euler equation was used. Each joint was actuated by a single motor separately to achieve relative motion of adjacent links. This work focused on the movements of the human (upper arm, shoulder, wrist, elbow and finger). Eventually, these analyzes have been extended to design a humanoid robot manipulator.

The authors [12] and [13] presented methods for estimating upper limb resistance in three-dimensional space with 7 degrees of freedom. The proposed methods utilized to measure the dynamic characteristics of the human arm in 3D space. The joint angles of the robot were calculated using the Jacobian method. On the other hand, [14] designed and developed a PC-based robotic arm with six degrees of freedom. Each joint was connected to a suitable servo motor. The robotic arm design and control solution were implemented by a self-developed computer program called Smart RAM. Similarly, [15] proposed a new PSO optimization model called Dual Particle Swarm Optimization (DPSO) for the reverse motion problem of the automated processor in the process of shutting down a nuclear power plant. The particle swarm was improved by splitting the particle swarm algorithm into two so that each algorithm can work separately. However according to [16], a complex case of a 76 kg Indian male in standing position was transformed into a

3D CAD model of the human body using different damping ratios by utilizing ANSES and Solid Work programs.

Whereas [17] and [18] conducted studies on hand-transmitted vibration exposure, biodynamic responses, and biological effects over the past 20 years, systematically reviewing the literature and identifying areas requiring further research. These studies collectively demonstrated the ongoing efforts in modeling and optimizing both robotic and prosthetic arms, employing a variety of advanced mathematical methods, computational tools, and design techniques to enhance their performance and functionality. The continued evolution of these models played a critical role in improving the capabilities of prosthetic devices, making them more effective and versatile for users.

Recently, exoskeletons designed for rehabilitating the human arm and hand, which rely on mechanical structures, have played a significant role in rehabilitation. In [19], the exoskeleton system was described using 3D mechanical solid software. Then the model was controlled using a sliding mode command to stabilize a 3D human arm exoskeleton due to uncertainties conditions through the rehabilitation approach. [20] presented a dynamic response to compute arm muscle torques based on musculoskeletal approaches due a wide range of shoulder joint motion. Then, the proposed system was validated using a simple dynamical model.

In this paper, the dynamic response of a prosthetic human arm with five degrees of freedom is studied. The system is modeled using the Lagrange-Euler method to develop a set of second-order equations, which are then linearized to optimize the system parameters using Lagrangian multipliers optimization method. A comparison between the linear and nonlinear responses is obtained to verify the accuracy of the linearized model in representing the nonlinear behavior. The system parameters are evaluated to achieve short settling time and low overshoot based on Lagrangian multipliers optimization method. Eventually, the human arm performance is analyzed while participating a wide range of initial excitations.

2 System Modeling

A mathematical model of the human arm system can indeed be exhibited using a Mass-Damper-Spring model. According to Fig. 1 the system has five degrees of freedom which are represented by the coordinates are: $z_1, z_2, z_3, y_1, \theta$. The angle $\tilde{\theta}$ shows the nominal position of the upper arm which we will analysis the vibrations. The coordinates z_1, z_2, z_3 represent the longtudinal motions of the hand, the forearm and the elbow respectively while the coordinate y_1, θ introduce the vertical and rotational motions of the elbow joint, respectively. The value z_0 describes the external displacement acting on the hand.

In our system, energy can be obtained from different forms: **kinetic energy, potential energy,** and **dissipated energy**. The human arm system generates kinetic energy due to the motion of the hand, forearm and the elbow so the total kinetic energy can be obtained as follows.

$$K_E = \frac{1}{2}m_1\dot{z}_1^2 + \frac{1}{2}m_2\dot{z}_2^2 + \frac{1}{2}m_3\left\{\begin{array}{c}\left[\dot{z}_3 - L_g\dot{\theta}\sin(\theta + \tilde{\theta})\right]^2 \\ +\left[\dot{y}_1 + L_g\dot{\theta}\cos(\theta + \tilde{\theta})\right]^2\end{array}\right\} + \frac{1}{2}J\dot{\theta}^2 \tag{1}$$

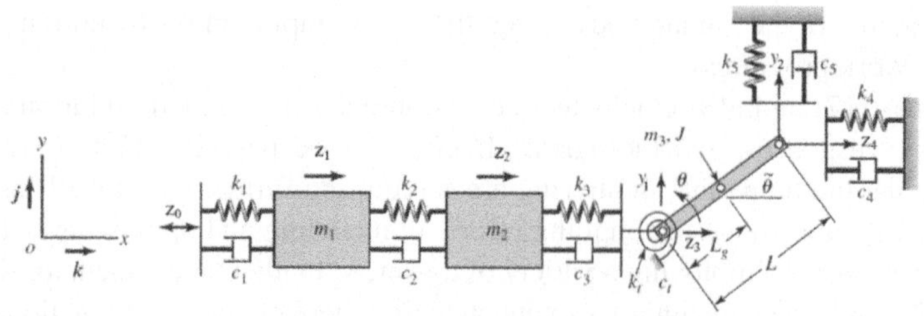

Fig. 1. Human arm system [21]

In this model, the kinetic energy of the hand and forearm components comes from the horizontal motion only, while the kinetic energy at the elbow results from both horizontal and vertical motions as illustrated in Eq. (1).

The second source of energy in this system is potential energy, which primarily depends on elastic and gravitational forces. The arm muscles demonstrate elastic behavior, enabling them to store and release energy, just like a spring. Furthermore, the elbow exhibits gravitational force due to its position relative to the reference point. The total potential energy can be extracted in the following shape:

$$\begin{aligned} P_E =& \frac{1}{2}k_1(z_1 - z_0)^2 + \frac{1}{2}k_2(z_2 - z_1)^2 + \frac{1}{2}k_3(z_3 - z_2)^2 + \frac{1}{2}k_t\left(\theta + \tilde{\theta}\right)^2 \\ &+ \frac{1}{2}k_4\left[z_3 + Lcos\left(\theta + \tilde{\theta}\right)\right]^2 + \frac{1}{2}k_5\left[y_1 + Lsin\left(\theta + \tilde{\theta}\right)\right]^2 + m_3g\left[y_1 + L_g sin\left(\theta + \tilde{\theta}\right)\right] \end{aligned} \tag{2}$$

Eventually, the arm muscles can resist movement, allowing them to absorb energy and reduce fluctuations, as with a damper. In addition, energy can be lost due to friction at the elbow joint. The total wasted energy can be presented as shown below:

$$\begin{aligned} D_E =& \frac{1}{2}c_1(\dot{z}_1 - \dot{z}_0)^2 + \frac{1}{2}c_2(\dot{z}_2 - \dot{z}_1)^2 + \frac{1}{2}c_3(\dot{z}_3 - \dot{z}_2)^2 + \frac{1}{2}c_t\dot{\theta}^2 \\ &+ \frac{1}{2}c_4\left[\dot{z}_3 - L\dot{\theta}\sin\left(\theta + \tilde{\theta}\right)\right]^2 + \frac{1}{2}c_5\left[\dot{y}_1 + L\dot{\theta}\cos\left(\theta + \tilde{\theta}\right)\right]^2 \end{aligned} \tag{3}$$

The proposed human arm model, like any mechanical engineering system, follows the principle of energy conservation. The Euler-Lagrange method will be used to describe the arm's motion by including the effects of kinetic, potential, and dissipated energies.

$$\frac{d}{dt}\left[\frac{\partial T}{\partial q_i}\right] - \frac{\partial T}{\partial q_i} + \frac{\partial D}{\partial q_i} + \frac{\partial v}{\partial q_i} = Q_i \tag{4}$$

where q represents the state variable and the index i refers to the number of degrees of freedom. To find the governing equations of motion, the human arm system is assumed to have five degrees of freedom, with no external forces (F) acting on it. The five nonlinear governing equations of motion can be derived as follows:

$$m_1\ddot{z}_1 + (c_1 + c_2)\dot{z}_1 - c_2\dot{z}_2 + (k_1 + k_2)z_1 - k_2z_2 = c_1\dot{z}_0 + k_1z_0$$

$$m_2\dot{z}_2 + (c_2 + c_3)\dot{z}_2 - c_2\dot{z}_1 - c_3\dot{z}_3 + (k_2 + k_3)z_2 - k_2z_1 - k_3z_3 = 0$$

$$m_3\left[\ddot{z}_3 - L_g\cos(\theta + \tilde{\theta})\dot{\theta}^2 - L_g\sin(\theta + \tilde{\theta})\ddot{\theta}\right] + (c_3 + c_4)\dot{z}_3 - c_3\dot{z}_2$$
$$- c_4 Lsin(\theta + \tilde{\theta})\dot{\theta} + (k_3 + k_4)z_3 - k_3z_2 + k_4 Lcos(\theta + \tilde{\theta}) = 0$$

$$m_3\left[\ddot{y}_1 - L_g\sin(\theta + \tilde{\theta})\dot{\theta}^2 + L_g\cos(\theta + \tilde{\theta})\ddot{\theta} + g\right] + c_5[\dot{y}_1 + L\dot{\theta}\cos(\theta + \tilde{\theta})]$$
$$+ k_5[y_1 + Lsin(\theta + \tilde{\theta})]$$
$$\left(J + m_3L_g^2\right)\ddot{\theta} + m_3L_g[-\ddot{z}_3\,sin(\theta + \tilde{\theta}) + \ddot{y}_1 cos(\theta + \tilde{\theta})] + c_t\dot{\theta} - c_4 L\sin(\theta + \tilde{\theta})$$
$$[\dot{z}_3 - L\dot{\theta}\sin(\theta + \tilde{\theta})] + c_5\, L\cos(\theta + \tilde{\theta})[\dot{y}_1 + L\dot{\theta}\cos(\theta + \tilde{\theta})]$$
$$+ k_t(\theta + \tilde{\theta}) - k_4\, Lsin(\theta + \tilde{\theta})[z_3 + Lcos(\theta + \tilde{\theta})] + k_5\, Lcos(\theta + \tilde{\theta})$$
$$[y_1 + Lsin(\theta + \tilde{\theta})] + m_3 g L_g\cos(\theta + \tilde{\theta}) \tag{5}$$

In our propped model, the human arm is described as a mechanical system so, mechanical properties allow the model to oscillate naturally while controlling it by actuator systems. During the oscillation, the system undergoes a nonlinear behavior which makes it difficult to find the transmissibility of arm. Therefore, a linearization method introduces to analyze the system's response while oscillating around an operating point to satisfy approximately a linear motion. To do that, the system has been linearized around an equilibrium elbow angle $(\tilde{\theta})$ predicting small perturbations. Applying the identities of sine and cosine addition formulas.

$$\begin{pmatrix} sin\left(\theta + \tilde{\theta}\right) = sin\theta cos\,\tilde{\theta} + sin\,\tilde{\theta}\,cos\theta \\ cos\left(\theta + \tilde{\theta}\right) = cos\theta cos\,\tilde{\theta} - sin\theta sin\,\tilde{\theta} \end{pmatrix} \tag{6}$$

Assuming that the arm oscillates with a small motion $(\sin\theta \simeq \theta, \cos\theta = 1)$

$$\begin{pmatrix} sin\left(\theta + \tilde{\theta}\right) = \theta cos\,\tilde{\theta} + sin\,\tilde{\theta} \\ cos\left(\theta + \tilde{\theta}\right) = cos\,\tilde{\theta} - \theta sin\,\tilde{\theta} \end{pmatrix} \tag{7}$$

Substituting Eq. (7) into Eq. (5), we obtain

$$m_1\ddot{z}_1 + (c_1 + c_2)\dot{z}_1 - c_2\dot{z}_2 + (k_1 + k_2)z_1 - k_2z_2 = c_1\dot{z}_0 + k_1z_0$$

$$m_2\ddot{z}_2 + (c_2 + c_3)\dot{z}_2 - c_2\dot{z}_1 - c_3\dot{z}_3 + (k_2 + k_3)z_2 - k_2z_1 - k_3z_3 = 0$$

$$m_3\ddot{z}_3 - L_g m_3 sin\tilde{\theta}\ddot{\theta} + (c_3 + c_4)\dot{z}_3 - c_3\dot{z}_2 - c_4 Lsin\tilde{\theta}\dot{\theta}$$
$$+ (k_3 + k_4)z_3 - k_3z_2 - k_4 Lsin\tilde{\theta}\theta = -k_4 Lcos\tilde{\theta}$$

$$m_3\ddot{y}_1 + L_g m_3 cos\tilde{\theta}\ddot{\theta} + c_5\dot{y}_1 + Lc_5 cos\tilde{\theta}\dot{\theta} + k_5 y_1$$

$$+k_5 L cos\tilde{\theta}\theta = -m_3 g - k_5 L sin\tilde{\theta}$$

$$\begin{aligned}\left(J + m_3 L_g^2\right)\ddot{\theta} - m_3 L_g sin\tilde{\theta}\ddot{z}_3 + m_3 L_g cos\tilde{\theta}\ddot{y}_1 + c_t\dot{\theta} - c_4 L sin\tilde{\theta}\dot{z}_3 + c_4 L^2 sin\tilde{\theta}^2\dot{\theta} \\ + c_5 L cos\tilde{\theta}\dot{y}_1 + c_5 L^2 cos\tilde{\theta}^2\dot{\theta} + k_t\theta - k_4 L^2 cos\tilde{\theta}^2\theta - k_4 L sin\tilde{\theta} z_3 \\ + k_4 L^2 sin\tilde{\theta}^2\theta + k_5 L cos\tilde{\theta} y_1 + k_5 L^2 cos\tilde{\theta}^2\theta - k_5 L^2 sin\tilde{\theta}^2\theta \\ - m_3 g L_g sin\tilde{\theta}\theta = -k_t\tilde{\theta} + k_4 L^2 sin\tilde{\theta} cos\tilde{\theta} - k_5 L^2 sin\tilde{\theta} cos\tilde{\theta} - m_3 g L_g cos\tilde{\theta}\end{aligned} \tag{8}$$

The **equations of motion** for the human arm system in matrix form, involving mass, damping, and stiffness, can be obtained in Eq. (9).

$$\begin{aligned}&\begin{bmatrix} m_1 & 0 & 0 & 0 & 0 \\ 0 & m_2 & 0 & 0 & 0 \\ 0 & 0 & m_3 & 0 & -L_g m_3 sin\,\tilde{\theta} \\ 0 & 0 & 0 & m_3 & L_g m_3 cos\,\tilde{\theta} \\ 0 & 0 & -L_g m_3 sin\,\tilde{\theta} & L_g m_3 cos\,\tilde{\theta} & J + L_g{}^2 m_3 \end{bmatrix}\begin{bmatrix}\ddot{z}_1 \\ \ddot{z}_2 \\ \ddot{z}_3 \\ \ddot{y}_1 \\ \ddot{\theta}\end{bmatrix} \\ &+\begin{bmatrix} c_1 + c_2 & -c_2 & 0 & 0 & 0 \\ -c_2 & c_2 + c_3 & -c_3 & 0 & 0 \\ 0 & -c_3 & c_3 + c_4 & 0 & -c_4 L sin\,\tilde{\theta} \\ 0 & 0 & 0 & c_5 & c_5 L cos\,\tilde{\theta} \\ 0 & 0 & -c_4 L sin\,\tilde{\theta} & c_5 L cos\,\tilde{\theta} & c_t + c_4 L^2 sin\,\tilde{\theta}^2 + c_5 L^2 cos\,\tilde{\theta}^2 \end{bmatrix}\begin{bmatrix}\dot{z}_1 \\ \dot{z}_2 \\ \dot{z}_3 \\ \dot{y}_1 \\ \dot{\theta}\end{bmatrix} \\ &+\begin{bmatrix} k_1 + k_2 & -k_2 & 0 & 0 & 0 \\ -k_2 & k_2 + k_3 & -k_3 & 0 & 0 \\ 0 & -k_3 & k_3 + k_4 & 0 & -k_4 L sin\,\tilde{\theta} \\ 0 & 0 & 0 & k_5 & k_5 L cos\,\tilde{\theta} \\ 0 & 0 & -k_4 L sin\,\tilde{\theta} & k_5 L cos\,\tilde{\theta} & \left(\left(k_t - k_4 L^2\left(cos\,\tilde{\theta}^2 - sin\,\tilde{\theta}^2\right)\right) + \left(k_5 L^2 \times \left(cos\,\tilde{\theta}^2 - sin\,\tilde{\theta}^2\right)\right) - m_3 g L_g sin\,\tilde{\theta}\right) \end{bmatrix}\begin{bmatrix} z_1 \\ z_2 \\ z_3 \\ y_1 \\ \theta\end{bmatrix} \\ &= \begin{bmatrix} c_1\dot{z}_0 + k_1 z_0 \\ 0 \\ -k_4 L sin\,\tilde{\theta} \\ -m_3 g - k_5 L sin\,\tilde{\theta} \\ -k_t\,\tilde{\theta} + k_4 L^2 sin\,\tilde{\theta}\, cos\,\tilde{\theta} - k_5 L^2 sin\,\tilde{\theta}\, cos\,\tilde{\theta} - m_3 g L_g cos\,\tilde{\theta} \end{bmatrix}\end{aligned} \tag{9}$$

In a static state, the system is in equilibrium; thus, the initial displacement of the human arm can be evaluated by solving Eq. (9), assuming no dynamic effects as follows:

$$
\begin{bmatrix} z_1 \\ z_2 \\ z_3 \\ y_1 \\ \theta \end{bmatrix}_0 = \begin{bmatrix} k_1+k_2 & -k_2 & 0 & 0 & 0 \\ -k_2 & k_2+k_3 & -k_3 & 0 & 0 \\ 0 & -k_3 & k_3+k_4 & 0 & -k_4 L sin\,\tilde{\theta} \\ 0 & 0 & 0 & k_5 & k_5 L cos\,\tilde{\theta} \\ 0 & 0 & -k_4 L sin\,\tilde{\theta} & k_5 L cos\,\tilde{\theta} & \left(\begin{array}{c} \left(\begin{array}{c} k_t - \\ k_4 L^2 \left(cos\,\tilde{\theta}^2 - sin\,\tilde{\theta}^2 \right) \end{array} \right) + \\ \left(\begin{array}{c} k_5 L^2 \\ \times \left(cos\,\tilde{\theta}^2 - sin\,\tilde{\theta}^2 \right) \end{array} \right) - \\ m_3 g L_g sin\,\tilde{\theta} \end{array} \right) \end{bmatrix}^{-1}
$$

$$
\times \begin{bmatrix} c_1 \dot{z}_0 + k_1 z_0 \\ 0 \\ -k_4 L sin\tilde{\theta} \\ -m_3 g - k_5 L sin\tilde{\theta} \\ -k_t \tilde{\theta} + k_4 L^2 sin\tilde{\theta} cos\tilde{\theta} - k_5 L^2 sin\tilde{\theta} cos\tilde{\theta} - m_3 g L_g cos\tilde{\theta} \end{bmatrix} \quad (10)
$$

As illustrated in Eq. (10), the human arm is subjected to a static load so that the system displacements can be computed based on equilibrium state under the static forces. Hence, these initial conditions will serve as starting points for solving the system's dynamic response for both linear and nonlinear simulations. Figures 2a–e illustrate the displacement responses of various segments of a human arm (hand, forearm, and elbow) considering both linear and nonlinear simulations. The system with both seniors oscillates to ensure similar underdamped patterns before capturing steady-states. Figures 2a–c, shows that the nonlinear hand, forearm, and elbow displacements exhibit a slightly large initial amplitude due to the nonlinear behavior in the transient phase. Then, the simulations are satisfactorily matched, indicating successful linearization. On the other hand, as shown in Figs. 2d–e, the nonlinear angular and vertical displacements of the elbow don't track the linear simulations as it should due to a significant role of the nonlinearity in the elbow joint. However, these differences in simulations might not be detrimental to create an appropriate linear model which can be used then to compute the system response using Lagrangian Optimization Method.

3 Optimization

Figure 2 presented the dynamic responses of the human arm under both linear and nonlinear simulations, using arbitrary system parameters. This comparison was conducted to ensure that the nonlinear model can be effectively represented by the proposed linear model. However, response parameters for each segment arm should match the proposed values and satisfy the system constraints. Thus, the system parameters of the human arm are evaluated to achieve the desired settling time and overshooting values. Lagrange multipliers optimization method is employed to minimize the objective function while involving system parameters constraints. To do that, the linear system of the human arm (Eq. (9)) is propagated numerically from the equilibrium state using estimated system parameters. The overshooting and settling time characteristics are evaluated after integrating the system using MATLAB program. In this study, we plan to optimize the system's parameters (masses, spring stiffness, and damper coefficients) while achieving the desired overshooting and settling time as illustrated in Table 1. In order to solve this optimization problem using the method of Lagrange multipliers, the objective function is defined as follows

$$f(x) = \sum\nolimits_{i=1}^{5} \left(os_i - os_{i,desired}\right)^2 + \left(t_{s,i} - t_{s,i,desired}\right)^2 \tag{11}$$

The human arm operates as a complex biomechanical structure, including multiple degrees of freedom (DOF) due to its diverse joints (shoulder, elbow, wrist) and muscles. In the arm, the damping ratio for each part is based on various factors such as mechanical and material properties of the joints and tissues. Therefore, knowing the human arm damping ratios is not simple and needs experimental information. However, it simply possible to perform the human arm system by improving the overshooting and reducing the settling time. Table 1 presents the desired overshooting and settling time parametrizes.

Table 1. The desired overshooting and settling time values

Response parameter	Constraints
Settling Time, t_s (s)	<0.1 s
Overshooting (100%), os	<15%

To apply the proposed optimization method, the Lagrangian function can be expressed as

$$\Im(x, \lambda_1, \lambda_2, \lambda_3, \lambda_4, \lambda_5, \lambda_6, \lambda_7, \lambda_8, \lambda_9, \lambda_{10}, \lambda_{11}, \lambda_{12}, \lambda_{13}, \lambda_{14}, \lambda_{15}) = \left(f(x) - \sum_{j-1}^{15} \hat{\lambda}_j g_j(x)\right) \tag{12}$$

where x is the vector of the optimal parameters $(m_1, m_2, m_3, k_1, k_2, k_3, k_4, k_5, k_t, c_1, c_2, c_3, c_4, c_5, c_t)$, f(x) and $g_i(x)$ are the objective and constraints functions, respectively, and λ_i are Lagrangian multipliers required to satisfy the optimization [22].

Figures 3a–e shows a 3D representation of iterations, stiffness vector, and objective function during the optimization of human arm stiffness (k_1-k_t), respectively, under the nominal conditions ($\dot{z}_0 = 0$, $z_0 = 15$ cm). The objective function surface evolves rapidly from a cost function value of 500 (dark red) to a small value (light blue), then gradually captures nearly zero (dark blue). As we can see from Fig. 3 that the quadratic objective function converges as the iterations increase fulfilling the optimal stiffness values.

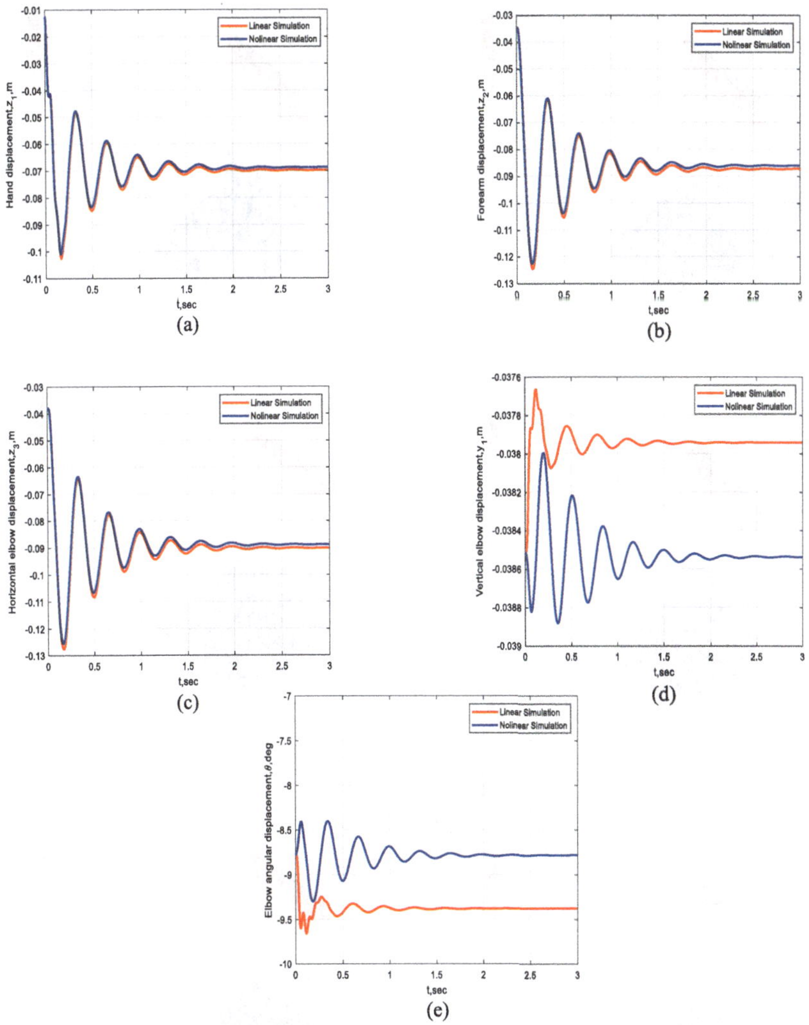

Fig. 2. Human arm states vs. time for linear and nonlinear cases; (a) human hand displacement vs. time; (b) human forearm displacement vs, time; (c) human elbow displacement vs. time; (d) vertical elbow displacement vs. time; (e) elbow angular displacement vs. time.

Likewise, Figs. 4a–e and Figs. 5a–e present a 3D plot of iterations vs. damper vector vs. objective function and a 3D plot of iterations vs. mass vector vs. objective function in the optimization of human arm dampers (c_1-c_t) and masses (m_1-m_3), respectively. Following the same way as Fig. 3, the Lagrangian multipliers method used in Figs. 4 and 5 can successfully minimize the proposed objective function to approach the target values while involving the system parameters constraints.

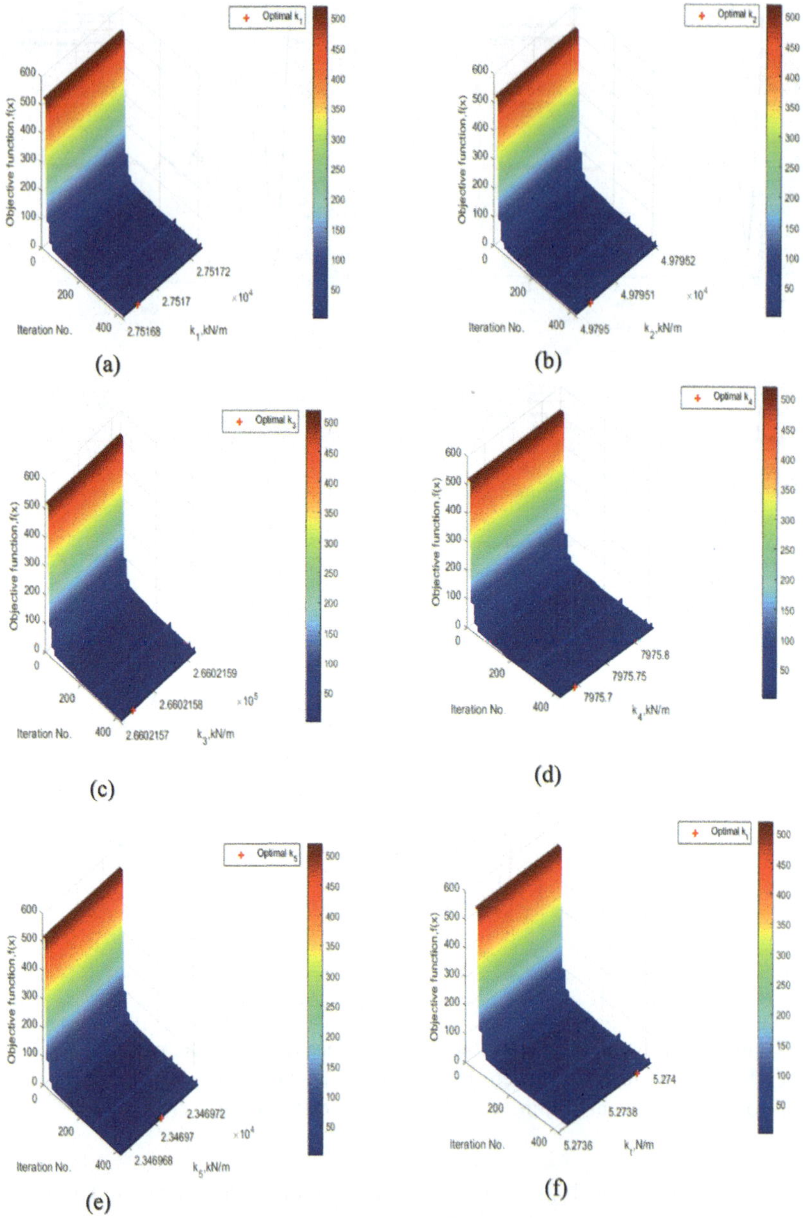

Fig. 3. 3D objective function surface using to optimize the human arm stiffness; (a) stiffness (k_1); (b) stiffness (k_2); (c) stiffness (k_3); (d) stiffness (k_4); (e) stiffness (k_5); (f) stiffness (k_t).

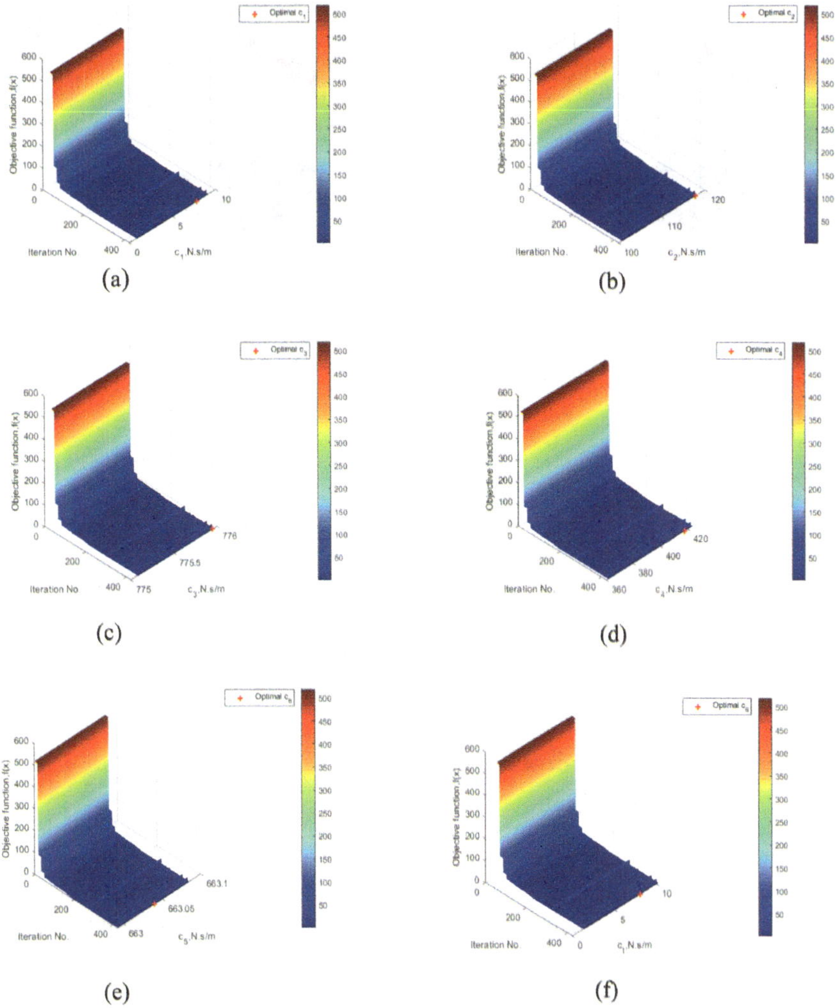

Fig. 4. 3D objective function surface using to optimize the human arm dampers; (a) damper (c_1); (b) damper (c_2); (c) damper (c_3); (d) damper (c_4); (e) damper (c_5); (f) damper (c_t)

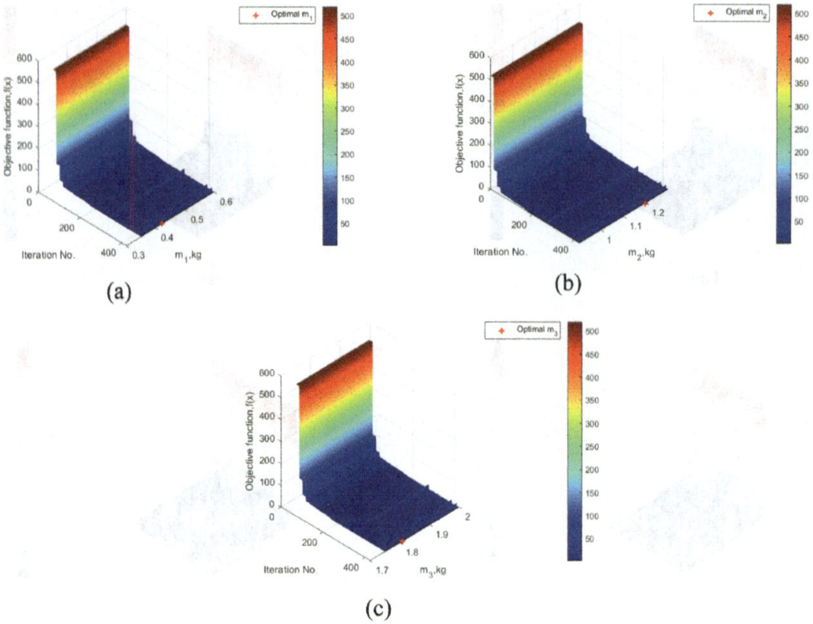

Fig. 5. 3D objective function surface using to optimize the human arm masses; (a) hand mass (m_1); (b) forearm mass (m_2); (c) elbow mass (m_3)

4 Results and Discussions

Thus, the optimization algorithm relying on the Lagrangian multipliers method can effectively stabilize the human arm model while ensuring a faster settling time and low overshoot under nominal conditions. However, it is unlikely that the human model will only encounter nominal conditions without experiencing severe off-nominal conditions. Hence, the optimal system parameters that computed under the nominal conditions will be used to evaluate the system performance while exhibiting the off-nominal conditions ($\dot{z}_0 = 0$, $z_0 = 5$ cm and $\dot{z}_0 = 0$, $z_0 = 25$ cm).

Figures 6a–e present the human hand, forearm, horizontal elbow, vertical elbow, and elbow angular displacements vs. time, respectively, for three different input values. Figures 6a–c show decaying behavior of displacements; they start at higher initial values and decrease rapidly towards the desired values maintaining rapid settling time with a maximum value of (0.0352 s) and minimal overshoot with a maximum amount of (3.345%). On the other hand, as shown in Fig. 6d, the elbow angle increases smoothly in the direction of the steady state elbow angle without exceeding the proposed constraints of settling time (<0.1 s) and overshoot (<15%). Eventually, Fig. 6e illustrates that the vertical elbow displacement decreases gradually from the higher initial condition to the target value with 0.0904 s and 5.4% as the maximum settling time and the maximum overshot, respectively. Table 2 lists the system response characteristics for various system input. Table 2 notes that all the results are acceptable and do not exceed the system's constraints, resulting in good system performance.

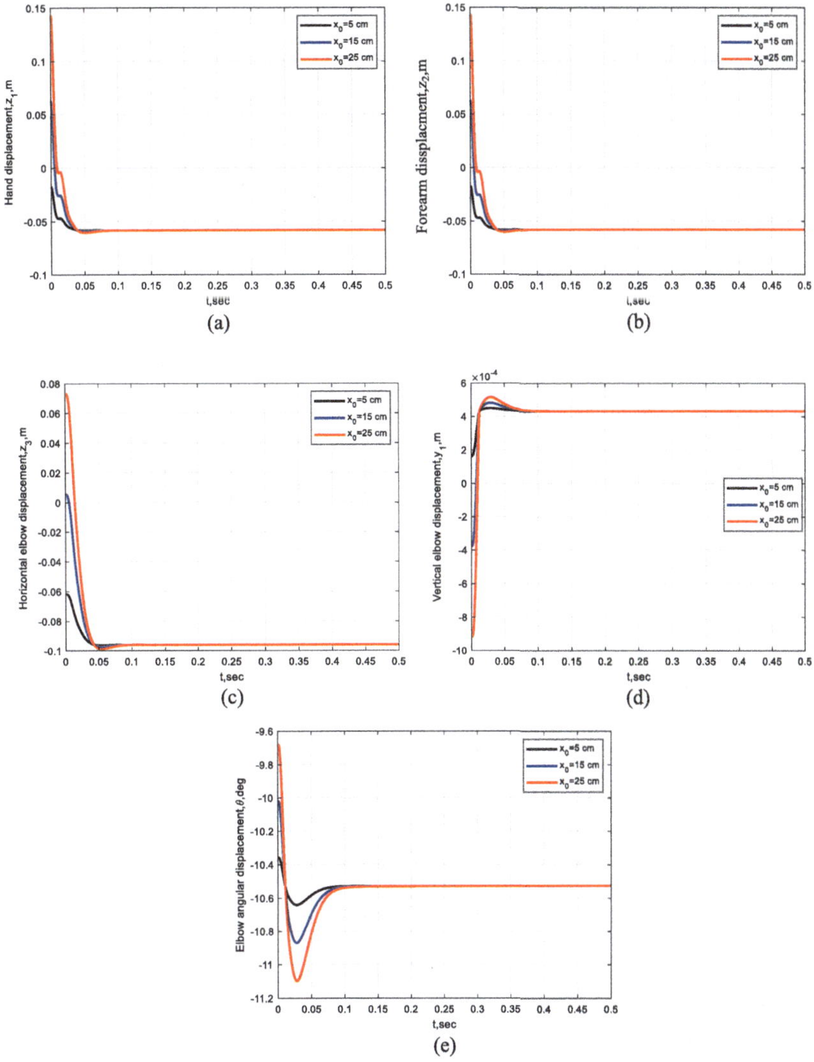

Fig. 6. Human arm responses vs. time for three input values; (a) hand displacement vs. time; (b) forearm displacement vs. time; (c) horizontal elbow displacement vs. time; (d) elbow angular displacement vs. time; (e) vertical elbow displacement vs. time.

Table 2. System response characteristics

No.	State	$z_0 = 5$ cm		$z_0 = 15$ cm		$z_0 = 25$ cm	
		Settling time (s)	Overshoot 100%	Settling time (s)	Overshoot 100%	Settling time (s)	Overshoot 100%
1	Hand displacement	0.0335	0.6721	0.0342	2.0143	0.0352	3.3450
2	Forearm displacement	0.0373	0.6224	0.0372	1.8666	0.0375	3.1428
3	Elbow displacement	0.0382	0.5885	0.0383	1.7713	0.0385	2.9628
4	Elbow angle	0.0563	3.4939	0.0581	9.4824	0.0563	14.4696
5	Vertical elbow displacement	0.0904	1.0778	0.0899	3.2331	0.0897	5.3905

5 Conclusion

A complicated dynamical model for a five-DOF human hand-arm exposed to step input was presented in this study. To decrease the complexity of the system, the system's non-linear and linear dynamic equations were compared, and the well comparison outcomes ensured that the linearized system can successfully represent the human arm model and can effectively compute the system parameters. The Lagrangian multipliers optimization method was able to accurately compute the arm parameters (masses, damper coefficients, and stiffness) while maintaining a short settling time and low overshoot. Under a higher initial displacement value, the hand, forearm and elbow displacements decreased promptly to capture the desired values while maintaining short settling time limited to (0.0385 s) and low overshoot without exceeding (3.345%). While the elbow angle increased smoothly towards the steady state elbow angle preserving rapid settling time with a maximum value of (0.0581 s) and minimal overshoot with an upper limit of (14.45%). The results indicated that the proposed optimization method can greatly improve the human arm dynamic performance despite significant input displacement conditions.

References

1. Shah, J., Rattan, S.S., Nakra, B.C.: Kinematic analysis of 2-DOF planer robot using artificial neural network. Int. J. Mech. Mechatron. Eng. **5**(9), 1720–1723 (2011). https://doi.org/10.5281/zenodo.1058563
2. Al-Zu'be, L.A., Al-Tamimi, A.A., Al-Momani, T.D., Alkarala, A.J., Alzawahreh, M.A.: Modeling and simulating human arm movement using a 2 dimensional 3 segments coupled pendulum system (2012)
3. Abbas, T.F.: Forward Kinematics Modeling of 5 DOF Stationary Articulated Robots (2013). https://doi.org/10.30684/etj.31.3a.8

4. Barakat, A.N., Gouda, K.A., Bozed, K.A.: Kinematics analysis and simulation of a robotic arm using MATLAB. In: 4th International Conference on Control Engineering and Information Technology, CEIT 2016, pp. 1–5. IEEE (2017). https://doi.org/10.1109/CEIT.2016.7929032
5. Serbest, K., Cilli, M., Yildiz, M.Z., Eldogan, O.: Development of a human hand model for estimating joint torque using MATLAB tools. In: 2016 6th IEEE International Conference on Biomedical Robotics and Biomechatronics (BioRob), pp. 793–797. IEEE (2016). https://doi.org/10.1109/BIOROB.2016.7523724
6. Agarana, M.C., Akinlabi, E.T.: Mathematical modelling and analysis of human arm as a triple pendulum system using Euler - Lagragian model. IOP Conf. Ser. Mater. Sci. Eng. 012010 (2018). https://doi.org/10.1088/1757-899X/413/1/012010
7. Mohanto, T., Talukde, A., Tasneem, Z.: Development of a 3DOF color sorting based robotic arm using MATLAB GUI. In: 2020 23rd International Conference on Computer and Information Technology (ICCIT), pp. 1–6. IEEE (2020). https://doi.org/10.1109/ICCIT51783.2020.9392750
8. Moses, S.O., Olalekan, E.I., Aniemeka, E.T., Lukman, A.: Development of a programmable robotics arm using MATLAB (2022). https://doi.org/10.33545/27076571.2022.v3.i1a.41
9. Salman, H.D., Hamzah, M.N., Bakhy, S.H.: Kinematics analysis and implementation of three degrees of freedom robotic arm by using MATLAB. IRAQI J. Mech. Mater. Eng. **21**(2), 118–129 (2021). https://doi.org/10.32852/iqjfmme.v21i2.547
10. Li, Y., et al.: Optimization of dynamic load distribution of a serial-parallel hybrid humanoid arm. Mech. Mach. Theory **149**, 103792 (2020). https://doi.org/10.1016/j.mechmachtheory.2020.103792
11. Sheng, B., Meng, W., Deng, C., Xie, S.: Model based kinematic & dynamic simulation of 6-DOF upper-limb rehabilitation robot (2016). https://doi.org/10.1109/ACIRS.2016.7556181
12. Bieda, R., Jaskot, K.: Analysis of human arm nonlinear and linear mathematical models (2016). https://doi.org/10.1007/978-3-319-21118-3_22
13. Artemiadis, P.K., Katsiaris, P.T., Liarokapis, M.V., Kyriakopoulos, K.J.: Human arm impedance: characterization and modeling in 3D space. In: 2010 IEEE/RSJ International Conference on Intelligent Robots and Systems, pp. 3103–3108. IEEE (2010). https://doi.org/10.1109/IROS.2010.5652025
14. Hao, W.G., Leck, Y.Y., Hun, L.C.: 6-DOF PC-based robotic arm (PC-ROBOARM) with efficient trajectory planning and speed control. In: 2011 4th International Conference on Mechatronics (ICOM), pp. 1–7. IEEE (2011). https://doi.org/10.1109/ICOM.2011.5937171
15. Khan, H., Kim, H.H., Abbasi, S.J., Lee, M.C.: Real-time inverse kinematics using dual particle swarm optimization DPSO of 6-DOF robot for nuclear plant dismantling. IFAC-PapersOnLine **53**(2), 9885–9890 (2020). https://doi.org/10.1016/j.ifacol.2020.12.2695
16. Dhaka, M., Sharma, V., Khetan, K.: Control of mechanisms and robots using LabVIEW and SolidWorks and Arduino. Int. J. Eng. Res. **V5**(02) (2016). https://doi.org/10.17577/IJERTV5IS020312
17. Hassan, M., Baqar, M.: Design and implementation of a DTMF based pick and place robotic arm. Int. J. Recent Res. Civ. Mech. Eng. **2**, 232–240 (2020). www.paperpublications.org
18. Mailah, M., Jahanabadi, H., Zain, M.Z.M., Priyandoko, G.: Modelling and control of a human-like arm incorporating muscle models. Proc. Inst. Mech. Eng. Part C J. Mech. Eng. Sci. **223**(7), 1569–1577 (2009). https://doi.org/10.1243/09544062JMES1289
19. Djoudi, D., Bouzid, Z.: Design of matlab/opensim robust controller of an exoskeleton for disuse muscular atrophy rehabilitation of a human arm
20. Zoufaly, O., Chadwick, E., Blana, D., Daniel, M.: Calculation of a force effect from muscle action to a quaternion-based musculoskeletal model. arXiv preprint arXiv:2502.03475 (2025)

21. Thomas, C., Rakheja, S., Bhat, R.B., Stiharu, I.: A study of the modal behavior of the human hand arm system. J. Sound Vibr. **191**(1), 171–176 (1996)
22. Cherian, T., Rakheja, S., Bhat, R.B.: An analytical investigation of an energy flow divider to attenuate hand-transmitted vibration. Int. J. Ind. Ergon. **17**(6), 455–467 (1996)

Modeling of Combined Heating-Ventilating System

Oleg Krol[1(✉)], Vladimir Sokolov[1], Dănuț-Iulian Stanciu[2], Marin Zhilevski[3], and Mikho Mikhov[3]

[1] Volodymyr Dahl East Ukrainian National University, 59a Central Avenue, Severodonetsk 93400, Ukraine
krolos@i.ua

[2] National Institute of Research and Development in Mechatronics and Measurement Technique INCDMTM, 6-8 Pantelimon, 2nd District, 021631 Bucharest, Romania

[3] Technical University of Sofia, 8 Kliment Ohridski Blvd, Sofia 1000, Bulgaria

Abstract. The mathematical model of aerothermodynamic processes for combined heating-ventilating system as an object of automatic control has been developed. In the heating-ventilating system, the supply air comes from the central ventilation unit at its constant temperature. Regulation of the supply air flow is carried out by the special terminal, and additional change of its temperature is made in the heat exchanger. The work processes taking place in the system are reflected on the functional scheme. The structural scheme of the two-circuit automatic control system for air heating has been constructed. In general, the system provides various regulators for the flow rate and temperature of the supply air. The calculation of the transient process in the system is performed when the set temperature changes in the serviced room. When modeling the system, MPC controller was used in the supply air flow control circuit, and PID controller was used in the temperature control circuit. By combination of controllers for supply air flow and its temperature, with optimal adjustment of their parameters, it is possible to improve the regulation quality indicators of the automatic air heating control system and, as a result, to ensure the increase of the efficiency of the heating-ventilating system.

Keywords: Heating-ventilating system · Structural scheme · Controller · Transient process · Quality indicators

1 Introduction

Heating, ventilation and air conditioning (HVAC) systems are an integral part of buildings and construction [1, 2]. The basic element of the HVAC system is a serviced room in which the air constantly transitions from one state to another. To maintain the set (required) parameters, supply air is supplied to the room, which has parameters that differ from the parameters inside the serviced room. The supply air mixes with the internal air and displaces it, while heating and humidifying the indoor air or assimilating excess heat and moisture. HVAC systems provide the required sanitary and technical standards

D. D. Cioboată and J. Machado (Eds.): ICoRSE 2025, LNNS 1592, pp. 99–109, 2025.
https://doi.org/10.1007/978-3-032-02508-1_9

in production premises, labor safety and compliance with technological processes [3, 4]. Ventilation and air conditioning play the crucial role in a number of enterprises in the machine-building, chemical, and textile industries, in mines, where it is impossible to conduct many technological processes without them [5, 6].

Important reserve for reducing energy consumption is the reduction of power consumption by reducing energy losses associated with the processes of regulating the aerothermodynamic characteristics of HVAC systems and, above all, heating-ventilating systems (HVS) [7, 8]. This can be achieved by developing automatic control systems (ACS) that ensure the rational use of energy resources by improving the quality indicators of HVS regulation [9, 10]. The solution of this problem can give a significant positive effect in the technical, economic and social aspects [11, 12].

2 Literature Review

In ACS for air heating (or HVS) is usually a variable thermal load, and humidity and gas loads are relatively constant, or require some small consumption of outdoor air [13, 14]. Therefore, three types of temperature control systems are possible for such control object: systems with constant flow and variable supply air temperature, systems with variable supply air flow of a certain temperature and mixed (combined or two-circuit) systems [15, 16]. The first two types of ACS for air heating are the most common [17, 18]. These systems are sufficiently fully considered and investigated as objects of automatic control, transfer functions are analytically obtained and experimentally confirmed for them, recommendations are given for determining transfer coefficients and time constants [19, 20].

Currently, new trends and approaches to the construction of ACS for HVAC have been developed, which provide opportunities to improve the control processes of HVAC in the organization of air heating of buildings and construction [21, 22]. For example, instead of traditional PID controllers, more complex algorithms and methods of optimal automatic control can be used to control the thermal regime of premises according to temperature schedules [23, 24, 32].

In the context of Industry 4.0 and digital transformation, the integration of digital tools and data exchange platforms in SMEs, particularly in the machine-building sector, also plays a vital role in improving system efficiency and adaptability under stress conditions [25]. Moreover, modern challenges such as force majeure conditions require resilient and adaptive approaches to system design and control, including the use of AI-based planning models to ensure sustainable operation and flexibility of enterprises [26].

In this regard, it is of undoubted interest to study the processes and analyze the regulation quality of aerothermodynamic characteristics for combined HVS to increase their efficiency. The purpose of this paper is to develop the mathematical model of combined HVS as control object, building on its basis ACS for air heating.

3 Mathematical Model of Combined Heating-Ventilating System

Let's consider the combined HVS system based on the supply ventilation system VAV (Variable Air Volume) [27, 28], which has the possibility of additional heating or cooling of the air. The operation scheme of HVS for variable air flow with regulation of its temperature is shown in Fig. 1.

In the considered HVS, the supply air comes from the central ventilation unit at its constant temperature. The supply air flow rate is regulated by the VAV terminal (controller or VAV valve) [28, 29], which is indicated in Fig. 1 as FC (flow controller). The change in air temperature, as shown in Fig. 1, occurs in the heat exchange apparatus using the temperature controller TC. Thus, in the serviced room, the maintenance of a set temperature regime in significant (working) zone, which is controlled by the temperature sensor TS, is ensured by both regulating the flow rate of supply air and its temperature.

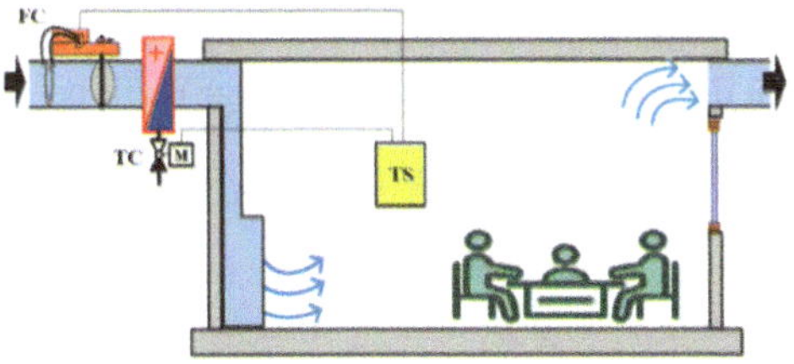

Fig. 1. The operation scheme of HVS for variable air flow with regulation of its temperature (FC – flow controller, TC – temperature controller, TS – temperature sensor).

The work processes in the combined HVS can be reflected in the functional scheme shown in Fig. 2. According to the functional scheme, the HVS is considered as the two-circuit ACS for air heating, in which the input signal is a set temperature (the law of its change in time t) in the serviced room $T_s(t)$. By the deviation $\varepsilon(t)$ of the set temperature $T_s(t)$ from the temperature in the significant (working) zone $T(t)$

$$\varepsilon(t) = T_s(t) - T(t) \tag{1}$$

the flow controller FC forms the value of the mass air flow rate $G(t)$ of the constant temperature coming from the central supply ventilation system. Next, the supply air enters the heat exchanger, in which, according to the temperature deviation $\varepsilon(t)$, the temperature controller TC changes the temperature of the supply air $T_h(t)$, which directly enters the serviced room. In the future, the temperature at the most significant (working) point $T(t)$ will be called the room temperature.

The flow rate of the supply air is equal to the flow rate of the removed air, therefore, heat gain from the ventilation system Q_{ah} to the serviced room (see Fig. 2) it will be determined by the difference between the supply air temperature $T_h(t)$ and the room temperature $T(t)$

$$Q_{ah}(t) = cG(t)(T_h(t) - T(t)), \tag{2}$$

where c – thermal capacity of the air.

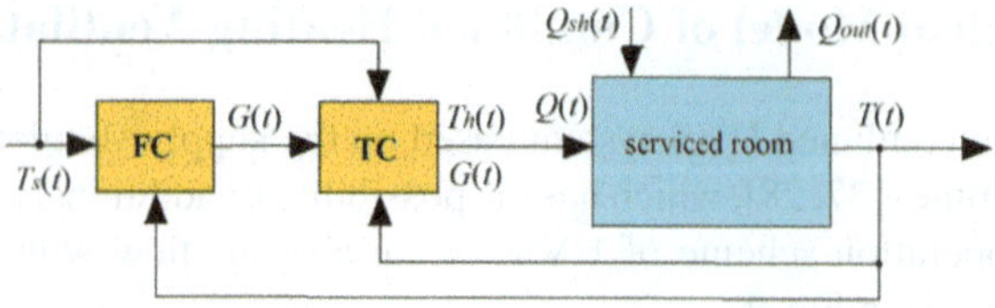

Fig. 2. The functional scheme of the combined HVS for the serviced room.

In addition, heat gain to the serviced room may be from other sources (additional excess heat) $Q_{sh}(t)$ that are not related to the supply ventilation system (see Fig. 2), and be from people, equipment, lighting, solar radiation, etc. Then the total heat gain $Q_{in}(t)$ in the serviced room can be determined by the dependence

$$Q_{in}(t) = Q_{ah}(t) + Q_{sh}(t). \tag{3}$$

The heat losses in the functional diagram (Fig. 2) are indicated as $Q_{out}(t)$. The total excess heat $\Delta Q(t)$ (heat deficiency with a negative sign) will be

$$\Delta Q(t) = Q_{in}(t) - Q_{out}(t). \tag{4}$$

It is quite appropriate to assume that HVS can be in a conditionally stationary mode, i.e., in a state for a relatively long period of time when its parameters are constant. This condition corresponds to the values of the mass flow rate of the supply air G_0, its temperature, indoor temperature T_0 and outdoor air. For the stationary mode, heat gain is equal to heat losses, these values are denoted by Q_0

$$Q_{in}^0 = Q_{out}^0 = Q_0. \tag{5}$$

It should be taken into account that the heat gain for the stationary mode consists of stationary values of heat gain from the supply ventilation system and additional excess heat

$$Q_{in}^0 = Q_{ah}^0 + Q_{sh}^0. \tag{6}$$

The operation scheme (Fig. 1) and the functional scheme of HVS (Fig. 2) allow to build the structural scheme of the two-circuit ACS for air heating.

The time change in excess heat $\Delta Q(t)$ in the serviced room, taking into account (5), is defined as the difference between the change in heat gain $\Delta Q_{in}(t)$ and heat losses $\Delta Q_{out}(t)$

$$\Delta Q(t) = \Delta Q_{in}(t) - \Delta Q_{out}(t). \tag{7}$$

Excess heat $\Delta Q(t)$ lead to change of the room temperature

$$\Delta T(t) \sim \Delta Q(t). \tag{8}$$

Considering the serviced room as the control object, we connect the dimensionless deviations of variables transformed according to Laplace, using transfer functions

$$T(s) = W_{TQ}(s)(Q_{in}(s) - Q_{out}(s)) \tag{9}$$

or taking into account (3)

$$T(s) = W_{TQ}(s)(k_{ah}Q_{ah}(s) + k_{sh}Q_{sh}(s) - Q_{out}(s)), \quad (10)$$

where s – the Laplace variable; $T(s)$ – the Laplace transformation of the dimensionless deviation of the room temperature from the stationary value T_0 (the deviation is divided by the difference between the stationary values of room temperatures T_0 and outside air T^0_{out})

$$\Delta\overline{T}(t) = \Delta T(t) \Big/ \left(T_0 - T^0_{out}\right); \quad (11)$$

$Q_{in}(s)$, $Q_G(s)$, $Q_{sh}(s)$ – the Laplace transformation of the dimensionless deviations of total heat gain, heat gain from the ventilation system and additional excess heat from their stationary values

$$\Delta\overline{Q}_{in}(t) = \Delta Q_{in}(t) \Big/ Q_0; \quad (12)$$

$$\Delta\overline{Q}_G(t) = \Delta Q_{ah}(t) \Big/ Q^0_{ah}; \quad (13)$$

$$\Delta\overline{Q}_{sh}(t) = \Delta Q_{sh}(t) \Big/ Q^0_{sh}; \quad (14)$$

$Q_{out}(s)$ – the Laplace transform of the dimensionless deviation of heat losses from the stationary value Q_0

$$\Delta\overline{Q}_{out}(t) = \Delta Q_{out}(t) \Big/ Q_0; \quad (15)$$

$W_{TQ}(s)$ – transfer function for the room temperature according to the heat flow.

By virtue of (5) and (6) we have

$$k_{ah} = \frac{Q^0_{ah}}{Q_0}; \quad k_{sh} = \frac{Q^0_{sh}}{Q_0}; \quad k_{ah} + k_{sh} = 1. \quad (16)$$

The coefficients k_{ah} and k_{sh} have a physical meaning, they show the proportion of responsible air heating and additional excess heat in the heat supply of the serviced room for stationary operation.

Let's perform linearization of the dependence (2) for heat gain from the ventilation system. The Taylor series expansion for deviations of variables has the form

$$\begin{aligned}\Delta Q_{ah}(t) &= \left.\frac{\partial Q_{ah}}{\partial G}\right|_0 \Delta G(t) + \left.\frac{\partial Q_{ah}}{\partial (T_h - T)}\right|_0 (\Delta T_h(t) - \Delta T(t)) \\ &= c\left(T^0_h - T_0\right)\Delta G(t) + cG_0(\Delta T_h(t) - \Delta T(t)). \quad (17)\end{aligned}$$

The index "0" for derivatives in expression (17) means calculating them for parameter values in the stationary operation mode of HVS. By writing Eq. (17) for the dimensionless deviations of variables.

$$\Delta\overline{G}(t) = \Delta G(t) \Big/ G_0; \quad (18)$$

$$\Delta\overline{T}_h(t) = \Delta T_h(t) \Big/ \left(T_h^0 - T_0\right), \tag{19}$$

and having performed the Laplace transformation for them, we get

$$Q_{ah}(s) = \frac{cG_0(T_h^0 - T_0)}{Q_{ah}^0} G(s) + \frac{cG_0(T_h^0 - T_0)}{Q_{ah}^0} T_h(s) - \frac{cG_0 T_0}{Q_{ah}^0} T(s). \tag{20}$$

It is easy to see that

$$cG_0\left(T_h^0 - T_0\right) = Q_{ah}^0, \tag{21}$$

therefore, the expression (20) can be reduced to the form

$$Q_G(s) = G(s) + T_h(s) - k_T T(s), \tag{22}$$

$$k_T = \frac{T_0 - T_{out}^0}{T_h^0 - T_0}. \tag{23}$$

The change in heat losses over time $Q_{out}(t)$ is very complex, with some approximation, it can be assumed that heat losses depend from the difference of the room temperature $T(t)$ and outdoor temperature $T_{out}(t)$. This dependence for dimensionless deviations of variables transformed by the Laplace is represented in the following form

$$Q_{out}(s) = W_{QT}(s)(T(s) - T_{out}(s)), \tag{24}$$

where $T_{out}(s)$ - the Laplace transform of the dimensionless deviation of the outdoor temperature from the stationary value

$$\Delta\overline{T}_{out}(t) = \Delta T_{out}(t) \Big/ \left(T_0 - T_{out}^0\right), \tag{25}$$

$W_{QT}(s)$ – transfer function for heat losses by the temperature difference in the room and outside.

Substituting (22) and (24) into (10) obtains the following expression

$$T(s) = W_{TQ}(s)[k_{ah}(G(s) + T_h(s) - k_T T(s)) + k_{sh} Q_{sh}(s) - W_{QT}(s)(T(s) - T_{out}(s))]. \tag{26}$$

After the transformations we have

$$\begin{gathered} T(s)\left[1 + k_{ah} k_T W_{TQ}(s) + W_{TQ}(s) W_{QT}(s)\right] = \\ = W_{TQ}(s)\left[k_{ah}(G(s) + T_h(s)) + k_{sh} Q_{sh}(s) + W_{QT}(s) T_{out}(s)\right]. \end{gathered} \tag{27}$$

Next, we convert (27) to the form

$$T(s) = W_0(s)\left[k_{ah}(G(s) + T_h(s)) + k_{sh} Q_{sh}(s)\right] + W_{out} T_{out}(s), \tag{28}$$

where $W_0(s)$ – the transfer function for the room temperature by total heat gain (heat gain from the supply system and excess heat)

$$W_0(s) = \frac{W_{TQ}(s)}{1 + k_{ah}k_T W_{TQ}(s) + W_{TQ}(s)W_{QT}(s)};\tag{29}$$

$W_{out}(s)$ – transfer function for indoor temperature by outdoor temperature

$$W_{out}(s) = \frac{W_{QT}(s)}{1 + k_{ah}k_T W_{TQ}(s) + W_{TQ}(s)W_{QT}(s)}.\tag{30}$$

Note that the transfer functions $W_0(s)$ and $W_{out}(s)$ are established either approximately using analytical approaches or using parametric identification methods.

Based on Eqs. (10), (28), we obtain the structural scheme of combined HVS as an object of automatic control, shown in Fig. 3.

The dimensionless deviation of the set room temperature from the stationary value T_0 is introduced similarly to the dimensionless deviations of the room temperature (11) and outdoor temperature (25)

$$\Delta\overline{T}_s(t) = \Delta T_s(t) \Big/ \left(T_0 - T^0_{out}\right).\tag{31}$$

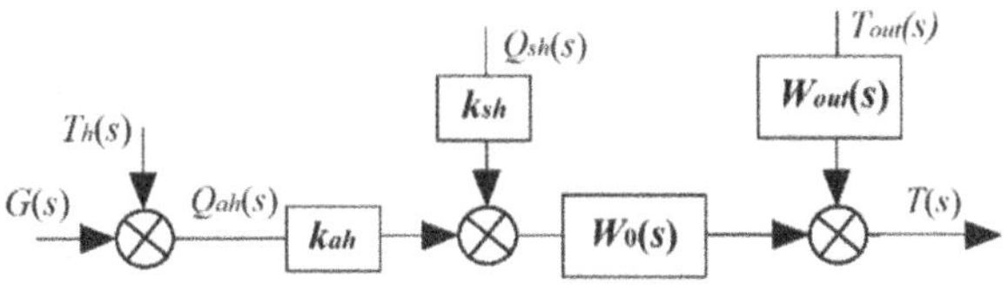

Fig. 3. The structural scheme of combined HVS as an object of automatic control for the room temperature $T(s)$.

According to (1) the difference of the dimensionless deviations of the set temperature and room temperature, transformed by Laplace,

$$E(s) = T_s(s) - T(s).\tag{32}$$

In the two-circuit ACS with air heating, in general, we consider various controllers for regulating the flow rate and temperature of the supply air

$$G(s) = W_{cG}(s) \cdot \varepsilon(s);\tag{33}$$

$$T(s) = W_{cT}(s) \cdot \varepsilon(s),\tag{34}$$

where $W_{cG}(s)$, $W_{cT}(s)$ – transfer functions of the flow controller and the temperature controller.

Based on (32–34) and Fig. 3, we obtain the structural scheme of the two-circuit ACS for air heating, shown in Fig. 4.

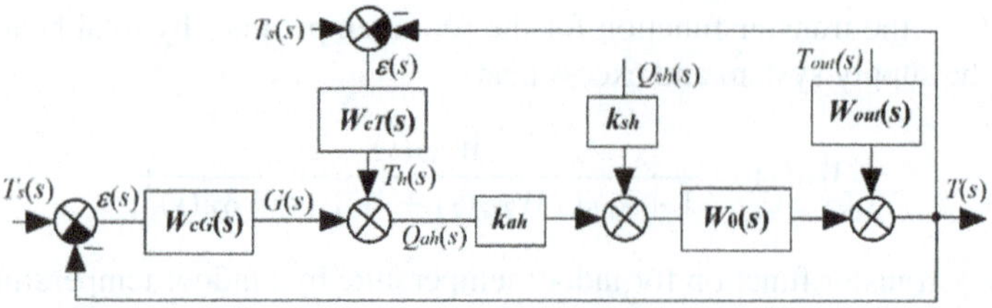

Fig. 4. The structural scheme of the two-circuit ACS for air heating ($W_{cG}(s)$, $W_{cT}(s)$ – transfer functions of the flow controller and the temperature controller).

4 Results

To simulate the two-circuit ACS for air heating in the Simulink environment of the MATLAB application software package, a block diagram was developed, the MPC controller (MPC – Model Predictive Control) was used in the control circuit of the supply air flow rate, and the PID controller was used in the control circuit of its temperature [30, 31].

The results of calculating the transient process in the two-circuit ACS for air heating at jump of the set temperature $T_s(s)$ in the serviced room are shown in Fig. 5.

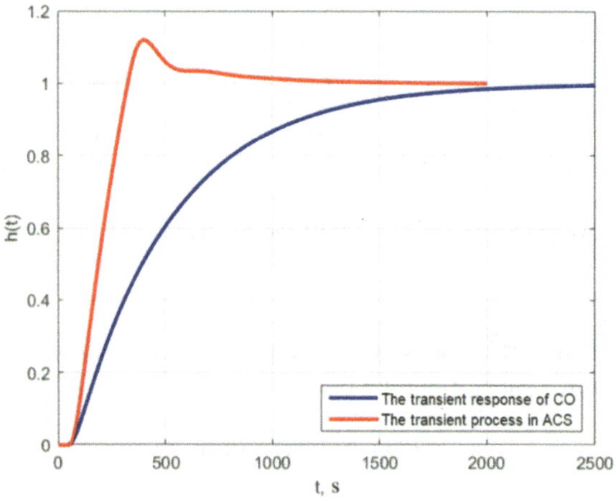

Fig. 5. The transient processes.

The transient characteristic of the control object is also presented here. During the simulation, the following transfer function $W_0(s)$ for the control object was considered

$$W_0(s) = \frac{e^{-49,7s}}{(25,8s+1)(457s+1)}. \tag{35}$$

The following parameters were set: coefficient $k_{ah} = 0.8$; the PID controller in the supply air temperature control circuit without the differentiating link $W_{cG}(s) =$

$1 + 1/200s$ and the minimum and maximum values at its output were limited by a non-linear link of the "saturation zone" type, respectively, with values -1 and $+1$.

During the synthesis of the MPC controller, the quantization period 10 *s* was taken, the prediction horizon 200 steps, the control horizon of 30 steps, the minimum value at the output was set -1, and the maximum value was 2.

As can be seen from Fig. 5, the ACS allows to increase the speed response of the system, and therefore increase the stability of the controlled parameters of HVS. A rational choice of controllers for the flow row of supply air and its temperature, with optimal adjustment of their parameters, can improve the quality indicators of ACS for air heating and ensure the efficiency increase of HVS.

5 Conclusions

Thus, the paper considers a combined HVS in which the supply air comes from a central ventilation unit at its constant temperature. The flow rate of the supply air is regulated by the special terminal, and additional change of its temperature is made in the heat exchanger. The work processes taking place in the system are reflected on the functional scheme.

The mathematical model of aerothermodynamic processes for combined HVS system as an object of automatic control has been developed. The structural scheme of the two-circuit ACS for air heating has been constructed. In general, the system provides various regulators for the flow rate and temperature of the supply air.

The calculation of the transient process in the system is performed when the set temperature changes in the serviced room. When modeling the system, MPC controller was used in the supply air flow control circuit, and PID controller was used in the temperature control circuit. By combination of controllers for supply air flow and its temperature, with optimal adjustment of their parameters, it is possible to improve the regulation quality indicators of the automatic air heating control system and, as a result, to ensure the increase of the efficiency of the heating-ventilating system.

In the future, it is planned to develop the presented results based on experimental implementation, as well as using alternative control strategies.

Acknowledgment. The authors would like to thank the Research and Development Sector at the Technical University of Sofia for the financial support.

References

1. Surin, V., Troyanovsky, Y., Maskaev, D., Tarasenko, I., Boldin, M.: Combined system development of heat ventilation air condition temperature control by fuzzy logic controller. IOP Conf. Ser. Mater. Sci. Eng. **869**, 022026 (2020)
2. Fomin, O., Sulym, A., Kulbovskyi, I., Khozia, P., Ishchenko, V.: Determining rational parameters of the capacitive energy storage system for the underground railway rolling stock. East.-Eur. J. Enterp. Technol. **2**(1(92)), 63–71 (2018)

3. Ifa, S., Dris, Z.: Experimental study of indoor air quality for cooling condition using ARDUINO system and PMV index testing the effect of heat sources. Int. J. Air-Cond. Ref. **29**(3), 2150021 (2021)
4. Krol, O.: Modeling of worm gear design with non-clearance engagement. In: Proceedings of the 6th International Conference on Industrial Engineering (ICIE 2020). Lecture Notes in Mechanical Engineering 1, pp. 36–46. Springer, Cham (2021). https://doi.org/10.1007/978-3-030-54814-8_5
5. Howell, R.: Principles of Heating Ventilating and Air Conditioning. ASHRAE, Atlanta (2017)
6. Shevchenko, S., Mukhovaty, A., Krol, O.: Modification of two-stage coaxial gearbox. In: Radionov, A.A., Gasiyarov, V.R. (eds.) ICIE2021. LNME, pp. 28–35. Springer, Cham (2021). https://doi.org/10.1007/978-3-030-54814-8_4
7. Esquinas, A.: Noninvasive Mechanical Ventilation in High-Risk Infections, Mass Casualty and Pandemics. Springer, Cham (2003)
8. Sokolova, Y., Azarenko, N., Sokolov, V.: The synthesis of system of automatic control of equipment for machining materials with hydraulic drive. East.-Eur. J. Enterp. Technol. **2**(2(68)), 56–60 (2014). https://doi.org/10.1015587/1729-4061.2014.23396
9. Awbi, H.: Ventilation Systems: Design and Performance. Routledge, London (2008)
10. Rogovyi, A., Korohodskyi, V., Neskorozhenyi, A., Hrechka, I., Khovanskyi, S.: Reduction of granular material losses in a vortex chamber supercharger drainage channel. In: Advances in Design, Simulation and Manufacturing V. Lecture Notes in Mechanical Engineering, pp. 218–226. Springer, Cham (2022)
11. Posohin, V., Safiullin, R., Broyda, V.: Ventilation. ACV, Moscow (2020)
12. Sulym, A, Fomin, O., Khozia, P., Mastepan, A.: Theoretical and practical determination of parameters of on-board capacitive energy storage of the rolling stock. Naukovyi Visnyk Natsionalnoho Hirnychoho Universytetu **2018**(5), 79–87 (2018)
13. Cao, S.J., Feng, Z.: Handbook of Ventilation Technology for the Built Environment: Design, control and testing. The Institution of Engineering and Technology, London (2021)
14. Rogovyi, A., Neskorozhenyi, A., Krasnikov, S., Tynyanova, I., Khovanskyi, S.: Improvement of vortex chamber supercharger performances using slotted rectangular channel. In: Advanced Manufacturing Processes IV. Lecture Notes in Mechanical Engineering, pp. 552–561. Springer, Cham (2023)
15. Ming, C., Wai, K., Ming, H.: Prediction and Control of Noise and Vibration from Ventilation Systems. CRC Press, Boca Raton (2017)
16. Romanchenko, O.: Principles of design of specialized technological equipment. Diagnostyka **23**(1), 2022109 (2022)
17. Nekrasov, A.: Mathematical modeling of a hybrid ventilation system with a technical floor. AIP Conf. Proc. **2701**, 020024 (2023)
18. Sokolov, V.: Increased measurement accuracy of average velocity for turbulent flows in channels of ventilation systems. In: Radionov, A.A., Gasiyarov, V.R. (eds.) ICIE2021. LNME, pp. 1182–1190. Springer, Cham (2021). https://doi.org/10.1007/978-3-030-54817-9_137
19. Sagin, S., et al.: Ensuring the environmental friendliness of drillships during their operation in special ecological regions of northern Europe. J. Mar. Sci. Eng. **10**(9), 1331 (2022)
20. McPherson, M.: Subsurface Ventilation and Environmental Engineering. Springer (2012)
21. Sokolov, V.: Hydrodynamics of flow in a flat slot with boundary change of viscosity. In: Radionov, A.A., Gasiyarov, V.R. (eds.) ICIE2021. LNME, pp. 1172–1181. Springer, Cham (2021). https://doi.org/10.1007/978-3-030-54817-9_136
22. Svistunov, V., Pushnyakov, N.: Heating, ventilation and air conditioning facilities of the agro-industrial complex and utilities. Polytechnic, St. Petersburg (2007)
23. Irfa, S., Driss, Z.: Ventilation system performance in a heated room testing the indoor air quality and thermal comfort. In: Indoor Air Quality: Control, Health Implications and Challenges, pp. 77–100. Springer, Cham (2022)

24. Krol, O., Sokolov, V.: Selection of worm gearing optimal structure for machine rotary table. Diagnostyka **22**(1), 3–10 (2021). https://doi.org/10.29354/diag/129949
25. Basova, Y., Dobrotvorskiy, S., Balog, M., Iakovets, A., Chelabi, M.A., Zinchenko, A.: Increasing SME supply chain resilience in the face of rapidly changing demand with 3D model visualisation. Int. J. Mechatron. Appl. Mech. **14**, 35–47 (2023). https://doi.org/10.17683/ijomam/issue14.5
26. Basova, Y., Dobrotvorskiy, S., Yakovenko, I.: Investigation of the impact of force majeure circumstances as a market instability factor on the flexibility and sustainability of engineering SMEs. EAI Endorsed Trans. Digital Transform. Ind. Process. **1**(1) (2025). https://publications.eai.eu/index.php/dtip/article/view/8951
27. Sudak, V., Hendiger, Y.: VAV systems. Short description. SMAY, Krakov (2009)
28. Tsankov, P.: Modeling of vertical spindle head for machining center. J. Phys. Conf. Ser. **1553**, 012012 (2020)
29. Andriichuk, N.: Aerohydromechanics. VDEUNU, Lugansk (2009)
30. Blaabjerg, F., Ionel, D.: Renewable Energy Devices and Systems with Simulations in MATLAB® and ANSYS®. CRC Press, Boca Raton (2017)
31. Dorf, R., Bishop, R.: Modern Control Systems. Prentice Hall (2011)
32. Mikhov, M., Zhilevski, M., Spiridonov, A.: Modeling and performance analysis of a spindle electric drive with adaptive speed control. J. Proc. Manufact. Syst. **7**(3), 153–158 (2012)

Structural and Parametric Identification of Mathematical Models of HVAC Systems

Oleg Krol[1(✉)], Vladimir Sokolov[1], and Dănuț-Iulian Stanciu[2]

[1] Volodymyr Dahl East Ukrainian National University, 59a Central Avenue, Severodonetsk 93400, Ukraine
krolos@gmail.com

[2] National Institute of Research and Development in Mechatronics and Measurement Technique INCDMTM, 6-8 Pantelimon, 2nd District, 021631 Bucharest, Romania

Abstract. The improvement of HVAC systems requires consideration of the work processes that take place in them. HVAC systems are complex dynamic systems with distributed parameters, and the work processes occurring in them are difficult to formalize clearly. Therefore, the task of identifying mathematical models of HVAC systems as object of automatic control is actual. The purpose of paper is to analyze the methods for structural and parametric identification of complex HVAC systems and to develop recommendations for the construction of mathematical models of HVS. It is shown that HVAC systems are inertial, in which the regulated variable is stabilized at a certain value, and transients have a non-oscillatory appearance and characteristic delay. The identification of mathematical models for working processes of HVAC systems by the transfer functions of the aperiodic links of the first and second orders with delay is recommended. The authors present the example of mathematical model for heating-ventilating system is presented in the form of the transfer function for the temperature at significant (working) zone of the room by the heat flow power of supply air.

Keywords: HVAC systems · Structural identification · Parametric identification · Transient process

1 Introduction

Heating, ventilation and air conditioning (HVAC) systems and, in particular, heating-ventilating systems (HVS), are one of the most important components of the design and development of industrial and administrative buildings and residential premises [1–4]. The main tasks of control HVS are: creating and maintaining the microclimate within a building, construction or room that is comfortable for humans or animals and plants, as well as material objects (equipment, substances, products, works art, etc.); energy savings spent on creating and maintaining a microclimate; technological safety of the system and cost reduction for its operation [5–8].

Important reserve for reducing energy consumption of HVS is the reduction of power consumption by decrease energy losses associated with the processes of regulating the

D. D. Cioboată and J. Machado (Eds.): ICoRSE 2025, LNNS 1592, pp. 110–119, 2025.
https://doi.org/10.1007/978-3-032-02508-1_10

aerothermodynamic characteristics. This can be achieved by developing automatic control systems (ACS) that ensure the rational use of energy resources by improving the quality indicators of HVS regulation. The solution of this problem can give a significant positive effect in the technical, economic and social aspects [9–12].

2 Literature Review

To improve ACS and control methods, it is necessary to have adequate description of the processes that occur in HVAC systems. From the point of view of the theory of automatic control, these systems are complex objects with distributed parameters, and the processes that occur in them are difficult to formalize clearly [13–16].

Identification is the process of constructing mathematical model based on observations of the inputs and outputs of the system [17–20]. The identification of dynamic objects is understood as the procedure for determining the structure and parameters of their mathematical models, which, with the same input signals of the model and the object, ensure the proximity of the output parameter of the model to the output parameter of the object in accordance with the given quality criterion [21–25].

In most cases, identification is the multi-step procedure. Its main stages are structural, parametric identification, as well as adequacy testing. Structural identification consists in determining by theoretical reasoning the structure of mathematical model. Parametric identification consists in conducting the identification experiment to find the values of the model parameters that have the properties of sufficiency and efficiency. The adequacy check is the quality assessment of the of the model using the accepted criterion of similarity of the object and the model [26–30].

In addition, approaches to modeling complex dynamic systems, such as those used in small and medium-sized machine-building enterprises (SMEs), provide valuable insights for HVAC modeling. For example, SMEs increasingly rely on digital tools and Industry 4.0 principles to integrate information across the product lifecycle, which can enhance the structural identification and simulation of real-world processes under variable and force majeure conditions [31]. Moreover, modeling the impact of external stressors using integral flexibility functionals, as applied in the machine engineering sector, highlights the importance of adaptive, AI-supported planning tools such as PPLAN. These approaches can also be extrapolated to the modeling of HVAC systems, particularly when evaluating system resilience and performance under stress or external influences [32].

The purpose of this paper is to analyze the methods for structural and parametric identification of complex HVAC systems and to develop recommendations for the construction of mathematical models of HVS.

3 Mathematical Structural and Parametric Identification of Mathematical Models of HVAC Systems

Two approaches are used to obtain a mathematical model: analytical and experimental. The analytical approach consists in the analytical derivation of the object equations based on the analysis of physical processes and laws, as well as the determination of the parameters included in these equations based on experimental data established by specially conducted experiments. This approach is often called modeling. The experimental approach has several names: the "black box" method, cybernetic, identification method, etc. In the case of solving the identification problem in the broad sense, there is no initial information about the object, and then the system is presented as a "black box". When the object structure and its model are known, and the parameters of the model can be determined by the results of experiment, then in this case the approach is called identification in the narrow sense or experimental-analytical. This approach is also commonly referred to as the "gray box" method or parametric identification.

The analytical approach usually requires a lot of time and provides the relatively low accuracy model. But at the same time, the resulting mathematical model has operability in the wide range of variable changes. The experimental approach makes it possible to quickly determine the parameters of the model. The found model can fairly accurately describe individual modes of operation of the object, however, the model's performance will be limited by narrower ranges of changes in the variables under study.

In practice, there are no purely analytical or purely experimental methods, usually the combination of them is resorted to. For each specific case, the reasonable combination of analytical and experimental approaches is found in the construction of a mathematical model.

It should be understood that for the same process occurring in the system, mathematical models may have practically nothing in common if they are developed for use for different purposes. In general, to build models of processes in the studied objects, it is necessary to establish the following parameters and characteristics: the boundaries or scope of the process under study; physical limitations, as well as restrictions imposed by safety conditions; depth of detail; required accuracy; nature of control: steady-state control or dynamic control; necessity and the nature of the improvement of the object model; state variables and control variables; basic disturbances and uncontrollable variables.

It should be noted that along with classical approaches to observation, systematization, hypothesis building and testing, intuition and common sense play the important role in the development of models of control objects. Intuition plays the important role for the formation and acceptance of basic assumptions when determining the dependencies between the basic variables, as well as for developing the initial approach to building model of the object under study. And to ensure the necessary balance, on the one hand, the accuracy and completeness of the description of mathematical model and, on the other hand, the complexity and laboriousness, common sense is required.

Identification methods are classified according to the certain number of characteristics, in particular, by the properties of the control object, by the type of input impact, by the degree of efficiency, etc. Methods are also distinguished depending on the accepted criteria for approximating the model and the object.

According to the properties of control objects, identification methods are distinguished: static and dynamic, linear and nonlinear, stationary and non-stationary, one-dimensional and multidimensional, continuous and discrete, with lumped and distributed parameters, deterministic and stochastic control objects. The most detailed methods are the identification of static and dynamic linear stationary control objects with lumped parameters.

According to the type of input exposure, methods of the active experiment and methods of the passive experiment are distinguished.

According to the degree of efficiency, there are methods of the operational identification (simultaneously with incoming information) and the retrospective (post-experimental) identification. For the most part, identification methods are retrospective.

The available methods of parametric identification make it possible to establish the parameters of control object with the given structure of its mathematical model for the experimental data obtained (meaning both the physical experiment and the numerical experiment on complex mathematical models, for example, models with distributed partial differential parameters). Under this condition, there is no need for information about the distribution laws for random values of the parameters of the object under study. Taking into account the complexity and distribution of HVAC systems, it is rational to carry out parametric identification of the object based on experimental transient characteristics.

Thus, in order to obtain the adequate mathematical model of aerothermodynamic processes in HVAC system, it is most appropriate to use an experimental method, it is then that the active and the passive experiments make it possible to obtain the reliable mathematical description of the processes that occur in the system. Moreover, the passive experiment can be conducted without the active participation of the researcher due to the independent course of the process. Note that the passive experiment is experiment in which the levels of factors in each experiment are recorded by the researcher, but not set. Performing the active experiment for HVAC system is possible, but only with the development of pre-set levels of certain factors that require time-consuming planning and conducting experiment.

To obtain mathematical models of dynamic systems based on measured input and output data of the real system, it is recommended to use the capabilities of the special set of software tools "System Identification Toolbox" (SIT) of the MATLAB application software package. Work with the SIT begins through the graphical interface with the help of the command "ident". With the help of the specialized SIT package, it is possible to identify the control object as the "black box" when evaluating the model parameters previously defined by the user. It is also possible to use the identified model to predict the reaction of the system, as well as to simulate the time and frequency characteristics of the system in the Simulink environment. The models built using the SIT package are discrete and can be represented in three forms of matrix model recording. The first form is representation in the so–called theta format (used for time models), the second form is representation in the frequency format (used for frequency characteristics) and the third form is representation in the form of zeros and poles.

To select the most optimal structure and parameters of mathematical models, the specialized SIT MATLAB package has special functions for parametric estimation, as well as for specifying, changing or refining the model structure. The evaluation of the model parameters for the selected structure is performed by minimizing the generally accepted criterion of model quality – the average square of the mismatch between the outputs of the object and its model.

For numerical evaluation of models in SIT MATLAB, the residual sum of squares of the error *LF* (Loss function), the final prediction error Akaike *FPE* (Akaike's Final

Prediction Error) and the standard deviation *BF* (Best Fits) are used. The definition of *LF* and *FPE* criteria is closely related.

The residual sum of squares of the error *LF* is called the theoretical Akaike Information Criterion, which is usually called simply the *AIC* – Akaike's Information Theoretic Criterion. The value of this criterion is determined by the expression of the form

$$LF = \log(L) + 2d/N, \tag{1}$$

where L – maximized value of the likelihood function of the model; d – number of estimated parameters; N – number of evaluation data.

The AIC information criterion is the quality measure of the model, which simulates a situation when the model is being tested for another data set. In fact, AIC is estimated measure of the quality of each of the available parametric models, since they relate to each other with the certain data set. This makes this criterion the ideal way to choose a model.

The final prediction error Akaike *FPE* is set according to the following expression

$$FPE = L\left(\frac{1 + d/N}{1 - d/N}\right). \tag{2}$$

After considering several different models, it is possible to compare them according to these criteria. For the most accurate model according to Akaike theory, the lowest values of *LF* and *FPE* are obtained.

As noted above, the set of software tools SIT MATLAB also offers the *BF* adequacy criterion, which is the standard deviation and which displays the measure of coincidence of experimental and predicted data as percentage. The value of $BF = 100\%$ indicates the perfect match of the parametric model, and $BF = 0\%$ corresponds to the model that does not fit at all.

Work processes in HVAC systems obey complex physical laws and are described by mathematical models with distributed parameters in the form of partial differential equations. The complete description of the aerothermodynamic characteristics in HVAC system requires consideration of spatial turbulent flow, heat transfer processes, which, in general, are carried out by thermal conductivity, convection and thermal radiation. Therefore, the realistic mathematical model is likely to be practically unsuitable for management purposes.

Taking into account the above, when developing ACS HVAC, it is necessary to replace the complex mathematical model of aerothermodynamic characteristics with distributed parameters in partial differential and with boundary conditions, which can also be partial differential equations, with the simpler adequate model with concentrated parameters in the form of ordinary differential equations. Such model should include variables that depend only on time and do not depend on coordinates.

Considering HVAC systems as control object, in general, the change in thermal power should be considered as the control signal, in particular cases this parameter may be the temperature of the heat carrier or its flow rate, the temperature or flow rate of the supply air for HVS, etc. Usually, the temperature in the main (working) zone of the room is considered as the controlled variable.

At the same time, despite the complexity of the mathematical description of work processes in HVS, they have common features. These systems are inertial and are objects with self-regulation, i.e., the regulated variable is stable at the certain value. Transient processes in HVS have the non-oscillatory appearance and the characteristic delay. In this regard, it would be most rational to build mathematical models of HVS as objects of automatic control by identifying them from experimental data in the form of the following transfer functions:

$$W_1(s) = \frac{ke^{-\tau s}}{T_0 s + 1}; \tag{3}$$

$$W_2(s) = \frac{ke^{-\tau s}}{(T_1 s + 1)(T_2 s + 1)}, \tag{4}$$

where s – the Laplace variable; k – transfer coefficient; T_0, T_1, T_2 – time constants; τ – delay time.

The transfer function (3) is the aperiodic link of the first order with delay, the transfer function (4) also defines the aperiodic link with delay, but already of the second order.

The transient characteristics for the transfer functions (3) and (4) have the following form

$$h(t) = \begin{cases} 0, & t \le \tau; \\ k\left(1 - e^{-\frac{t-\tau}{T_0}}\right), & t > \tau. \end{cases} \tag{5}$$

$$h(t) = \begin{cases} 0, & t \le \tau; \\ k\left(1 - \frac{T_1}{T_1 - T_2} e^{-\frac{t-\tau}{T_1}} - \frac{T_2}{T_2 - T_1} e^{-\frac{t-\tau}{T_2}}\right), & t > \tau. \end{cases} \tag{6}$$

The presence of analytical expressions for the transfer functions (4) and (5) makes it possible to reduce the problem of parametric identification to optimization problem: to find the minimum function of several variables: The sum of the quadratic deviations of the calculated and experimental values of the output variable for control object is taken as the minimized function. Note that the experimental value, in general, should be understood as the result of both a physical experiment and the result of a numerical experiment based on the complete mathematical model. You can automate the process by using the built-in function "fminsearch" in the MATLAB application software package.

The value of the transfer coefficient k is conveniently determined by the steady-state value of the output variable. One can also consider dimensionless variables, i.e., take $k = 1$. Then, in the case of identification of control object by the aperiodic link of the first order with delay, the problem of parametric identification is reduced to the problem of optimizing the function of two variables: τ and T_0, and in the case of identification of control object by the aperiodic link of the second order with delay, to the problem optimization of the function of three variables: τ, T_1, and T_2.

4 Results

Let's show the above with example. When applying the heat flow with the power 2.52 kW from HVS to the room with the volume 134.4 m^3, temperature measurements T in time t in the significant (working) zone with the step 60 s were carried out, shown in Fig. 1.

Structurally, we identify the system by the aperiodic link of the second order with delay (4). To simplify the parametric identification procedure, we reduce the experimental data obtained to the dimensionless form

$$\bar{T}(t) = \frac{T(t) - T(0)}{T_{ssv} - T(0)}, \tag{7}$$

where $T(0)$, T_{ssv} – initial and steady–state temperature value in the working zone (for experimental conditions $T(0) = 15.8$ °C, $T_{ssv} = 21.1$ °C).

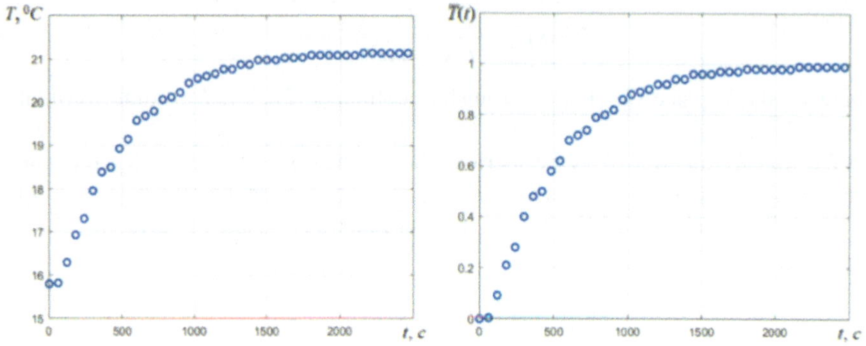

Fig. 1. The temperature change in the work zone.

The change of the dimensionless temperature in the significant (working) zone is shown in Fig. 1.

For the obtained dependence $k = 1$, and the problem of parametric identification is reduced to the problem of finding the minimum for function the sum of the quadratic deviations of the calculated (6) and experimental values (7) for three variables τ, T_1, and T_2. The calculation results in the environment of the MATLAB application software package give the following results: $T_1 = 25.8$ s; $T_2 = 457$ s; $\tau = 49.7$ s.

The transient characteristic $h(t)$ for control object with the transfer function

$$W(s) = \frac{e^{-49,7s}}{(25,8s + 1)(457s + 1)} \tag{8}$$

shown in Fig. 2.

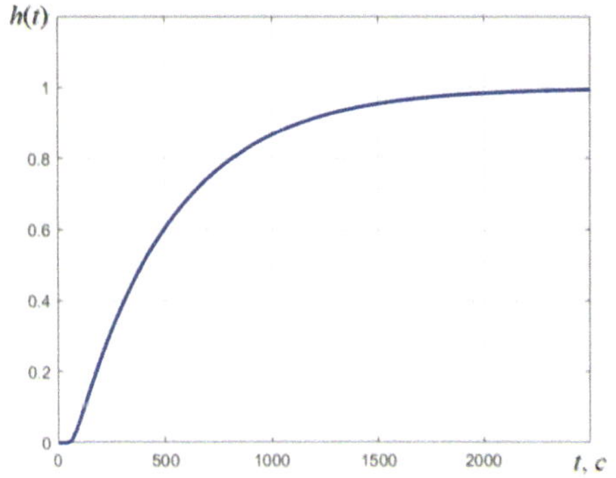

Fig. 2. The transient characteristic.

5 Conclusions

The improvement of HVAC systems requires consideration of the work processes that take place in them. HVAC systems are complex dynamic systems with distributed parameters, and the work processes occurring in them are difficult to formalize clearly. Therefore, the task of identifying mathematical models of HVAC systems as object of automatic control is actual.

The analysis of methods for structural and parametric identification of complex dynamic systems is carried out and recommendations for the construction of mathematical models for HVAC systems are developed. It is noted that in order to obtain the adequate mathematical model of aerothermodynamic processes, it is most appropriate to use the experimental method that allows obtaining the reliable mathematical description of the processes that occur in the system. In this case, the experiment can be passive, when the levels of factors in each experiment are recorded by the researcher, but not set.

It is shown that HVAC systems are inertial and are self-regulating objects, i.e., the regulated variable is stabilized at the certain value. Transient processes in such systems have the non-oscillatory appearance and the characteristic delay. In this regard, it would be most rational to construct mathematical models for HVAC systems as objects of automatic control by identifying them according to experimental data in the form of transfer functions of the first and second orders with the delay.

The example of mathematical model for HVS is presented in the form of transfer function for the temperature in significant (working) zone of the room by the heat flow power of supply air.

References

1. Howell, R.: Principles of Heating Ventilating and Air Conditioning. ASHRAE, Atlanta (2017)
2. Sokolov, V.: Increased Measurement accuracy of average velocity for turbulent flows in channels of ventilation systems. In: Radionov, A.A., Gasiyarov, V.R. (eds.) ICIE 2021. LNME, pp. 1182–1190. Springer, Cham (2021). https://doi.org/10.1007/978-3-030-54817-9_137
3. Sudak, V., Hendiger, Y.: VAV systems. Short description. SMAY, Krakov (2009)

4. Krol, O.: Modeling of Worm Gear Design with Non-clearance Engagement. In: Radionov, A.A., Gasiyarov, V.R. (eds.) ICIE 2021. LNME, pp. 36–46. Springer, Cham (2021). https://doi.org/10.1007/978-3-030-54814-8_5
5. Awbi, H.: Ventilation Systems: Design and Performance. Routledge, London (2008)
6. Esquinas, A.: Noninvasive Mechanical Ventilation in High-Risk Infections, Mass Casualty and Pandemics. Springer, Cham (2003). https://doi.org/10.1007/978-3-031-29673-4
7. Shevchenko, S., Mukhovaty, A., Krol, O.: Modification of Two-Stage Coaxial Gearbox. In: Radionov, A.A., Gasiyarov, V.R. (eds.) ICIE 2021. LNME, pp. 28–35. Springer, Cham (2021). https://doi.org/10.1007/978-3-030-54814-8_4
8. Sagin, S., et al.: Use of biofuels in marine diesel engines for sustainable and safe maritime transport. Renew. Energy **224**, 120221 (2024)
9. Awbi, H.: Ventilation of Buildings. Routledge, London (2003)
10. Posohin, V., Safiullin, R., Broyda, V.: Ventilation. ACV, Moscow (2020)
11. Rogovyi, A., Korohodskyi, V., Neskorozhenyi, A., Hrechka, I., Khovanskyi, S.: Reduction of granular material losses in a vortex chamber supercharger drainage channel. In: Ivanov, V., Pavlenko, I., Liaposhchenko, O., Machado, J., Edl, M. (eds.) DSMIE 2022. LNME, pp. 218–226. Springer, Cham (2022). https://doi.org/10.1007/978-3-031-06044-1_21
12. Sulym, A, Fomin, O., Khozia, P., Mastepan, A.: Theoretical and practical determination of parameters of on-board capacitive energy storage of the rolling stock. Naukovyi Visnyk Natsionalnoho Hirnychoho Universytetu **2018**(5), 79–87 (2018)
13. Cao, S.J., Feng, Z.: Handbook of Ventilation Technology for the Built Environment: Design, Control and Testing. The Institution of Engineering and Technology, London (2021)
14. Stefanov, Y.: Ventilation and Air Conditioning. AVOK NORTH-WEST, St. Petersburg (2005)
15. Rogovyi, A., Neskorozhenyi, A., Krasnikov, S., Tynyanova, I., Khovanskyi, S.: Improvement of vortex chamber supercharger performances using slotted rectangular channel. In: Tonkonogyi, V., Ivanov, V., Trojanowska, J., Oborskyi, G., Pavlenko, I. (eds.) InterPartner 2022. LNME, pp. 552–561. Springer, Cham (2023). https://doi.org/10.1007/978-3-031-16651-8_52
16. Romanchenko, O.: Principles of design of specialized technological equipment. Diagnostyka **23**(1), 2022109 (2022)
17. Svistunov, V., Pushnyakov, N.: Heating, Ventilation and Air Conditioning Facilities of the Agro-Industrial Complex and Utilities. Polytechnic, St. Petersburg (2007)
18. Ifa, S., Dris, Z.: Experimental study of indoor air quality for cooling condition using ARDUINO system and PMV index testing the effect of heat sources. Int. J. Air-Cond. Refrig. **29**(3), 2150021 (2021)
19. Tsankov, P.: Modeling of vertical spindle head for machining center. J. Phys. Conf. Ser. **1553**, 012012 (2020)
20. Sokolov, V.: Hydrodynamics of flow in a flat slot with boundary change of viscosity. In: Radionov, A.A., Gasiyarov, V.R. (eds.) ICIE 2021. LNME, pp. 1172–1181. Springer, Cham (2021). https://doi.org/10.1007/978-3-030-54817-9_136
21. McPherson, M.: Subsurface Ventilation and Environmental Engineering. Springer, Luxembourg (2012). https://doi.org/10.1007/978-94-011-1550-6
22. Irfa, S., Driss, Z.: Ventilation system performance in a heated room testing the indoor air quality and thermal comfort. In: Indoor Air Quality: Control, Health Implications and Challenges, pp. 77–100. Springer, Cham (2022)
23. Pupkov, K., Egupov, N.: Methods of Classical and Modern Theory of Automatic Control. MSTU, Moscow (2004)
24. Krol, O., Sokolov, V.: Selection of worm gearing optimal structure for machine rotary table. Diagnostyka **22**(1), 3–10 (2021)
25. Fomin, O., Sulym, A., Kulbovskyi, I., Khozia, P., Ishchenko, V.: Determining rational parameters of the capacitive energy storage system for the underground railway rolling stock. Eastern-Eur. J. Enterpr. Technol. **2**(1(92), 63–71 (2018)

26. Dorf, R., Bishop, R.: Modern Control Systems. Prentice Hall, Upper Saddle River (2011)
27. Etheridge, D., Sandberg, M.: Building Ventilation: Theory and Measurement. Wiley, Chichester (1997)
28. Blaabjerg, F., Ionel, D.: Renewable Energy Devices and Systems with Simulations in MATLAB® and ANSYS®. CRC Press, Boca Raton (2017)
29. Sokolova, Y., Azarenko, N., Sokolov, V.: The synthesis of system of automatic control of equipment for machining materials with hydraulic drive. Eastern-Eur. J. Enterpr. Technol. **2**(2(68)), 56–60 (2014)
30. Lurie, B., Enright, P.: Classical Methods of Automatic Control. BHV-Petersburg, Saint-Petersburg (2004)
31. Basova, Y., Dobrotvorskiy, S., Balog, M., Iakovets, A., Chelabi, M.A., Zinchenko, A.: Increasing SME supply chain resilience in the face of rapidly changing demand with 3D model visualisation. Int. J. Mechatron. Appl. Mech. **14**, 35–47 (2023). https://doi.org/10.17683/ijomam/issue14.5
32. Basova, Y., Dobrotvorskiy, S., Yakovenko, I.: Investigation of the impact of force majeure circumstances as a market instability factor on the flexibility and sustainability of engineering SMEs. EAI Endorsed Trans. Digit. Transf. Ind. Processes **1**(1) (2025). https://publications.eai.eu/index.php/dtip/article/view/8951

Dynamic 3D Mapping of Airborne Pollutants Using UAV with Multiple Data Sensing Instrument

Karolina Macúchová[1(✉)], Tomáš Tichý[1], Dominik Pilnaj[2], and Jan Macúch[3]

[1] HiLASE Centre, Institute of Physics CAS, Za Radnicí 828, 25241 Dolní Břežany, Czech Republic
karolina.macuchova@hilase.cz

[2] Faculty of Electrical Engineering, Czech Technical University in Prague, Technická 2, 166 27 Prague, Czech Republic

[3] University of Creative Communication (VŠKK/PSCC), Na Pankráci 420/54, 140 00 Prague, Czech Republic

Abstract. Unmanned Aerial Vehicles (UAVs) are increasingly utilized for air quality monitoring due to their ability to access challenging environments and provide three-dimensional pollution profiles. This study explores UAV-based measurements of particulate matter (PM), CO_2, volatile organic compounds (VOCs), and other pollutants across diverse terrains in the southeastern part of Prague, including fields, highways, and forests. A novel sensor shielding design was implemented to mitigate rotor-induced airflow interference, ensuring measurement accuracy. Data acquisition involved multiple altitude profiles and vertical sampling near pollution sources, with meteorological conditions factored into the analysis. Results revealed pollutant distributions influenced by terrain and wind direction, such as elevated PM concentrations near highways and forests. The study confirms the effectiveness of UAVs for dynamic airborne sampling while addressing challenges like rotor effects and spatial data processing. These findings highlight UAVs' potential for environmental monitoring and underscore their role in advancing air quality research.

Keywords: UAV · Air quality monitoring · Particulate matter · PM2.5 · Carbon dioxide · CO_2 · Volatile organic compounds · VOCs · Topology · CFR print · Environmental monitoring · Sensor technology · Rotor-induced airflow interference · Spatial pollutant distribution · Meteorological factors · Three-dimensional pollution profiling · Real-time data acquisition · Pollution source tracking · Terrain influence on air quality · Air sampling · Aerial flexibility · Volumetric cloud

1 Introduction

The assessment of air quality has been extensively studied in numerous articles and research papers [1]. Various methodologies exist for measuring airborne particles as well as sampling techniques [2]. UAVs (Unmanned Aerial Vehicles) offer unparalleled aviation flexibility for pollution profile measurements at different distances and heights from

D. D. Cioboată and J. Machado (Eds.): ICoRSE 2025, LNNS 1592, pp. 120–132, 2025.
https://doi.org/10.1007/978-3-032-02508-1_11

a source - typically between 2–100 m above the terrain. UAVs enable the creation of pollution maps by facilitating the measurement of pollution profiles around specific objects [3]. Their ability to operate in three-dimensional space allows for targeted localization of pollution sources [4, 5].

A significant advantage of UAV-based atmospheric data collection lies in their ability to gather data from otherwise inaccessible or hazardous areas. For instance, UAV-based atmospheric measurements above volcanic sites are highlighted in [6]. These features of UAV-based measurements can be advantageously applied to monitor PM (particulate matter), NO_2, volatile organic compounds (VOC) and other polluting emissions from various sources. Among the primary sources of emissions, high-traffic roads are significant contributors [7]. Monitoring these sources and analyzing the dispersion of pollutants into the surrounding environment, both horizontally and vertically, is essential.

Site-specific environmental variability – such as terrain features and vegetation - significantly influence dispersion of pollutants in the air. For example, study [8] examines the effect of roadside vegetation on PM distribution and recommends optimal green infrastructure designs to mitigate PM spread. However, this study focuses exclusively on low-altitude measurements. In contrast, the research presented in [9] explores vertical PM profiles using UAVs, analyzing the influence of roadside vegetation on PM distribution up to a height of 100 m.

Despite their potential, the use of UAVs for such measurements faces several challenges. A prominent issue is the influence of rotor-induced airflow on measurement accuracy [10, 11]. Nevertheless, UAV-based atmospheric sensing is promising and increasingly popular, with significant potential for future advancements.

2 Instrumentation

2.1 UAV Instrumentation

An 8″ quadcopter, constructed on a MARK 4 carbon fiber frame, is utilized for the experiments. The measurement instrumentation is mounted on a custom nylon composite platform, reinforced with carbon fibers for strength and durability while maintaining low weight. Additional reinforcement is provided by continuous glass fiber using advanced CFR printing technology, achieving an optimal balance between weight and stiffness. A topological study optimized the design for efficient material distribution and minimized weight. This setup is versatile, designed for various scientific and technical experiments, with a instrumentation base plate featuring a matrix of M6 threads at a 25×25 mm pitch for easy attachment of experimental hardware. This modularity allows adaptation for sensors, cameras, or light sources, ensuring structural integrity and flight safety (Fig. 1).

2.2 Sensors and Data Logging

A Raspberry Pi 3B+ was selected as the primary controller for sensor management and data logging due to its versatile connectivity, which facilitates the integration of sensors from various manufacturers using multiple communication protocols (UART, SPI, and ADC). The software architecture, written in Python 3.7, utilizes asynchronous

Fig. 1. Overview of the UAV (battery removed for better visibility of the payload instrumentation)

data reading and timestamping to enhance system robustness and minimize latency in real-time data collection.

The UAV's position was tracked using a GEP-M10 module. Spatial accuracy depended on atmospheric conditions and satellite availability, with at least 15 satellites monitored before takeoff. This ensured 3D position accuracy within 0.35 m and an HDOP factor below 2 during flight. The timestamp was synchronized with the satellite signal and further aligned with the controller at UAV launch.

Aerosols were detected using the Sensirion SPS30 light-scattering sensor, which can resolve PM 1.0, 2.5, 4, and 10 μm within a range of 0–1000 $\mu g/m^3$. This low-power sensor detects airborne particles like pollen, smoke, and dust from traffic or combustion processes.

The Sensirion SCD30 sensor module measured CO_2 concentration, temperature, and humidity. It uses NDIR spectroscopy for CO_2 detection within 400–10,000 ppm, with 3% accuracy. Humidity and temperature are detected using CMOS technology, with accuracies of 3% rH and 0.1 °C. These parameters provide crucial insights into environmental conditions affecting pollutant interactions. The sensor was mounted beneath the UAV to minimize rotor-induced airflow interference.

Volatile organic compounds (VOCs) were detected using the Alphasense PID-AH2 photoionization sensor, with detection limits around 10 ppb and responses normalized to toluene equivalent. Inorganic gases (NO_x, SO_2, O_3) were detected using Alphasense electrochemical sensors. Electrochemical sensors were powered and warmed up for 60 min prior measurement to obtain a stable baseline.

2.3 UAV Air Pollutant Measurements

While UAVs offer many advantages for measuring air pollutants, challenges remain, particularly the influence of UAV rotors on airflow measurement. This effect can significantly impact PM particle measurements, as sensors with internal fans are affected by rotor turbulence, potentially compromising accuracy. The UAV's presence also disturbs the sampling environment, affecting aerosol measurements.

A common solution is adding a tube upstream of the sensor intake to distance it from rotor influence, but this introduces measurement delays and uncertainties, making it more suitable for semi-stationary measurements. Our research focused on dynamic airborne sampling, so we created a shield for the PM sensor to prevent direct propeller airflow impact. The shield's effectiveness was confirmed by comparing PM 2.5 concentrations at different rotor speeds, showing no clear dependence, thus validating the shield design.

3 Results

3.1 Experimental Methodology

The experimental methodology was designed to systematically capture the spatial and vertical distribution of airborne pollutants across diverse terrains in southeastern Prague (Czech Republic), including fields, a highway, and a spinney. This area was selected for its heterogeneous landscape and the presence of distinct pollution sources, enabling the assessment of how terrain and meteorological factors influence pollutant dispersion.

Flight paths were planned to maximize spatial coverage and data representativeness while adhering to airspace regulations valid in Czech Republic. Altitude profiles were selected based on the expected vertical extent of pollutant dispersion and the physical features of each site. For horizontal mapping, semi-constant altitude flights were conducted at increments of 10 m, ranging from 10 to 60 m above ground level, to capture horizontal pollutant gradients over the field and the southern side of the highway.

Above the forest canopy, flights ranged from 40 to 75 m, targeting pollutant accumulation zones and vertical mixing processes. Two additional flights near the highway implemented vertical profiling, with ascents performed at three distinct points on each side of the road. At each sampling altitude or profile node, the UAV hovered for a minimum of 60 s to allow sensor readings to stabilize and to minimize transient effects from UAV movement. The UAV remained stationary during these intervals to reduce measurement artifacts caused by airflow.

To avoid contamination from take-off and landing, initial readings during rotor spin-up were discarded, and designated take off zones were selected away from known pollution sources. All sensor data were timestamped and synchronized with UAV telemetry using the onboard GPS module, ensuring accurate spatial and temporal alignment. Sensor measurements were logged at varying frequencies depending on sensor type, with subsequent resampling and interpolation to achieve a uniform, one-second time base for integrated analysis.

Meteorological conditions were closely monitored before and during each flight using forecasts from the ECMWF model and real-time data from the Copernicus Atmosphere Monitoring Service (CAMS). Flights were scheduled during periods of stable weather to minimize confounding effects from sudden changes in wind speed or direction Table 1 summarizes the meteorological parameters recorded during the main sampling campaign; these values represent averages measured immediately prior to and during the flights.

Table 1. Meteorological conditions

Marker	Value
Atmospheric pressure	1028 hPa
Temperature	4 °C
Wind speed	$6\ m{\cdot}s^{-1}$
Wind direction	117°
PM 2.5 concentration	$13\ \mu g/m^3$
NO_2 concentration	$5.7\ \mu g/m^3$
O_3 concentration	367 DU
SO_2 concentration	$12.9\ mg/m^2$
CO_2 concentration	292 ppbv

Prior to each flight session, all sensors underwent calibration using reference standards or baseline checks to ensure measurement accuracy. Electrochemical sensors were powered and warmed up for at least 60 min before deployment. Data reliability was further assessed by comparing readings from replicate flights and by evaluating consistency with ground-based reference measurements where available.

A schematic map of the study area and flight paths is provided (see Fig. 4), illustrating the spatial relationship between sampling transects, pollution sources, and terrain features. All methodological decisions-including flight altitudes, sampling intervals, and quality control procedures-were guided by established best practices in UAV-based environmental monitoring and are documented in detail to facilitate replication by other researchers.

Access to the highway was restricted to its protection zone, which may have limited proximity sampling near the most intense pollution sources. Additionally, while efforts were made to standardize meteorological conditions, some variability in wind and temperature may have influenced pollutant dispersion during the campaign.

3.2 Methods of Data Visualization

Sensor readings were acquired using both asynchronous and synchronous sampling modes, with varying sampling intervals. To integrate the data into a cohesive dataset, the UAV telemetry timestamp was used as the primary temporal reference. Sensor measurements were collected at varying frequencies, requiring resampling and interpolation (at one-second intervals) to create a uniform, time-aligned dataset.

Data evaluation was conducted using Python-based scripts, MATLAB 4D scatter plot functions, the Plotly framework, and volumetric cloud rendering via the Three.js library. Each visualization method offers distinct advantages tailored to specific needs.

MATLAB provides highly accurate visualizations that remain true to the measured data. Its robust functionality supports further data processing and modeling, making it an ideal choice for detailed analysis (see Fig. 7).

Plotly introduces flexibility by enabling interactive 4D charts that can be displayed directly in web browsers. This feature facilitates sharing results with a broader audience. Plotly supports vector rendering through its SVG API, producing visuals like MATLAB while integrating additional capabilities such as support of map-based visualizations, allowing users to overlay data on OpenStreetMap. By pre-processing data into JSON format and embedding a simple selector in the web page, users can easily switch between different measured values or combinations thereof. This approach allows pairing measured data with a UAV's flight path and its relative position to potential pollutant sources. Data points are represented as discrete markers, with measured values indicated by a color scale and optional measurement labels.

Volumetric cloud rendering utilizes Three.js leverages WebGL for 3D visualization directly in browsers. This method provides a visually intuitive representation of measured data (see Fig. 2).

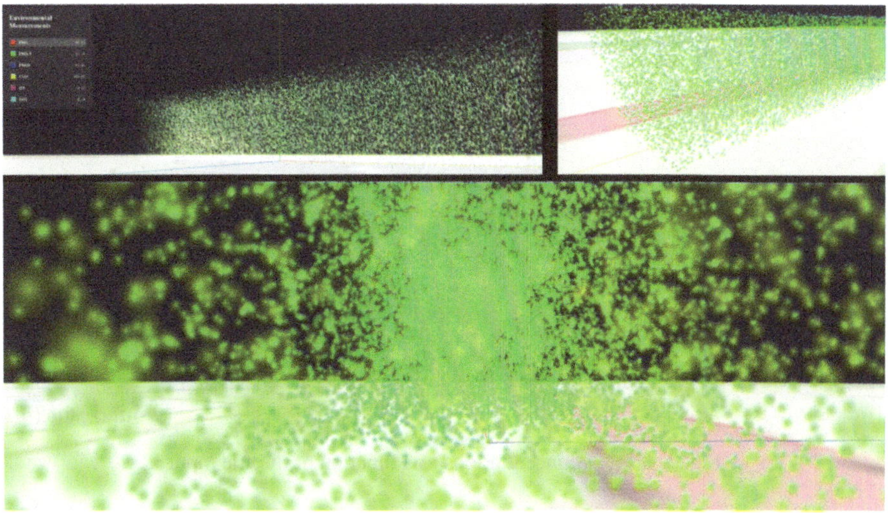

Fig. 2. Volumetric cloud representation of the PM2.5 particle cloud

The bounding box envelope of the cloud is defined using minimum and maximum values across three axes, with individual measurement points serving as nodes in a 3D gradient mesh that forms the volumetric cloud. For this experiment, up to 50,000 particles were rendered per pollutant class, distributed according to area density gradients via rejection sampling. Each pollutant is assigned a unique hue, and particle number is reduced by enlarging size while increasing opacity and color intensity in highest density areas and using occlusion culling. These optimizations do not impact the fidelity of visualized data, yet they ensure swift rendering and smooth orbiting around and zooming through the cloud. This efficient rendering method can also enable live low latency updates during live measurements where data stream from measuring device is reliable. Such capability would allow spotters to guide UAV pilots effectively, ensuring comprehensive coverage of the area of interest. Volumetric cloud rendering does not provide single-point data analysis, yet it excels at illustrating the distribution and correlation of pollutants across multiple regions using density and color variations. Additionally, this visualization method is ideal communication of air quality findings to the public by presenting results in an accessible format compared to traditional scatter plots with numerical values.

3.3 Verifying Data Reliability

The initial measurements confirmed the UAV's influence and potential distortion. Research indicates that measurements will always be affected to some extent, with the light-scattering sensor for particulate matter being most influenced by rotor effects. This effect is illustrated in Fig. 3 below.

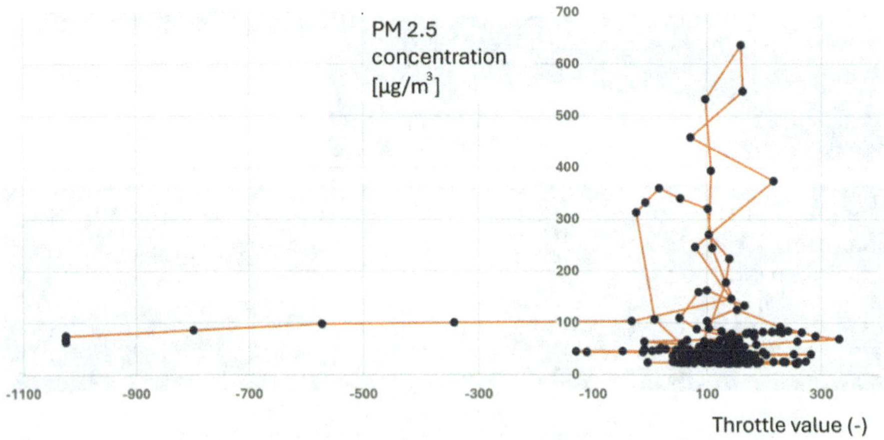

Fig. 3. $PM_{2,5}$ concentrations as a function of throttle input

Data acquisition begins at a throttle value of −1000, corresponding to initial rotor spin-up (arming). The throttle setting required for take-off is approximately 0. Flight operations typically occur within the throttle range of 0 to 500, where the relevant measurements are situated. No significant anomalies are observed in the sensor output, even during the gradual increase in throttle prior to take-off (from −1000 to 0). The VOC, NO_x, SO_2, O_3, and CO_2 sensor signals were evaluated using the same approach, and no correlation was found between their readings and throttle values. Accordingly, the measured data can be regarded as reliable in this context, unaffected by rotor-induced airflow.

To preserve measurement dynamics, a shield was used and found effective. The data show that solid particle measurements are unaffected by rotor flow, maintaining measurement dynamics, with the take-off moment visible in the datasets – see Fig. 4.

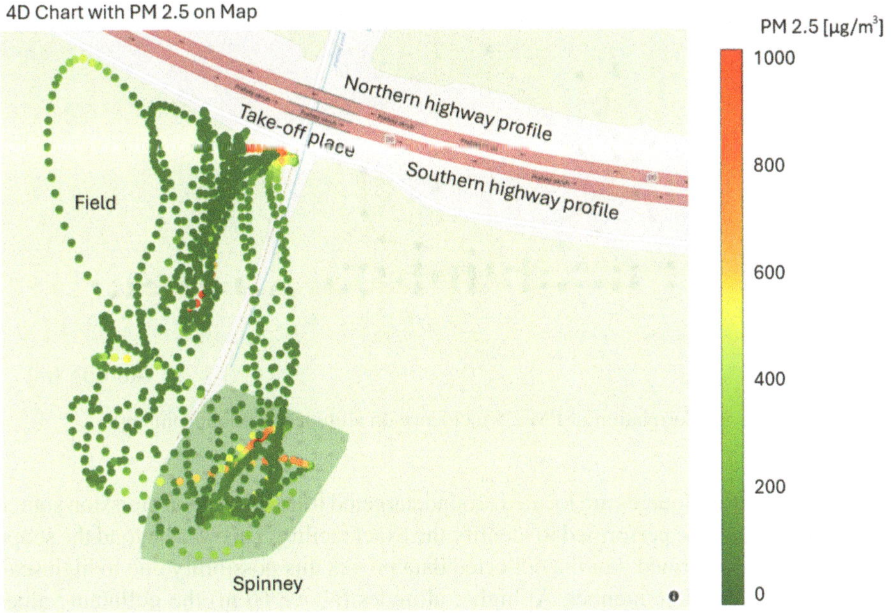

Fig. 4. Visible take-off moment captured in the dataset, showcasing $PM_{2.5}$ concentration levels

It is assumed that the sensor detected dust particles resuspended by rotor-induced airflow during take-off. However, as the UAV gained altitude, these airborne particulates appeared to disperse rapidly and did not influence the remaining measured variables.

3.4 Environmental Variability and Pollutant Distribution

The measured values revealed significant environmental variability across all defined regions: a spinney, field and near highway. For example, PM 2.5 concentration values were observed with measurements over the spinney, reaching 25 $\mu g/m^3$ for the 55 m flight dataset and above 500 $\mu g/m^3$ in isolated areas. The highest levels of aerosol presence were recorded in the 40–60 m altitude band above the forest canopy. See Fig. 5 for more details.

At higher altitudes (60–70 m), $PM_{2.5}$ concentrations consistently decreased across all flights within this area. This phenomenon is likely attributable to pollen release from early blooming species within the Betulaceae family, such as hazel (Corylus sp.), alder (Alnus sp.), or hornbeam (Carpinus sp.), which are common pioneer tree species in this region. Alternatively, it might be caused by emission from significant remote source. This hypothesis is supported by higher concentration in CO_2 and NO_2, detected by photoionization and electrochemical sensor, and slight increase in temperature. Many

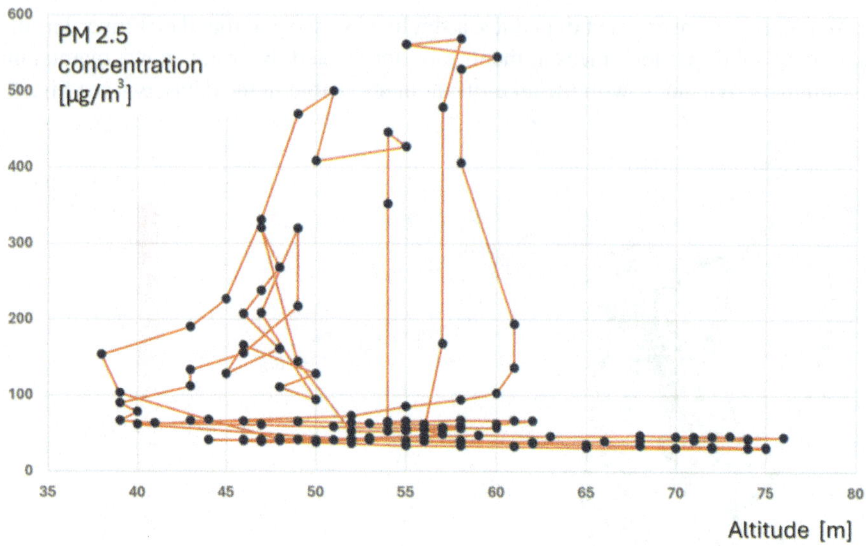

Fig. 5. Correlation of PM 2.5 incidence on altitude (location: spinney)

industrial pollution sources are located around, targeted following of the emission source during flight could be performed to identify the exact facility. This was beyond the scope of experiment performed, but the collected data proves this possibility due to high scan rate in nearly real-time manner. At higher altitudes (above 60 m) the pollutant values already decrease, mostly due to their sedimentation while the wind blew from the higher altitudes with lower contamination level.

Regarding the measurement of other variables, the NO_x concentration is of particular interest. Over the forested area, an increase in NO_x levels was observed between 50 and 60 m. A comparable trend was also identified in the CO_2 measurements. In the light of these variations in trace gas concentrations, ambient air temperature was also evaluated. A similar pattern emerged, with increase temperatures detected within the 40–60 m range (see Fig. 6).

Aerosol concentrations above the field were comparatively low, as no direct emission source was present. Sporadic increases in particulate matter were observed in the dataset and are attributed to distant pollution sources, such as the adjacent forest or highway, since they predominantly occurred near these areas or following UAV overflights nearby. The measurements were conducted outside the vegetation season, with tillage having been completed at the end of the previous year.

Additionally, the experiment took place during a relatively cold and humid period, under which wind-driven soil erosion—detectable by the light-scattering sensor—was not expected. A distinct decrease in air temperature with increasing altitude was observed over the field area, which is a typical pattern in a relatively homogeneous environment. Aside from boundary effects near the edges, no dominant landscape features were present that would induce vertical temperature anomalies. UAV-based monitoring during active field operations—such as ploughing or pesticide application—could offer valuable insights into pollutant dispersion and environmental impact.

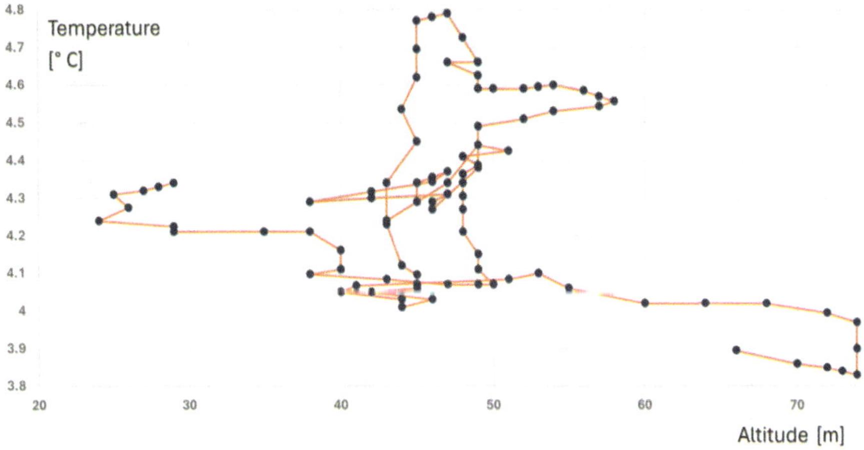

Fig. 6. Temperature variation with altitude (location: spinney)

Vertical profiling flights were conducted near the highway on both its southern and northern sides. This approach was selected based on previous horizontal measurements, which had proven challenging to interpret due to limited spatial context. In contrast, vertical flight paths enable clearer visualization of spatial trends, as illustrated in Fig. 7 and Fig. 8, which shows the vertical distribution of $PM_{2.5}$ concentrations.

Generally, for $PM_{2.5}$ concentrations measured across the field, most values fall within the lower tens of $\mu g/m^3$. However, there are occasional instances where concentrations reach the higher hundreds of $\mu g/m^3$. These are attributed to the dust blown by rotors during UAV take-off, proximity of potential sources of particulate matter (e.g. vegetation or remote industrial sources) as well as the direction of wind flowing from the forest to the field in the study area. Thus, it is very likely that the cloud of particulate matter located over the forest moved over the field area causing a local increase in the measured values. This is mainly noticeable at elevations under 40 m, which is approximately the height of trees. As expected, there is a noticeable decrease in temperature with increasing height, which is not affected by spatial inhomogeneity as in the case of forest cover.

Regarding the measured values, a general trend of decreasing aerosol concentration with altitude was observed, albeit with localized deviations. In both southern and northern profiles, intermittent increases or decreases appear throughout the data. However, across the dominant portion of the altitude range, a gradual decline in $PM_{2.5}$ concentration with height is evident (see Fig. 8).

It can also be stated that on the southern and northern sides, the measured $PM_{2.5}$ values approach the values predicted in the meteorological forecast - prediction 13 $\mu g/m^3$, subsequently measured range 10–20 $\mu g/m^3$. Despite the wind direction (from south to north), no fundamental difference was found between the values on the southern and northern sides in all monitored quantities. The concentrations of VOCs, NO_x, SO_2, and O_3 remain nearly constant as the UAV ascends, with only a few variations that are likely due to random effects or noise. In the case of measured particulate matter, there is a

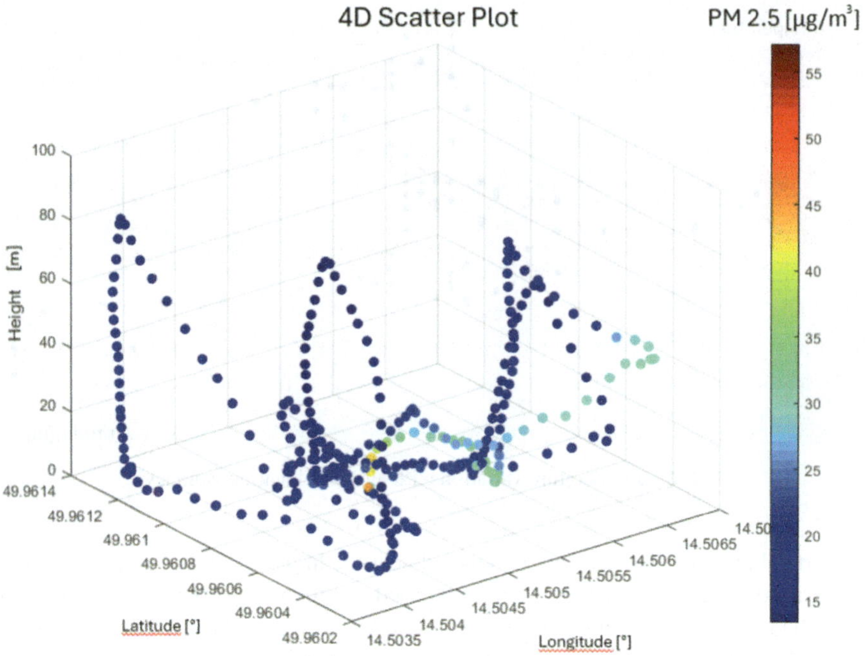

Fig. 7. 4D representation of distribution of PM 2.5

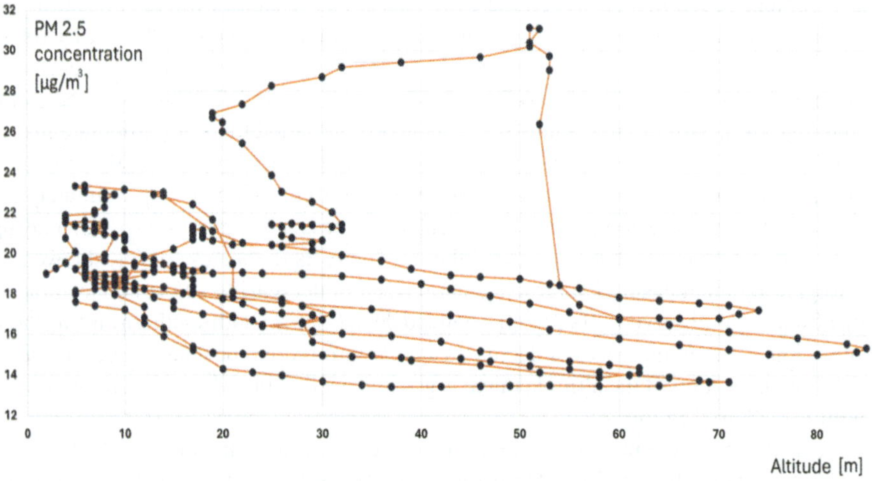

Fig. 8. 2D representation of concentration of PM 2.5 on altitude

slight decrease in values with increasing altitude. It can be concluded that there is a rapid dispersion of particles into the surroundings.

4 Conclusion

The integration of UAV-based dynamic 3D mapping with advanced sensor technology and adaptive flight strategies, as demonstrated in this study, offers transformative potential for environmental monitoring. By combining rigorous experimental design with innovative technical solutions, the research addresses critical challenges in airborne pollutant measurement while advancing methodologies for spatial and temporal data acquisition.

One of the challenges addressed is mitigating the influence of UAV rotor-induced airflow on both the measuring instruments and the surrounding environment. This paper introduces a custom-designed protective shield for the particulate matter sensor, which, according to the evaluated data, effectively prevents measurement distortion while preserving the dynamics of the readings, ensuring measurement validity even during takeoff and dynamic flight maneuvers. As a result, the sensed data can be considered valid immediately upon takeoff.

The findings underscore the complex interplay between terrain, meteorological conditions, and pollutant dispersion. Elevated $PM_{2.5}$ and CO_2 levels above forest canopies (40–60 m) suggest contributions from both biological sources (e.g., pollen) and industrial emissions, with wind direction acting as a critical transport mechanism. Vertical profiling near highways revealed rapid particle dispersion at higher altitudes, consistent with simulations of UAV-driven spatial pattern analysis. These results highlight the necessity of 3D mapping to capture heterogeneous pollution gradients, particularly in topographically diverse regions.

Advanced visualization techniques-including 4D scatter plots, interactive web-based charts, and volumetric cloud rendering-enabled intuitive interpretation of multi-dimensional datasets. The integration of real-time machine learning models (hybrid 3D AQI mapping systems) could further enhance the immediacy and precision of pollution tracking. Such tools not only support scientific analysis but also improve public communication of air quality risks.

Acknowledgement. The authors gratefully acknowledge the financial support from the Technology Agency of the Czech Republic (TAČR Naccas) under grant TN02000009/11 for the "Compact Optical Hyperspectral Camera FREYA II for CubeSat Missions, " and from the European Union and the Czech Republic's state budget under project LasApp CZ.02.01.01/00/22_008/0004573.

References

1. Mannucci, P.M., Harari, S., Martinelli, I., et al.: Effects on health of air pollution: a narrative review. Intern. Emerg. Med. **10**, 657–662 (2015)
2. Jońca, J., Pawnuk, M., Bezyk, Y., Arsen, A., Sówka, I.: Drone-assisted monitoring of atmospheric pollution—a comprehensive review. Sustainability **14**(18) (2022)
3. Cichowicz, R., Dobrzański, M.: Spatial analysis (measurements at heights of 10 m and 20 m above ground level) of the concentrations of particulate matter (PM10, PM2.5, and PM1.0) and gaseous pollutants (H2S) on the university campus: a case study. Atmosphere **12**(1) (2021)

4. Ho, Y.H., Lin, Y.C.: Searching for the fine particulate matter (PM 2.5) pollutant emission source using a drone. Measurement **232**(15) (2024)
5. Duangsuwan, S., Prapruetdee, P., Subongkod, M., Klubsuwan, K.: 3D AQI mapping data assessment of low-altitude drone real-time air pollution monitoring. Drones **6**(8) (2022)
6. Rüdiger, J., Tirpitz, et al.: Implementation of electrochemical, optical and denuder-based sensors and sampling techniques on UAV for volcanic gas measurements: examples from Masaya, Turrialba and Stromboli volcanoes. Atmos. Meas. Tech. (11), 2441–2457 (2018)
7. Belkacem, I., Khardi, S., Helali, A., Slimi, K., Serindat, S.: The influence of urban road traffic on nanoparticles: roadside measurements. Atmos. Environ. (242) (2020)
8. Abhijith, K.V., Kumar, P.: Field investigations for evaluating green infrastructure effects on air quality in open-road conditions. Atmos. Environ. (201) (2018)
9. Dubey, R., Patra, A.K., Joshi, J., Blankenberg, D., Penchala, A., Lee, X.: Vertical distribution of particulate matter near a national highway and influence of roadside tree canopy: a drone/UAV based study. Atmos. Environ. (336) (2024)
10. Hedworth, H.A., Sayahi, T., Kelly, K.E., Saad, T.: The effectiveness of drones in measuring particulate matter. J. Aerosol Sci. (152) (2021)
11. Villa, T.F., Salimi, F., Morton, K., Morawska, L., Gonzalez, F.: Development and validation of a UAV based system for air pollution measurements. Sensors **16**(12) (2016)
12. Samad, A., Diego, A.F., Ioannis, Ch., Ulrich, V.: Concept of using an Unmanned Aerial Vehicle (UAV) for 3D investigation of air quality in the atmosphere—example of measurements near a roadside. Atmosphere **13**(5), 663 (2022)

Controlling a Magnetorheological (MR) Fluid Brake by Simultaneously Adjusting the Electrical Current and the Fluid-Film Thickness

Salwan Obaid Waheed Khafaji[1,2(✉)], Hasan H. Ali[3], Fawaz F. Al-Bakri[4], and Noah D. Manring[5]

[1] Mechanical Power Technical Engineering Department, College of Engineering and Technologies, Al-Mustaqbal University, 51001 Babylon, Iraq
salwan.obaid@uomus.edu.iq

[2] Al-Mustaqbal Energy Research Center, Al-Mustaqbal University, 51001 Babylon, Iraq

[3] Medical Instrumentation Techniques Engineering, College of Engineering and Technology, Al-Mustaqbal University, 51001 Babylon, Iraq

[4] Department of Biomedical Engineering, College of Engineering, University of Babylon, Babylon, Iraq

[5] Mechanical and Aerospace Engineering, University of Missouri at Columbia, Columbia City, MO, USA

Abstract. In this paper, a method of controlling a MR fluid brake by simultaneously altering both the input current to the electric coil, and the MR fluid thickness between the stationary and rotating part of a single-disk brake was presented. In the development of the nonlinear torque-model for the brake, nondimensional analysis was used to generalize the problem for any brake configuration with similar attributes, while providing design guidance for making the energized and non-energized brake components comparable in strength. In order to control the brake speed, two saturating proportional-integral (PI) controllers were used in parallel: one for adjusting the MR fluid thickness, and the other for adjusting the input current to the electric coil. It was shown in this paper that the controller always achieves a steady-state output with zero error; however, the combination of fluid thickness and current is non-unique and depends upon initial conditions and saturation events that occur during the transient response. In conclusion, the control method proposed in this paper is shown to extend the range of torque capacity for the brake without increasing the radial envelope for the brake itself.

Keywords: MR Fluid Brake · Torque Capacity · Brake Radial Envelope · Torque Control

D. D. Cioboată and J. Machado (Eds.): ICoRSE 2025, LNNS 1592, pp. 133–149, 2025.
https://doi.org/10.1007/978-3-032-02508-1_12

1 Introduction

1.1 Background

Rotating mechanical systems are typically designed to include a brake or a rotational damping device for the purposes of controlling speed. Historically, the braking device has been constructed using a friction brake design, which has often been accompanied with adverse wear and premature failure. In order to prevent material wear, and to apply direct electrical control to the brake, magnetorheological (MR) fluid brakes have been introduced for rotational systems. MR fluid brakes are designed based upon principles that govern viscous dampers where fluid shear-stress is used to resist the motion of mechanical parts. In these systems the strength of the viscous damper is proportional to the fluid viscosity. The novel property of MR fluids is that their apparent viscosity can be significantly altered by exposing the fluid to an electrically induced magnetic field. By controlling the strength of the magnetic field, the MR fluid viscosity is easily controlled with a very fast response time and a high torque-to-volume ratio. Magnetorheological fluid technology holds great promise for many electrically controlled, mechanical systems including brakes and clutches [1, 2], dampers for vibration control [3], robotic applications [4], and haptic devices [5].

1.2 Literature Review

A comprehensive literature review for MR fluid brakes has been presented by Rossa and colleagues [6]. In their review, the authors presented the most commonly used models for estimating MR fluid shear-stress which is known as the Herschel-Bulkley and Bingham model [7, 8]. They also describe the many applications for MR fluid technologies and they discuss past areas of research which include optimal design configurations. Those optimal design studies were focused on minimizing the rotating inertia of the system [9] and the required power of operation [10], and increasing the torque density by using multiple disks [11]. The reader is encouraged to peruse these references for other examples. An adaptive control for an MR fluid brake was designed in [12]. A dynamic model was created first and the effect of the temperature and speed was taken into account. The results were validated experimentally. It was shown that the velocity has an impact on the current and torque gain while the temperature only affects the fluid viscosity. In [13], a closed loop control system was designed for an MR fluid brake. Proportional controller along with Gauss function was applied for controlling the angular position.

Within the body of literature known to the authors, the control for MR fluid brakes has been achieved by altering the input current to the electrical coil. This is the obvious control input, however, an additional option exists for controlling the brake by simultaneously altering both the input current and the MR fluid thickness. This simultaneous control may be used to achieve a wider range of torque without increasing the radial envelope, and to compensate for the saturation limits that occur within electromagnetic fields. This current paper has been written to demonstrate this novel control method for a single disk design, and to show that good dynamic response may be achieved using two proportional-integral (PI) controllers in parallel.

1.3 Objectives

In this paper, a new technique of controlling a MR fluid brake by simultaneously altering both the input current to the electric coil, and the MR fluid thickness between the stationary and rotating part of a single-disk brake is presented. In the development of the nonlinear torque-model for the brake, nondimensional analysis was introduced to generalize the problem for various brake configurations with similar attributes. It also provides a design procedure for making the energized and non-energized brake components comparable in strength. In order to control the brake speed, two saturating PI controllers are used in parallel. the first one is used for controlling the MR fluid thickness, while the other is used for adjusting the input current to the electric coil. It is shown in this paper that the controller always achieves a steady-state solution with zero error; however, the combination of fluid thickness and current is non-unique and depends upon initial conditions and saturation events that occur during the transient response. In conclusion, the control method proposed in this paper extends the range of torque capacity for the brake without increasing the radial envelope and the number of frictional discs for the brake itself. That can be considered as a braking torque optimization way keeping brake volume constant.

1.4 Descriptions

A schematic for a single disk, MR fluid brake is shown in Fig. 1. The left-hand-side of this figure shows a non-energized rotating disk connected to a shaft. The angular velocity of the disk is given by ω and the torque generated by the brake is shown by the symbol T. Magneto-rheological (MR) fluid is shown to exist between the rotating disk and a stationary surface. The thickness of the MR fluid is given by the dimension h. As it turns out, in the absence of a magnetic field, MR fluid sustains a shear stress that is proportional to its strain rate which means that the non-energized MR fluid is Newtonian and characterized by a certain fluid viscosity. In this non-energized state the shear stress in the MR fluid generates a brake torque that is proportional to the fluid viscosity and the angular velocity, while being inversely proportional to the MR fluid thickness h.

The right-hand-side of Fig. 1 shows the same brake geometry but with a magnetic field that has been generated by passing current through the electric coil which is shown to be wrapped around the brake. The current is depicted in Fig. 1 using the symbol which flows through the coil into the paper at the top, and out of the paper at the bottom. The novel property of a MR fluid is that the viscosity of the fluid is increased when exposed to a magnetic field. This additional viscosity of the fluid increases with increased magnetic-flux density which is proportional to the current passing through the electric coil. In other words, increasing the current through the coil increases the viscosity of the MR fluid, which then increases the resistive torque generated by the brake.

In this paper, a closed-loop control for the MR fluid brake was designed by simultaneously adjusting the MR fluid thickness h and the coil current i. The actuator for adjusting the fluid thickness and the driver for adjusting the coil current are not depicted in Fig. 1, and will be idealized in this research for illustrating the control method. Note: the linear actuator for adjusting the fluid thickness will be fairly small (no additional radial space required) due to the fact that the force requirements will be low. There are

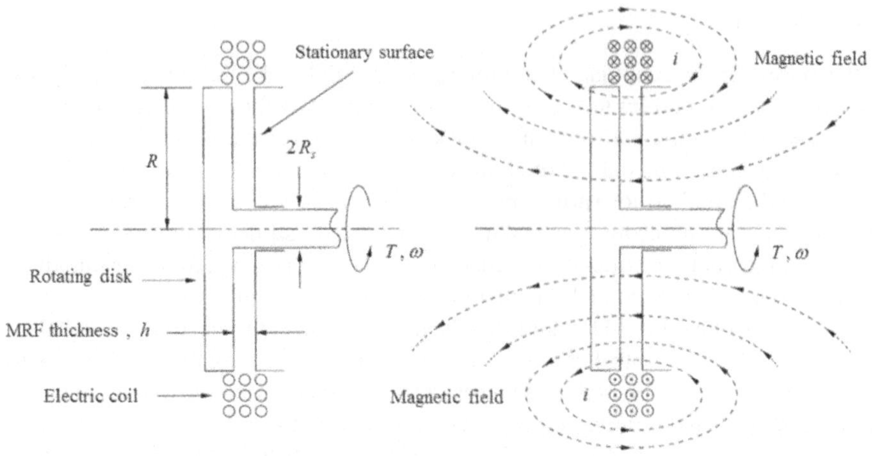

Fig. 1. Schematic of the single disk, MR fluid brake

three reasons for designing a controller that adjusts both h and i as opposed to making only one adjustment. First, an adjustment for the fluid thickness cannot be used to fully stop the brake, and therefore a control for the electric current is necessary if zero velocity is desired. Second, the fluid thickness and the electric current are both limited in their range of operation and when one parameter reaches a saturation point the other parameter continues to regulate the brake speed. Third, by adjusting both parameters a wider range of braking torque may be achieved. For most operating conditions, the controller simultaneously adjusts both h and i; however, a simplified way to think about this control is to consider the adjustments for h to be applied primarily for controlling high angular-velocities while adjustments for i are applied primarily for controlling low angular-velocities. This control design will be clarified in the sections that follow.

2 Analysis

2.1 Dynamic System

Figure 2 shows a schematic of the rotating system to be controlled by the MR fluid brake. In this schematic, the left-hand-side represents the MR fluid brake which exerts a control torque T on the rotating flywheel. The flywheel is shown to have a mass moment-of-inertia given by J. Using a flywheel is an analogy which can be used for different practical application. On the right-hand-side of this schematic, the input torque to the rotating system is given by the load torque T_L. The system rotates about the central axis with an instantaneous angular-velocity ω.

According to Newton's second law, the equation of motion for the rotating system shown in Fig. 2 is given by

$$J\frac{d\omega}{dt} = T_L - T \tag{1}$$

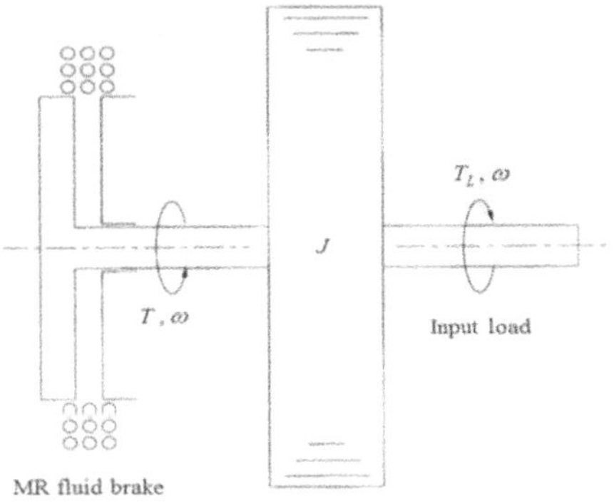

Fig. 2. Schematic of the rotating system being controlled by the MR fluid brake

where t is time. In the following paragraphs an equation for describing the brake torque T will be developed. In a subsequent section of this paper, a feedback control system will be designed for regulating the rotational speed ω by altering the brake torque itself.

2.2 Brake Equations

In this section the mechanical and electrical models for the MR fluid brake were derived and presented. Figure 3 shows a sectional view of the single disk where an infinitesimal torque dT is generated on the disk surface by an infinitesimal fluid-shear force df. Using Fig. 3 this torque may be expressed as:

$$dT = r\,df \tag{2}$$

where r is the radial location of the infinitesimal shear force. Recognizing that the infinitesimal shear force is given by $df = 2\pi\tau r dr$, where τ is the shear stress of the fluid at the surface of the disk, the total torque exerted on the brake disk by the MR fluid may be represented as:

$$T = 2\pi \int_{R_s}^{R} \tau\, r^2\, dr \tag{3}$$

where R is the outer radius of the disk and is the radius of the shaft.

In order to solve Eq. (3), it is necessary to present a shear stress model for the MR fluid. As it turns out, MR fluid is most analogous to a Bingham viscoplastic material that behaves as a solid during low stress conditions, but flows like a fluid during high stress conditions [8]. The most common model used in the literature for describing the shear stress of a Bingham material is known as the Herschel-Bulkley model [1, 6, 7] and when applied to MR fluids this shear-stress model is given by:

$$\tau = \frac{\mu\, r\, \omega}{h} + k\, B^{\alpha} \tag{4}$$

where μ is the absolute viscosity of the non-energized MR fluid, B is the magnetic-flux density generated by the electric coil, and k and α are material properties for a specific MR fluid. As it turns out is near unity for many MR fluids [1, 6] and the magnetic-flux density is proportional to the electric current for most operating conditions of the coil [15]. If we use a MR fluid with $\alpha = 1$, and while operating within the linear range of the magnetic-flux density as it varies with current, then the induced shear-stress resulting from the magnetic field may be written as:

$$k\ B^{\alpha} = \frac{\tau_B}{i_{max}}\ i \tag{5}$$

where τ_B is the maximum shear yield-stress that may be added to the MR fluid when the maximum allowable current i_{max} is applied to the coil. Estimates for τ_B are readily obtained from MR fluid suppliers [16].

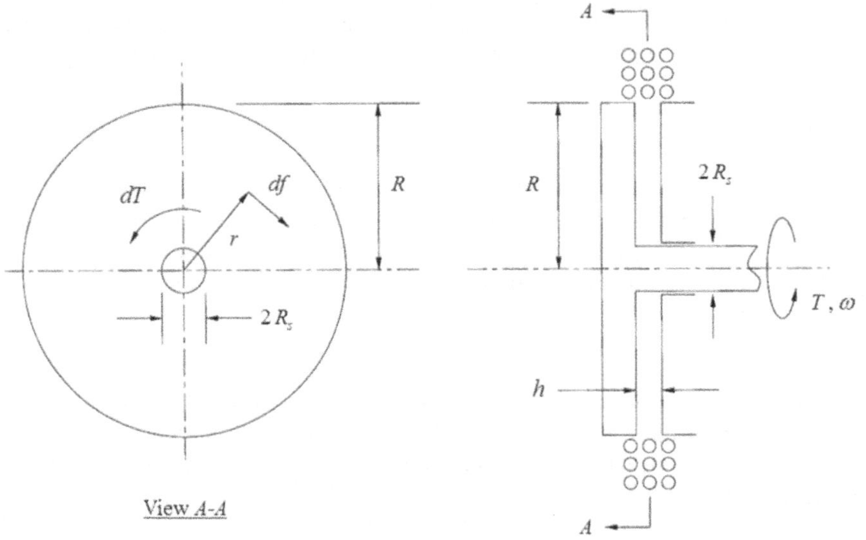

Fig. 3. Geometry and infinitesimal force diagram for the single disk, MR fluid brake

Substituting Eqs. (4) and (5) into Eq. (3) and evaluating the integral produces the following result for the total torque that is generated by the MR fluid brake:

$$T = \frac{\pi}{2}\frac{\mu\,\omega}{h}R^4\left[1-\left(\frac{R_s}{R}\right)^4\right] + \frac{2\pi}{3}\tau_B\frac{i}{i_{max}}R^3\left[1-\left(\frac{R_s}{R}\right)^3\right]. \tag{6}$$

The first term on the right-hand-side of Eq. (6) represents the resistance applied by the non-energized brake while the second term represents the additional resistance that is applied by exposing the MR fluid to the magnetic field.

2.3 Nondimensional Analysis

Using dimensionless quantities reduces the number of variables and makes the model more general [17]. It is convenient to nondimensionalize the governing equations by using the following definitions:

$$\begin{aligned} h &= h_o\,\hat{h}\ , \quad i = i_{max}\,\hat{i}\ , \quad R_s = R\,\hat{R}_s\ , \\ T &= T_o\,\hat{T}\ , \quad T_L = T_o\,\hat{T}_L\ , \quad t = t_o\,\hat{t} \text{ and } \omega = \omega_o\,\hat{\omega} \end{aligned} \tag{7}$$

where h_O is the nominal MR fluid thickness, T_O is the nominal load torque, ω_O is the nominal angular velocity, t_O is a characteristic time constant for the dynamic response, and all other parameters have been previously defined. In Eq. (7) the symbols with carets are dimensionless and of order one. Substituting Eqs. (6) and (7) into Eq. (1) produces the following nondimensional equation of motion for the rotating system shown in Fig. 2:

$$\frac{d\hat{\omega}}{d\hat{t}} = \hat{T}_L - \frac{\hat{\omega}}{\hat{h}} - \hat{i}. \tag{8}$$

In order to produce this equation, the following definitions were used for the nominal load torque and the characteristic time:

$$T_o = \frac{\pi}{2}\frac{\mu\,\omega_o}{h_o} R^4\left(1-\hat{R}_s^4\right) \text{ and } t_o = \frac{J\,\omega_o}{T_o}. \tag{9}$$

In addition to using these definitions, the following design constraint was enforced:

$$\frac{4}{3}\frac{\tau_B}{\mu\,\omega_o}\frac{h_o}{R}\frac{1-\hat{R}_s^3}{1-\hat{R}_s^4} = 1. \tag{10}$$

Physically speaking, this design constraint ensures that the non-energized braking torque and the magnetic-field braking torque are similar in magnitude when applied. As shown in the Appendix, this design constraint is easily satisfied using realistic numbers.

Equation (8) represents evidence on the simplicity of the governing equation. In this nondimensional form, which captures all of the physics of the problem, the rotational speed of the system depends upon only three things: 1) the nondimensional load torque $\hat{T}_L$, 2) the nondimensional MR fluid thickness$\hat{h}$, and 3) the nondimensional current$\hat{i}$. Furthermore, all of these parameters are of order one for reasonable brake operation. In the section that follows, a control law will be presented for simultaneously adjusting and in order to regulate the angular velocity $\hat{\omega}$.

3 Control Design

3.1 General

The objective of the proposed controller is to regulate the speed of the rotational system shown in Fig. 2 by altering the MR fluid brake torque $\hat{T}$, which is adjusted by simultaneously altering the MR fluid thickness $\hat{h}$ and the electric coil current $\hat{i}$. A block diagram for this controller is shown in Fig. 4 with two saturating PI controllers and another non-linear block which describes the inverse proportionality of the MR fluid thickness for the non-energized torque that is developed by the brake. The desired speed for the rotating system is shown in Fig. 4 by the symbol $\hat{\omega}_d$ and the system error is given by $\hat{\varepsilon}$. Again, the carets indicate that these are nondimensional quantities. The Laplace operator *1/s* is shown in Fig. 2 to denote integration.

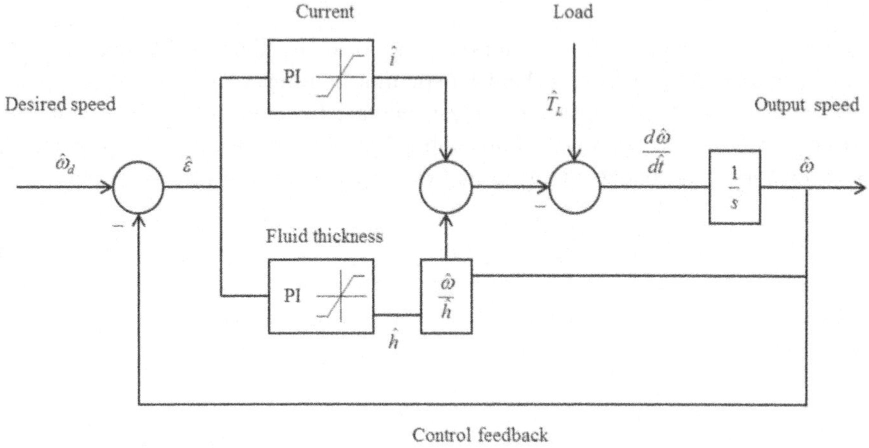

Fig. 4. Block diagram for the MR fluid brake control showing two PI controllers with saturation and a nonlinear relationship for $\hat{h}$

Due to physical constraints, the fluid thickness $\hat{h}$ and the coil current $\hat{i}$ are likely to saturate during extreme speed conditions. For instance, a linear actuator will be used to control the MR fluid thickness between a maximum and minimum value allowed by the brake. For very slow speeds it will be desirable for $\hat{h}$ to be as small as possible in order to maximize the non-energized brake torque; however, the fluid thickness cannot be allowed to approach zero. A minimum non-zero value for $\hat{h}$ is required to keep the rotating disk from making metal-to-metal contact with the stationary surface during operation. Similarly a maximum value for the fluid thickness is needed during high speed operation, however, the maximum value for is limited by the maximum stroke length for the linear actuator. In this work, the nondimensional MR fluid thickness was bounded between 0.1 and 3. As previously noted, we want the input current to remain within the linear region of the magnetic-flux density curve and should therefore not exceed the dimensional value of i_{max} where the nonlinearity in this curve begins [12]. Furthermore, we will limit the minimum allowable current to zero since there is no physical advantage of reversing the direction of current through the coil. In other words, the physical constraints for the fluid thickness and the electric current are given by

$$0.1 < \hat{h} < 3 \quad \text{and} \quad 0 < \hat{i} < 1. \tag{11}$$

These saturation characteristics are shown graphically in the PI control blocks of Fig. 4 and will be used in the following paragraphs to describe the overall control design.

PI Controllers. In order to simultaneously adjust the fluid thickness and the electric current for regulating system speed, a saturating PI control is chosen for both adjustments. The proportional control is the basic controller, while the integral control is added to ensure a zero steady-state error while regulating a constant speed.

Mathematically, the PI control law for adjusting the fluid thickness is given by

$$\hat{h} = \hat{K}_p \, \hat{\varepsilon} + \hat{K}_i \int \hat{\varepsilon} \, d\hat{t} \tag{12}$$

where $\widehat{K}_p$ and $\widehat{K}_i$ are the nondimensional proportional gain and integral gain respectively, and the nondimensional error is given by

$$\hat{\varepsilon} = \hat{\omega}_d - \hat{\omega}. \tag{13}$$

As previously noted, Eq. (12) is used to calculate the controlled fluid-thickness within the bounds noted in Eq. (11). If the fluid thickness computed from Eq. (12) that exceeds 3, then the fluid thickness is simply set to 3. On the other hand, if Eq. (12) computes a fluid thickness that is less than 0.1 then the fluid thickness is set to 0.1. In order to prevent integrator windup within the PI control during saturation, the integral gain is turned on and off as follows:

$$\hat{K}_i = \begin{cases} 0 & \text{Eq.}(12) < 0.1 \\ \hat{K}_i & 0.1 < \text{Eq.}(12) < 3 \\ 0 & 3 < \text{Eq.}(12) \end{cases}. \tag{14}$$

As previously noted, the overall output of the fluid-thickness control is limited by the bounds shown in Eq. (11).

Similarly, the PI control law for adjusting the electric current is given by

$$\hat{i} = -\hat{\Phi}_p \, \hat{\varepsilon} - \hat{\Phi}_i \int \hat{\varepsilon} \, d\hat{t} \tag{15}$$

where $\widehat{\phi}_p$ and $\widehat{\phi}_i$ are the nondimensional proportional gain and integral gain respectively, and the nondimensional error is given in Eq. (13). Again, Eq. (15) is used to calculate the controlled current within the bounds noted in Eq. (11). If Eq. (15) computes a current that exceeds unity then the current is simply set to 0. On the other hand, if Eq. (15) computes a current that is less than zero, then the current is set to. Once again, in order to prevent integrator windup within the PI control during saturation, the integral gain is turned on and off as follows:

$$\hat{\Phi}_i = \begin{cases} 0 & \text{Eq.}(15) < 0 \\ \hat{\Phi}_i & 0 < \text{Eq.}(15) < 1 \\ 0 & 1 < \text{Eq.}(15) \end{cases}. \tag{16}$$

And again, the overall output of the current control is limited by the bounds in Eq. (11).

3.2 Non-Unique Solutions

By setting the left-hand-side of Eq. (8) to zero, the steady-state solution for the speed of the system may be determined as a function of the MR fluid thickness and the current that is applied to the electric coil. This solution is given by

$$J\frac{d\omega}{dt} = T_L - T. \tag{17}$$

The purpose for presenting the steady-state result is to illustrate that there is no unique combination of MR fluid thickness and electric current that establishes the steady-state condition for the brake. Rather, an infinite number of combinations exist within the bounds of operation that are stipulated in Eq. (11). Subsequent discussions relative to Figs. 6 and 7 show that the proposed controller always finds a solution with zero steady-state error; however, the exact combination of $\widehat{h}$ and $\widehat{i}$ depends upon the initial conditions from which the controller begins seeking this solution, and whether or not any saturation points are encountered along the way.

3.3 Linearization and Control-Gain Selection

The controlled system presented in this paper describes a nonlinear system that is partly discontinuous. In order to check for some level of stability and to provide guidance for selecting control gains, it is helpful to linearize the system and to examine cases when no saturation limits are reached. For linearized conditions where $\widehat{\omega} \approx 1$ and $\widehat{h} \approx 1$, and where the two PI controls in Eqs. (12) and (15) are used, it may be shown that the dynamic response for the regulated system is governed by the following second-order differential equation:

$$\frac{d^2\widehat{\omega}}{d\hat{t}^2} + (1 + \hat{K}_p + \hat{\Phi}_p)\frac{d\widehat{\omega}}{d\hat{t}} + (\hat{K}_i + \hat{\Phi}_i)\,\widehat{\omega} = (\hat{K}_i + \hat{\Phi}_i)\,\widehat{\omega}_d \,. \tag{18}$$

Equation (18) shows that the integral controllers successfully eliminate the steady-state error. This is shown by setting the time derivatives in Eq. (18) to zero and noting that $\widehat{\omega} = \widehat{\omega}_d$. Although not proven here, this property holds true even for the nonlinear system as demonstrated later in the simulation studies.

Another thing to observe from Eq. (18) is that all of the coefficients on the left-hand-side are positive, which means that the linearized and continuous system is stable. Although we present no proof of stability for the nonlinear system, the exercised nonlinear model appears to be robust and has demonstrated no instabilities in our study thus far. Again, this will be demonstrated in the simulation studies that follow.

Finally, by inspecting Eq. (18) it may be shown that the nondimensional undamped natural-frequency and the damping ratio for the the linearized system are given by

$$\widehat{\omega}_n = \sqrt{\hat{K}_i + \hat{\Phi}_i} \quad \text{and} \quad \zeta = \frac{1 + \hat{K}_p + \hat{\Phi}_p}{2\sqrt{\hat{K}_i + \hat{\Phi}_i}} \,. \tag{19}$$

In order to provide an adequate rise time for the transient response, the nondimensional undamped natural-frequency was designed so that $\widehat{\omega}_n = 4$. Similarly, design the damping ratio was chosen such that $\zeta = \sqrt{2}/2$. The following gain selections satisfy these conditions and will be used in the simulations which follow: $\widehat{K}_p = 2.33$, $\widehat{K}_i = 4$, $\widehat{\phi}_p = 2.33$, and $\widehat{\phi}_i = 12$.

4 Results and Discussion

4.1 High Speed Control

Figure 5 represents a transient response for the controlled brake system where the applied load torque is given by $\hat{T}_L = 1$ and the desired shaft speed is given by $\hat{\omega}_d = 2$. The initial shaft speed is given by $\hat{\omega} = 1$. In order to increase the shaft speed, the controller seeks to provide a larger MR fluid thickness to reduce the non-energized brake torque. The initial error in the control signal is so large that the linear actuator soon saturates at the point where $\hat{h} = 3$, which is the maximum allowable thickness that can be provided by the linear actuator due to a limited actuator stroke. While the fluid thickness remains saturated, the current in the electric coil is also saturated at zero which ensures that no energized torque is applied to the brake.

As shown in Fig. 5 the system is stable and the settling time occurs within 4 time constants. The transient response for speed exhibits very little overshoot. It is also significant to note that this high-speed condition could not have been achieved without the ability to control the fluid thickness $\hat{h}$.

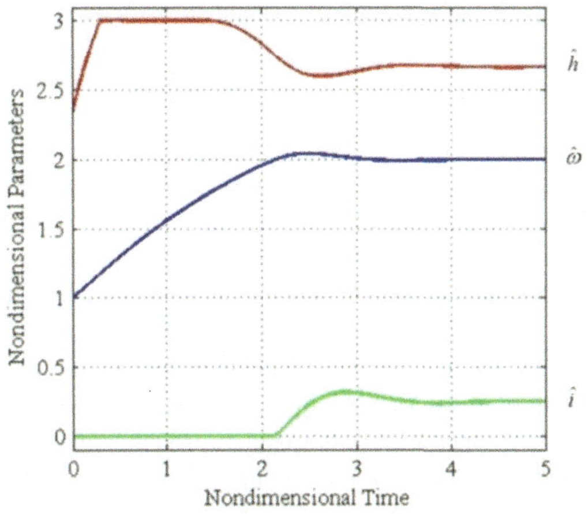

Fig. 5. Transient response illustrating high speed control for $\hat{T}_L = 1$ and $\hat{\omega}_d = 2$. The blue line is shaft speed, the red line is MR fluid thickness, and the green line is current.

4.2 Moderate Speed Control

Figure 6 represents a transient response for the controlled brake system where the applied load torque is given by $\hat{T}_L = 1$ and the desired shaft speed is given by $\hat{\omega}_d = 1$. The initial shaft speed is given by $\hat{\omega} = 0$. In order to increase the shaft speed, the controller seeks to provide a larger MR fluid thickness to reduce the non-energized brake torque. Since the desired speed is moderate, no saturation occurred in the linear actuator. During the initial increase of speed, the current in the electric coil was saturated at zero to ensure that no energized torque is applied to the brake. During these minimal braking conditions, the shaft speed increases to the desired level $\hat{\omega}_d = 1$. Once the desire speed is reached, the fluid thickness and the electric current adjust themselves to create stronger braking conditions in order to hold the desired shaft speed. As shown in Fig. 6, the system is stable and the settling time occurs within **4-time** constants. The response shown in Fig. 6 is very similar to that of Fig. 5 where the transient response for speed exhibits acceptable overshoot.

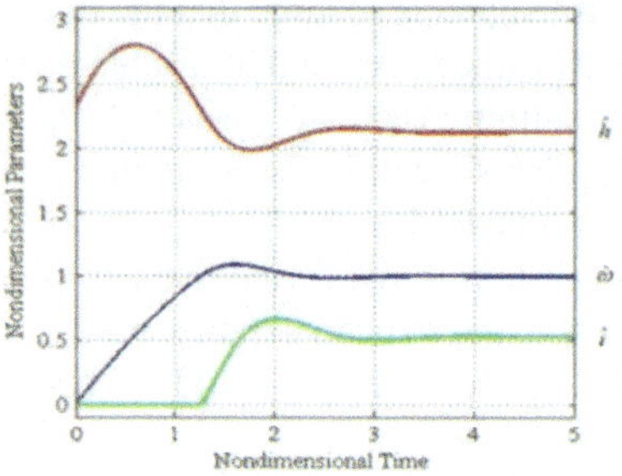

Fig. 6. Transient response illustrating moderate speed control with a low-speed initial condition, and where $\hat{T}_L = 1$ and $\hat{\omega}_d = 1$. The blue line is shaft speed, the red line is MR fluid thickness, and the green line is current.

Figure 7 represents the transient response for the controlled brake system where the applied load torque is given by $\hat{T}_L = 1$ and the desired shaft speed is given by $\hat{\omega}_d = 1$; however, the initial shaft speed is now given by $\hat{\omega} = 2$. This +. In other words, the controller needs to make initial adjustments to slow the shaft down rather than speed the shaft up. In order to slow the shaft down, the controller starts with a small, saturated MR fluid thickness to increase the non-energized brake torque. At the same time, the electric current is saturated at its maximum level in order to apply a maximum energized brake torque. With both controllers applying a maximum brake torque to the shaft, the shaft speed drops quickly. Once the shaft speed approaches its desired value, the fluid-thickness control relaxes to its nominal operating condition and the magnetic brake de-energizes. Like the previous two examples, this response is stable and timely with a settling time of near 4 time constants. The speed response exhibits a bit more overshoot (undershoot) compared to the previous examples.

The important thing to note about the responses shown in Figs. 6 and 7 is that while the control objective is the same for both examples, the steady-state solutions for $\hat{h}$ and $\hat{i}$ are different. This non-unique solution was discussed relative to Eq. (17) and has now been demonstrated with the same controller using different initial conditions. At this point one might question which of the two solutions is more advantageous from, say, an energy point of view. In order to consider this question, a model for the energy

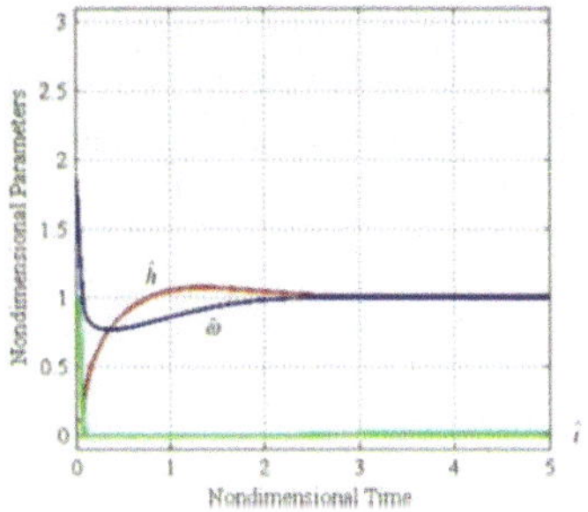

Fig. 7. Transient response illustrating moderate speed control with a high-speed initial condition, and where $\hat{T}_L = 1$ and $\hat{\omega}_d = 1$. The blue line is shaft speed, the red line is MR fluid thickness, and the green line is current.

consumption of the linear actuator and the electric coil would need to be evaluated. although not pursued here, based upon these energy models an optimal control solution may be pursued.

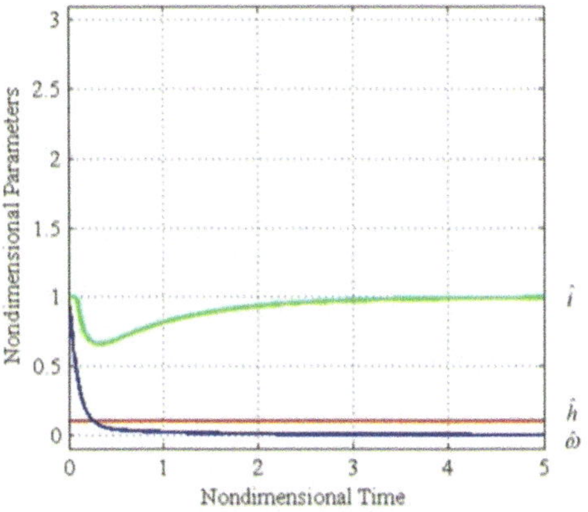

Fig. 8. Transient response illustrating low speed control for $\hat{T}_L = 1$ and $\hat{\omega}_d = 0$. The blue line is shaft speed, the red line is MR fluid thickness, and the green line is current.

4.3 Low Speed Control

Figure 8 illustrates a transient response for the controlled brake system where the applied load torque is given by $\hat{T}_L = 1$ and the desired shaft speed is given by $\hat{\omega}_d = 0$. The initial shaft speed is given by $\hat{\omega} = 1$. In order to decrease the shaft speed, the controller seeks to provide the smallest MR fluid thickness possible in order to apply the maximum non-energized brake torque. As such, the fluid thickness saturates at its minimum value $\hat{h} = 0.1$. It is important to note that the non-energized brake torque is insufficient for stopping the shaft since the development of this torque requires a certain amount of shaft speed. In order to stop the shaft, the electric coil must be energized to its maximum value $\hat{i} = 1$. As shown in Fig. 8, the maximum current is developed after 4 time constants at which point the desired shaft speed of zero has been achieved. Again, the response for the shaft speed is very timely and exhibits no overshoot (undershoot).

4.4 Tracking Control

Figure 9 represents a tracking response for the controlled brake system where the applied load torque is given by $\hat{T}_L = 1$ and the desired shaft speed is given by $\hat{\omega}_d = 1 + 0.2\ \sin(2\pi\,\hat{t})$. This desired shaft speed was arbitrarily selected to illustrate the tracking behavior of the control system. As shown in this response, both the fluid thickness and the electric current are oscillating in order to track the desired shaft speed. The fluid thickness oscillates without saturation, while the electric current saturates at both maximum and minimum values. As shown in Fig. 9 the system demonstrates good stability; however, the desired tracking frequency clearly exceeds the bandwidth frequency of the system with an amplitude ratio of 0.5 and a phase lag of 90°.

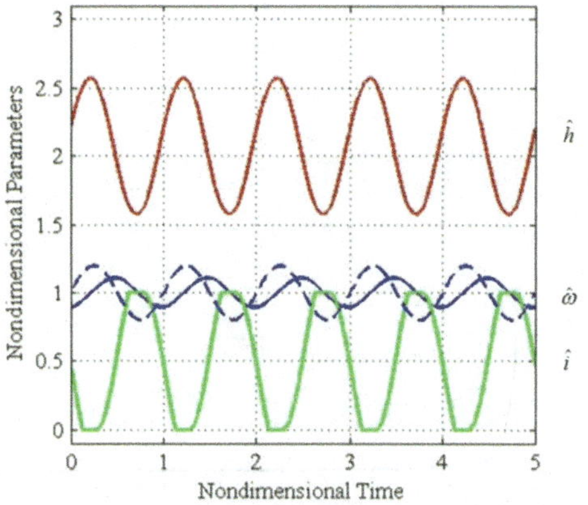

Fig. 9. Tracking response illustrating moderate, sinusoidal speed control for $\hat{T}_L = 1$ and $\hat{\omega}_d = 1 + 0.2\ \sin(2\pi\,\hat{t})$. The dashed blue line is the desired speed while the solid blue line is actual speed, the red line is MR fluid thickness, and the green line is current.

5 Conclusion

The following conclusions are supported by the analysis, results, and discussion presented in the paper:

1. The MR fluid brake is capable of generating torque from two sources: 1) a non-energized source resulting from the Newtonian viscosity, angular velocity, and MR fluid thickness; and 2) an energized source resulting from the MR fluid which increase its viscosity when exposed to a magnetic field. The non-energized brake torque is

controlled by altering the MR fluid thickness. The energized brake torque is controlled by altering the current passing through the electric coil, which then alters the strength of the magnetic field.

2. By simultaneously controlling the MR fluid thickness and the electric current, a large range of brake torque is achieved without increasing the radial envelop for the brake, and saturation conditions in one controller are compensated for by the other controller. High angular velocities of the brake are primarily controlled by increasing the MR fluid thickness, while low angular velocities are primarily controlled by increasing the electric current.
3. By enforcing the design constraint of Eq. (10), the strength of the non-energized torque and the energized torque can be made comparable. The ability to satisfy this design constraint has been illustrated in the Appendix.
4. As shown by Eq. (8), the governing equation for describing the angular velocity of the brake can be made quite elegant using nondimensional analysis. The physics of this problem are captured by three things: 1) the applied nondimensional load-torque, 2) the nondimensional MR fluid thickness, and 3) the nondimensional electric current. An alteration in any one of these three things will impact the dynamic and steady-state response of the brake. Using the definitions in Eq. (7) the final results may be re-dimensionalized for the practicing engineer.
5. For this paper, two PI controls have been used to simultaneously adjust the MR fluid thickness and the electric current. These linear controllers have been selected because they are known to produce an easily tuned response with zero steady-state error. These linear controllers have been shown to perform well for controlling the nonlinear system described in this paper. Figures 5, 6, 7, 8 and 9 show good transient responses for regulating a constant speed (high, moderate, and low), and good stability while seeking to track a sinusoidal input.
6. Because both the fluid thickness and the electric current are being simultaneously controlled to achieve a single speed objective, a unique solution does not exist for the combination of fluid thickness and current that will achieve our goal. Rather, the combination of fluid thickness and electric current is dependent upon the initial conditions, and whether or not any saturation has been experienced along the way. In the end, however, the control objective is always achieved in a timely and stable way.

In summary, the control system proposed in this paper demonstrated good controllability for the MR fluid brake. The nondimensional form of this analysis has been used to generalize our results for any MR fluid brake design that is similar to the system shown in Figs. 1 and 2. It is important to note that all nondimensional solutions may be re-dimensionalized using the definitions in Equation.

Appendix

The following design parameters are presented as typical characteristics for a MR fluid brake. These values are physically realistic and may be used to satisfy the design constraint shown in Eq. (10). Note: design parameters for the electric coil have not been included here.

Description	Symbol	Value	Units
Nominal MR fluid thickness	h_o	50	μm
Mass moment-of-inertia	J	0.064	$Kg.m^2$
Disk radius	R	95	mm
Shaft radius	R_s	20	mm
Nominal brake torque	T_o	53	N.m
Characteristic time constant	t_o	0.25	s
MR fluid viscosity	μ	0.1	Pa s
MR fluid maximum shear stress	$4\ \tau_B$	30	kPa
Nominal shaft speed			rpm

References

1. Li, W.H., Du, H.: Design and experimental evaluation of magnetorheological brake. Int. J. Adv. Manuf. Technol. **21**(508–515), 508–515 (2003)
2. Haung, J., Zhang, J.Q., Yang, Y., Wei, Y.Q.: Analysis and design of a cylindrical magneto-rheological fluid brake. J. Mater. Process. Technol. **129**, 559–562 (2002)
3. Duan, Y.F., Ni, Y.Q., Ko, J.M.: Cable vibration control using magnetorheological dampers. J. Intell. Mater. Syst. Struct. **17**(4), 321–325 (2006)
4. Kikuchi, T., Oda, K., Furusho, J.: Leg-robot for demonstration of spastic movements of brain-injured patients with compact magnetorheological fluid clutch. Adv. Robot. **24**(5/6), 671–686 (2010)
5. Chen, J.Z., Liao, W.H.: Design, testing, and control of magnetorheological actuator for assistive knee braces. Smart Mater. Struct. **129**(3), 035029 (2010)
6. Rossa, C., Jaegy, A., Lozanda, J., Micaelli, A.: Design considerations for magnetorheological brakes. IEEE/ASME Trans. Mechatron. **19**(5), 1669–1680 (2014)
7. Lee, D., Wereley, N.M.: Analysis of electro- and magneto-rheological flow mode dampers using Herschel-Bulkley model. Proc. SPIE **3989**(1), 244–255 (2000)
8. Lindler, J., Wereley, N.M.: Quasi-steady Bingham plastic analysis of an electrorheological flow mode bypass damper with piston bleed. Smart Mater. Struct. **12**(3), 305–317 (2003)
9. Kikuchi, T., Kobayashi, K.: Development of cylindrical magnetorheological fluid brake for virtual cycling system. In: IEEE International Conference on Robotics and Biomimetics, pp. 2386–2392 (2011)
10. Shiao, Y.J., Chang, C.Y.: Design of an innovative high-torque brake. Adv. Mater. Res. **339**, 84–87 (2011)
11. Nikitczuk, J., Weinberg, B., Mavroidis, C.: Control of electro-rheological fluid based resistive torque elements for using in active rehabilitation devices. Smart Mater. Struct. **16**(2), 418–428 (2007)
12. Russo, R., Terzo, M.: Design of an adaptive control for a magnetorheological fluid brake with model parameters depending on temperature and speed. Smart Mater. Struct. **20**(11), 115003 (2011). https://doi.org/10.1088/0964-1726/20/11/115003
13. Rocha-Gómez, I., et al.: Position control based on Gaussian function applied on magneto-rheological fluid disc brake of a hybrid mechanical device (MR brake - DC motor). Int. J. Appl. Electromagn. Mech. **73**, 1–15 (2023). https://doi.org/10.3233/JAE-220302

14. Cao, Z., Yang, X., Zhu, X.: Xia, Optimal design and experimental study of comb-type disc magnetorheological brake. Measurement **229**, 114458 (2024)
15. Carlson, A.B., Gisser, D.G.: Electrical Engineering Concepts and Applications, 2nd edn. Addison-Wesley Publishing Company, New York (1990)
16. LORD Technical Data. Accessed 2017. MRF-132DG Magneto-Rheological Fluid. https://www.lord.com/sites/default/files/Documents/TechnicalDataSheet/DS7015_MRF-132DGMRFluid.pdf
17. Ali, H.H., Fales, R.C., Manring, N.D.: Design of a velocity control system using an inlet metered pump. In: Bath/ASME Symposium on Fluid Power and Motion Control (FPMC 2017), Sarasota, FL, 16–19 October 2017

Design and Simulation of Mechatronic Learning Modules for Solving Mathematical Problems Using Arduino and Tinkercad

Iulian Sorin Munteanu[1], Liviu Marian Ungureanu[1], Mircea Daniel Popescu[1], Florentina Badea[2](✉), Sorin Ionuț Badea[2], and Cristian Radu Badea[2]

[1] National University of Science and Technology Politehnica, Splaiul Independentei, 313 Bu-Charest, Romania
[2] National Institute of Research-Development in Mechatronics and Measurement Technique - INCDMTM, Sos. Pantelimon, 6-8, Sector 2, 021631 Bucharest, Romania
mihaiflori@yahoo.com

Abstract. This paper presents a modular mechatronic learning environment based on Arduino Uno and simulated in Tinkercad, developed within the Master's level Mechatronics laboratory. The study highlights three C++ applications that solve fundamental mathematical-logical problems: (1) a quadratic equation solver, (2) a multi-operand arithmetic calculator, and (3) a statistical mean calculator (arithmetic and geometric). These digital programs run on an embedded system with LCD and keypad input. The versatility of the Basic Mechatronic Assembly (BMA) encourages interdisciplinary thinking, promotes algorithmic reasoning, and introduces expandable programming models that can be adapted to various STEM educational contexts. The system supports further enhancements with touchscreen HMI displays or sensor-based data acquisition modules. This educational best-practice model underlines the role of simulation and programming in developing student creativity, engagement, and technical competence in modern engineering education.

Keywords: Arduino Uno · Tinkercad · mechatronic simulation · C++ · educational application · Modeling and simulation · Mathematical problems

1 Introduction

In today's landscape of higher technical education [1–4], teaching strategies based on simulation, prototyping, and programming are increasingly important for building competencies in the field of mechatronics [5, 6].

This paper presents a practical and innovative example developed within the Mechatronics laboratory [7–9] of the IMST Master's program – an interactive mechatronic project based on the Arduino Uno platform and virtually simulated using Tinkercad [10], generically referred to as BMA – Basic Mechatronic Assembly.

The project focuses on designing a numeric input and processing system for solving second-degree equations, where coefficients are entered through a keypad and solutions

D. D. Cioboată and J. Machado (Eds.): ICoRSE 2025, LNNS 1592, pp. 150–163, 2025.
https://doi.org/10.1007/978-3-032-02508-1_13

are displayed on an LCD screen. Beyond this core function, the BMA can be expanded with a broad variety of interdisciplinary applications, especially in mathematics and complementary STEM fields.

Examples include:

- a basic arithmetic calculator (+, −, ×, ÷) for integers and decimals;
- even/odd number verification;
- computation of GCD and LCM;
- geometric calculations for area and perimeter (triangle, square, trapezoid, etc.);
- trigonometric and inverse function evaluation;
- arithmetic and geometric mean calculation;
- basic probability problems and unit conversions (temperature, pressure, etc.);
- scrolling messages with customizable graphical effects.

Students are encouraged to explore the synergy between basic electronics, C++ programming, and algorithmic reasoning, fostering logical thinking and technical creativity.

The project demonstrates a strong educational impact through active, hands-on STEM learning, where every component—from hardware to software—is designed to stimulate innovation and personal initiative.

This educational approach aligns with current trends in engineering education, as emphasized by Hernández-de-Menéndez et al. (2020) [11], who advocate for integrating hands-on methods and experiential learning to enhance student engagement and skill acquisition.

2 Implementation in TINKERCAD of a C++ - Based Mechatronic Application for Solving Second-Degree Equations Using the Basic Mechatronic Assembly (BMA)

Within the Mechatronics laboratories, including the IMST Master's program, from the Faculty of Industrial Engineering and Robotics, within the Department of Theory of Mechanisms and Robots, we encourage students to make applications in the Tinkercad virtual environment, and we emphasize programming concepts, the possibility of developing in our students the desire to perform in C++, and to find new solutions for optimizing or creatively developing basic mechatronic applications, i.e. selected for educational purposes, for their training potential.

Thus, on a minimal configuration of high-performance electronic components, such as those presented below, students can develop a series of distinct and interesting applications, based on arithmetic mathematics, trigonometric mathematics and even mechatronic applications from primary to advanced level, by increasing the primary configuration presented below.

The fact that a variety of common problems can be solved, through a 3D modeled configuration in Tinkercad and with a significant contribution of programming, requires students to have an active, challenging, STEM-type learning, which channels their creative effort towards new innovations and new outstanding results.

In terms of the components required for the mechatronic mechanism, we use the following basic components – already modeled in 3D in the Tinkercad software component database: an Arduino Uno R3 board that already has a programmable ATmega328P type microcontroller built into it, on which the following are connected in the appropriate slots: Arduino Uno R3, 32 (0 × 20) LCD 16 × 2 (I2C) - MCP23008-based, 4 × 4 Matrix keyboard (Keypad), connection cables. These electronic components will be connected according to an electronic scheme (see Fig. 1) that is given to students in the explanatory documentation of the specific laboratory work, and they will later test the functionality of the mechatronic assembly made by manually entering commands through the Keypad (keyboard).

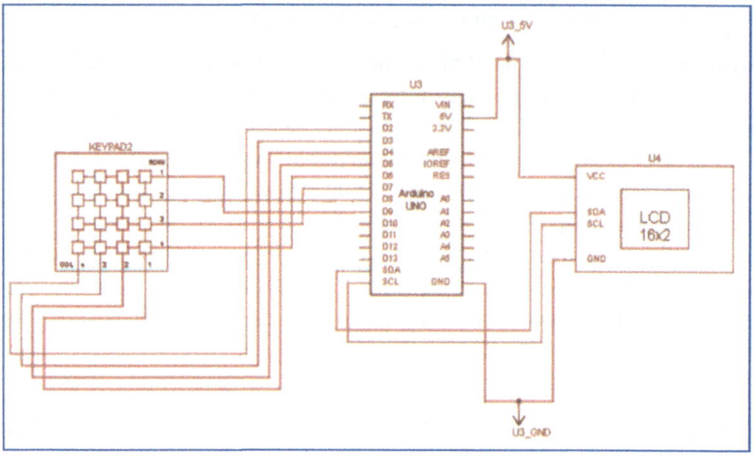

Fig. 1. View of an electronic scheme

Physical connection method of the basic mechatronic assembly - BMA:

- LCD I2C: Connecting the LCD pins on the Arduino board (pins with the same name will be connected by cables, as follows: SDA - > SDA, SCL - > SCL).
- 4 × 4 Matrix Keyboard: Connect each keyboard pin to a free digital pin on the Arduino board (D2 ÷ D9).

We make the wiring with metal wires between the Arduino Uno R3 board and the LCD, then we make the wiring between the Arduino Uno R3 board and the Keypad, in the TINKERCAD software application, and we obtain the assembly of 3D modeled components - functional electronics, as in the figures below Fig. and Fig..

Step 1 for assembly – We make the wire connection between the ARDUINO UNO board and the LCD in the workspace, by accessing the WIRE TYPE - Normal command, and connecting the pins 5V- VCC (red wire), GND- GND (black wire), 16 Digital - SDA (green wire), 17 Digital - SCL (green wire) (see Fig. 2).

Step 2 for assembly – We make the wire connection between the ARDUINO UNO board and the KEYPAD in the workspace (see Fig. 3), by accessing the WIRE TYPE - Normal command and joining the pins with wires as follows: D2 - Column 4 (first wire

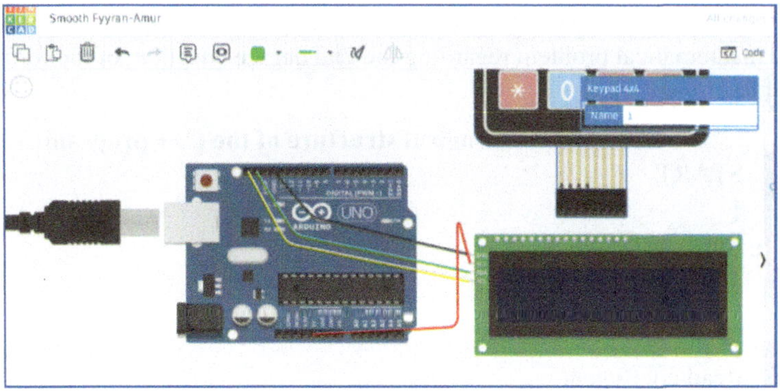

Fig. 2. Arduino Uno board wiring with LCD

from the right on the Keypad), D3 - Column 3 (second Keypad wire), D4 - Column 2, D5 - Column 1, D6 - Row 4, D7 - Row 3, D8 - Row 2, D9 - Row 1, according to Fig. 3. At this point, we have exhausted all the circuit elements in the statement and ensured their necessary electrical connection, obtaining the final electronic assembly below, following which we will design a specific program in C + +, which will correctly solve the mathematical problem regarding the quadratic equation: ax^2 + bx + c = 0.

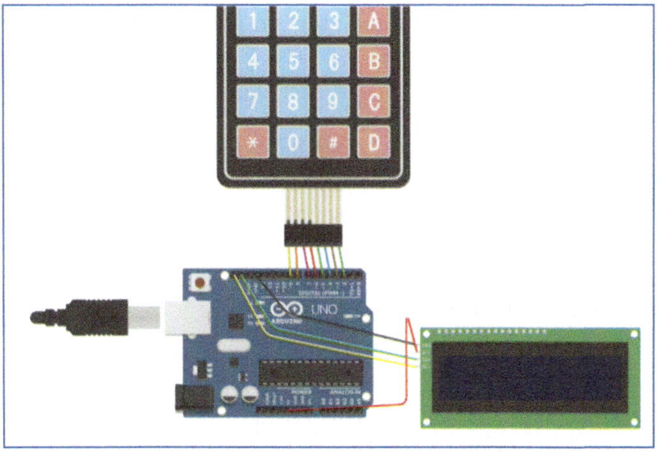

Fig. 3. Final electronic assembly in Tinkercad - Basic Mechatronic Assembly

When the circuit is powered and the Simulation Module is started, we can enter various data for the variable parameters of the 2nd degree equation, namely the coefficients (a), (b) and (c). Once the coefficients are confirmed, the program calculates and displays the roots of the equation on the LCD screen, formatted with two decimal places for better readability.

Step 3 for assembly – We create the specific program in C++, which will correctly solve the mathematical problem regarding the 2nd degree equation, of the form ax^2 + bx + c = 0.

Logical structure of the C++ program:

START
↓
Read coefficient a
↓
Read coefficient b
↓
Read coefficient c
↓
Compute discriminant D = b^2 - 4ac
↓
IF D > 0 → 2 real roots → display x1 and x2
ELSE IF D == 0 → 1 real root → display x1
ELSE → display "Complex solutions"
↓
END

Notes: User input is handled via the 4x4 matrix keypad. Each coefficient (a, b, and c) is entered manually and confirmed by pressing the '#' key, which simulates an ENTER command. This method ensures that the program processes one coefficient at a time before proceeding to the next step.

Testing Methodology

To validate the proper execution of the program, we performed test cases with various combinations of coefficients, including positive, negative, and fractional values.

The LCD correctly displayed the corresponding real or complex results based on the value of the discriminant.

Example of a Solved Numerical Case

- Enter the coefficient of (a) = 101.88 (see Fig. 4)
- Enter the coefficient of (b) = 403.66 (see Fig. 4)
- Enter the coefficient of (c) = −402.33 (see Fig. 5)

Show solutions to equation gr. 2:

x1 = 0.82 (see Fig. 6)
x2 = −4.79 (see Fig. 6.)

Explanations in brief: This C++ code reads the coefficients (a), (b), and (c) using a matrix keyboard and calculates the solutions of the equation type like: ax2 + bx + c = 0, displaying final solutions on an LCD screen. First of all, special programms checks the discriminant to determine whether the solutions are real or complex. If the discriminant

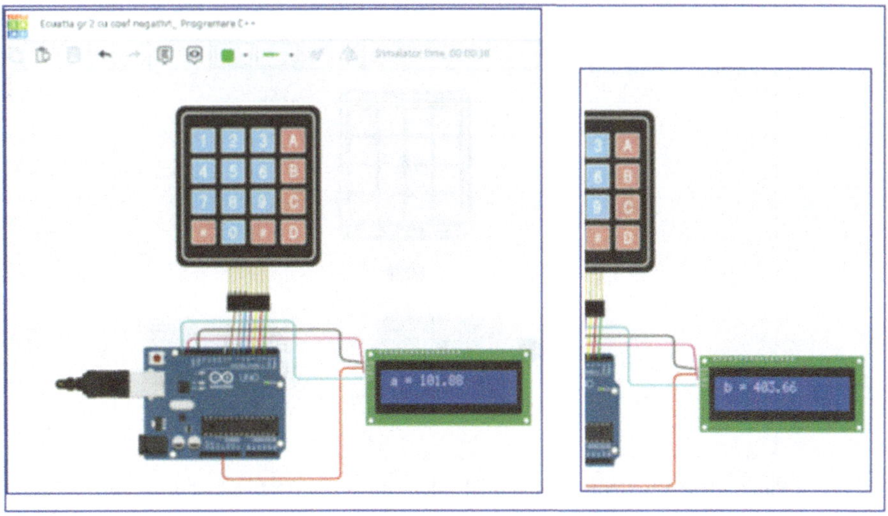

Fig. 4. BMA with input values: (a) = 101.88 and for (b) = 403.66

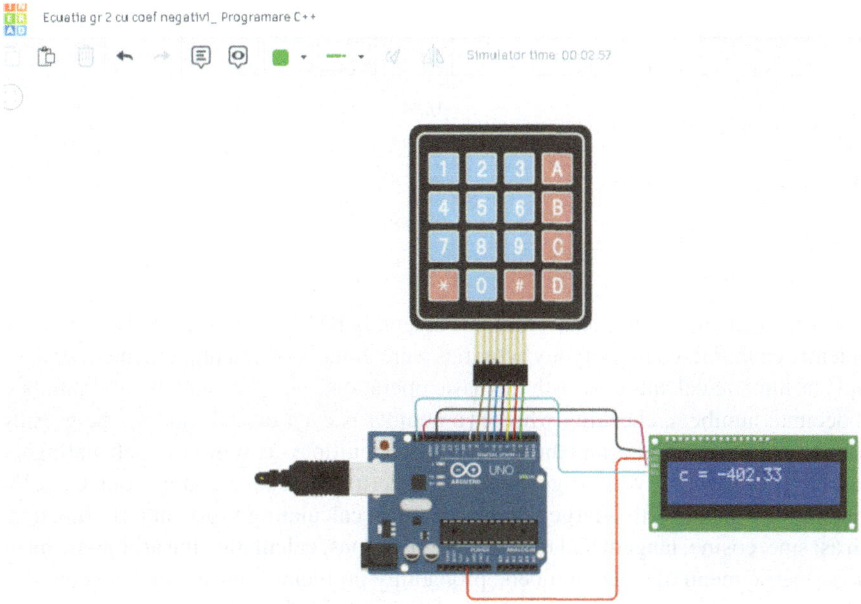

Fig. 5. BMA with input value for (c) = −402.33

is positive, it calculates and displays two real solutions (see Table 1). If the discriminant is zero, it calculates and displays the unique solution. If the discriminant is negative, it informs the user that the solutions are complex.

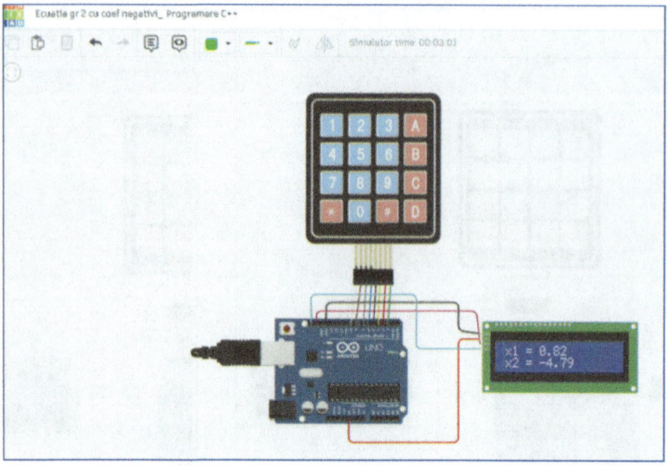

Fig. 6. BMA with displaying final values for: x1 = 0,82 and x2 = −4,79

Table 1. Examples of more calculated solutions of this equation

Enter coefficient values			Display calculated values	
coefficient (a)	coefficient (b)	coefficient (c)	x1	x2
0,22	0,33	−0,54	0,99	−2,49
30,41	20,22	−80,36	1,33	−1,99
211,48	309,35	−521,56	1,00	−2,46
33,11	23,33	−81,58	1,26	−1,96

With the help of this simple mechatronic assembly BMA, a wide range of applications frequently encountered in everyday activities were thought of and implemented, such as: simple arithmetic calculations with the basic operations (+,−,*,/) with natural numbers and decimal numbers, checking whether a number is even or odd, finding the greatest common divisor c.m.m.d.c. and the least common multiple - c.m.m.m.c., calculating the areas and perimeters of various geometric shapes (triangle, square, trapezoid, etc.), the perimeter of a shape with a large number of sides, calculating trigonometric functions such as: sine, cosine, tangent and their inverse functions, calculating the arithmetic mean and geometric mean of some numbers, probability problems, conversion between units of measurement (examples: temperature in Celsius and Fahrenheit, pressure in bar and atm, etc.), viewing scrolling messages - with different graphic effects, etc.

2.1 Arithmetic Calculator with Flexible Operand Input and Real-Time Display Using BMA

Another possible application, using the Arduino Uno configuration - LCD display - 4 × 4 keyboard (i.e. BMA) achieved by changing only the programming script, modified to accept any number for the operations of addition, subtraction, multiplication and division.

Essential modifications required:

- it displays the instructions for choosing the operation in the required format.
- it reads an unlimited number of values until the user presses # to complete the list.
- it applies the chosen operation in a chain (for example, for addition: n1 + n2 + n3 +…).
- it displays the final result on the LCD.

Logical structure (pseudocode) for the multi-number arithmetic calculator:

```
START
↓
Display available operations: A=+, B=-, C=*, D=/
↓
Wait for user to press key A, B, C, or D → store operation
↓
Display instruction: "Enter numbers, press # to stop"
↓
WHILE true:
    → Read number from keypad
    → IF input is "#" and no number entered: terminate
    → IF input is "#" and at least one number: stop loop
    → ELSE store number
    → IF first number: set as initial result
    → ELSE apply operation to running result
↓
IF division and user entered 0 → display "Divide by 0 error"
↓
Display final result on LCD
↓
END
```

Libraries included in this code:

1. Include < Adafruit_LiquidCrystal.H >

This library allows controlling an I2C compatible LCD screen with MCP23008 chip, specifying the addressing mode via I2C.

It is essential for:

- initializing the LCD (lcd.begin)
- controlling the cursor and displaying it on the screen (lcd.print, lcd.setCursor, lcd.clear)

2. Include < Keypad.H >

Provides the interface between the Arduino and the 4 × 4 matrix keyboard.

Allows:

- defining the keymap (makeKeymap)
- reading individual keys (keypad.getKey())

Structure and Functionality of the Arithmetic Calculator Program

This second application expands the functionality of the Basic Mechatronic Assembly by enabling basic arithmetic operations over a dynamic number of operands, entered manually via keypad. The operation is selected at the beginning ($A = +$, $B = -$, $C = \times$, $D = \div$), after which users can input a sequence of decimal or integer values, terminated with the # symbol.

A dedicated subroutine readFloat() collects user input one character at a time and builds a floating-point number, while readOperation() ensures only valid operation keys are accepted. The calculation is performed incrementally: the first number initializes the result, and each subsequent value updates it based on the selected operation. A division-by-zero is safely detected and handled with a specific message on the LCD.

This program structure reflects a robust and flexible approach to embedded programming, using modular functions and clear control flow to enable real-time data processing and user interaction.

Benefits of the Program Structure:

- Allows flexible input of any number without fixed limit (up to pressing #)
- Division by zero error is handled explicitly with message
- Clear separation of reading and calculation functions makes the code reusable

2.2 Implementation of a Statistical Mean Calculator with User-Defined Input Range and Dual-Line LCD Output Using BMA

A third possible application of BMA achieved by changing only the programming script accordingly, so that:

- can be entered between 2 and 4 real numbers (with decimals).
- the Arithmetic Mean and Geometric Mean are calculated depending on the initial option chosen.
- the LCD screen displays the result on two lines - first it prints the Arithmetic Mean/ or Geometric Mean, then on line 2 it prints the calculated value - clearly and with 4 decimals.

Logical structure (pseudocode) for the calculating Arithmetic and Geometric Mean:

```
  START
  ↓
  Display: "Calculate Mean"
  ↓
Ask user to choose mean type:
  1 → Arithmetic
  2 → Geometric
↓
Ask user how many numbers (2 to 4)
↓
FOR each number (i = 1 to N):
  → Prompt: "Enter number i"
  → Read number via keypad using readFloat()
  → Store number in array
↓
IF Arithmetic:
  → Compute sum of all numbers
  → Divide by N → Display result
ELSE IF Geometric:
  → Compute product of all numbers
  → Take N-th root of product → Display result
↓
Display result on LCD:
  Line 1: Mean type
  Line 2: Value (4 decimal places)
↓
Wait 6 seconds → Reset to initial menu
↓
END
```

Subroutines and structure in code
☑ readFloat() – reused function
Identical or similar to the one in previous applications.
It can also include the introduction of negative signs (for expand).
☑ loop() structure
The main code contains 3 parts:
- ✓ Choosing the average type (averagetype = 1 or 2)
- ✓ Choosing the number of terms (numberofNumbers = between 2 and 4)
- ✓ Reading, calculating and displaying results

☑ Calculation:
- ➢ Arithmetic Average:

arithmeticaverage = sum / numberofNumbers;
- ➢ Geometric Average:

geometricaverage = pow(product, 1.0 / numberofNumbers);
We use pow() from the <math.h> library for the n-ary root (i.e. product of 1/n)

geometricaverage = pow(product, 1.0 / numberofNumbers);
We use pow() from the < math.h > library for the n-ary root (i.e. product of 1/n).
Statistical mean calculator – program structure and analysis.

The third application within the BMA project calculates the arithmetic or geometric mean for 2 to 4 real numbers, based on user input. The interaction is structured as a step-by-step selection process, where the type of mean and the number of values are first selected via the keypad. The user then enters the required values individually.

The program relies on the readFloat() subroutine to read decimal input, and uses conditional logic to perform the appropriate calculation. The arithmetic mean is computed as the sum divided by the count, while the geometric mean uses the pow() function from <math.h> to evaluate the n-th root of the cumulative product.

Results are displaed on two lines of the LCD screen: the first line shows the type of mean selected, and the second displays the calculated value with four decimal precision. This approach enhances clarity and user feedback. The modular structure of the program supports future expansion to more complex statistical operations or variable-length datasets.

3 Conclusions

The work presented in this paper demonstrates not only a successful educational experiment but also a highly effective platform for fostering creativity, innovation, and hands-on learning within engineering education.

The Basic Mechatronic Assembly (BMA), developed on the Arduino Uno platform and virtually modeled in the Tinkercad environment, has proven to be a versatile and powerful pedagogical tool. Through its accessible configuration and flexible architecture, the BMA enables students to go beyond theory and directly engage with practical concepts via real-time programming and interactive simulations.

By tackling a diverse range of mathematical-logical tasks—from solving quadratic equations to performing multi-operand arithmetic calculations and computing statistical means—the BMA system highlights its wide-ranging capabilities. Students acquire essential competencies in C++ programming, logical structuring, user input control, and LCD output display. These skills not only cultivate algorithmic thinking but also help develop engineering fluency in a highly applied and motivating context.

What makes this approach particularly impactful is its alignment with contemporary STEM education principles [7–9].

Students are encouraged to explore, test, iterate, and improve. They are allowed—and even encouraged—to make mistakes, learning from each challenge as they optimize both code and logic structures. This nurtures not only technical mastery but also the creative autonomy vital for tomorrow's engineers.

As supported by relevant literature, such as Atmatzidou and Demetriadis [12], simple and modular mechatronic assemblies can be key in developing algorithmic reasoning across student populations.

Furthermore, Martínez Ortiz and Coyle (2017) [13] underline the essential role that Arduino boards play in developing computational thinking and structured reasoning among students—core competencies embedded in the BMA's logic.

Likewise, Sahin, Ayar, and Adiguzel [14] affirm the value of extracurricular STEM activities in enhancing both motivation and academic outcomes. These findings validate the BMA as a tool not only for structured laboratory sessions but also for student-led projects, creative workshops, and even competitive engineering challenges. Fleming's (2016) [15] findings also support the use of Tinkercad as an effective, visual platform for introductory electronics education.

The project also opens numerous opportunities for continued development and real-world application. Within our mechatronics laboratories—at both Master's and adapted Bachelor's levels—we strongly encourage students to transition from simulation to physical prototyping, using real electronic components such as sensors, actuators, low-voltage motors, and communication modules. This closes the loop between design and physical realization and brings STEM learning to life through tangible, exploratory engineering.

Looking ahead, one of the most promising directions involves the integration of advanced HMI (Human-Machine Interface) touchscreen modules, including models from Nextion, DWIN, or STONE HMI. These modern alternatives to the classic MCP23008-based I2C LCDs allow for direct and intuitive user interaction with application parameters on screen, greatly enriching the educational experience and expanding the range of programmable features.

Alongside our Master's students, we are also exploring new areas of development, including:

- probabilistic simulations and random number distributions,
- sensor-based physical measurements (e.g., temperature, humidity, distance),
- real-time data visualization on OLED or e-paper displays,
- wireless integration (Bluetooth, WiFi) for IoT-based educational applications,
- and gamified interfaces to enhance student engagement and interactive learning.

These new developments position the BMA platform not merely as a teaching aid, but as a catalyst for innovation, interdisciplinary teamwork, and student-led research. It has the potential to serve as the foundation for pre-incubation projects, smart device prototypes, and innovative educational competitions.

We further emphasize the critical importance of collaboration between the academic environment and the scientific research ecosystem. Institutions such as INCDMTM, INMA, or INOE 2000 IHP, as well as research-oriented SMEs, are essential partners in forming multidisciplinary consortia with universities.

These collaborations are vital for the development of advanced, scalable, and socially relevant technological solutions—both at the national and European levels, and even beyond.

In conclusion, we extend our sincere appreciation to the students, faculty, and partner institutions who continuously contribute to these educational innovations. This work stands as evidence of what is possible when passion for teaching, technical ingenuity, and accessible digital tools come together.

We remain fully committed to guiding our students toward designing both virtual and physical mechatronic models, embracing the latest technologies, and cultivating the skills, curiosity, and boldness needed to become competitive and inventive engineers from the earliest stages of their academic journe.

References

1. Munteanu I.S.: Mechatronics Course Support (2024). https://curs.upb.ro/2024/course/view.php?id=3069. Accessed 28 Mar 2025
2. Munteanu, I.S., Ungureanu, L.M., Petrescu, C., Badea, F., Apostolescu, TC., Văleanu, D.M.: Realization of virtual laboratory application and prototype for a mechatronic mechanism with Arduino Uno board programmed for a customized security application. In: Cioboată, D.D. (ed.) ICoRSE 2024. LNNS, vol. 1129, pp. 273–283. Springer, Cham (2024). https://doi.org/10.1007/978-3-031-70670-7_24
3. Munteanu, I.S., Ungureanu, L.M., Caraiman, CC., Crişan, RG., Niculae, E., Sorin, B.: Modelling and simulation of some mechatronics assembly realized on Arduino Uno board, through the Tinkercad application. In: Cioboată, D.D. (ed.) ICoRSE 2023. LNNS, vol. 762, pp. 374–384. Springer, Cham (2023). https://doi.org/10.1007/978-3-031-40628-7_31
4. Niţicală V., Munteanu I. S., Ştefănescu A., Munteanu A.M., Popescu D.M., STEM Education and Mechatronic Applications: Mechanisms, Devices, and Systems in the Development of Youth and Students, "HIDRAULICA" Magazine (No. 3/2024), ISSN 1453 – 7303, pp. 80–87. Hydraulics and Pneumatics Research Institute https://doi.org/10.6084/m9.figshare.27679410.v1
5. https://www.arduino.cc/education. Accessed 01 Apr 2025
6. www.robofun.ro Arduino Accessed 02 Apr 2025
7. Caraiman, C.C., Munteanu I.S.: Dissertation - Research on the development of smart greenhouses with multisensors through virtual simulations, Faculty of Industrial Engineering and Robotics (Bucharest Center), Program Management Engineering of Technical Systems -IMST, Graduation year (2023)
8. Barbu, A.M., Munteanu, I.S.: Dissertation - development of original applications in Tinkercad through the contribution of green energy generated by photovoltaic panels, Faculty of I.E.R. (Bucharest Center), Program IMST, Graduation year (2024)
9. Petrescu, C.G., Munteanu, I.S.: Dissertation - mini meteorological station for educational purposes - physical realization through virtual modeling and simulation using specialized software Arduino and Kicad, Faculty of of I.E.R. (Bucharest Center), Program Modeling and Simulation of Mobile Mechanical Systems - MSSMM, Graduation year (2024)

10. Jacob, F., Alberto, A., Guimaraes, P.: Use of tinkercad platform for teaching electronics subject in post-secondary technical courses, In: TEEM 2021: Ninth International Conference on Technological Ecosystems for Enhancing Multiculturality (TEEM 2021) October 2021, pp. 543–547 (2021). https://doi.org/10.1145/3486011.3486517
11. Hernández-de-Menéndez, M., Morales-Menendez, R., Escobar, C.A., Montes, J.M.: Engineering education: changes and challenges for innovative teaching. J. Eng. Educ. **109**(2), 163–189 (2020). https://doi.org/10.1002/jee.20347
12. Atmatzidou, S., Demetriadis, S.: Advancing students' computational thinking skills through educational robotics: a study on age and gender relevant differences. Robot. Auton. Syst. **75**, 661–670 (2016). https://doi.org/10.1016/j.robot.2015.10.008
13. Martínez Ortiz, A., Coyle, H.: Teachers and students using Arduino boards in STEM education: development of computational thinking skills. J. Sci. Educ. Technol. **26**(5), 516–527 (2017). https://doi.org/10.1007/s10956-017-9700-y
14. Sahin, A., Ayar, M. C., Adiguzel, T. (2014). STEM related after-school program activities and associated outcomes on student learning. Educational Sciences: Theory Pract, **14**(1), 309–322 (2024). https://files.eric.ed.gov/fulltext/EJ1038787.pdf
15. Fleming, J.: Tinkercad for Classroom Electronics Projects: A Guide for Educators (2016)

Selective Paint Removal from Plastic Components Using Laser Advanced Techniques

Alexandru Sover(✉) and Markus Zink

Faculty of Technology, Ansbach University of Applied Science, Residenzstraße 8, 91522 Ansbach, Germany
a.sover@hs-ansbach.de

Abstract. Laser technology presents a compelling alternative to conventional methods for removing coatings from plastic and metallic parts, offering advantages over chemical solvents and media blasting for component reuse and recycling. This study investigates laser paint removal from thermoplastic materials, particularly those with complex 3D geometries. Utilizing a 1064 nm pulsed fiber laser, experiments were conducted to analyze paint ablation from plastic substrates and characterize the resulting surfaces. Experimental results on plastic substrates demonstrate the feasibility of complete paint removal while preserving the integrity of the thermoplastic. The laser's energy density, scanning speed, and spot size are identified as key parameters influencing removal efficiency and substrate integrity. The study presents the necessary steps and process conditions and analyzes the resulting quality of the processed surfaces. The study concludes that laser paint removal facilitates the reuse or high-quality recycling of plastic parts and can be an environmentally friendly, flexible, and highly efficient method for paint removal.

Keywords: Paint Removal · Laser Advanced Techniques · Paint Stripping · Polymer Recycling

1 Introduction

The widespread use of painted plastic parts stems from their advantageous combination of inexpensive thermoplastic substrates and the high resistance provided by a thin thermoset coating. This pairing results in reliable and cost-effective products. Particularly in the automotive industry, thermoplastic parts with thermoset coatings are favored for achieving appealing aesthetics through varied colors, alongside excellent resistance to diverse surface stresses [1–3].

Not only plastics, but also metallic components commonly receive coatings for protection against wear or corrosion, as well as for design and color differentiation, especially in the aviation and automation sectors [3, 4].

The increasing global focus on circular economy principles necessitates efficient and effective recycling processes for end-of-life materials. Painted thermoplastic components represent a significant waste stream across various industries, including automotive, aerospace, and consumer electronics. The paint layers on these materials pose a

D. D. Cioboată and J. Machado (Eds.): ICoRSE 2025, LNNS 1592, pp. 164–172, 2025.
https://doi.org/10.1007/978-3-032-02508-1_14

considerable challenge to their successful recycling and to the recovery of high-quality secondary resources. This scientific overview highlights the critical importance of paint removal technologies in enhancing the recyclability of both thermoplastic and metallic materials [2, 4].

The presence of paint coatings on thermoplastics can impede recycling processes in several key ways:

- **Contamination of the Recyclate:** During the mechanical recycling of plastics, residual paint can contaminate the melt by way of unmelted thermoset particles, thus reducing the mechanical properties and compromising the aesthetics of the recycled product (Hopewell, Dvorak, & Kosior, 2020) [5, 6].
- **Process Incompatibility:** Paint layers can be incompatible with certain recycling technologies. For instance, coatings and base polymers may degrade at different temperatures, causing processing difficulties and material degradation during the melting process (Awaja & Pavel, 2005) [7].
- **Reduced Material Value:** The presence of paint (usually a thermoset) often diminishes the economic value of the recovered recycled polymer. Contaminated plastic regrind may require additional processing steps or be relegated to lower-value applications, hindering the economic viability of recycling initiatives (Ragaert, Delva, & Van Geem, 2017) [8]. Unlike thermosets, thermoplastics can be melted for recycling by being mechanically broken down for remoulding [9].

Implementing efficient paint removal technologies offers significant benefits for the recycling of thermoplastics:

- **Enhanced Recyclate Quality:** Removing paint prior to recycling yields cleaner and higher-quality secondary materials. For thermoplastics, this translates to regrind with improved mechanical strength, and processability, expanding its potential applications (Pickering, 2010) [9].
- **Improved Recycling Efficiency:** Clean materials are often easier and more energy-efficient to process. Removing paint can reduce the energy required for melting plastics and minimize the formation of undesirable byproducts or emissions (Habib et al., 2019) [10].
- **Increased Material Value:** High-quality recyclates command better prices in the secondary materials' market, enhancing the economic feasibility and sustainability of recycling programs. This incentivizes the collection and processing of painted components that might otherwise end up in landfills [11].
- **Facilitating Advanced Recycling Technologies:** Some advanced recycling technologies for plastics, such as chemical recycling (e.g., pyrolysis, solvolysis), are particularly sensitive to contaminants like paint. Effective pre-treatment through paint removal is often a crucial step for the successful implementation of these advanced methods, enabling the recovery of monomers or other valuable chemicals (Rahimi & Garcia, 2017) [12].

Various technologies are being explored and implemented for paint removal from plastics and metals, each with its own advantages and limitations:

- **Mechanical Abrasion:** Techniques like media blasting or grinding can physically remove paint layers. While relatively simple, these methods can be energy-intensive and may generate secondary waste (e.g., abrasive media and paint debris) requiring further management [13].

- **Chemical Stripping:** Solvents can be used to dissolve or detach paint coatings. However, the use of harsh chemicals raises environmental and safety concerns regarding solvent disposal and worker exposure [14, 15].
- **Thermal Methods:** Controlled heating can be employed to decompose or volatilize paint layers. This requires careful temperature control to avoid damaging the underlying material and necessitates effective emission control systems [16].
- **Laser Ablation:** Laser technology offers a precise and potentially environmentally benign method for selectively removing paint layers without damaging the substrate. This technology is particularly promising for complex geometries and delicate materials (Ouyang et al., 2020) [17–19].

2 Laser Technology: A Precise Tool for Paint Removal

Laser technology has emerged as a highly effective and increasingly popular method for removing paint and coatings from various substrates, including metals, plastics, and composites [20]. Unlike traditional methods such as chemical stripping or abrasive blasting, laser ablation offers a precise, non-contact, and often environmentally friendlier alternative. [18–20].

The process involves directing a focused laser beam onto the painted surface. The intense energy from the laser is absorbed by the paint layer, causing it to rapidly heat up, vaporize, or ablate [21]. By carefully controlling the laser parameters – such as wavelength, pulse duration, energy density, and scanning speed – the paint can be selectively removed layer by layer, leaving the underlying substrate largely unaffected [20, 22].

Advantages of Laser Paint Removal

- **Precision and Selectivity:** Lasers can be precisely controlled to remove paint from specific areas without damaging the surrounding material. This is particularly beneficial for intricate parts or delicate substrates [21].
- **Non-Contact Process:** Eliminates physical contact with the surface, reducing the risk of mechanical damage or contamination [23].
- **Environmental Friendliness:** Often reduces or eliminates the need for harsh chemicals and the generation of secondary waste associated with traditional methods. The primary byproduct is often vaporized paint particles that can be captured and filtered [24].
- **Versatility:** Applicable to a wide range of paint types and substrate materials [24, 25].
- **Automation Potential:** The process can be easily automated and integrated into industrial production lines, increasing efficiency and reducing labor costs [25].

Principles of Laser Paint Removal

Laser paint removal, also known as laser ablation or laser cleaning, relies on the interaction between a high-energy pulsed laser beam and the coating material and support material. The process involves:

- **Absorption:** The paint layer absorbs the laser energy. The efficiency of absorption depends on the laser wavelength and the optical properties of the paint [21].
- **Rapid Heating:** The absorbed energy leads to a rapid increase in the temperature of the paint layer [26].
- **Ablation/Vaporization:** When the energy density exceeds the ablation threshold of the paint, the material undergoes rapid vaporization, sublimation, or decomposition into plasma and particulate matter.
- **Ejection:** The ablated material is ejected from the surface, often assisted by the pressure wave generated during the process.

The key to successful laser paint removal lies in carefully selecting and controlling the laser parameters (wavelength, pulse duration, frequency, power density, and scanning speed) to selectively remove the paint while minimizing damage to the underlying substrate [20–22]. The damage threshold of the substrate is a critical factor in determining the optimal laser parameters.

Metals are a common and often ideal substrate for laser paint removal due to their relatively high thermal conductivity and damage threshold compared to some plastics [27].

Ceramics present a unique challenge due to their brittle nature and potential sensitivity to thermal shock. Laser ablation of paint from ceramics relies on similar principles as with metals. Ceramics have a lower thermal conductivity than metals, potentially leading to localized heating if laser parameters are not optimized. However, careful control of the laser energy input is crucial to prevent cracking or damage to the ceramic surface or glaze [28].

Plastics exhibit a wide range of thermal and chemical properties, making laser paint removal more complex than with metals or ceramics. The success of laser paint removal from plastics depends heavily on the laser wavelength and the absorption characteristics of both the paint and the plastic substrate. Ideally, the laser wavelength should be strongly absorbed by the paint and poorly absorbed by the plastic to ensure selective removal (KEYENCE International Belgium, [29]).

Plastics have low melting and degradation temperatures, making them susceptible to thermal damage if laser parameters are not optimized. Different types of plastics (e.g., PP, ABS, PC) have varying absorption spectra and damage thresholds, requiring tailored laser parameters. Additives and pigments in the plastic can affect its laser absorption characteristics [30].

Research into laser-based paint removal from plastics indicates that the separation of the paint layer from the plastic carrier is achievable through optimized laser parameters [18]. The following figure shows the paint removal tests with a pulsed fiber laser from Coherent-Rofin (1064 nm, 50 W) with different parameters on a PP plastic support material. Experiments were performed on square surfaces measuring 10×10 mm.

The microscopic analysis of the processed plastic surface illustrates the successful removal of the coating layer (paint and/or protective layer) while preserving the integrity of the support material's surface.

A complete separation of the layer was observed, and further surface analysis confirmed these results.

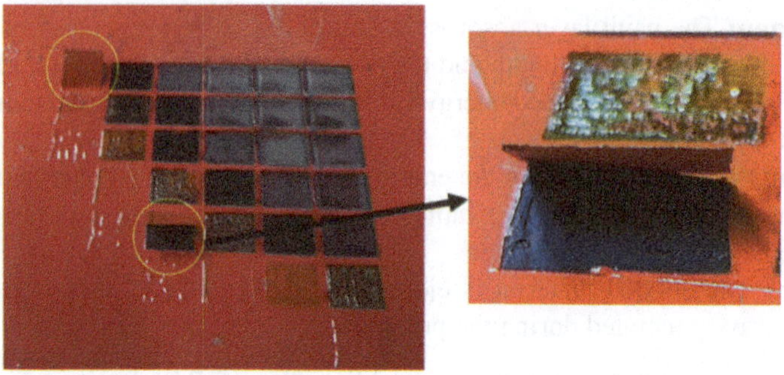

Fig. 1. Paint layer separation from the laser processed thermoplastic surface [18]

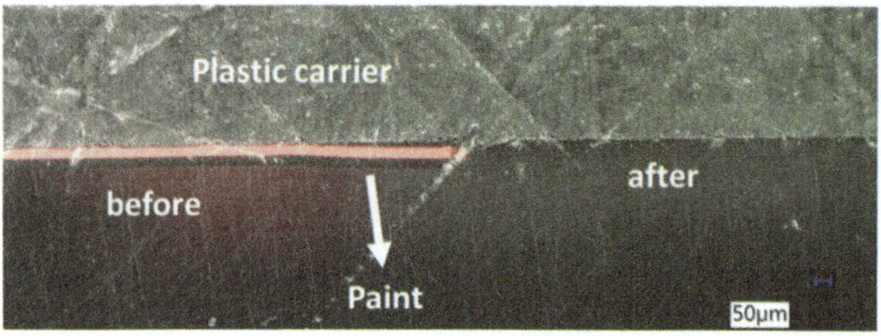

Fig. 2. Microscopy cross section of the laser processed plastic part [18]

Optimal results were achieved at a current intensity of 15 A and a frequency of 60 kHz. Any remaining residuals between the coating layer and the carrier material were also easily removed. To assess the scalability of the laser paint removal process, a larger coated part (200 × 200 mm) from an automotive spoiler was treated using the previously determined optimal parameters (Fig. 3).

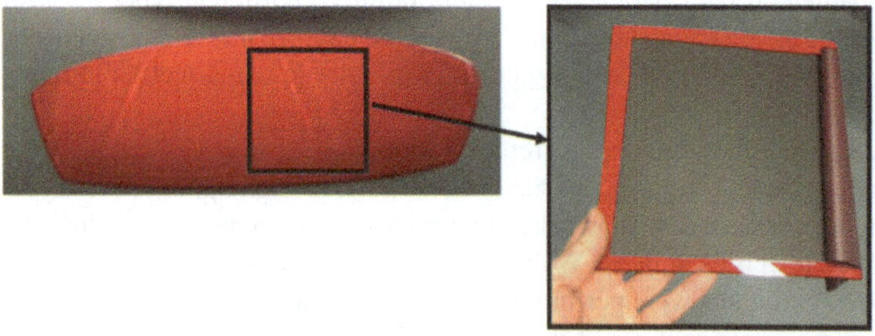

Fig. 3. Laser processing of a larger coated part from an automotive spoiler

This application demonstrated the feasibility of the technology for the removal of the coating layer from more substantial and geometrically complex automotive components.

The technology was also successfully employed on plastic support material coated with white paint, where the painted layer was effectively removed by laser processing, Fig. 4.

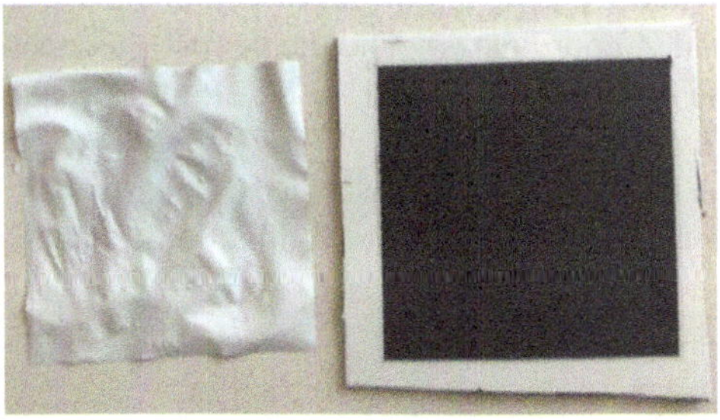

Fig. 4. Laser paint stripping of a larger part from an automotive plastic component coated with white paint [31]

In further tests, different plastic support materials commonly used for coated automotive products with varying coating colours were processed, potentially with adjustments to laser parameters, Fig. 5. It was possible to observe that the coating layer could be separated after the laser processing, suggesting the adaptability of the laser system to effectively remove coatings from a range of different plastic and paint combinations.

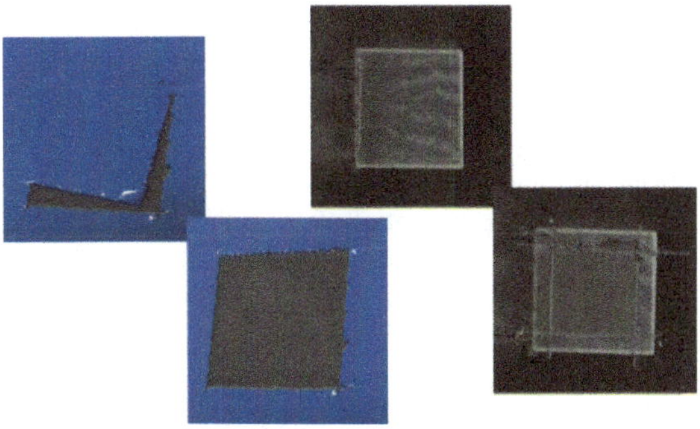

Fig. 5. Laser paint stripping of a PP/EPDM plastic carrier coated with blue paint (left) and a PC/PBT plastic carrier coated with black paint (right) [33]

The obtained results presented in Figs. 1, 2, 3, 4 and 5 were achieved on flat surfaces or those with only slight 3D geometry variations. This allowed the laser to operate within a 2D plane, maintaining the processed surfaces in optimal focus and ensuring precise and consistent energy delivery throughout the processing time. However, many coated plastic parts possess complex 3D geometries, which involve slightly different processing conditions since the laser deviates from its optimal focus. New laser technology offers the capability to dynamically adjust the focus in Z-direction during the processing of a 3D surface. This is made possible by the scan head of the laser, where the focus shift is performed with the aid of the lenses positioned upstream of the mirrors. The movable lens alters the divergence of the laser beam, thereby causing the focus to move upwards or downwards. The focusing lens and its position determine the field size and the working distance. This "intelligent" dynamic focus control is provided for by the software, which synchronizes the X, Y, and Z positions using a correction file (assigning height and angle values to the X, Y, and Z coordinates). The theoretical processing speeds can exceed 5000 mm/s. To enable this process, an STL or STEP file is required, as it supplies the necessary geometric data of the 3D part.

Figure 6 shows the laser processing of a coated 3D plastic part. The figure compares laser processing with and without dynamic Z-direction focus adjustment of the laser beam.

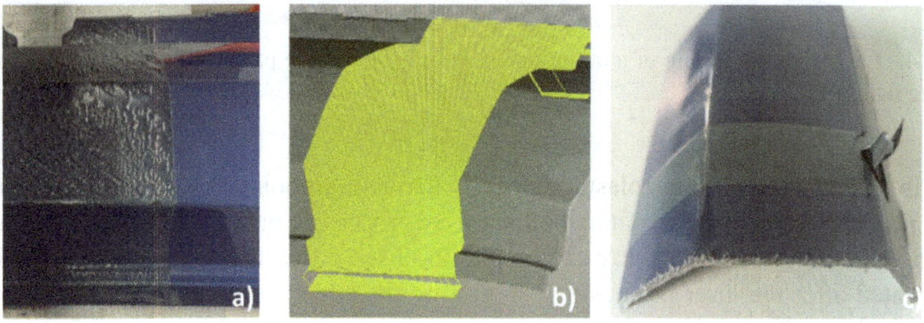

Fig. 6. Laser processing of a 3D geometry without dynamical focusing a), software mapping of the geometry b) and processing with dynamical focus c)

The laser used for the processing of the 3D geometries was built by OPTOGON Deutsche Industrielaser GmbH and is a 200-Watt fibre laser (1064 nm) offering increased performance and pulse width adjustment.

These results demonstrate the potential of laser technology to "clean" coated thermoplastic parts for the reuse in a new coating process or for recycling without the coating layer, which is typically a thermoset and cannot be melted. Further investigation into the laser paint removal technique will be continued to enable new applications for large, coated parts.

3 Conclusions

The efficient removal of paint from end-of-life thermoplastic is a critical step towards achieving an intensified circular economy. By addressing the challenges posed by paint contamination, effective paint removal technologies enhance the quality, efficiency, and economic viability of recycling processes. The laser paint removal is a promising technology, its application on plastics with high pigment loads is limited by the risk of thermal damage and residue formation, and on highly curved geometries by the challenges of maintaining consistent focus and achieving uniform coverage. As this paper proves, these challenges are often overcome by employing advanced laser systems that feature sophisticated control, dynamic focusing, and robotic integration.

Continued research and development in this area, focusing on environmentally friendly and cost-effective solutions like advanced mechanical methods and laser ablation, are essential to unlock the full potential of material recovery and reduce our reliance on virgin resources.

Laser technology offers a compelling and increasingly viable solution for paint removal from metals, ceramics, and plastics. Its precision, non-contact nature, and environmental benefits make it an attractive alternative to traditional methods. However, the optimal application of laser ablation requires a thorough understanding of the laser-material interaction, careful selection of laser parameters tailored to the specific substrate and paint, and consideration of the unique challenges associated with each material category. Continued advancements in laser technology and application knowledge will further expand the use of this precise tool in various industrial and restoration processes.

References

1. Sover, A., et al.: Paint removal from thermoplastic materials and its influence on the physical-mechanical properties for the recycling of the polymer. IOP Conf. Ser. Mater. Sci. Eng. **1037**, 012032 (2021)
2. Zambrano, C., Tamarit, P., Fernandez, A.I., Barreneche, C.: Recycling of plastics in the automotive sector and methods of removing paint for its revalorization: a critical review. Polymers **16**, 3023 (2024). https://doi.org/10.3390/polym16213023
3. Patil, A.; Patel, A.; Purohit, R.: An overview of polymeric materials for automotive applications. Mater. Today Proc. **4**, 3807–3815 (2017). http://www.springer.com/lncs. Accessed 21 Nov 2016
4. Ragaert, K., Delva, L., Van Geem, K.: Mechanical and chemical recycling of solid plastic waste. Waste Manag. **69**, 24–58 (2017)
5. Ventosinos Louzao, V., et al.: Impact of recyclability on the mechanical properties of coated plastic materials for the automotive and electronic sectors [version 1; peer review: 2 approved with reservations]. Open Res Europe. **4**, 51 (2024). https://doi.org/10.12688/openreseurope.16888.1
6. Hopewell, J., Dvorak, R., Kosior, E.: Plastics recycling: challenges and opportunities. Phillos. Trans. Ser. B, Biol. Sci. **364**, 2115–2126 (2009). https://doi.org/10.1098/rstb.2008.0311
7. Awaja, F., Pavel, D.: Recycling of PET. Eur. Pol. J. **41**(7), 1453–1477 (2005). https://doi.org/10.1016/j.eurpolymj.2005.02.00
8. Ragaert, K., Delva, L., Van Geem, K.: Mechanical and chemical recycling of solid plastic waste. Waste Manage. **69**, 24–58 (2017). https://doi.org/10.1016/j.wasman.2017.07.044
9. Pickering, K., Beg, H.: Quality and durability of recycled composite materials (2010). https://doi.org/10.1533/9781845697662.3.303
10. Zeng, X., Qin, W., Li, J., Wang, Z.: Research on mechanism and process of paint removal with pulsed fiber laser. MATEC Web Conf. **189**, 01008 (2018). https://doi.org/10.1051/matecconf/201818901008
11. https://www.eea.europa.eu/en/analysis/publications/measuring-the-quality-of-recycling, last accessed 2025/04/28
12. Rahimi, A., Garcia, J.: Chemical recycling of waste plastics for new materials production. Nat. Rev. Chem. **1**, 0046 (2017). https://doi.org/10.1038/s41570-017-0046
13. Momber, A.: Blast Cleaning Technology. Springer Science & Business Media, Berlin (2007). https://doi.org/10.1007/978-3-540-73645-5

14. Agarwala, V.S., Rajeshwar, K.: Paint and coating removal technologies: past, present and future. In: CORROSION 2001, Houston, Texas, March 2001
15. Operowsky, R.M.: Chemical immersion paint stripping. Met. Finish. **98**(6), 477–480 (2000)
16. https://www.arenacomet.eu/page/cleaning-shops-metals-paintstripping-organic-coatings-removal/f1.html. Accessed 28 Apr 2025
17. Li, X., et al.: Laser paint stripping strategy in engineering application: a systematic review. Optik **241**, 167036 (2021). https://doi.org/10.1016/j.ijleo.2021.167036
18. Sover, A.: Research on the process of paint removal from thermoplastic materials by laser. IOP Conf. Ser. Mater. Sci. Eng. **564**, 012028 (2019). https://doi.org/10.1088/1757-899X/564/1/012028
19. Zink, M., Sover, A.: Study on the process of paint removal from thermoplastic materials by laser technology. In: Proceedings of the 2nd International Conference of the University of Applied Sciences Ansbach 3rd - 4th of July 2020
20. https://www.cleanlaser.de/en/applications/application-fields/paint-removal-and-decoating/. Accessed 28 Apr 2025
21. https://www.laserax.com/applications/laser-cleaning-paint-removal. Accessed 28 Apr 2025
22. https://www.laserflux.com/en/about-laser-cleaning/what-is-laser-cleaning/. Accessed 28 Apr 2025
23. https://www.trumpf.com/en_INT/solutions/applications/surface-processing-with-the-laser/laser-paint-removal. Accessed 28 Apr 2025
24. https://www.boeing.com/innovation/innovation-quarterly/2021/03/light-solution. Accessed 28 Apr 2025
25. https://looptechnology.com/automated-laser-cleaning-on-a-large-scale. Accessed 28 Apr 2025
26. Lu, Y., Yang, L., Wang, M.L., Wang, Y.: Improved thermal stress model and its application in ultraviolet nanosecond laser cleaning of paint. Appl. Optics. **59**, 7652–7659 (2020). https://doi.org/10.1364/AO.398101
27. https://www.mactrontech.net/advantages-of-laser-paint-remover/. Accessed 28 Apr 2025
28. https://dplaser.com/how-to-use-laser-cleaning-machine-for-paint-stripping/. Accessed 28 Apr 2025
29. https://www.keyence.com/products/marker/laser-marker/applications/laser-processing/laser-paint-removal.jsp. Accessed 28 Apr 2025
30. https://sabreen.com/library/smart-additives-enhance-plastics-laser-marking/. Accessed 28 Apr 2025
31. Sover, A., Zink, M.: Lackschichten umweltfreundlich entfernen. JOT Journal für Oberflächentechnik **61**(6), 50–53 (2021). https://doi.org/10.1007/s35144-021-1188-4

Development of a Framework to Coordinate Capacity with Market Demand

Eduardo Pereira[1], André S. Santos[2,3,4], João Bastos[1,3], Paulo Ávila[1,3], Leonilde Varela[3], Nuno Leal[5], and José Machado[6(✉)]

[1] ISEP - School of Engineering, Polytechnic of Porto, 4200-072 Porto, Portugal
{jab,psa}@isep.ipp.pt

[2] Interdisciplinary Studies Research Center (IRSC), ISEP - School of Engineering, Polytechnic of Porto, 4200-072 Porto, Portugal
abg@isep.ipp.pt

[3] Tecnologia E Ciência, INESC TEC - Instituto de Engenharia de Sistemas E Computadores, Campus da FEUP, Rua Dr. Roberto Frias, 4200 - 465 Porto, Portugal
leonilde@dps.uminho.pt

[4] Algoritmi Research Centre, University of Minho, Guimarães, Portugal

[5] Sonae Arauco Portugal, S.A, Oliveira Do Hospital, Portugal

[6] MEtRICs Research Centre, University of Minho, Guimarães, Portugal
jmachado@dem.uminho.pt

Abstract. This document addresses and develops a framework tool to solve reliability issues in the calculation of processing times for components, using their dimensions. This framework was implemented in a real industrial setting, specifically in a multinational company that manufactures highly customizable electric motors according to customer requirements. After identifying the most critical components and their respective process diagrams, a prototype of the proposed framework was developed to calculate production time. Additionally, another prototype was developed to aid in visualizing the company's workload. As a result of this work, various improvements were observed in the company, including a 42% reduction in the time required to create workflows and an increase in the reliability and dependability of process times. The framework significantly enhanced operational efficiency, streamlined production processes, and provided a robust solution for managing the complexities of custom manufacturing, demonstrating its effectiveness in a real-world industrial environment. Furthermore, this approach has the potential to be adapted for use in other industries facing similar challenges.

Keywords: Process Diagram · Operation Plan · Operation Process

1 Introduction

In today's market, the high demand for customization and the significant variability in products make it challenging to predict the manufacturing time for various types of equipment, particularly when they are custom-made. These products often consist of different components, and in some cases, standard components with modified dimensions

D. D. Cioboată and J. Machado (Eds.): ICoRSE 2025, LNNS 1592, pp. 173–180, 2025.
https://doi.org/10.1007/978-3-032-02508-1_15

and tolerances. In the company where this tool was developed, many components are large and need to be machined, and the process time tends to be lengthy. Therefore, it is crucial to accurately calculate manufacturing times to inform production planning and control, enabling effective factory management and reducing the lead time for component delivery and, consequently, the motor itself.

To achieve this, two tools were developed. The first tool calculates the manufacturing time of the component using machine parameters and time measurements. The second tool compiles production orders recorded in the SAP ERP software and delays in Power BI to assist in sequencing production orders on the factory floor. This document consists of a brief introduction to the work, followed by two theoretical sections that support the practical developments, and finally, a third chapter demonstrating and explaining both tools.

2 Literature Review

2.1 Study of Methods

The study of methods can be defined as the process of meticulously examining each operation of a specific task to eliminate any unnecessary elements or steps, and determine the most efficient way to perform each essential element or operation [9]. Some authors define the study of times and methods as: "The study of times and methods involves the organized recording and rigorous evaluation of the ways tasks are performed, with the aim of promoting improvements" [1, 2].

2.2 Operation/Process Diagram

To obtain a clear understanding of the manufacturing process of a product, detailed sheets, outlining the description and sequence of operations for a specific task or product, were developed. These sheets are known as operation or process diagrams. An operation diagram describes all the operations required to manufacture each component, their assemblies, and the packaging of the final product [3].

A process diagram is more comprehensive than an operation diagram, as it includes universally accepted symbols as described by Abbasian-Hosseini S. in [4]. According to [5], a process diagram explicitly details all the stages necessary to produce a product, from its initial state to the final product. It also contains information regarding time, materials used, machines, and tools. As stated by [6], this diagram helps identify steps that do not add value to the process. Additionally, there is a process flow diagram, which, according to [3], results from the combination of the process and operation diagrams. Currently, there are several themes to be developed and refined, such as: a time estimation approach based on Machine Learning (ML) methods, aimed at obtaining standard times for manual operations [7, 8]; methods based on Virtual Reality for automatically evaluating manual work processes using MTM-Methods Time Measurement [9–12]; intelligent work measurement systems with energy monitoring devices based on the Internet of Things [13]; systems for analyzing the physical demands on operators and the impact of these factors in the workplace [14, 15]; and finally, the development of methods using image processing to obtain process times [16].

3 Development of the Work

3.1 Methodological Approach

Initially, a Pareto diagram was created with the aim of identifying which components have the longest processing times in the machining section. Subsequently, the machining process for these components was analyzed and studied to determine the necessary machines. Thereafter, a prototype tool was developed in VBA Excel, capable of calculating the processing times of the parts based on CNC machine parameters. Finally, an additional tool was developed in Power BI, capable of using these previously calculated times to demonstrate the existing workload in the company.

3.2 Identification of the Most Critical Components and Definition of the Process Diagram

Using a cost deployment file from 2023, a Pareto diagram was constructed to identify which components require the longest manufacturing times in the machining section. The diagram highlights housings as the most critical component. Notably, it shows that 80% of the machining workload is attributed to housings, covers, inner fixing rings, junction boxes, intermediate bases, and rotor pressing rings.

Since the company had acquired a new machine for manufacturing junction boxes, a study was requested to determine the manufacturing time for the other components produced by this machine. The current times being used by the company for these components were based on an older machine. Due to this request and the limited timeframe in which the study was conducted, it was only possible to develop the calculation tool for the following components: Housing; Junction Boxes; Injected Rotors; Shperical Supports. After identifying the most critical components, their production process was studied and analyzed. The process diagram is presented in Fig. 1.

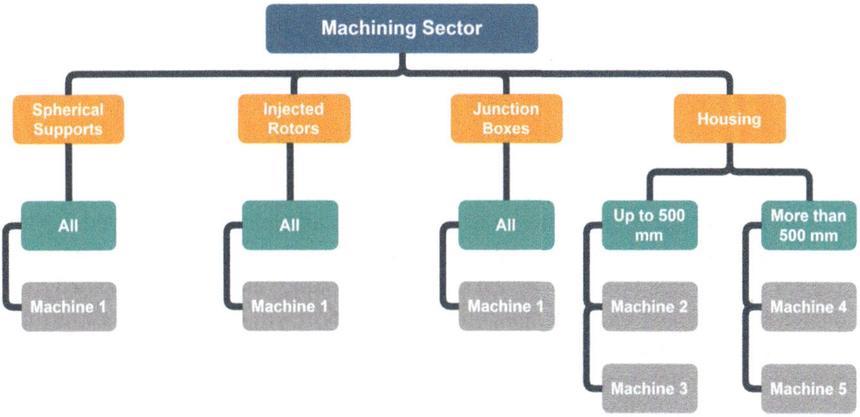

Fig. 1. Process Diagram of Machining Sector

As shown in Fig. 1, the diagram details the machines required for the critical components, but it lacks manufacturing times for each machine. To address this, a time calculation tool will be introduced in the next subsection. It is important to note that calculating manufacturing time requires studying and identifying the machines used in the process.

3.3 Time Calculation Tool

To enhance the reliability of process times for future components to be manufactured in the factory, a tool was developed. In the initial phase, a database was created, documenting the parameters of each tool on each machine, as detailed in Table 1.

Table 1. Excerpt from the database of tool parameters

Tool	(mm)	F-Feed	S-rot/min	Speed- m/min	Index	Material	Machine
Barra	379	100		X	WTBX	Ferro	Machine 1
Barra	380	50	500	X	WTBX	Ferro	Machine 1
Barra	264	100	250	X	WTBX	Ferro	Machine 1

Once the tool parameters have been obtained, it is necessary to identify the machining strategies, that is, all the operations required to machine a part completely. Once these operations are identified, it is possible to extract the perimeters of each part based on their dimensions. Consequently, in the second phase, a software application was developed in VBA, capable of compiling this information to determine the time required for each operation. The Eq. (1) demonstrates the fundamental principle of the calculation that underpins the VBA software for machines considered as machining centers, where: Distance: Refers to the distance to be traveled by the tool in millimeters; Tool feed rate: Refers to the machine parameter, indicating how many millimeters the tool advances per minute; Number of passes: Refers to the number of times the cutting tool must traverse a specific path.

$$Time = \frac{Distance * Number\ of\ passes}{\text{Tool feed rate}} \tag{1}$$

On the initial page (Fig. 2), the parts available for calculation are displayed. It should be noted that by the completion date of this document, the software had only been developed for junction boxes, spherical supports, housings, and injected-molded rotors.

When selecting, for example, junction boxes, a form will be displayed as illustrated in Fig. 3. In this form, the user must complete the fields according to the characteristics of the part and the provided technical drawing. If any crucial field for the calculation is left incomplete, or if a dimension is entered for which there is no corresponding tool in the database, the program will issue a warning message.

To simplify and expedite the form filling process, the time for all fixed operations, whose dimensions do not vary, has been pre-calculated in the software ("Perimeters" and

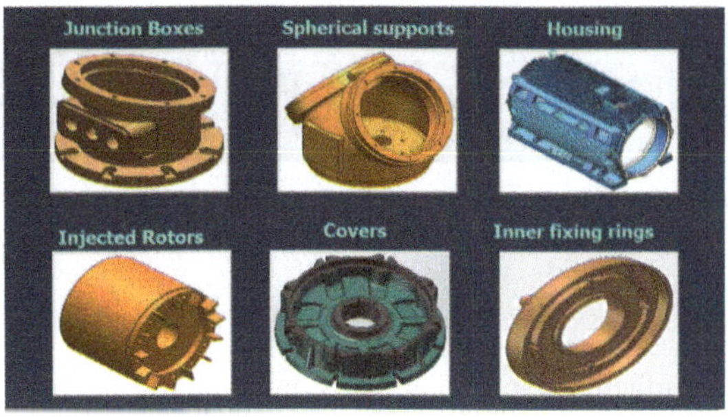

Fig. 2. Initial Page of the VBA Software

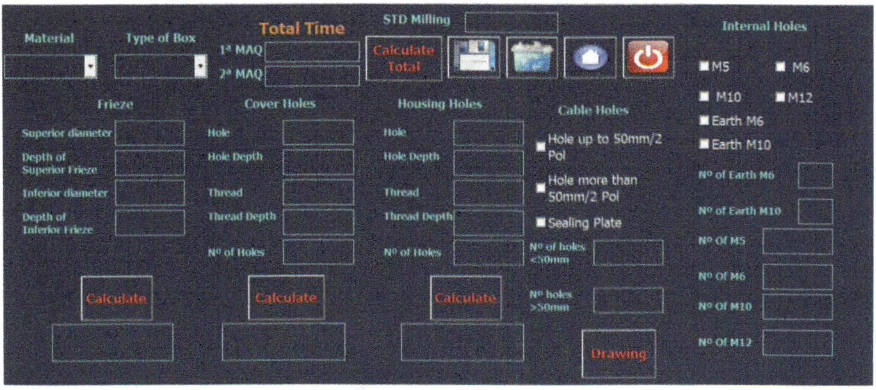

Fig. 3. Junction Boxes Userform

"Timing for Housings"). Upon executing the calculation commands in the VBA form, these time values are automatically imported. Operations where dimensions may vary based on customer orders, such as holes in covers and housings, must be filled in by the user according to the technical drawing provided in the SAP reference of the respective component. It is important to note that tool changeover time and tolerance measurement time have been timed and incorporated into the code.

After entering all data and performing the calculation, the user needs only to click to save the calculations in a sheet. In this sheet, calculation results are presented in a predefined layout for each component or part. Several SAP scripts are associated with this sheet, which automatically create the routing and input the previously calculated values. Therefore, the operator simply needs to enter the SAP component reference and click on the blue-bordered icon as illustrated in Fig. 4.

All other parts function in an identical manner to the junction boxes; the latter was used solely to illustrate the operational procedure.

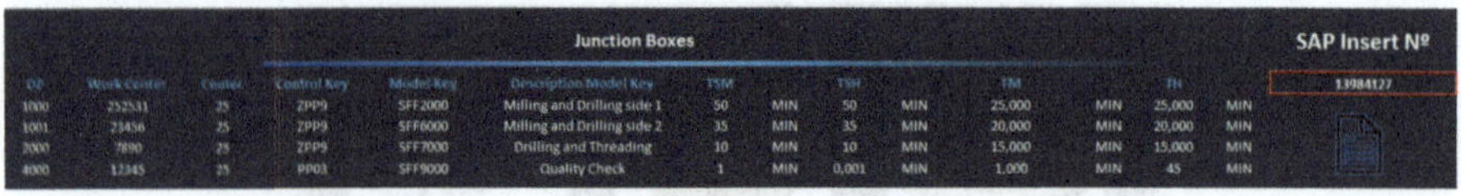

OP	Work Center	Center	Control Key	Model Key	Description Model Key	TSM		TSH		TM		TH	
			Junction Boxes										
1000	252531	25	ZPP9	SFF2000	Milling and Drilling side 1	50	MIN	50	MIN	25,000	MIN	25,000	MIN
1001	23456	25	ZPP9	SFF6000	Milling and Drilling side 2	35	MIN	35	MIN	20,000	MIN	20,000	MIN
2000	7890	25	ZPP9	SFF7000	Drilling and Threading	10	MIN	10	MIN	15,000	MIN	15,000	MIN
4000	12345	25	PP03	SFF9000	Quality Check	1	MIN	0,001	MIN	1,000	MIN	45	MIN

SAP Insert Nº: 13984127

Fig. 4. Routing Layout

3.4 Load Control Tool

Since the factory manufactures special and customized motors, there is no standardized routine in production; in other words, adaptations to the production system are necessary daily, and production sequencing is therefore manually handled by an associate. The primary objective of this tool is to simplify and enhance the efficiency of the production sector, specifically production sequencing, by visualizing the workload imposed by Production Planning and Control (PPC) and associated delays. This approach aims to enable operators to have a consolidated view and facilitate adjustments as mentioned earlier.

The workload control tool involves creating a Power BI dashboard (Fig. 5) and working with existing data in SAP, including previously calculated manufacturing times for components imported into SAP. Accordingly, every day at 7:30 AM, the scheduled workload planned by Production Planning and Control (PPC) and delays from the previous day's second shift are automatically exported from SAP to Power BI. At 3:30 PM, delays from the first shift are also automatically exported to assist the second shift.

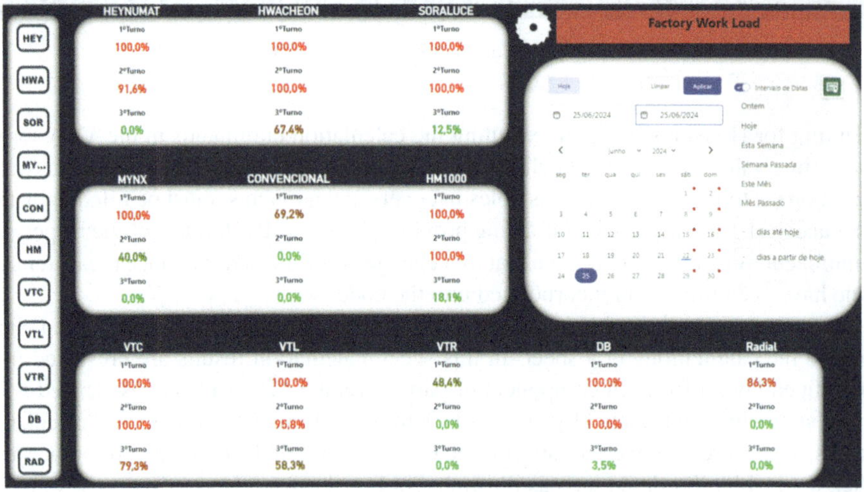

Fig. 5. Load Control Dashboard

In the initial dashboard (Fig. 5), an overview of the workload for the 11 machining machines in the factory is observed. Additionally, there is a column displaying machine abbreviations; clicking on each abbreviation directs users to a page (Fig. 26) where more detailed information is available regarding the components scheduled for production within the selected timeframe and their respective production delays. Figure 6 presents two primary pieces of information: firstly, operators can visualize delays by filtering the "Dt.Mot.Atr" field to the previous day for the first shift and to the current day for the second shift.

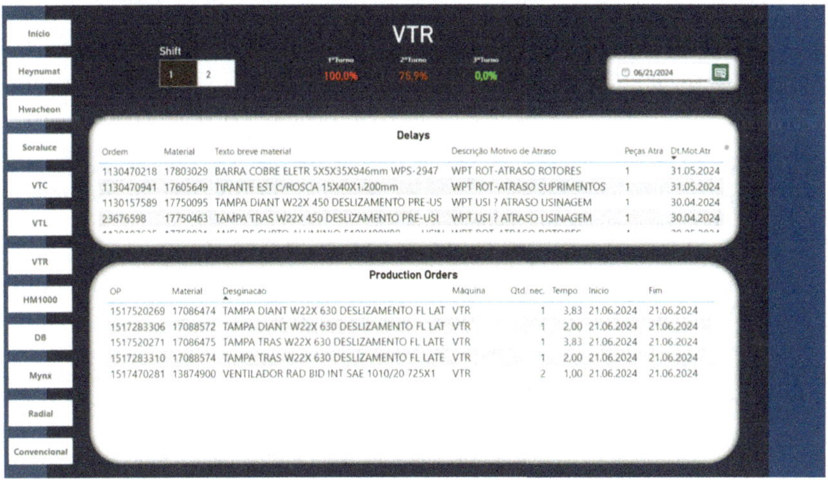

Fig. 6. Detailed View of each Machine

4 Conclusions

Through this work, significant improvements in the efficiency of process diagram execution have been noted, with a 42% reduction in the time required to create new routings and consequently a decrease in costs. For the production sequencing manager, tasks were substantially simplified. Consequently, there were improvements in the management of production sequencing, an acceleration of the process, and an optimization in the grouping of components. There has also been an increase in the reliability and accuracy of calculated times, as they have been validated on the factory floor.

Acknowledgments. This work has received support from the European Union under the Next Generation EU, through a grant of the Portuguese Republic's Recovery and Resilience Plan (PRR) Partnership Agreement, as part of the PRODUTECH R3 project – "Agenda Mobilizadora da Fileira das Tecnologias de Produção para a Reindustrialização", aimed at the mobilization of the production technologies industry to promote the reindustrialization of the manufacturing sector (Project ref. nr. 60 - C645808870-00000067; Total project investment: 166.988.013,71 Euros; Total Grant: 97.111.730,27 Euros).

References

1. Patel, M.B., Thakkar, H.R., Santosh, M.R.: Reducing manufacturing cycle time of milk tanks by work study technique in small scale fabrication industry. Int. J. Innov. Res. Sci. Eng. Technol. **4**, 12532–12541 (2007). https://doi.org/10.15680/IJIRSET.2015.0412163
2. George. Kanawaty and International Labour Office. Introduction to work study. International Labour Office (1992)
3. Meyers, F.E., Stewart, J.R.: Motion and Time Study for Lean Manufacturing (2002)
4. Abbasian-Hosseini, S.A., Nikakhtar, A., Ghoddousi, P.: Verification of lean construc-tion benefits through simulation modeling: a case study of bricklaying process. KSCE J. Civ. Eng. **18**(5), 1248–1260 (2014). https://doi.org/10.1007/s12205-014-0305-9
5. Pratiwi, A.I., et al.: Development of operation process chart and analysis of inventory control based on material requirement planning in assembly line. JIEHIS. **1**(1), 30–38 (2020)
6. Czégé, L.: Process analysis and optimization
7. Çaklt, E., Dağdeviren, M.: Comparative analysis of machine learning algorithms for predicting standard time in a manufacturing environment. Artif. Intell. Eng. Des. Anal. Manuf. AIEDAM **37**, e2 (2023). https://doi.org/10.1017/S0890060422000245
8. Rocha, A., et al.: Online scheduling: a survey (2023)
9. Gorobets, V., Holzwarth, V., Hirt, C., Jufer, N., Kunz, A.: A VR-based approach in conducting MTM for manual workplaces. Int. J. Adv. Manuf. Technol. **117**(7–8), 2501–2510 (2021). https://doi.org/10.1007/s00170-021-07260-7
10. Benter, M.: Derivation of MTM-UAS® analyses from virtual reality tools using MTMmotion®. In: Proceedings of the Conference on Production Systems and Logistics, Publishing in cooperation with TIB - Leibniz Information Centre for Science and Technology University Library, pp. 916–923 (2023). https://doi.org/10.15488/13510
11. Andreopoulos, E., Gorobets, V., Kunz, A.: Automatic MTM-Transcription in Virtual Reality Using the Digital Twin of a Workplace (2022). https://doi.org/10.21203/rs.3.rs-2110194/v1
12. Silva, H., Santos, A.S., Varela, L.R., Trojanowska, J., Ivanov, V.: Virtual and augmented reality: past, present, and future. In: Trojanowska, J., Kujawińska, A., Pavlenko, I., Husar, J. (eds.) Advances in Manufacturing IV: Volume 2 - Production Engineering: Digitalization, Sustainability and Industry Applications, pp. 46–59. Springer Nature Switzerland, Cham (2024). https://doi.org/10.1007/978-3-031-56444-4_4
13. Jung, W.K., Kim, H., Park, Y.C., Lee, J.W., Ahn, S.H.: Smart sewing work measurement system using IoT-based power monitoring device and approximation algorithm. Int. J. Prod. Res. **58**(20), 6202–6216 (2020). https://doi.org/10.1080/00207543.2019.1671629
14. Laring, J., Forsman, M., Kadefors, R., Örtengren, R.: MTM-based ergonomic workload analysis. Int. J. Ind. Ergon. **30**(3), 135–148 (2002). https://doi.org/10.1016/S0169-8141(02)00091-4
15. Faccio, M., Ferrari, E., Gamberi, M., Pilati, F.: Human factor analyser for work measurement of manual manufacturing and assembly processes. Int. J. Adv. Manuf. Technol. **103**(1–4), 861–877 (2019). https://doi.org/10.1007/s00170-019-03570-z
16. Sim, E.S., Lee, H.G., Lee, J.C., Park, J.W.: Efficient work measurement system of manufacturing cells using speech recognition and digital image processing technology. Int. J. Adv. Manuf. Technol. **29**(7–8), 772–785 (2006). https://doi.org/10.1007/s00170-005-2557-5

Traffic Analysis and Optimization of Network Equipment Using Artificial Intelligence

Marius Toderici[1] and Aurel Mihail Titu[2,3](✉)

[1] Faculty of Industrial Engineering and Robotics, National University of Science and Technology POLITEHNICA Bucharest, Splaiul Independentei No. 313, 6th District, Bucharest, Romania
marius@toderici.ro

[2] Faculty of Engineering, Industrial Engineering and Management Department, Lucian Blaga" University of Sibiu, 10 Victoriei Street, 550024 Sibiu, Romania
mihail.titu@ulbsibiu.ro

[3] Academy of Romanian Scientists, 3 Ilfov Street, Bucharest, Romania

Abstract. In the hybrid cloud infrastructure, data networks play a very important role, ensuring the connection between the local infrastructure and the public or private cloud infrastructure. Data traffic increases greatly when using the cloud and this will be achieved by increasing bandwidth and by segregating data on multiple connections. IT management must analyze the types of data and the quantities that are transmitted to the cloud and received from the cloud and, based on this analysis, establish the necessary connections and technical specifications to support communications within the cloud. Depending on the location where the on-premises or cloud data is stored, connections will be established so that communications ensure the data flow necessary for the proper functioning of IT services. IT services and applications used by the organization will use separate, encrypted communication channels to ensure data security. If the cloud email service is used, it is ideal to ensure this on a separate communication channel. Internet access for browsing and information will be provided separately from the communication channels necessary for IT services serving different applications. Modern communication equipment allows the definition of software networks that have the characteristics of physical networks, share the same physical communication environment but are separate and governed by their own communication rules. Software-defined networks are very versatile and have the great advantage of allowing the implementation of all these requirements in terms of flexibility, versatility and security, but the great advantage is given by the possibility of their automation.

Keywords: Network management · Hybrid cloud · Traffic optimization · Artificial intelligence · Network cost management

1 Introduction

In hybrid cloud infrastructure (Fig. 1), data networks play a very important role, providing the link between the on-premises infrastructure and the public or private cloud infrastructure. Data traffic increases greatly when using the cloud and this will be achieved by

D. D. Cioboată and J. Machado (Eds.): ICoRSE 2025, LNNS 1592, pp. 181–190, 2025.
https://doi.org/10.1007/978-3-032-02508-1_16

increasing bandwidth and segregating data across multiple connections. IT management needs to do an analysis of the types of data and how much is being sent to the cloud and how much is being received from the cloud, and based on this analysis determine the connection requirements and technical specifications to support communications within the cloud [1].

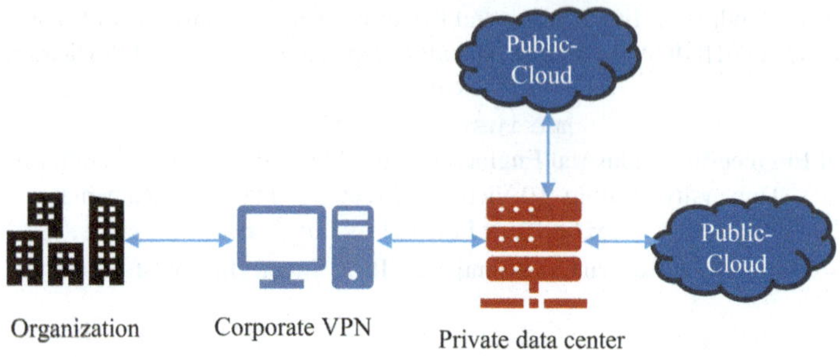

Fig. 1. Hybrid cloud infrastructure

Depending on where the data is stored on-premises or in the cloud, the connections will be determined so that the communications ensure the necessary data flow for the smooth operation of the IT services. IT services and applications used by the organization will use separate, encrypted communications channels to ensure data security. If the email service is used in the cloud, it is ideally secured on a separate communication channel. Access to the internet for browsing and information will be provided separately from the communication channels required by the IT services serving the different applications.

A requirement of the hybrid cloud is to ensure data and process flexibility, i.e. to allow data and processes to move seamlessly between the IT work environments provided by the hybrid cloud infrastructure. This requirement can be achieved very well using software defined networks (SDWANs) [2].

Modern communication equipment allows the definition of software defined networks that have the characteristics of physical networks, share the same physical communication environment but are separate and governed by their own communication rules. Software-defined networks are very versatile and have the great advantage that they allow the implementation of all these requirements on flexibility, versatility and security but the big advantage is the possibility to automate them [3].

There are different network equipment management software systems on the market. Large manufacturers of active network equipment provide users with different solutions for equipment management, traffic analysis and traffic management. With the help of these solutions, IT specialists, have an overview of the organization's network and their equipment. Through equipment monitoring and traffic analysis, data flow can be optimized both by establishing shorter routes and by optimizing equipment. By monitoring a month's data traffic in the organization, analysis data is created for traffic, equipment and IT services. Thus, from the analysis of this data, the network architecture can be optimized providing much better network scalability, increased security and ensuring the necessary capacity for the organization's data traffic [4].

The big challenge of the moment is the use of artificial intelligence to monitor the organization's network infrastructure and even optimize it. New network infrastructure monitoring and management tools that use artificial intelligence algorithms need a learning period within the network after which they can be used both to identify problems occurring in the network and to optimize or secure them [5].

2 Data Traffic Management

In a hybrid cloud infrastructure, reliable connectivity to both the internet and cloud services is essential, especially as data traffic continues to grow exponentially in both directions—to and from the cloud. For large enterprise organizations with distributed infrastructures, maintaining multiple data connections is critical to ensure communication redundancy and minimize the risk of downtime.

Each data connection should include both a primary and a secondary communication path, ideally sourced from different communication service providers. This approach helps to ensure continuity of service in the event of a provider outage or infrastructure failure.

Implementing redundant communication paths is particularly vital for all critical services within the organization. It is the responsibility of the IT management team, in collaboration with organizational leadership, to identify which IT systems and services are deemed critical to business operations. Once identified, appropriate redundancy measures—including communication path diversification—should be put in place to support those services [6].

Modern communications equipment may use one or more communications service providers and may be configured to switch between connections when a connection is very close to its maximum limit. In this case, policies can be defined to allow traffic splitting and maximum utilization of available active connections. Network traffic management plays a very important role in ensuring data flows and security. Network traffic rules need to be prepared in such a way that traffic and equipment load balancing, proactive monitoring can quickly identify traffic anomalies and quickly respond to potential threats that may be a sign of malicious activity or vulnerability. In this case by defined policies should isolate equipment that is making abnormal traffic to protect the integrity of the data in the organization [8].

Network traffic management is very important for delivering efficient and reliable IT services in large and very large organizations, and traffic control is at the heart of this process. By implementing traffic shaping and traffic control strategies, network administrators will ensure that networks are resilient, resilient and able to manage all communication flows, enabling them to avoid bottlenecks and reduce communication outages while delivering high quality communication services.

On the traffic optimization side, it is essential to prioritize different types of data or IT services according to their importance to the organization. Depending on the level of importance of a service it will be prioritized more on the bandwidth allocation side and will be routed to equipment that is less loaded in order to get the maximum performance required. For example, voice calls will be prioritized over video calls, but if the organization has a video conferencing solution this will also be prioritized. From this point of view, it is very important to prioritize communication services in order to ensure their optimization. The IT management has to establish the priority level for each type of service and then to ensure through continuous monitoring of traffic and equipment the necessary data flow for the smooth running of IT operations in the organization.

A proactive approach will not only ensure improved service performance but will also provide a better experience for internal and external users of the organization. By monitoring traffic, some parameters can be set such as the limits of data packets downloaded or uploaded by a user or a machine and when these limits are breached without warning, it is ideal for network administrators to be able to intervene by limiting access or even blocking the machine or user until the reasons for exceeding the limits are established in a way that minimizes data loss. Forming a set of rules on the data traffic side both on the downloading or uploading to the cloud helps to minimize data loss or data theft [9].

In this case traffic management software based on artificial intelligence is very useful. Applications of this type are installed on the network, prioritization policies, traffic limits are loaded and these are additionally left on the network to learn for a period of time. Afterwards they can be used for traffic management and traffic optimization by being able to make real-time decisions based on historical data, learning capabilities as well as on the set policies. The great advantage of these systems is that they are active 24/7 and can isolate a machine that is making abnormal traffic during off-hours, being a very versatile and useful tool in network traffic management.

3 Automatic Network Equipment Management

Network equipment management (Fig. 2) in an enterprise organization means managing tens, hundreds or thousands of active network devices such as switches, routers, firewall equipment, access points, servers, etc., which need to be maintained, patched, tested and then put into production [10].

Fig. 2. Network automation benefits

The primary objective of network management is to ensure that the network infrastructure is running efficiently, without disruption and smoothly under secure conditions. Any unplanned downtime translates into financial losses for the organization, long-term reputation damage, loss of customers, etc. Network management also means adding new equipment to the network and integrating it.

The network maintenance strategy should include an inventory of all network equipment as they are at a given point in time T0, a plan for upgrading equipment taking into account equipment dependencies, compatibility, critical equipment, equipment utilization.

The first thing that needs to be done is to inventory all the equipment, identifying the versions of firmware or software running on it. This is best done with the help of an equipment management application, in the absence of a management tool it is difficult to keep track of and access settings, remotely update, test or put updates into production. Once the inventory has been completed a backup of all the equipment on the network should be made so that if the need arises and an update does not complete successfully or if a piece of equipment does not behave well after the update it can be reverted to its previous settings. This step is mandatory before any update [11].

Table 1. Key AI Applications for Automatic Network Equipment Management

AI Technique	Use Case
Reinforcement Learning	Dynamic Routing, Load Balancing, Resource Allocation
Machine Learning	Traffic Classification, Anomaly Detection, Predictive Maintenance
Deep Learning (LSTM, CNN)	Traffic Forecasting, Intrusion Detection
Autoencoders	Network Fault and Anomaly Detection
Graph Neural Networks	Network-wide Decision Making, Topology-aware Optimization
Natural Language Processing (NLP)	AI-powered network management assistants (self-service)

Depending on the heterogeneity of the equipment in the network infrastructure, the manufacturer's recommendations for each update will be taken into account, pre-required prerequisites, dependencies related to compatibility with certain versions, etc. At this point, the files with the new firmware images will be downloaded and uploaded to the test equipment. Testing the new configurations or software updates is a very important step because at this step possible problems can be identified and fixed. The whole process will be documented and the upgrade procedure will be created for each type of equipment and the procedure for testing and validating the upgrade. In case the upgrade does not complete successfully it will revert to the initial state using the backup images/settings. In case not all equipment is available for testing, equipment on the network that is less used during certain periods of time can be identified and tested [12].

Once the upgrade scenario is prepared, the most appropriate time for the upgrade will be identified, most often during off-hours when the upgrades will be applied and tested according to the above procedure. Testing of the equipment will be done using various software applications that will validate the proper functioning of the equipment taking into account the most important parameters such as presence (ping), routing, latency and packet loss.

This patch management procedure must be applied for each type of equipment in the network. The patch management strategy must have as little impact as possible on the smooth operation of the data network, so the patch management strategy will define the time during which updates can be made, the time during which a device can be unavailable without affecting the smooth operation of the network. Depending on these parameters for certain equipment, specific network configurations can be established to allow them to be upgraded without impact, or traffic configurations can be established to avoid a piece of equipment if it is unavailable, which creates upgrade and test windows without impact on production [13].

4 Cost Management and Resource Optimization

An enterprise network is a complex system that generates a very large amount of data: logs, network health data, equipment health data, equipment configuration, telemetry and performance data. Collecting and analyzing them manually is impossible due to the multitude of equipment, so managing them can only be done with software management tools.

Proactive network management allows to reduce network failure rates. Downtime in which certain, often critical, network equipment is not working has a negative impact on the organization causing financial and image losses for the organization.

The use of equipment that allows the configuration of software-defined networks brings flexibility in use, deployment and allows long-term cost reduction as it allows the reuse of equipment by simply programming it. Another feature of SD-WAN is the high degree of automation, which means lower maintenance costs. Deployment of SD-WAN and virtual networks enables the automation of network processes which makes the work of IT administrators easier by enabling easier monitoring of network traffic, traffic balancing and routing, and resource optimization. Network virtualization enables the creation of logical networks that use the same physical network infrastructure inside the hybrid cloud but these networks can be highly controlled, segregated and network isolated ensuring the highest level of communication security.

This translates into serious savings both on the equipment side and on the information propagation media side because if we need a separate network for a service or an application, we no longer need to create physical links, they will be created quickly virtually using the same physical propagation medium for multiple subnets. Maintenance costs for these virtual networks are also practically non-existent, as they are done with the maintenance of physical equipment, which is one of the great advantages of virtual networks: low creation and maintenance costs.

If the organization uses redundant communications connections these can be exploited to their full potential by using automatic traffic balancing and traffic routing so that there is no need to purchase very large bandwidths at high cost, instead using traffic balancing systems and existing connections. Efficient traffic distribution will ensure network performance, application reliability and improve the user experience by reducing latency and bandwidth overhead of the main connection. At the same time it is recommended to segregate the communications taking into account the main IT services used as well as the critical level of usage for the organization. The hybrid cloud needs connection services for the main applications hosted in the public cloud. Email and business critical applications critical to the organization will have different redundant connections so that they are resilient.

The use of network monitoring and configuration software tools based on artificial intelligence algorithms and/or agents enables real-time network management by quickly identifying congestion, configuration issues, abnormal network traffic or security breaches. All this coupled with the use of virtual networks, encryption of data flowing to and from the cloud allows to quickly isolate and block potential security issues at an early stage.

5 Predictive Resource Analysis Using AI

The integration of artificial intelligence in hybrid cloud management brings major changes in the way organization can manage and optimize IT infrastructure. The hybrid cloud is a complex IT environment due to the various interconnected public and private integrated platforms that need advanced knowledge for administration and operation. Operating and managing a hybrid cloud system cannot be done without proper software tools.

To be able to do predictive resource analysis using artificial intelligence we need historical data or a model where the artificial intelligence system needs to be trained (Fig. 3). Thus, I think we have to start from the premise that we have a public cloud component with scalable on-demand resources where we host data and IT services consumed by different users. In the private cloud we host sensitive data and critical workloads for the organization as well as data that needs to be processed as close as possible to the source where it is produced (with low latency). Inter-connectivity between the two cloud systems is ensured by high-speed, redundant, end-to-end encrypted connections so that moving data between the two environments is fast and secure.

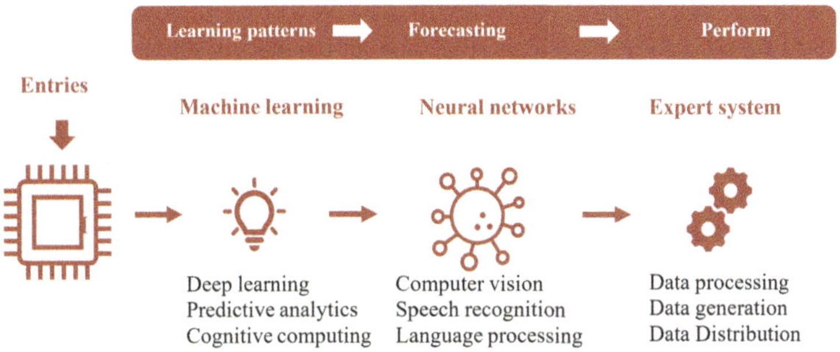

Fig. 3. Learning model using artificial intelligence

An artificial intelligence-based system can be added on top of this infrastructure using machine learning algorithms to analyze the data in some cases:

- Storage systems - It is essential to analyze the storage environments required to support data processing across various applications. An artificial intelligence (AI)-based system can evaluate data stored and processed by the organization's primary applications over a monitoring period of 1 to 3 months. Based on this analysis, the AI model can then forecast future storage and backup requirements, allowing for more accurate capacity planning and resource optimization. For data storage and retrieval, AI-powered solutions from providers such as Veeam and Rubrik are already in use. These platforms leverage machine learning to enhance data protection strategies, optimize backup scheduling, detect anomalies, and improve recovery processes, contributing to a more intelligent and resilient storage infrastructure.
- Data Processing - AI-based systems can be employed to analyze the processing power required by the organization's production IT applications over a period of several months. By examining usage patterns and performance metrics, these systems can generate accurate predictions of future processing requirements for both applications and operational processes. This predictive capability allows for proactive resource planning and optimization, ensuring that compute resources are allocated efficiently while avoiding overprovisioning or performance bottlenecks. AI-driven platforms such as Turbonomic and IBM Densify are commonly used in this context. These tools dynamically adjust workloads and optimize compute resources in real time, improving overall system performance and cost efficiency.
- Network optimization - On the communications network side, AI-based systems can analyze the full range of network components, including physical and virtual equipment, traffic patterns, and overall data volume. After a learning period, these systems can provide intelligent insights that help optimize network traffic flows, improve the performance of network infrastructure, and detect potential issues before they impact operations. AI systems can proactively identify anomalies, anticipate system failures, and detect possible security threats in real time, enabling faster incident response and minimizing downtime. In the area of network monitoring and optimization, platforms like Cisco DNA Center offer AI-driven capabilities for intelligent automation, policy enforcement, and continuous performance optimization.
- Security - AI-based security platforms leverage machine learning to identify zero-day vulnerabilities, detect abnormal behavior, and perform real-time threat analysis. By continuously collecting and analyzing data across the organization's environment, these platforms enable proactive detection and response to potential security incidents before they escalate. AI-driven security solutions can adapt to evolving threat landscapes, reduce false positives, and provide automated remediation recommendations, enhancing the overall cybersecurity posture. Leading AI-based platforms in this area include Prisma Cloud, CrowdStrike Falcon, and Palo Alto Networks, all of which offer advanced threat intelligence, behavioral analytics, and real-time protection across cloud and on-premise environments.

Although these systems are in their infancy, artificial intelligence-based and generative AI systems are moving fast and the pace is accelerating. Furthermore, machine learning capabilities enable levels of automation and resource management like never before and as AI capabilities evolve, cloud IT infrastructure management in the cloud will become automated, predictive and autonomic capabilities.

Hybrid solutions, deployed in the Enterprise using Artificial Intelligence (AI)-based technology combine cloud-based AI capabilities with security and control of on-premise systems. By leveraging the strengths of both environments, this method addresses business challenges by providing a wide range of tools to promote, innovate, increase productivity and facilitate better decision making. The adoption of hybrid AI applications across all industries emphasizes the flexibility and strategic significance of these cases and the advantages merit much further exploration. The benefits of using hybrid AI applications that incorporate AI technologies in enterprise environments bring a number of strategic benefits tailored to meet the specific requirements of today's organizations [7].

6 Conclusions

In today's rapidly evolving digital landscape, the demand for efficient, reliable, and scalable networks is higher than ever. Traditional rule-based and static network management techniques are no longer sufficient to cope with the dynamic and complex nature of modern traffic patterns. Artificial Intelligence (AI) has emerged as a powerful tool to address these challenges by enabling intelligent traffic analysis and optimization of network equipment.

Through advanced AI models such as machine learning, deep learning, reinforcement learning, and graph-based neural networks, network operators can now gain deeper insights into traffic behavior, predict future trends, and proactively resolve performance bottlenecks. AI-based traffic analysis allows for accurate classification of traffic types, detection of anomalies, and precise forecasting of traffic demands. Moreover, AI-driven optimization techniques enhance resource utilization, reduce congestion, improve Quality of Service (QoS), and contribute to energy-efficient networking.

The integration of AI into network equipment and management systems paves the way towards fully automated, self-optimizing, and self-healing networks, capable of adapting to changing conditions without human intervention. This not only leads to better network performance and user experience but also significantly reduces operational costs [14].

In conclusion, artificial intelligence plays a vital role in the future of traffic analysis and optimization of network equipment. Continued research and development in this area will further enhance the intelligence, automation, and resilience of next-generation networks.

References

1. Miliano, I. S.: Machine learning-based automated problem categorization in a helpdesk ticketing applica-tion. In: 8th International Conference on Orange Technology (ICOT), pp. 1–6. Daegu (2020)
2. Agarwal, S.B.: Automated assignment of helpdesk email tickets: an artificial intelligence life-cycle case study. AI Mag. **41**, 45–62 (2020)
3. Asma, A., Badr, B.: Intelligent Network Management and Control: Intelligent Security, Multi-criteria Optimization, Cloud Computing, Internet of Vehicles, Intelligent Radio. Willey (2021)

4. Gooley, J.D.Y.: Cisco Software-Defined. Cisco Press (2021)
5. Leng, T.: SD-WAN Solutin. Huawei Technologies Co. (2024)
6. Segeč, P., Moravčik, M.: SD-WAN - architecture, functions and benefits. In: 18th International Conference on Emerging eLearning Technologies and Applications (ICETA), pp. 593–599. Košice (2020)
7. Sagi, S.: Hybrid AI: harnessing the power of cloud and on-premise datacenter for enterprise AI use cases. J. Artif. Intell. Cloud Comput. SRC/JAICC-246 **230**, 2-4 (2024)
8. Sarwar, M.I., Abbas, Q.: Digital transformation of public sector governance with IT Service management–a pilot study. IEEE Access **11**, 6490–6512 (2023)
9. Uddin, M., Islam, S.: A dynamic access control model using authorising workflow and task-role-based access control. IEEE Access **7**, 166676–166689 (2019)
10. Majumdar, S., et al.: Security compliance auditing of identity and access management in the cloud: application to OpenStack. In: International Conference on Cloud Computing Technology and Science (CloudCom), pp. 58–65. IEE, Canada, Vancouver (2015)
11. Cheng, S., Jie, B.: Software-Defined Wide Area Network Architectures and Technologies, CRC Press (2021)
12. Miller, L.: Secure Access Service Edge (SASE), Palo Alto Ntetworks 2nd Special Edn. (2022)
13. Eiers, W., Sankaran, G.: Quantifying permissiveness of access control policies. In: 2022 IEEE/ACM 44th International Conference on Software Engineering (ICSE), pp. 1805–1817. Pittsburgh, USA (2022)
14. Carretero, J., Izquierdo-Moreno, G., Vasile-Cabezas, M., Garcia-Blas, J.: Federated Identity Architecture of the European eID System, vol. 6. IEEE Access (2018)

Some Considerations on Microfabrication Methods and Simulation of Reliable Si-Based MEMS Devices

Georgeta Ionaşcu[1], Edgar Moraru[1(✉)], Lucian Bogatu[1], Tudor Cătălin Apostolescu[2], Elena Manea[3], and Vlad-Andrei Stănescu[1]

[1] National University of Science and Technology, POLITEHNICA Bucharest 313, Splaiul Independentei, 060042 Bucharest, Romania
edgar.moraru@upb.ro
[2] Titu Maiorescu University 189, Calea Vacaresti, 0400511 Corp M- Bucharest, Romania
[3] National Institute for Research and Development in Microtechnologies, 126A, Erou Iancu Nicolae Street, 077190 Bucharest, Romania

Abstract. The paper describes the investigations of the authors on the methods of microfabrication to obtain important silicon MEMS (microelectromechanical systems) devices met in applications of microsensors and microactuators from micromechanics and microrobotics. There are considered an acceleration piezoresistive microsensor performed by bulk micromachining technique using crystalline anisotropy property of the material, monocrystalline silicon wafer, and as well as some thermal microtweezers made from silicon polycrystalline by surface micromachining using sacrificial layer technique. Simulations of modal analysis, and thermal and static structural analysis of the investigated structures are also performed. We comment on the reliability aspects from the perspective of the performance of the devices under study, imposed by the applications and influenced by the material, the fabrication method, the dimensions of the structures, the modulus of elasticity, the fundamental frequency, the vibration modes, and the thermo-mechanical characteristics, presented through simulations.

Keywords: Bulk micromachining · Surface micromachining · Microsensors · Microactuators · Reliable MEMS devices · Fusion 360 package

1 Introduction

Microelectromechanical systems, or more recently nanoelectromechanical systems, abbreviated MEMS or NEMS, are devices that incorporate mechanical and electrical components at the micro or nanoscale and play a crucial role in various fields of major importance including miniaturized sensors and actuators, telecommunications, automotive, medical devices, aerospace or microfluidic devices and so on [1–10]. Figure 1 shows some uses of MEMS in various fields of application. The importance of these microdevices derives not only from the perspective of the field of use, but also from the following considerations: MEMS technology is able to create very small devices

D. D. Cioboată and J. Machado (Eds.): ICoRSE 2025, LNNS 1592, pp. 191–204, 2025.
https://doi.org/10.1007/978-3-032-02508-1_17

that solve complex tasks; sensors, actuators and other electronic components can be integrated on a single chip, increasing functionality and streamlining costs; the manufacturing technologies for MEMS structures are similar to semiconductor fabrication, which can lead to reduced production costs and increased scalability [1, 11, 12].

The fundamental feature of MEMS applications is that they detect and influence phenomena appropriately, where micro-scale interactions are associated with macro-scale effects. Small size, low energy consumption, high reliability, low cost, and the ability to process and deliver undistorted signals in a form that the user can easily exploit, are important features of MEMS [1].

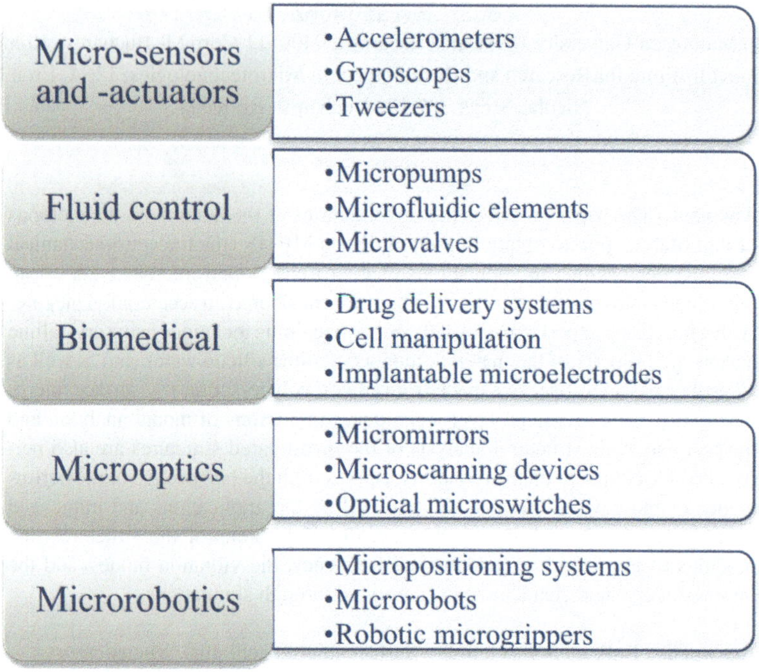

Fig. 1. Examples of MEMS uses in various fields

As in any other macromechanical applications, reliability aspects are extremely important when it comes to MEMS structures and their optimal operation. These can be discussed and analyzed from several perspectives, among which can be highlighted: durability and degradation over time aspects – material fatigue and wear over time can cause reliability problems; manufacturing and material aspects – reliability of MEMS devices is very closely related to materials and microfabrication techniques in order to ensure continuous performance; MEMS structures can be susceptible to environmental changes (e.g. pressure, humidity, temperature and so on) and implicitly affect reliability and longevity, and to mitigate these influences, appropriate packaging and design are considered; standards and testing protocols are constantly being developed and improved to guarantee that MEMS devices meet rigorous reliability and safety criteria (peculiarly medical or automotive areas) [13, 14].

As specified and mentioned above, to ensure adequate reliability and durability, the material and microfabrication technology used is extremely important. For the fabrication of MEMS structures, the following micromachining techniques are most often used: bulk micromachining, surface micromachining and LIGA process (mostly for metal and polymeric micromechanical structures) [1, 15–17], but also additive technologies with concepts similar to the macro domain could be considered [18–21]. The first two of these are the most suitable for processing silicon and for obtaining MEMS devices (due to the integrability of MEMS into common structures with electronic signal processing systems). Bulk micromachining allows the production of cavities of various shapes, closed or open holes, flat/profiled membranes, elastic elements such as leaf springs or planar spiral springs. The creation of these shapes is based on selective and anisotropic wet chemical etching and influencing the shape through controlled doping of the material from which they are made. On the other hand, surface micromachining or the sacrificial layer technique is based on the deposition and configuration of thin or thick layers whose profile in plane or space represents the negative of spaces or cavities necessary to obtain microstructures and which are ultimately "sacrificed", that is, removed (usually photoresists or silicon dioxide are used as sacrificial layers) [1]. A more detailed comparison between the two micromachining techniques will be presented in Sect. 2.

Considering the above discussed and the need of ensuring MEMS device reliability, which requires careful consideration of materials, manufacturing processes and environmental factors, the authors investigate some particularities on microfabrication methods of some MEMS Si-based devices and namely bulk micromachining for acceleration piezoresistive microsensor realization (single-crystal silicon) [2, 22–24] and surface micromachining for electrothermal microtweezers (polycrystalline silicon) [3, 25]. Also, the simulation of these microdevices with the help of finite element analysis is considered using simulation modules suitable for each structure according to its application.

2 Bulk Micromachining and Surface Micromachining Techniques for MEMS

Bulk micromachining and surface micromachining are two Si-based fundamental techniques that are used in the field of microfabrication to realize MEMS components or to obtain other microstructures and in each technique, specific manufacturing methods and approaches and materials are used to create the microstructure [1–3, 16]. Figure 2 presents a comparison of silicon based technological approaches for MEMS realization in terms of the main features. Practically, bulk micromachining creates microstructures by selectively etching within (in the volume of) the substrate (usually single-crystal silicon wafer or glass and other ceramic materials), while surface micromachining consists of creating microstructures by deposition and etching the structural layers (usually polycrystalline silicon) followed by removing structures from the substrate through etching the sacrificial layers. Regarding the process technology, in the case of bulk micromachining, anisotropic wet etching is used (dry etching can also be used), which exploits the specific properties of monocrystalline silicon and can result in characteristic shapes and cavities depending on the crystallographic planes. In surface micromachining the main technology is represented by thin film depositions, correlated with chemical etching and layer configuration through lithographic processes.

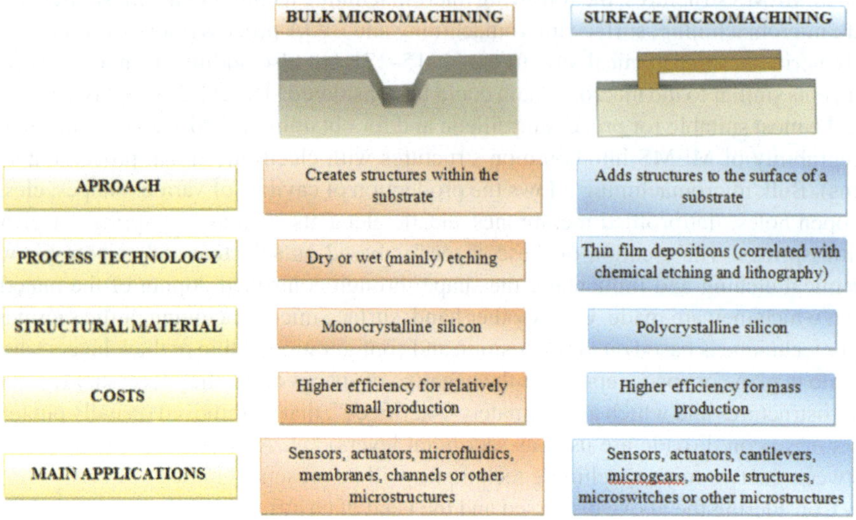

Fig. 2. Comparison between bulk micromachining and surface micromachining

In addition to the main single-crystalline silicon and polycrystalline silicon materials, other auxiliary materials are also used in these processes: in bulk micromachining - SiO_2 (insulating or sacrificial layer), Si_3N_4 (passivation layer), polymers (sacrificial layers), metals (electrical contacts or even structural layers); in surface micromachining – single-crystalline silicon (substrate wafer), SiO_2 (sacrificial layer), Si_3N_4 (structural or passivation layer), metals (contacts or structural layers), resists. The main technology of the bulk micromachining process is anisotropic wet chemical etching, but in some cases DRIE (Deep Reactive Ion Etching) or IBE (Ion Beam Etching), i.e. dry etching methods are also used for structures with a high aspect ratio. In the case of surface micromachining, a wider and more complex range of technologies is used: thin film depositions, for example chemical vapor deposition methods (CVD) for polycrystalline silicon or physical vapor depositions (PVD) for metals (sputtering or thermal evaporation), photolithography or e-beam lithography for layer configuration, isotropic wet and dry chemical etching to remove the sacrificial layers and to define the structures [1, 16, 26–29].

In terms of geometric complexity, bulk micromachining has the capability to achieve relatively larger and deeper three-dimensional structures with high aspect ratio and higher complexity (when deep etching and significant material removal are required), while surface micromachining results in less deep and usually (more) flat structures, in general structures with greater complexity can be obtained due to the additive approach in layers (including from different materials).

If we talk about manufacturing costs, bulk micromachining is more suitable and efficient for relatively small production (relatively larger structures and a small number of identical structures), but for large production volumes the costs increase due to the removal of a significant amount of material and the use of complex equipment and conditions for deep etching. In contrast, surface micromachining is more suitable and efficient for mass production (less material is used, and it is compatible with semiconductor device manufacturing technologies), but generally involves higher costs because it has more technological stages and implicitly uses more materials and equipment.

The main applications include: bulk micromachining – sensors, actuators, membranes, various complex channels and cavities and other microstructures [1, 30–32]; surface micromachining – sensors, actuators, cantilevers, resonant structures, microtweezers and other microstructures [1, 30, 33, 34].

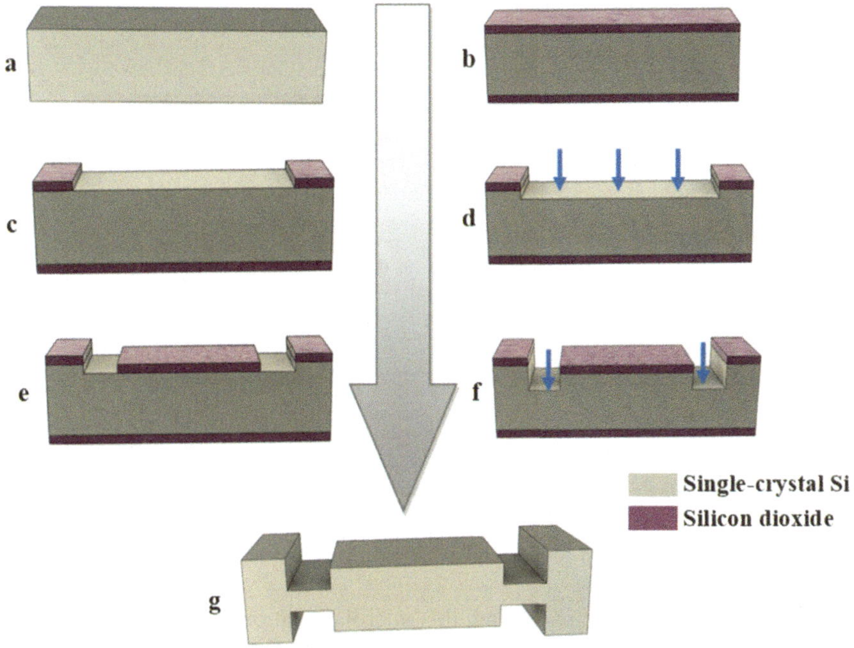

Fig. 3. Schematics of bulk micromachining process

The processing schemes for the two discussed methods are presented below. Figure 3 illustrates the main steps for obtaining a microstructure by means of bulk micromachining [1, 2, 16], namely:

a. – substrate preliminary processing steps which consist in bringing the substrate to the technical conditions imposed by the subsequent processes (in the given example, double side polished p-type silicon wafer);
b. – silicon dioxide thin film deposition of silicon dioxide which serves as a mask. The deposition is done through CVD processes (e.g. thermal oxidation), with thicknesses that usually do not exceed 2 μm;

c. – configuring the oxide layer to create the desired shape through lithography and chemical etching (wet isotropic type with specific etchants, for example buffered HF);
d. – chemical etching of the substrate in silicon dioxide areas patterned in previous step. Anisotropic wet chemical etching (for example with KOH) is used, but in some cases dry etching techniques (DRIE or IBE) are used, especially when specific structures with a high aspect ratio or deeper configurations are necessary;
e. – depending on the desired final shape, step b) can be repeated: in this case another deposition of silicon dioxide;
f. – depending on the desired final shape, step d) can be repeated: in this case another chemical etching of the substrate in areas patterned in step e) to obtain deeper trenches/grooves;
g. – final obtained shape (suspended structure) after silicon dioxide removing (the same steps a)-f) were applied on the bottom side).

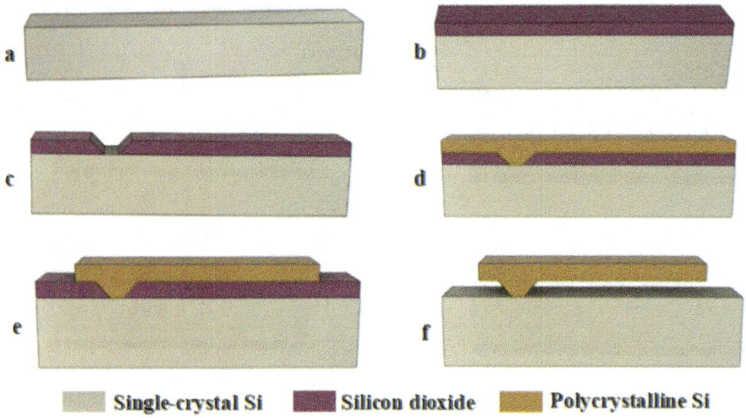

Fig. 4. Schematics of surface micromachining process

Regarding surface micromachining, this microfabrication method usually involves the use of more technologies and work steps, Fig. 4 [1, 3, 16] presenting a simplified graphical schematization of the sacrificial layer technique:

a. – substrate preliminary processing steps which consist in bringing the substrate to the technical conditions imposed by the subsequent processes (mechanical and mechano-chemical processing of silicon wafer);
b. – oxidation and doping to obtain the sacrificial layer – deposition of a thin layer of silicon dioxide (by thermal oxidation). The thickness of this layer determines the distance from the microstructure to the substrate surface;
c. – the sacrificial layer is patterned by lithography and etching through by selective removal of material from the substrate surface;
d. – the polycrystalline silicon (structural layer) is deposited (usually by CVD techniques) with a thickness of a few micrometers. Depending on the structuring material,

other thin layer deposition techniques can be used – thermal evaporation, sputtering or even electrodeposition;

e. – patterning and defining the shape of the structural layer through a second combined lithography - etching process;
f. – removing (etching) the sacrificial layer and obtaining the final structure (cantilever beam).

The discussed microfabrication techniques have a special applicability in the MEMS field, each of them having specific processing particularities for obtaining certain configurations depending on the geometry, application and materials involved, and the following chapters exemplify and detail an application for each method.

3 Some Considerations Regarding Fabrication and Simulation of MEMS Accelerometers

This section describes the fabrication process that can be applied to obtain a piezoresistive microaccelerometer (Fig. 5) using the bulk micromachined technique [2, 22–24]. The preliminary stage consists of mechanical and mechano-chemical processing of the silicon substrate to ensure the conditions for subsequent processing stages, in this case obtaining a double side polished p-type silicon wafer (450 μm thickness).

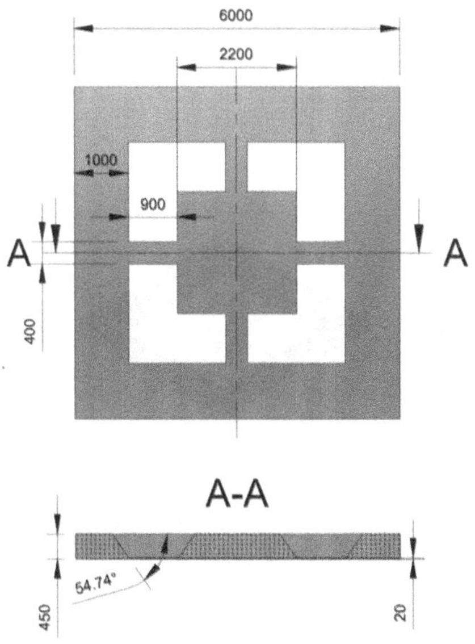

Fig. 5. Designed piezoresistive MEMS accelerometer

A technological approach follows that consists of combining technological processes applicable to integrated circuits and silicon bulk micromachining, namely (mentioned in order): firstly CVD process is applied (thermal oxidation) on the two sides of the silicon substrate (as a masking layer with a thickness of 180 nm); partial des-oxidation (on the front side) in order to counteract the stresses that may appear in the thin microbeams due to the too thick oxide layer; piezoresistor diffusion process (combine photolithographic and etching processes in order to create windows opening in SiO_2 on the front side of the silicon substrate and diffuse boron); same as the previous process for windows opening for metal contacts; thin film deposition of Cr-Au through PVD process (thermal evaporation) on the front side of the wafer; combined photolithography and wet etching for delimitation of metallization tracks; photolithography and etching for windows opening for anisotropic chemical etching on the front side of the silicon substrate; same as the previous process on the back side of the silicon substrate; anisotropic etching (front-back) of the silicon substrate with the help of potassium hydroxide; the last main step represents the wafer front side covering with a special resin with a protection role [2].

For the ideal structure of accelerometer (Fig. 5, dimensions are expressed in micrometers), considering it entirely made from silicon, as an isotropic model, a numerical simulation by FEM (Finite Element Method) was performed by means of Autodesk Fusion (2601.1.37) software, respectively *Modal Frequencies* module Fusion 360 package, using 218575 elements of 3D solid interconnected in 327707 nodes. The material data used in computation were: density $\rho_{Si} = 2330\ kg/m^3$, Poisson' ratio $\nu_{Si} = 0.22$, and $E_{dynamic} = 103.9$ GPa representing the dynamic Young's modulus, analytically computed with formula of beam model assimilated to the suspension arms of the inertial mass and the experimentally values of the natural frequencies measured on MSA-500 Micro System Analyzer by Laser-Doppler Vibrometry (LDV) method [23]. The main general steps of simulation were applied: material properties assignment (single-crystal silicon) [2, 22–24], boundary conditions setting, meshing and solving the problem. The structure was fixed on the lateral surfaces, cancelling all degrees of freedom in these areas.

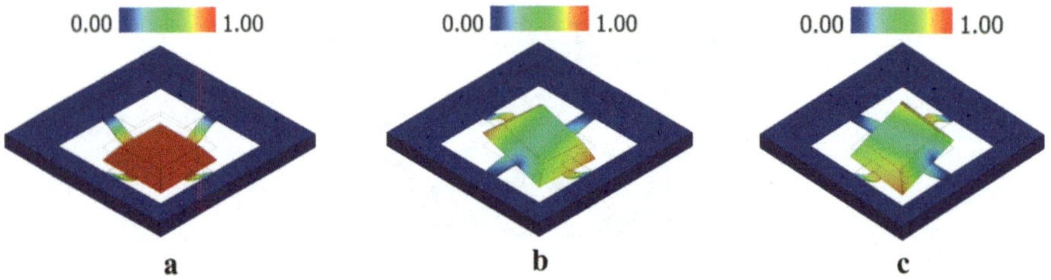

Fig. 6. The first three modes of vibration and the resonance frequencies of the analyzed piezoresistive MEMS accelerometer, determined by FEM with Fusion 360 package. a) ode 1: 2668.68 Hz - translation of the inertial mass along the Oz axis; b) Mode 2: 5050.782 Hz - rotation of the inertial mass relative to the Ox axis; c) Mode 3: 5052.203 Hz - rotation of the inertial mass relative to the Oy axis

After running the numerical model, a total of eight vibration modes were obtained, the most important ones being illustrated in Fig. 6 with the following frequencies obtained: 2668.68 Hz, 5050.782 Hz and 5052.203 Hz respectively. It can be seen that the first vibration mode has an integral mass participation on the Z axis, and the last two modes have close values due to symmetry aspects. The natural frequencies have high values, which is due to a very small oscillating mass in relation to the fairly high rigidity. The other vibration modes are less representative, some of them exceeding frequencies of 100 kHz.

The natural frequencies from the first to the third vibration mode, obtained by FEM were compared with the resonance frequencies measured with MSA-500 Analyzer [2], Table 1. Deviations of 11.9% have been obtained for all the three modes of vibration, considering the average value for the second and the third modes. After measuring, significant differences between the similar modes of vibration (2, 3) were found. Due to the structure symmetry of accelerometer, they would be equal or very close, as for the computed values. These differences can be explained by various behaviors of the four suspension arms due to technological execution errors and material anisotropy.

Table 1. The first three vibration modes of the MEMS accelerometer structure, experimentally determined and computed with FEM, Fusion 360 package

Vibration mode	f (Hz) measured with MSA-500 Analyzer	FEM with $E_{dynamic} = 103.9$ GPa		
		f (Hz)	Deviation (%)	
1	2384.4	2668.68	11.9	
2.	4232.8	5050.782	19.3	11.9 (average value)
3.	4836.7	5052.203	4.4	

These bulk micromachined MEMS devices that utilize the piezoresistive effect to measure acceleration can find multiple and diverse extremely important applications that are continuously growing due to their indisputable advantages, such as small size, efficient energy consumption or high sensitivity. Among these applications it can be mentioned: a variety of applications in automotive and aerospace fields, seismic sensors, medical and biological fields, robotics and many other examples.

Simulating these microdevices represents a powerful tool and is essential to ensure their reliability and performance over time from several possible reasons: interpreting the vibrational behaviour of the structure and avoiding unwanted vibrations that can cause major problems in the applications in which it is integrated; helps in the appropriate choice of material that can respond differently to vibrations, thus contributing to the optimization of reliability and performance of the system; information is received about the need to modify the geometric structure to reach the necessary modal frequencies; verifying the influence of external stresses (thermal, mechanical, etc.) on the vibrational behaviour of the device under different operating conditions, an extremely important aspect for ensuring the reliability, efficiency and durability of MEMS structures.

4 Some Considerations Regarding Fabrication and Simulation of MEMS Microgrippers

Regarding the structure that can be achieved by surface micromachining, this section exemplifies the fabrication process for a microtweezer (Fig. 7) that is based on two U-type electrothermal microactuators [3] with a symmetrical arrangement and a prehension mechanism [25, 35]. As with most MEMS structures, the first stage consists of applying mechanical and mechano-chemical processes to the silicon wafer, i.e. preparing the substrate for the following technological steps.

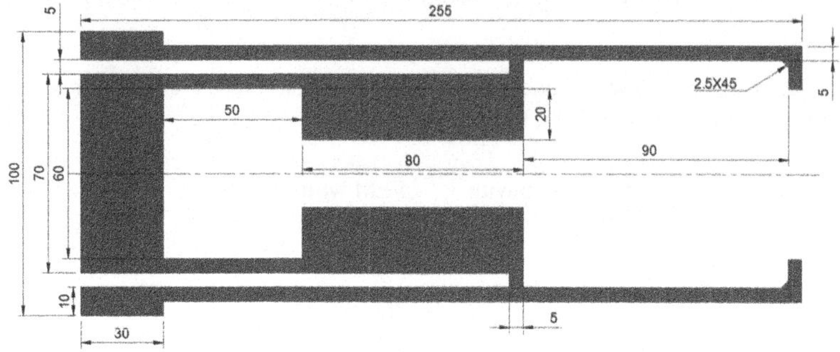

Fig. 7. Designed MEMS tweezer

After this, the work stages of the sacrificial layer technique begin, namely (main steps are mentioned in order): deposition of electrical insulator layer (silicon nitride – Si_3N_4 through LPCVD process – low pressure chemical vapor deposition) between single-crystal silicon wafer and polycrystalline silicon; deposition of sacrificial layer – in this case silicon dioxide through CVD process; photolithography and etching process in order to pattern the silicon dioxide for subsequent polysilicon deposition; deposition of structural layer – in this case polycrystalline silicon through LPCVD process; the same patterning (as in the case of SiO_2) to create the desired configuration for polysilicon structural layer; in order to release residual and inherent stresses that can be appear in structural layer, a special heat treatment process (RTP – Rapid thermal process) can be applied; removing the sacrificial layer by means of wet etching (HF) [3]. A thin metallic layer can be modeled using the lift-off technique above the structural layer before removing the sacrificial layer in order to improve operating characteristics (increase thermal expansion), but at the same time some problems can be caused (out-of-plane displacement), creating bimorph structure with different thermal properties [25].

To simulate the behavior of the microtweezer, the *Thermal stress* module (thermo-mechanical simulation) within the Fusion 360 program was used. The structure is based on two symmetric U-shaped electrothermal actuators [3] and a prehension device [25, 35], and the dimensional data of the simulated structure are presented in Fig. 7 (dimensions are expressed in micrometers). The microdevice is fixed (boundary condition) in the lateral anchors (left side of the figure). The assigned material is polycrystalline silicon with the appropriate material properties required in the simulation [3]. In terms of loads – it was applied a thermal load represented by a temperature of 200 °C on the hot arm (thinner arm) of the microactuators and 100 °C on the cold arm (thicker arm), respectively. Due to the variation of temperature and geometrical difference in the structure, a bending moment will be created in the microactuators which will lead to the "closing" of the prehension device, i.e. the approaching of the gripping elements. Figure 8 presents the obtained results: maximum of 122.321 MPa in terms of equivalent von Mises stress with its components of principal stresses (1st and 3rd principal stresses); displacement of 0.673 μm that occurs at a gripping element, i.e. almost 1.35 μm in total if the entire gripping mechanism is considered.

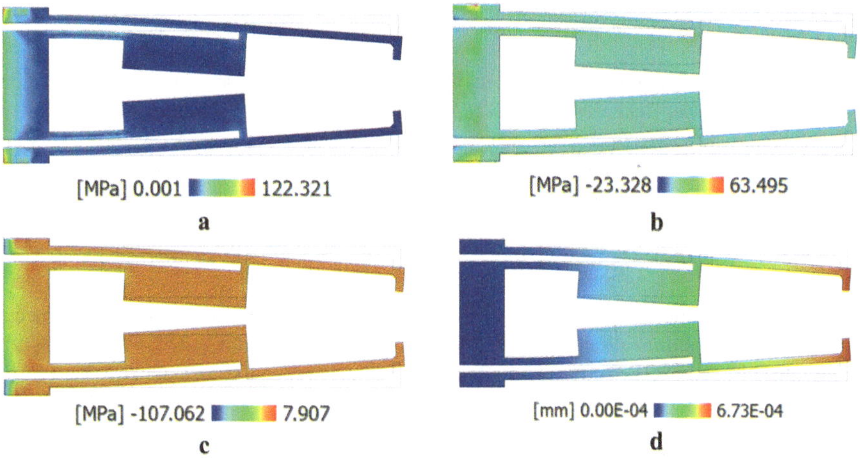

Fig. 8. Thermomechanical simulation results of MEMS tweezer: a – equivalent von Mises stress; b – 1st principal stress; c – 3rd principal stress; d – displacement

Simulating these microdevices is very important to anticipate the mechanical behavior of the structure and to choose the material, design and optimal operating conditions in order to ensure reliability and performance over time. These MEMS structures can be found in various micro scale applications such as biomedical fields (manipulating cells or other biological tissues or specimens), microrobotic systems or in material science and engineering to manipulate and characterize/test some micro-sized samples [3, 25, 35–40].

5 Conclusions

This paper exemplified the fabrication and simulation of some MEMS devices – piezoresistive sensor that can be realized by bulk micromachining technique and electrothermal microtweezer that can be realized by surface micromachining technique. The fabrication process through the appropriate choice of materials and technological conditions during manufacturing and pre- and post-processing steps are vital to ensure the reliability and performance of systems over time. At the same time, finite element simulation is an important tool that can help engineers and researchers to predict the behavior of a MEMS structure under certain loading or operating conditions. Practically, FEM analysis of MEMS structures can be used as a preceding stage to the fabrication process to reduce costs, time and increase system reliability.

Future research aim to develop the analysis and simulation of MEMS structures in terms of applying different materials and geometrical configurations (to study the influence of material, geometric and load variation on the structure) to investigated devices in order to improve the mechanical behavior and characteristics of the structures and to optimize their general design.

References

1. Ionascu, G.: Technologies of Microtechnics for MEMS (in Romanian). Cartea Universitara Publishing House, Bucharest (2004)
2. Comeaga, C.D., Bogatu, L., Ionascu, G., Alexandrescu, N., Sandu, A., Manea, E.: Manufacturing and testing technology for a silicon piezoresistive microaccelerometer. In: 2010 IEEE International Conference on Automation, Quality and Testing, Robotics (AQTR), vol. 2, pp. 1–6. IEEE (2010)
3. Moraru, E., Selagea, M.: Design, modeling and simulation of thermal microactuators. U.P.B. Sci. Bull., Series D, **80**(3), 19–30 (2018)
4. Percy, J.J., Kanthamani, S.: Revolutionizing wireless communication: a review perspective on design and optimization of RF MEMS switches. Microelectron. J. **139**, 105891 (2023)
5. Hommel, M., Knab, H., Yousef, S.G.: Reliability of automotive and consumer MEMS sensors-an overview. Microelectron. Reliab. **126**, 114252 (2021)
6. Bhatt, G., Manoharan, K., Chauhan, P.S., Bhattacharya, S.: MEMS sensors for automotive applications: a review. Sens. Autom. Aerospace Appl. 223–239 (2019)
7. Chircov, C., Grumezescu, A.M.: Microelectromechanical systems (MEMS) for biomedical applications. Micromachines **13**(2), 164 (2022)
8. Javed, Y., Mansoor, M., Shah, I.A.: A review of principles of MEMS pressure sensing with its aerospace applications. Sens. Rev. **39**(5), 652–664 (2019)
9. Ren, X., Yu, M., Kim, I., Nam, W., Qin, Y.: Advanced microfluidics for synthetic biology and/or biomedical applications. Front. Bioeng. Biotechnol. **12**, 1440206 (2024)
10. Hossain, N., et al.: Advances of materials science in MEMS applications: a review. Resul. Eng. 102115 (2024)
11. Fischer, A.C., et al.: Integrating mems and ICS. Microsyst. Nanoeng. **1**(1), 1–16 (2015)
12. Fedder, G.K., Howe, R.T., Liu, T.J.K., Quevy, E.P.: Technologies for cofabricating MEMS and electronics. Proc. IEEE **96**(2), 306–322 (2008)
13. Huang, Y., Vasan, A.S.S., Doraiswami, R., Osterman, M., Pecht, M.: MEMS reliability review. IEEE Trans. Device Mater. Reliab. **12**(2), 482–493 (2012)
14. Iannacci, J.: Reliability of MEMS: a perspective on failure mechanisms, improvement solutions and best practices at development level. Displays **37**, 62–71 (2015)
15. Moraru, E., Dontu, G.O., Cananau, S., Stanescu, V.A.: Approaches and processing technologies for medical devices: considerations from micro-and macroscale perspectives. In: International Conference on Reliable Systems Engineering, pp. 345–362. Springer, Cham (2023)
16. Antonello, R., Oboe R.: MEMS gyroscopes for consumers and industrial applications. Microsensors. InTech, 253–280 (2011)

17. Ionaşcu, G., Moraru, E., Bogatu, L., Apostolescu, T.C., Manea, E., Stănescu, V.A.: Research on using hot embossing technology for fabrication of polymer microstructures. In: International Conference on Reliable Systems Engineering, pp. 220–229. Springer, Cham (2024)
18. De Pasquale, G.: Additive manufacturing of micro-electro-mechanical systems (MEMS). Micromachines **12**(11), 1374 (2021)
19. Teh, K.S.: Additive direct-write microfabrication for MEMS: a review. Front. Mech. Eng. **12**, 490–509 (2017)
20. Besnea, D., Avram, M., Constantin, V., Spanu, A., Andrei, C.L.: Additive micromanufacturing technologies for composite materials based on nanotubes and polymers. Romanian Review Precision mechanics. Opt. Mechatron. (50), 16 (2016)
21. Ciobanu, R., Rizescu, C.I., Rizescu, D., Gramescu, B.: Surface durability of 3D-printed polymer gears. Appl. Sci. **14**(6), 2531 (2024)
22. Ionascu, G., Sandu, A., Bogatu, L., Comeaga, C. D., Manea, E., Besnea, D.: Simulation aspects of performance and experimental procedures for a microaccelerometer. UPB Sci. Bull., Series D **74**(4), 137–148 (2012)
23. Ionascu, G., Sandu, A., Bogatu, L., Comeaga, C. D., Manea, E., Besnea, D.: Computation models for studying the dynamic behaviour of mems structures. determination of silicon dynamic young's modulus (Edynamic). Romanian J. Inf. Sci. Technol. **14**(4), 365–379 (2011)
24. Ionascu, G., Comeaga, C.D., Bogatu, L., Sandu, A., Manea, E., Besnea, D.: Modelling of material properties for MEMS structures. J. Optoelectron. Adv. Mater. **13**(7), 998 (2011)
25. Cauchi, M., Grech, I., Mallia, B., Mollicone, P., Sammut, N.: Analytical, numerical and experimental study of a horizontal electrothermal MEMS microgripper for the deformability characterisation of human red blood cells. Micromachines **9**(3), 108 (2018)
26. Ionaşcu, G., Manea, E., Gavrilă, R., Moraru, E.: Perspectives on the use of thin films technologies in precision mechanics and mechatronics. Int. J. Mechatron. Appl. Mechan. **8**, 204–208 (2020)
27. Ionascu, G., Manea, E., Gavrila, R., Moraru, E.: Technologies for thin layers on ceramics substrate. In: International Conference on Reliable Systems Engineering, pp. 250–265. Springer, Cham (2021)
28. Ionascu, G.: Utilization of the Thin Film Structures Technologies in Precision Engineering and Mechatronics (in Romanian). Printech Publishing House, Bucharest (2004)
29. Ionascu, G., Bogatu, L., Manea, E., Moraru, E.: Defining the geometric configurations in thin films. Int. J. Mechatron. Appl. Mechan. **12**, 140–147 (2022)
30. Hierlemann, A., Baltes, H.: CMOS-based chemical microsensors. Analyst **128**(1), 15–28 (2003)
31. Pal, P., Swarnalatha, V., Rao, A.V.N., Pandey, A.K., Tanaka, H., Sato, K.: High speed silicon wet anisotropic etching for applications in bulk micromachining: a review. Micro Nano Syst. Lett. **9**, 1–59 (2021)
32. Paul, N., Bonam, S., Vanjari, S.R.K., Singh, S.G.: Silicon membranes for microbolometers: simulation and fabrication using frontend bulk micromachining. In: 2024 IEEE International Conference of Electron Devices Society Kolkata Chapter (EDKCON), pp. 623–628. IEEE (2024)
33. Pernu, T., Saarilahti, J., Kyynäräinen, J., Sillanpää, T.: Ultra-high sensitivity surface-micromachined capacitive differential pressure sensor for low-pressure applications. J. Microelectromech. Syst. **32**(1), 74–81 (2022)
34. Nabavi, S., Ménard, M., Nabki, F.: Surface micromachined out-of-plane electrostatic MEMS actuator integrated with displacement sensor. In: 2021 IEEE Sensors, pp. 1–4. IEEE (2021)
35. Kolahdoozan, M., Rouhani, E.A., Hassani, M.: Experimental and numerical investigation of the arms displacement in a new electrothermal MEMS actuator. Int. J. Adv. Des. Manuf. Technol. **10**(2), 71–81 (2017)

36. Pan, P., Wang, W., Ru, C., Sun, Y., Liu, X.: MEMS-based platforms for mechanical manipulation and characterization of cells. J. Micromech. Microeng. **27**(12), 123003 (2017)
37. Tabak, A.F.: Multi-scale robotics: a numerical investigation on mobile micro-tweezers for micro-manipulation with extreme requirements. Micromachines **16**(1), 40 (2024)
38. Botta, F., Rossi, A., Belfiore, N.P.: A feasibility study of a novel piezo MEMS tweezer for soft materials characterization. Appl. Sci. **9**(11), 2277 (2019)
39. Hopcroft, M.A., Nix, W.D., Kenny, T.W.: What is the Young's Modulus of Silicon? J. Microelectromech. S. **19**, 229 (2010)
40. Ionascu, G., Sandu, A., Manea, E., Bogatu, L.: Aspect regarding the elastic properties of Silicon and its influence of the behavior of micromechanical components. Int. Res. J. Eng. Technol. **3**(1), 16–20 (2016)

Co-Simulation Techniques for the Dynamic Testing of New Advanced Sandwich Structures

Roberto DI Bona, Erika Ottaviano(✉), Gabriel Testa, Alessandro Silvestri, and Andrea Checcacci

University of Cassino and Southern Lazio, Via Di Biasio 43, 03043 Cassino, Italy
ottaviano@unicas.it

Abstract. This work proposes the co-simulation concept applied to the dynamic testing of a new material produced by 3D printing and realized with sandwich structures for being used in the luxury sectors of Made in Italy industries, such as boat interiors, as well as in other key areas of the Italian economy, including cultural heritage restoration and preservation. In particular, this paper explores the material's behavior, which can be produced as the floor of a boat, in response to the dynamic action of the human walking on the surface for assessing the resistance to trampling. The co-simulation technology involves the interaction between two simulations working in sync, namely Multibody Dynamics (MBD) and Finite Element Method (FEM). It can be considered an emerging tool for adopting multidisciplinary integration. Co-simulation technology is essential for improving the connection between physical and virtual tests, optimizing processes, and efficiently managing data.

Keywords: Co-Simulation Techniques · Dynamic Simulation · Walking · Sandwich Structures · 3D Printing

1 Introduction

To achieve and sustain a competitive edge that relies on innovation and technology, it is essential to integrate digital processes into the design cycle of components, subsystems, and complex systems [1, 2]. This enables virtual exploration of all aspects of product development, including feasibility, performance, durability, cost, machining, and manufacturing [3]. Adopting multidisciplinary integration and co-simulation technology is crucial to reach this ambitious goal. This approach enhances the correlation between physical and virtual tests while optimizing processes and managing data effectively. Designing furnishings (e.g., for boats) presents unique challenges due to the stringent requirements for weight, mechanical strength, durability, aesthetics, and sustainability. While marble and wood are highly valued for their visual appeal and luxurious feel, they are also significantly heavy. Incorporating these materials directly into vehicle designs can result in undesirable consequences, such as reduced performance, higher fuel consumption, and compromised maneuverability. To address this issue, sandwich structures have become a popular solution. These structures combine thin layers of marble

D. D. Cioboată and J. Machado (Eds.): ICoRSE 2025, LNNS 1592, pp. 205–216, 2025.
https://doi.org/10.1007/978-3-032-02508-1_18

or wood with lightweight core materials, such as honeycomb or various foams, effectively reducing weight while preserving their aesthetic value. The strength-to-weight and stiffness-to-weight ratios of sandwich structures enable weight reduction in components and assemblies without sacrificing strength [4]. However, this approach introduces its complexities, requiring more intricate construction processes and specialized bonding techniques. In addition to aesthetics, these materials should possess adequate mechanical properties to be functional. Adopting those new materials can reduce greenhouse gas emissions and, consequently, the carbon footprint expressed in CO2 (carbon dioxide). The present research is developed thanks to the 3D-EcoCore project (founded by the European Union under the NextGenerationEU (PNRR) program), whose focus is the creation of new advanced sandwich structures using biodegradable and biocompatible materials produced through 3D printing. The project explores opportunities to use innovative technologies and sustainable materials to address these challenges. Challenges like delamination of layered structures, fractures within core materials, and the need for compatible adhesives across different materials demand advanced engineering solutions.

These solutions must effectively balance the complex requirements of aesthetic appeal, durability, and practical functionality, ensuring that the furnishings endure harsh environments while maintaining luxury and comfort standards.

Advanced architectural design is utilized across various materials, including metals, inorganic non-metallics, polymers, and more. To enhance the performance of these advanced materials, the architectural design can be refined through bio-inspired structures, optimization of structural parameters, and integration of multiple structural types [5, 6]. In advanced sandwich structures, composites with marble skins are particularly promising. These materials are used in high-end applications, such as nautical components, in which the visual appeal of marble is combined with lightweight cores to meet stringent weight requirements. They are also employed in exterior building panels, where durability and resistance to environmental conditions are essential.

Bio-inspired sandwich structures are starting to be studied for several diverse applications in a variety of industries, including automotive, structural, and biomedical, such as energy absorption [7–9], cushioning properties for engineering protective equipment [10], thermal insulation [11], and tuneable mechanical properties [12–16].

This paper aims to discuss a co-simulation test of the new advanced sandwich structures realized by a 3D-printing technique for producing furnishing and floors of boats. When dealing with new materials and applications, simulating the systems' behavior is crucial to optimize further and test engineering significance. The proposed idea and model can be extended to other materials and applications.

2 Multibody Modeling of Human Walking

Human walking has been studied using various approaches based on optimization techniques to develop controllers for simulation. Several software solutions, both open-source and commercial, offer biomechanical models. Notable examples include OpenSim [17], developed by Delp et al. in 2007 at the NCSRR at Stanford University, USA; BoB [18], created by Biomechanics in Bromsgrove, UK, and described by Shippen and May in 2016; and AnyBody [19], from AnyBody Technology A/S in Aalborg, Denmark, as reported by Damsgaard et al. in 2006. While these tools are highly effective, they are designed for specific applications and cannot simulate more general, complex, and diverse models. Additionally, they do not offer integrated multidisciplinary solutions. In this work, MSC Adams was used for the MBD modeling, MSC Marc Mentat dealt with the FEM simulation, and MSC CoSim provided the integrated co-simulation interface between the two solvers. The biomechanical model consists of 18 rigid bodies connected by 17 ideal spherical joints, resulting in 57 degrees of freedom in Cartesian space, as shown in Fig. 1. The computational model is defined by 228 mixed (natural+angular) coordinates, which include, for each time step, the natural coordinates of the joint centers, optical markers, and the center of mass (CoM) of the body segments.

The model is actuated with revolute joints, each following the Euler body-fixed XYZ formulation. Motion data were sourced from a Motion Capture (MoCap) environment reported in [20]. Inertia values were assigned to the body parts through an MSC Software plug-in, scaled to a 1.75 m tall human male and 75 kg.

PID-like torques were applied to the torso body segment to mimic the human body's equilibrium system. These acted by minimizing the error between the angular displacement measured at the current time increment and the corresponding value of the MoCap measurements.

The angular displacement error around Z was then measured (Fig. 2), peaking at 0.65°, so the model was confirmed to act consistently with the source data. The model was then prepared to run in a co-simulation by deactivating the body parts that are to be substituted by a FEM flexible body counterpart and selecting the markers responsible for the data exchange between the simulations. Two auxiliary parts were then modeled in correspondence with the ankle joints, and another set of PID forces was added. Detailed information on human body modeling is proposed in [21, 22].

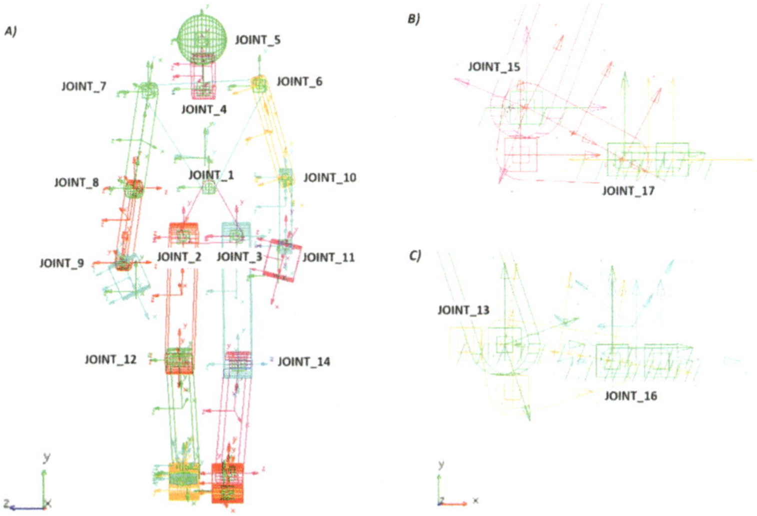

Fig. 1. The human body model in Adams with joint enumeration: ***A)*** Frontal view of the ***B)*** Side view of the left foot ***C)*** Side view of the right foot

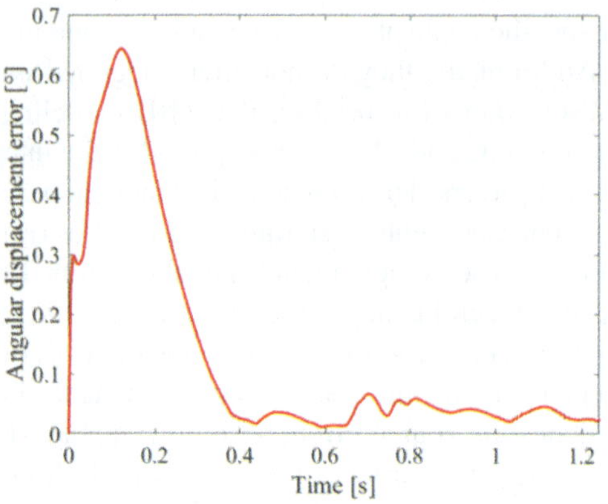

Fig. 2. Angular displacement error in Adams simulation

3 An Introduction to Co-Simulation

The current state of virtual prototyping for complex systems (such as aircraft, vehicles, machine tools, and robots) demonstrates impressive capabilities in predicting trajectory, force, torque, power, deformation, stress, noise, fatigue life, stability, vibrations, and more. However, some available software packages are limited in scope, addressing only specific features like kinematics, dynamics, structure, or noise. These tools may be model-based or data-driven, but often fail to accommodate new challenges, such as integrating lightweight materials for modern systems or higher dynamics requirements. Additionally, they can demand excessive computational resources—such as powerful hardware, extensive software, significant engineering time, and considerable computer time—and may not be user-friendly enough for a broader audience.

The main limitation today is the absence of a single tool capable of addressing all these varied problems using a unified mathematical model, and perhaps this is not the ultimate goal to strive for. In Sect. 4, a dynamic simulation will be considered.

4 Dynamic Simulations

The bio-degradable material considered for the activity was Polylactic acid, also known as PLA, a thermoplastic monomer derived from renewable, organic sources such as corn starch or sugar cane. It is known to be easily 3D printed using the Fused Filament Fabrication technique. The project aims to improve on sandwich composites through bio-inspired cores. A Voronoi geometry, based upon the sea urchin's microstructure, was designed through Rhinoceros' Grasshopper visual coding environment. The parametric 3D model was prepared to be comparable with the weight of common sandwich structures, as shown in Fig. 3. By changing the count and seeding of the cells and varying the strut diameter, the relative density of the core can be assigned. After iterations, a relative density of 13.5% was considered, corresponding to 160 kg/m^3 (H160).

The compression and 3-point bending tests, Fig. 4 and 5, were used to calibrate the mechanical behavior of the core to be used to model an equivalent FEM element. Since no plasticity should develop in the structure during walking, a simple power law flow rule was assumed to model the non-linear behavior of the core, neglecting the softening effect in the compression test related to the collapse of the strut.

The calibrated curve is reported in Fig. 6, and Table 1 summarizes the elastic parameters and the yield stress.

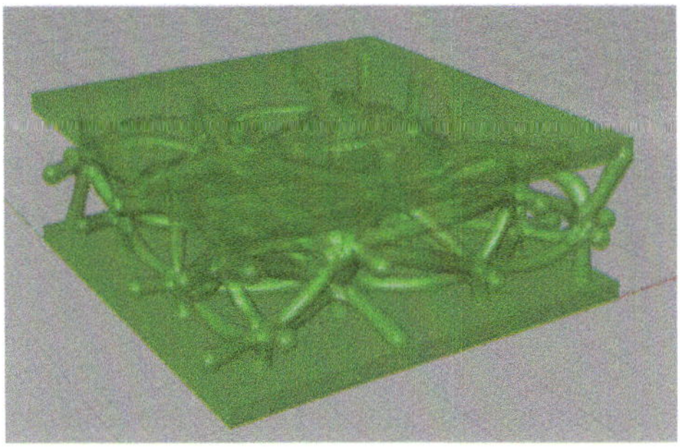

Fig. 3. 3D model of the bio-inspired core

Fig. 4. 3-point bending sample

In the scope of Computer Aided Engineering (CAE), the Finite Element Method (FEM) is a powerful tool at the disposal of engineers enabling them to study the behavior of materials through various analyses, such as structural, thermal, electrostatics, and magneto-dynamics. The FEM software used in this paper is MSC Marc Mentat, capable of efficiently and accurately solving multi-physics analyses. Marc is the implicit solver, while Mentat is the Graphical User Interface (GUI) for pre- and post-processing.

Given the complex geometry of the problem and the necessity to be coupled with the MBD android model, a 3D structural quasi-static FEM simulation was designed to be ready to run directly with the co-simulation. For this reason, the coordinate and unit systems are coincident.

Two simplified feet CAD models were realized starting from an open-access model [23], and a floor was made in a parallelepiped solid shape. The floor length and width were modeled according to the distance traveled by the android in the co-simulation, while the thickness was set at 20 mm. The CAD geometry was imported into the Mentat interface, as shown in Fig. 7.

Two material properties were created for the floor and feet, respectively. The PLA Eco-Core floor was assumed to be an isotropic elastoplastic material, featuring the fitted

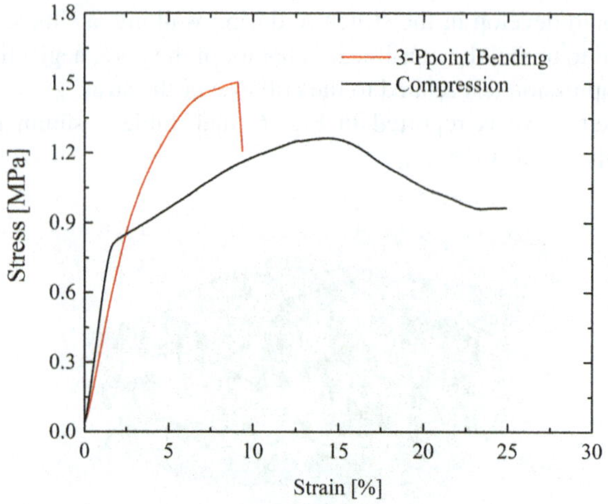

Fig. 5. Mechanical response of the PLA in compression and bending.

Table 1. Core elastic properties.

CHARACTERISTIC	VALUE
Young's Modulus	52 GPa
Poisson's Ratio	0.3
Yield stress	1MPa

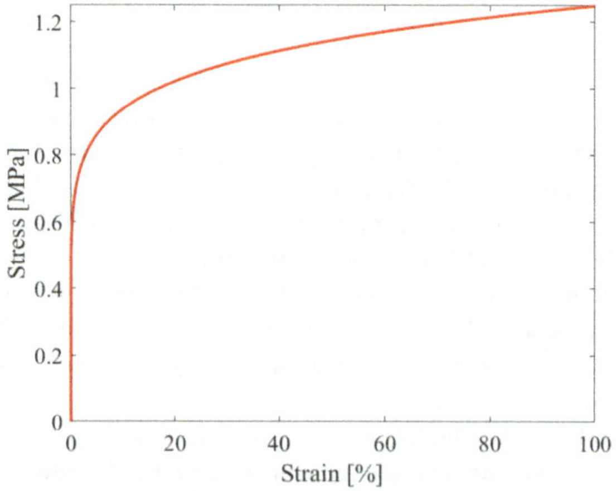

Fig. 6. Fitted stress-strain law for the core

stress-strain law in Fig. 6 and the elastic properties in Table 1. The parameters are valid since the floor is expected to experience mainly compressive loads.

Fig. 7. Front view of the 3D models featured in the FEM analyses.

The human foot is made of different tissues, each requiring extensive mechanical testing to be properly characterized, also depending on the individual age and health, thus posing significant challenges for FEM modeling. As this is not the focus of the work, a Mooney-Rivlin elastomer model was implemented for the feet, featuring material constants of a hyperelastic rubber sourced from [24]. The formulation is given in Eq. (1)

$$W^{TOD}_{deviatoric} = C_{10}(I_1 - 3) + C_{01}(I_2 - 3) \quad (1)$$

where:

- $W^{TOD}{}_{deviatoric}$ is the third-order degree - deviatoric strain energy
- C_{10} and C_{01} are material constants, in this case 1.8MPa and 1.57MPa respectively
- I_1 and I_2 are the first two invariants of the strain tensor

This material is expected to perform more rigidly globally than the real human feet. However, as a walking person typically wears shoes, the elastomer parameters are considered acceptable for modeling the surface of contact with the floor.

The mesh was made using the volume auto-mesher feature in the software. The resulting mesh, featuring bi-linear tetrahedral elements, is shown in Fig. 8. The limitations of the current co-simulation suite prevented the use of adaptive mesh techniques. Mooney-Rivlin materials require the use of elements computed through the Herrmann formulation. These features provide an additional degree of freedom in negative pressure and result in negative hydrostatic pressure. Three contact bodies were set to model the contact between the feet and the floor surfaces. Marc features various contact formulation implementations, either impact or restitution-based, the choice of which depends on the specific problem at hand. It was found that a very high friction coefficient was needed to prevent slip; further studies concerning the implementation of FEM contact in an MBD-FEM co-simulation are to be carried out. The communication between the two solvers in the co-simulation was modeled in the two markers/nodes corresponding to the ankle joints. In the FEM model, two isolated nodes were added at the exact coordinates as the MBD auxiliary parts and were rigidly linked to the respective upper feet surfaces' nodes. An imposed displacement boundary condition ensured the proper exchange of dynamic quantities through the nodes.

A detail of the boundary condition (in purple) and nodal links (in red) is shown in Fig. 9. The co-simulations featured a simulation time of 1.24 s and a time step of 5 ms. The results are shown in Figs. 10, 11, 12, 13, 14 and 15.

The co-simulation is a valuable tool for modeling the walk on the considered surface. Weight transfer during the gait phases can be evaluated. There is noticeable residual stress after lift-off due to the small plastic deformation computed numerically. However, the

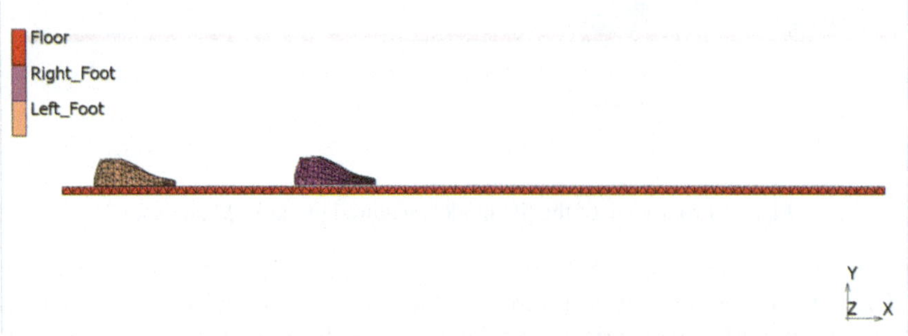

Fig. 8. Front view of contact bodies in the FEM analyses

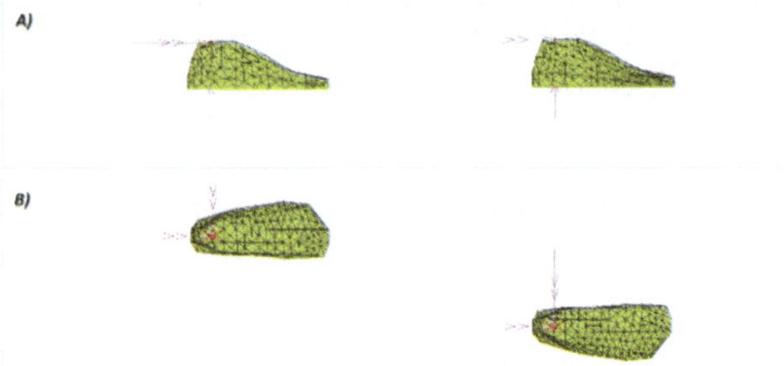

Fig. 9. A) Front view and ***B)*** Top view of the FEM boundary conditions and links

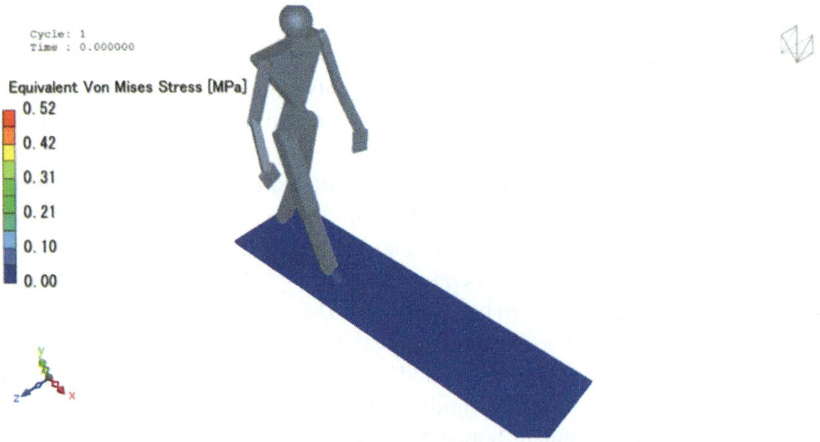

Fig. 10. 3D view of the co-simulation results at t = 0 s

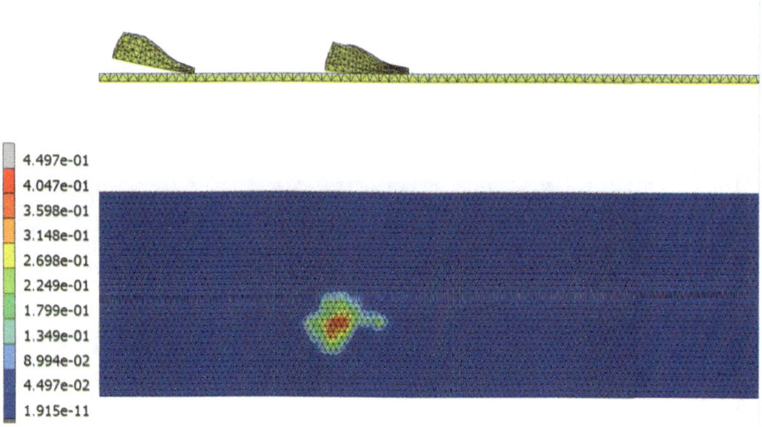

Fig. 11. Front and top view of the mesh and the Von Mises Stress on the floor at t = 0.095 s.

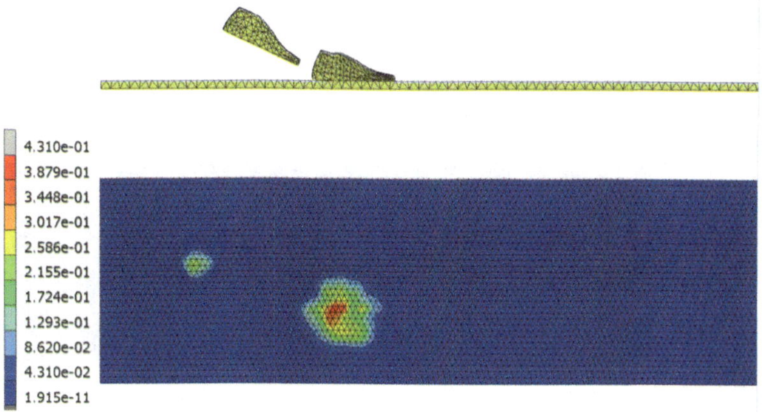

Fig. 12. Front and top view of the mesh and the Von Mises Stress on the floor at t = 0.315 s.

value remains well below the actual yield limit, as shown in Fig. 15. The computed Equivalent Von Mises stress remains throughout the simulation below the yield limit, with a maximum value of 0.52 MPa measured at t = 0.68 s (Fig. 13). However, the safety coefficient is <2, even for a fit person weighing 75 kg. Therefore, as a result of the proposed research, it has been verified that using this material is unadvisable for the intended purpose, but it can be used for paneling, ceilings, and walls.

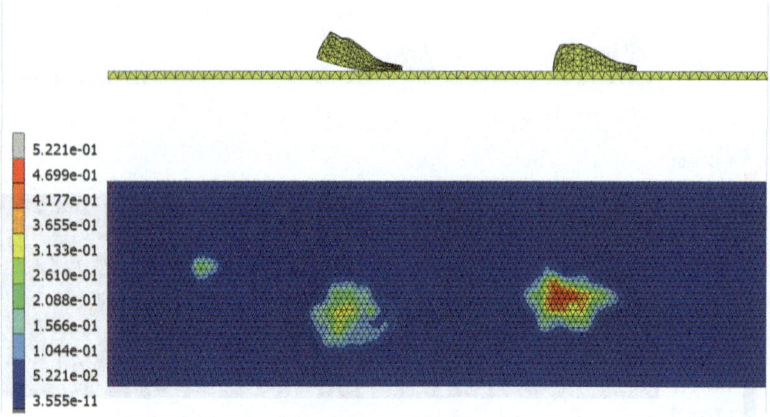

Fig. 13. Front and top view of the mesh and the Von Mises Stress on the floor at t = 0.68 s.

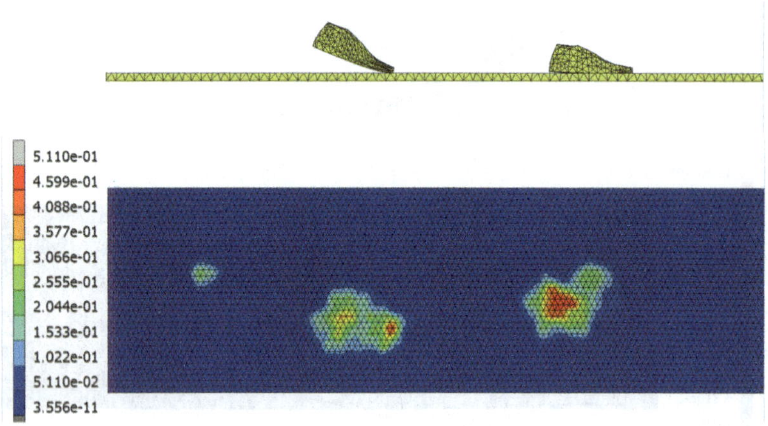

Fig. 14. Front and top view of the mesh and the Von Mises Stress on the floor at t = 0.73 s.

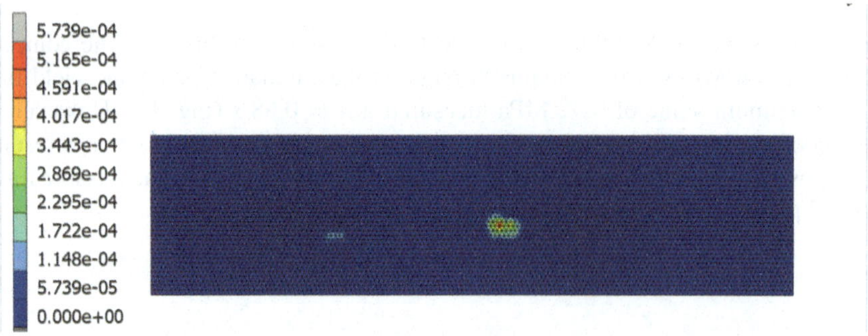

Fig. 15. Detailed top view of the equivalent plastic strain on the floor at t = 1.24 s.

5 Conclusions

We have proposed a co-simulation of MBD and FEM for the dynamic analysis of a new material produced by 3D printing and realized with sandwich structures for being used in specific applications, such as boat interiors. In particular, co-simulation techniques accomplish requests related to the practical use of the material, which has been designed and tested but has to be used in realistic scenarios. The paper answers a possible question related to the effects of dynamic walking in evaluating resistance to trampling. Still, it can be extended to any other type of dynamic load acting on it, and assessing the material behavior is a real case study, enabling an efficient model and tools for trampling resistance assessment.

Acknowledgments. This research was funded by the 3D ECO CORE project of MICS - Made in Italy Circolare e Sostenibile, an Extended Partnership between Universities Research Centers and Enterprises financed by MUR – Ministero dell'Università e della Ricerca thanks to funds by the European Union under the NextGenerationEU (PNRR) program.

References

1. Rea, P., Ottaviano, E., Antosz, K., Kozlowski, E., Lopes, H., Machado, J.: Simulation of robotic inspections based on systematic data acquisition and analysis. Int J. Simul. Model **24**(1), 29–40 (2025). https://doi.org/10.2507/IJSIMM24-1-708
2. Borràs, J., Thomas, F., Ottaviano, E., Ceccarelli, M.: A reconfigurable 5-DoF 5-SPU parallel platform. In: Proceedings of the ASME/IFToMM International Conference on Reconfigurable Mechanisms and Robots, ReMAR 2009, London, Article 5173892, 617–623 (2009)
3. Di Bona, G., Duraccio, V., Silvestri, A., Forcina, A.: Productive line reengineering through simulation techniques. In: Proceedings of the IASTED International Conference on Modelling, Identification and Control, pp. 291–296 (2014)
4. Kunzmann, C., Aliakbarpour, H., Ramezani, M.: Biomimetics design of sandwich-structured composites. J. Compos. Sci. **7**(8), 315 (2023). https://doi.org/10.3390/jcs7080315
5. Dai, H., Dai, W., Hu, Z., Zhang, W., Zhang, G., Guo, R.: Advanced composites inspired by biological structures and functions in nature: architecture design, strengthening mechanisms, and mechanical-functional responses. Adv. Sci. (Weinh). **10**(14), e2207192 (2023). https://doi.org/10.1002/advs.202207192
6. Che, S., Qu, G., Wang, G., Hao, Y., Sun, J., Ding, J.: A review of the biomimetic structural design of sandwich composite materials. Polymers **16**(20), 2925 (2024). https://doi.org/10.3390/polym16202925
7. Ha, N.S., Lee, T.-U., Ma, J., Li, J., Xie, Y.: Energy absorption of a bio-inspired cylindrical sandwich structure. Thin-Walled Struct. **195**, 111378 (2023). https://doi.org/10.1016/j.tws.2023.111378
8. Siddique, S.H., Hazell, P.J., Wang, H., Escobedo, J.P., Ameri A.A.H.: Lessons from nature: 3D printed bio-inspired porous structures for impact energy absorption – a review. Add. Manuf. **58** (2022). https://doi.org/10.1016/j.addma.2022.103051
9. Ghanbari, J., Panirani, P.N.: A hybrid bio-inspired sandwich structures for high strain rate energy absorption applications. Sci. Rep. **14**, 2865 (2024). https://doi.org/10.1038/s41598-024-53521-2

10. Xu, X., et al.: Cushioning performance design of metasandwich structures inspired by Kirigami. Virtual Physical Prototyp. **18**(1), e2285894 (2023). https://doi.org/10.1080/17452759.2023.2285894
11. Koštial, P., Rusnák, V., Malinarič, S., Jančíková, Z., Valíček, J., Harničárová, M.: Thermal properties of sandwiches for applications in transportation. In: Advanced Structured Materials, pp. 179–184. Springer, Heidelberg (2017)
12. Harhash, M., Palkowski, H., Sguazzo, C., Hartmann, S.: Experimental and numerical investigations ofmetal/polymer/metallaminates: thermo-mechanical and forming characteristics. Clausthal. Zent. Mater. **6**, 31–45 (2016)
13. Nayak, G.S., Palkowski, H., Carradò, A.: Precepts for designing sandwich materials. J. Exp. Theoret. Anal. **2**(1), 31–45 (2024). https://doi.org/10.3390/jeta2010003
14. Pelanconi, M., Ortona, A.: Nature-inspired, ultra-lightweight structures with Gyroid cores produced by additive manufacturing and reinforced by unidirectional carbon Fiber ribs. Materials **12**(24), 4134 (2019). https://doi.org/10.3390/ma12244134
15. Peng, X., Zhang, B., Wang, Z., et al.: Bioinspired strategies for excellent mechanical properties of composites. J. Bionic Eng. **19**, 1203–1228 (2022). https://doi.org/10.1007/s42235-022-00199-9
16. Solak, A., Aşçıoğlu Temiztaş, B., Bolat, B.: Wavy spider web honeycomb sandwich structures with PLA: Bending performance. https://doi.org/10.21203/rs.3.rs-4442290/v1 (2024)
17. Delp, S.L., et al.: OpenSim: open-source software to create and analyze dynamic simulations of movement. IEEE Trans. Biomed. Eng. **54**, 1940–1950. https://doi.org/10.1109/TBME.2007.901024. (2007)
18. Shippen, J., May, B. BoB – biomechanics in MATLAB. In: Proceedings of 11th International Conference Biomdlore 2016. VGTU Technika (2016)
19. Damsgaard, M., Rasmussen, J., Christensen, S.T., Surma, E., de Zee, M.: Analysis of musculoskeletal systems in the AnyBody modeling system. Simul. Model. Pract. Theory **14**, 1100–1111 (2006). https://doi.org/10.1016/j.simpat.2006.09.001
20. Michaud, F., Lamas, M., Lugrís, U., Cuadrado, J.: A fair and EMG-validated comparison of recruitment criteria, musculotendon models and muscle coordination strategies, for the inverse-dynamics based optimization of muscle forces during gait. J. Neuroeng. Rehabil. **18**, 17 (2021). https://doi.org/10.1186/s12984-021-00806-6
21. Di Bona, R., Catelani, D., Ottaviano, E., et al.: Cosimulation-based biomechanics of the human body with hip prosthesis. Multibody Syst. Dyn. (2025). https://doi.org/10.1007/s11044-025-10058-0
22. Gentile, D., Di Bona, R., Vanoli, G.P., Testa, G., Ricci, S.: Integrated FEM-multibody co-simulation of additively manufactured hip prosthesis containing cracks. Fract. Struct. Integr. **19**(71), 108–123 (2024). https://doi.org/10.3221/IGF-ESIS.71.09
23. Sketchfab. https://sketchfab.com/3d-models/simplified-foot-for-artists-13d7eba5e435462bba41351bb94e3935. Accessed 15 Apr 2025
24. Ning, H.: Experimental determination of the Mooney–Rivlin model for hyperelastic materials like rubber. Int. J. Acad. Res. Dev. **7**(4), 15–18 (2022). ISSN: 2455-4197

A Prototype Test Stand for Gear Transmissions Obtained Through Additive Technologies

Alexandra-Gabriela Vasilescu, Ionela-Mihaela Popescu(✉), Victor Constantin, Mihai Avram, and Bogdan Grămescu

National University of Science and Technology POLITEHNICA Bucharest, 313, Splaiul Independentei, 060042 Bucharest, Romania
ionela.baciu@upb.ro

Abstract. Additive manufacturing is increasingly used in mechanical engineering due to its significant advantages, such as reduced material consumption, the ability to create complex geometries, and shorter production times. This paper explores a method for designing and manufacturing gear transmissions using these technologies. To validate the selected transmission model, a dedicated test stand was first developed, providing a controlled environment for performance evaluation. Following the construction of this test stand, future improvements will focus on the integration of high-precision sensors and advanced monitoring systems. These additions will enable accurate measurement and analysis of key operational parameters, including torque, rotational speed, vibrations, and temperature. Monitoring these parameters is essential, as they directly affect mechanical efficiency, dimensional accuracy, operational stability, and system durability. The data obtained from future experimental tests will offer valuable input for optimizing the design, selecting appropriate materials for the additive manufacturing process, and enhancing the overall reliability of the transmission system.

Keywords: Additive manufacturing · Gear transmissions · Test stand

1 Introduction

The need for an innovative system, such as the one presented in this article (Chapter 3), is evident, as there are currently no accessible solutions for testing gear transmissions produced using additive manufacturing technologies. Most existing test stands (Chapter 2) are tailored for conventionally manufactured components, which limits the accurate assessment of systems developed through 3D printing.

Following the design and fabrication of the transmission system using additive manufacturing, a dedicated test stand was developed. This included custom-designed supports for precise alignment and secure mounting of the components, as well as the integration of necessary mechanical and control elements. The stand was intended to provide a controlled and repeatable environment for testing, while also ensuring ease of assembly and access for measurement and future system upgrades.

D. D. Cioboată and J. Machado (Eds.): ICoRSE 2025, LNNS 1592, pp. 217–228, 2025.
https://doi.org/10.1007/978-3-032-02508-1_19

This sequential approach—comprising the creation of the transmission, construction of the test rig, and its functional evaluation at various speeds—formed the foundation for more advanced investigations. Early experimental results indicated the need for real-time data acquisition to gain deeper insights into system performance. Consequently, subsequent sections of the article focus on the integration of advanced monitoring systems, including sensors and transducers capable of measuring torque, rotational speed, temperature, and position, thereby enabling comprehensive dynamic analysis and system optimization.

Incorporating these elements into the test stand will not only enhance the accuracy of the results but also provide a basis for the improvement of additive manufacturing techniques used in the production of gear transmissions. Additionally, this study contributes to the growing body of knowledge regarding the performance and reliability of 3D-printed mechanical systems.

2 Literature Review

Over time, numerous studies have investigated different types of test stands designed for gear transmissions, whether produced from plastic or metal materials [1, 2]. While some of these configurations remained at the conceptual level, others have been patented and are currently available on the market. Several examples of such systems are discussed in the following sections to highlight existing approaches and their limitations.

According to previous researchers, the development of technologies and materials has led to significant evolution in the field of test benches for gear transmissions. These are essential in the automotive industry and other industrial sectors for evaluating the performance, reliability, and durability of transmission components. In this context, recent research has made significant contributions to the design and implementation of advanced test benches that integrate modern monitoring and analysis technologies.

A notable example is the test bench for analyzing the dynamic behavior of planetary transmissions [3]. This system is equipped with an electric motor, a coupling, a transmission system, a torque sensor, a vibration sensor, and a data acquisition system. The main purpose of this test bench is to evaluate the dynamic behavior of planetary transmissions under varying operating conditions. The arrangement of sensors and components allows detailed monitoring and analysis of their performance.

Another example is the experimental bench for measuring the efficiency of spur gears [4]. This bench includes an electric motor, a transmission system, a torque sensor, a power measurement system, and a data acquisition system. The primary purpose is to evaluate the energy efficiency of spur gears under specific operating conditions. By integrating these components, the bench allows for precise data regarding the energy performance of transmission components.

Additionally, one of the researchers proposed a monitoring system based on "digital twin" technology for gear test benches [5]. This system integrates a virtual model of the test bench with real-time data collected from sensors such as accelerometers, torque sensors, and thermocouples. In this way, the behavior of the test stand under normal operating conditions can be simulated, and potential malfunctions or deviations from normal parameters can be identified, improving the reliability and efficiency of the tests.

Modern measurement and monitoring systems integrated into test benches are crucial for obtaining accurate data and evaluating the performance of transmission components. For example, in the test bench [5], sensors are used to measure torque, vibration, and temperature, providing essential data for evaluating the dynamic behavior of the transmission and for optimizing its design.

Similarly, in the experimental bench for measuring the efficiency of spur gears, power measurement systems and torque sensors are used to assess the energy efficiency of transmission components. These systems allow continuous monitoring of performance and the identification of potential energy losses or inefficiencies [6, 7].

In conclusion, the development and implementation of advanced test benches for gear transmissions are essential for improving the performance and reliability of transmission components. By integrating modern monitoring and analysis technologies such as "digital twin," it is possible to obtain precise and detailed data regarding their dynamic behavior [8, 9]. This information is crucial for optimizing design and ensuring efficient and reliable operation of transmission systems.

3 Proposed Solution

The proposed solution consists of developing a prototype model of a test stand for simple gear transmission. This approach aims to create an experimental system that allows for the analysis and evaluation of transmission performance in a controlled environment. To ensure the efficient operation of the test stand, it is essential to develop a clear conceptual diagram that illustrates the main components of the assembly and the interaction between them.

The novelty of the proposed stand is given by the possibility of testing various gear transmissions made using additive technologies and not only. During the testing, intelligent monitoring systems will be used to determine in real time its technical performance. The tested transmission model can be made of both metallic and non-metallic materials. Non-metallic ones find their utility in the construction of printers, toys, household appliances, but also in corrosive environments, where metal degrades, such as humid environments, or with chemicals.

3.1 General Description of the Stand's Principle of Operation

The principle scheme of the test stand comprises a structured arrangement of shafts, gears, and couplings, enabling realistic simulation of operational conditions. This setup facilitates the assessment of previously developed gear transmission in a controlled environment. The design ensures accurate motion transmission and supports efficiency and reliability analysis.

Figure 1 presents the principle diagram of the experimental test stand, which illustrates the functional interconnection between all key components involved in the operation and evaluation of a simple gear transmission system.

The electrical drive motor (2) functions as the primary source of mechanical energy, generating rotational motion that drives the entire assembly. This motion is transmitted through the elastic coupling (3), which connects the motor to the input shaft (4). The coupling plays a crucial role in damping mechanical shocks and vibrations, thus ensuring smooth operation and protecting the integrity of the connected components.

The input shaft (4) delivers the rotational motion to the gear transmission system (6), where the mechanical parameters—such as speed and torque—are modified in accordance with the gear ratio. The gear assembly is enclosed within the gear transmission housing (7), which serves both as a protective casing and as a structural constraint, maintaining proper alignment of internal elements and ensuring safe operation during testing.

Following the transmission stage, the modified rotational motion is carried by the output shaft (8) to the braking system (10). The braking mechanism introduces a controlled resistive torque, thereby simulating different loading conditions. This allows for real-time analysis of the transmission's behavior under variable mechanical stresses.

To reduce friction and ensure smooth, low-resistance rotation, both the input and output shafts are supported by bearings (5 and 9). These are mounted within dedicated bearing supports (12 and 15), which provide rigid structural alignment and minimize energy losses due to mechanical friction.

Central to the functionality of the stand is the electronic control and monitoring system (1). This system continuously tracks critical performance parameters, including speed, torque, and power. The data collected supports the accurate evaluation of system efficiency and offers valuable insights into design optimization and fault detection.

Mechanical stability and structural rigidity are achieved through the integration of several support elements: the motor support (11) ensures the correct positioning of the drive motor; the gear transmission box support (13) maintains the transmission housing in place; and the braking system support (16) secures the braking mechanism. All these elements are fixed to the main experimental stand support (14), which provides the foundational structure for the entire test setup.

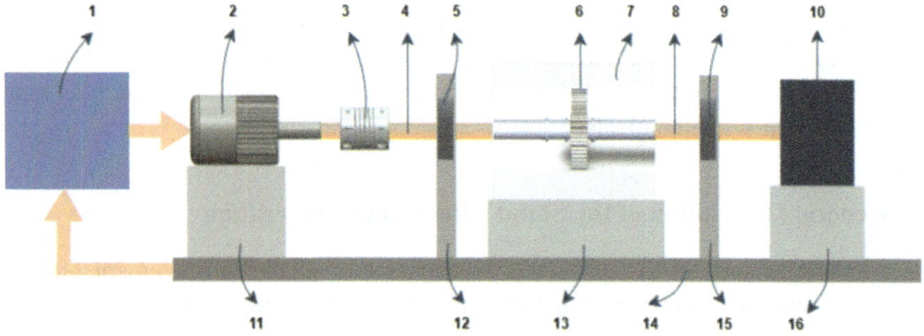

Fig. 1. Principle schematic of the test stand.

3.2 Prototype of Test Stand for Gear Transmission

This section presents the stages in the development of the test stand, from designing the basic structure and selecting the components to the actual assembly and functional testing of the system. Each stage is technically substantiated to ensure the coherence and relevance of the final model in the context of current engineering requirements.

To determine the dimensions of the test stand components, the design of the gear transmission was first required. The process began with the creation of a 3D model of a spur gear with straight teeth, as shown in Fig. 2.a. This model was based on an existing design used in a already deployed mechanism and developed in accordance with ISO 54 and ISO 21771 standards [10, 11], which define the geometrical parameters of gears, as detailed in Table 1. Furthermore, the design of the gears was also based on research conducted in a previous article, which provided valuable insights into the optimization of gear geometry, the creation of complex geometries, and their production using various additive technologies [12].

Table 1. Spur Gear Dimensions According to ISO 54/ISO 21771 [10, 11].

Parameter	Symbol	Value	Unit	Description
Module	m	2	mm	Defines tooth size; standard ISO gear module
Number of teeth	z	30	–	Determines pitch diameter
Pitch diameter	d	$m \times z = 60$	mm	Reference diameter of the gear
Outside diameter	d_a	$d + 2m = 64$	mm	Diameter including addendum
Base diameter	d_b	$d \times \cos(\alpha) \approx 56{,}57$	mm	For involute profile generation, $\alpha = 20°$
Root diameter	d_f	$d - 2.5m = 55$	mm	Includes dedendum extension
Face width	b	10	mm	Width of gear tooth face
Pressure angle	α	20	degrees	Standard value for involute gears
Circular pitch	p	$\pi \times m \approx 6.283$	mm	Distance between teeth along the pitch circle
Total tooth height	h	$2.25 \times m = 4.5$	mm	Addendum + dedendum
Addendum (tooth height above pitch circle)	h_a	$m = 2$	mm	
Dedendum (tooth depth below pitch circle)	h_f	$1.25 \times m = 2.5$	mm	

Figure 2.b) illustrates that the designed gear was fabricated using additive technologies, employing plastic materials such as PLA (Polylactic Acid).

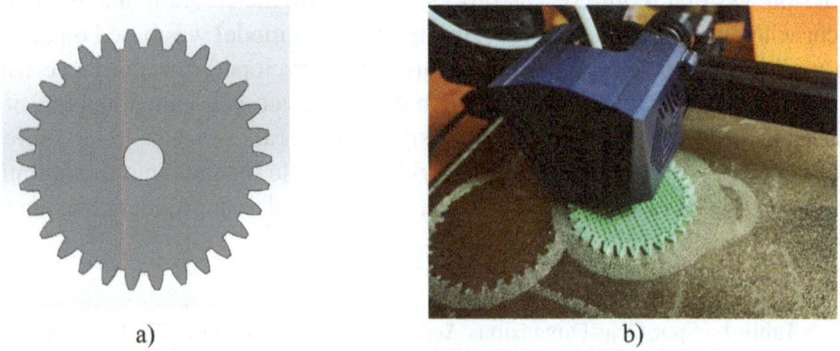

Fig. 2. a) 3D model of gear; b) Gear printing process.

To increase the durability and strength of the gears that form the tested transmission, the parameters set in the slicer were modified so that the wheels have a filling degree of 100%. In terms of the shape of the printing layers, a gyroidal structure was chosen that offers almost isotropic mechanical properties.

In the next stage, the input and output shafts were designed and printed using additive manufacturing technologies, like gear. These components were dimensioned according to the geometry of the gear transmission and the spatial constraints of the test stand used for assembly. In creating the 3D models of the shafts (Fig. 3.a), the coaxial condition was considered to ensure proper alignment, support, and transmission of motion between the motor and the braking system. Additionally, four dedicated supports were designed and fabricated for the input and output shafts, aiming to minimize friction between the moving components.

After printing, these shafts were assembled with the gears and bearings and mounted on the test stand, forming the complete transmission subassembly. This is shown in Fig. 3.b, where the correct positioning and integration of the transmission components into the structure of the stand can be clearly observed.

Based on the conceptual diagram, a physical model of a test stand for gear transmissions was created (see Fig. 4). For the functional testing of the transmission, it was necessary to select a drive motor capable of precise control over speed and torque, to observe and analyze the behavior of the mechanism under various operating conditions. The ECMA-C20604RS electric motor (1), along with its dedicated driver (14), was selected for this purpose. The motor drives a gear mounted on the input shaft (2), which meshes with a second gear mounted on the output shaft (4).

During initial testing, unwanted vibrations, displacements, and friction were observed. To improve the system's stability and efficiency, bearings were added and mounted in custom-designed supports (3, 3', 12, 12').

To stop the transmission after a certain operating period, a braking system was implemented, consisting of a shoe-brake (5) actuated by a stepper motor (10). A pulley (11) mounted on the motor shaft winds a high-strength steel wire connected to the brake.

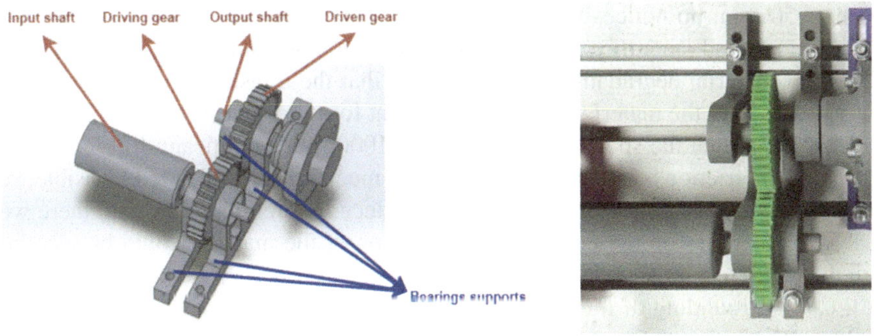

a) b)

Fig. 3. a) 3D Model of gear transmission; b) Gear transmission obtained after printing process.

Upon activation, the pulley tensions the wire, engaging the braking mechanism, which generates a counteracting force to gradually decelerate the transmission.

This system enables precise and repeatable braking control, while maintaining low complexity and high reliability. The stepper motor is controlled via a dedicated driver (9), and the electrical connections between components were established using a breadboard (8) and an Arduino development board (7), ensuring both operation and braking functions. To maintain coaxial alignment and proper component fixation, custom supports were designed for the electric motor, its driver, and the stepper motor (16, 13, 6).

Fig. 4. Prototype of test stand for gear transmission.

The gear transmission assembled using additive manufacturing was tested on the custom-designed test stand to evaluate its performance at different speeds. Two empirical speeds were selected for the test: 50 rpm and 1000 rpm. (see Fig. 5 a) and b)). At the lower speed of 50 rpm, no noticeable displacements or excessive vibrations were observed. The gears meshed smoothly, and the system ran without issues, with no visible signs of wear or material degradation. This indicated that the transmission operated well at lower speeds, and the material used was sufficient for this range of operations.

However, when the speed was increased to 1000 rpm, significant differences were observed. The friction between the gears became much higher, which led to visible wear on the gear teeth. At this speed, the system produced more vibrations, and there were clear signs of stress on the gears. The degradation of the material could be seen both with the naked eye, as shown in Fig. 6.a), and under the microscope in Fig. 6.b), where the wear and rounding of the gear teeth were evident. This suggests that the material used for 3D printing the gears may not be suitable for high-speed applications.

These observations highlight the need for further material optimization and testing at higher speeds to ensure better performance and durability. Additionally, the testing interface for controlling the electric motor provided real-time feedback on the transmission's behavior, which proved crucial in understanding the impact of speed on gear performance.

3.3 Identification of the Parameters to Be Monitored During the Development of the Test Stand

To ensure accurate performance evaluation and real-time diagnostics of gear transmissions, the integration of sensors and transducers is essential. These devices enable the continuous monitoring of key operational parameters such as torque, vibration, temperature, and position. As summarized in Table 2, each parameter is associated with a specific unit of measurement and sensing method that reflects its importance in transmission dynamics. Torque is typically measured in Newton-meters (Nm) to assess the mechanical load applied to rotating components. Vibration, which can indicate misalignment or wear, is quantified in meters per second squared (m/s2) using accelerometers. Temperature, critical for identifying overheating or lubrication issues, is measured in degrees Celsius (°C) with thermocouples or RTD sensors. Lastly, position tracking, often necessary for motion control and synchronization tasks, is measured in angular degrees (°) using optical or magnetic encoders. These measurements are crucial for the development of intelligent monitoring systems aimed at improving performance, extending component lifespan, and enabling predictive maintenance strategies [13, 14].

4 Conclusion

Based on existing research in the specialized literature regarding gear transmission test benches, this study aimed to design and manufacture such a test stand using additive manufacturing technologies.

A simple gear transmission consisting of two gears was selected, with dimensions defined according to an industrial standard commonly used in similar applications. Depending on these dimensions, the adjacent components, including the input and output shafts and the bearing supports, were subsequently designed and fabricated.

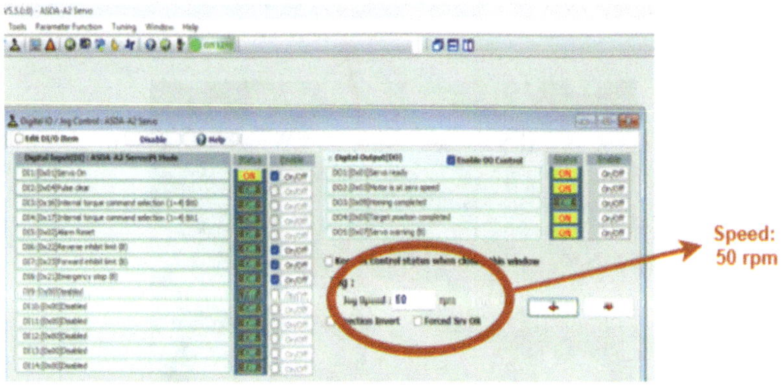

a)

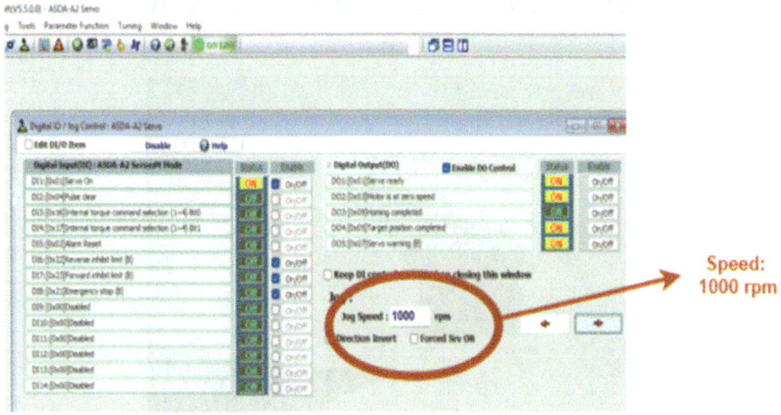

b)

Fig. 5. Interface of the electric motor control program for tested transmission at various speeds: a) 50 rpm; b) 1000 rpm.

Following the assembly process, an electric motor was selected to enable precise control of both torque and rotational speed. Measurements were conducted to develop custom mounts for the motor and its driver. For braking control, a shoe-type mechanism like those found in bicycles was implemented, driven by a stepper motor mounted on a dedicated support structure. This configuration allows precise control of the brake force and replicates real operating conditions within the test stand. The system enables detailed analysis of brake response, efficiency, and wear behavior under controlled actuation.

Despite these challenges, the assembled gear system demonstrated stable performance during testing at rotational speeds up to 1000 rpm, without evident damage to the gears themselves. However, other structural components within the assembly experienced signs of wear and mechanical degradation, emphasizing the need for material optimization in future iterations. These results confirm that both gear transmissions and their corresponding test stands can be effectively produced through additive manufacturing, highlighting their potential for industrial applications.

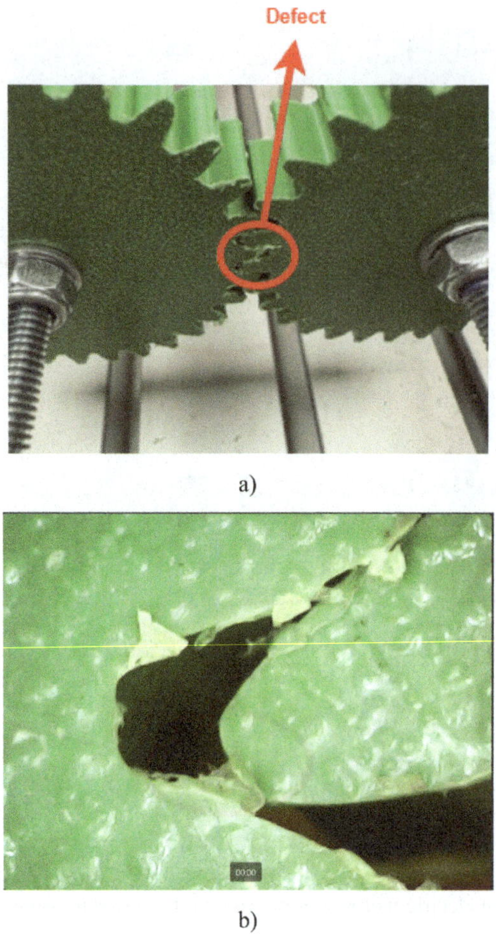

Fig. 6. Tooth profile of gear after testing transmission process: a) defect observed; b) microscopic image with the defect.

Future developments include the integration of advanced sensors to monitor torque, vibration, temperature, and position. These sensors will enable detailed performance analysis under varying operational conditions, facilitating design optimization and early detection of faults or inefficiencies.

In conclusion, the use of additive manufacturing for producing plastic gear transmission, together with a dedicated test bench and integrated monitoring systems, represents an innovative approach in mechanical engineering. This methodology contributes to enhancing the reliability of transmission systems and broadens the applicability of additive manufacturing in the production of functional mechanical components.

Table 2. Types of sensors and transducers for parameters measurement [15–18].

Parameter	Unit of measurement	Symbol	Description	An exemple of sensor/transducer
Torque	Newton-meter	Nm	Measures the twisting force applied to a shaft.	
Vibration	Acceleration	m/s^2	Typically measured using accelerometers mounted on the housing or bearings.	
Temperature	Degrees Celsius	°C	Measured use thermocouples or RTD sensors to monitor thermal conditions.	
Position	Angular degrees	°	Measured use optical or magnetic encoders to detect shaft position or motion.	

References

1. Besnea, D., Rizescu, C., Moraru, E., Dinu, E., Panait, I.: Comparative study regarding the wear of gearwheels manufactured through additive technologies. In: MATEC Web of Conferences, vol. 290, p. 08001. EDP Sciences (2019)
2. Moraru, E., Dontu, O., Petre, A., Vaireanu, D., Constantinescu, F., Besnea, D.: Some technological particularities on the execution of dental prostheses are realized by selective laser deposition. J. Optoelectron. Adv. Mater. **20**(3–4), 208–213 (2018)
3. Fernández del Rincón, A., Cerdá, R., Iglesias, M., de-Juan, A., García, P., Viadero, F.: Test bench for the analysis of dynamic behavior of planetary gear transmissions. In: Davim, J.P. (eds.) Advances in Condition Monitoring of Machinery in Non-Stationary Operations. Lecture Notes in Mechanical Engineering. Springer, Heidelberg (2013)
4. Dumitru, N., Drăguț, E., Crăciunoiu, N., Geonea, I.: Experimental bench for spur gears efficiency measurement. In: Zapciu, M., Amza, C., Coman, G., Popa, C. (eds.) New Advances in Mechanism and Machine Science. Mechanisms and Machine Science, vol. 57. Springer (2018)
5. Li, J., Wang, S., Yang, J., Zhang, H., Zhao, H.: A digital twin-based state monitoring method of gear test bench. Appl. Sci. **13**(5), 3291 (2023)
6. Pal, S., Saha, S.: Design and development of a gear test rig for performance evaluation. Mat. Today Proc. **45**, 2042–2047 (2021)
7. Zhang, Y., Li, X., Liu, Z.: Development of a high-speed gear test bench for evaluation of gear performance. Proc. Inst. Mech. Eng. C J. Mech. Eng. Sci. **235**(9), 2549–2558 (2021)
8. Kumar, A., Singh, R.: Design and fabrication of gear test rig for load distribution analysis. Mat. Today Proc. **45**, 2053–2058 (2021)
9. Wang, L., Li, J., Zhang, H.: Development of a gear test bench for noise and vibration analysis. J. Sound Vib. **496**, 230–240 (2021)

10. ISO 54:1996. Cylindrical gears for general engineering and for heavy engineering — Modules. International Organization for Standardization (1996), Confirmed (2022)
11. ISO 21771:2007. Gears — Cylindrical involute gears and gear pairs — Concepts and geometry. International Organization for Standardization (2007)
12. Baciu, I., Vasilescu, A.G., Avram, M., Grămescu, B., Constantin, V., Moraru, E.: A review of additive technologies applied to gear transmission. In: Machado, J., et al. (eds.) Innovations in Mechanical Engineering III, 44–55. Springer (2024)
13. Zhang, Y., Liu, Y.: Real-Time Torque Monitoring for Gear Systems Using Piezoelectric Sensors. Springer (2022)
14. Li, J., Chen, Z.: Vibration Analysis and Monitoring of Gearboxes with Integrated Accelerometers. Springer (2023)
15. https://ro.tst-balancas.com/torque-sensor-torque-transducer/dynamic-torque-sensor-torque-transducer.html. Accessed 17 Apr 2025
16. https://www.ifm.com/ro/ro/product/VVB010. Accessed 17 Apr 2025
17. https://aplisens.ro/apt-28.html. Accessed 17 Apr 2025
18. https://www.autonics.com/glb/series/EPM50S. Accessed 17 Apr 2025

The Role of Digitalization and 3D Information in the Technological Preparation of Manufacturing for Engineering Components in SMEs

Yevheniia Basova[1(✉)], Sergey Dobrotvorskiy[1,2], and Rafał Talar[3]

[1] National Technical University "Kharkiv Polytechnic Institute", 2, Kyrpychova Street, Kharkiv 61002, Ukraine
Yevheniia.Basova@khpi.edu.ua

[2] Poznan University of Life Sciences, Wojska Polskiego 38/42, 60637 Poznan, Poland

[3] Poznan University of Technology, 2 Piotrowo Street, 61-138 Poznan, Poland

Abstract. This article addresses the crucial role of digitalization and the integration of 3D modeling in enhancing the competitiveness of small and medium-sized enterprises (SMEs) within the engineering component manufacturing sector. Adopting advanced digital technologies, including digital twins and 3D information, enables efficient production planning, improved product quality, and cost reduction. The study highlights the significance of 3D models at multiple levels – from nano to global scales - in optimizing production processes and ensuring the sustainability of enterprises under fluctuating market conditions. Special attention is given to developing complex components, such as turbine parts, where implementing digital solutions improves aerodynamics, load resistance, and overall performance. The article also presents a new technological chain incorporating 3D concept tools for creating high-tech engineering products in response to evolving market demands. Through real-time analysis, predictive maintenance, and precise design modifications, this approach significantly enhances production efficiency, accelerates innovation, and strengthens the adaptability of SMEs in a dynamic global market.

Keywords: Flexibility · Adaptability · Modeling · Sustainability · Innovation

1 Introduction

Ensuring the competitiveness of modern engineering component manufacturing requires the intensification of processes for developing knowledge-intensive products. One of the key ways to achieve this is by implementing advanced manufacturing technologies, particularly digital solutions, at all stages of the product lifecycle. Digitalization significantly optimizes the technological preparation of production, enhancing efficiency and reducing manufacturing costs.

Small and medium-sized enterprises (SMEs) play a crucial role in this transformation, as their flexibility allows them to quickly adapt to changing market conditions. The

D. D. Cioboată and J. Machado (Eds.): ICoRSE 2025, LNNS 1592, pp. 229–239, 2025.
https://doi.org/10.1007/978-3-032-02508-1_20

adoption of digital technologies in production planning and execution provides SMEs with several advantages, including:

- reduced time-to-market for new products,
- improved precision and quality of components,
- lower production costs,
- more efficient management of complex manufacturing processes.

This is particularly relevant for modern engineering component manufacturing, which faces a growing demand for high-tech products such as thin-walled and low-rigidity parts, whether as standalone components or as parts of larger assemblies. However, many SMEs, particularly in Ukraine, encounter challenges that limit their ability to produce geometrically complex and technologically advanced components.

Developing complex-profile and high-tech components, such as turbine parts, necessitates the involvement of high-tech small and medium-sized enterprises (HTSMEs). These companies focus on integrating cutting-edge scientific advancements, innovative technologies, and Industry 4.0 tools, enabling them to remain competitive in an evolving market.

Key characteristics of HTSMEs include:

- a high share of investment in R&D and innovation-driven manufacturing processes,
- the use of advanced digital tools for production planning and technological preparation,
- high flexibility in manufacturing, allowing for rapid adaptation to market demand.

Implementing digital solutions in technological production preparation enables HTSMEs to minimize production cycles, optimize resource utilization, and dynamically adjust manufacturing processes. This, in turn, accelerates the release of innovative products and strengthens their global competitiveness.

Thus, integrating digital technologies into production planning is a key factor in improving SME efficiency and a critical condition for advancing high-tech engineering manufacturing in today's economic landscape.

The aim of this article is to substantiate the significance of digitalization in the technological preparation of engineering component manufacturing and to introduce a new 3D concept as a key tool for optimizing production processes.

2 Literature Review

Technological preparation of production (TPP) is a crucial stage in the production of machine-building components, directly influencing the efficiency of manufacturing processes, resource optimization, and product quality improvement. Traditionally, TPP has relied on empirical methods that ensure the basic functions of preparing production processes; however, these approaches often fail to respond swiftly to market changes and technological advancements [1–3]. With the shift toward digitalization in business and introducing new technologies, the essence of TPP has significantly changed, opening up new opportunities for optimizing production processes through integrating digital technologies.

Digital transformation in business is one of the most important trends in contemporary research. It provides high flexibility and adaptability in manufacturing systems, especially for small and medium-sized enterprises (SMEs). Specifically, digital technologies allow for rapidly adapting production processes to market changes, improving resource management efficiency and ensuring high business competitiveness [4–8]. Digital transformation is a key development factor for engineering SMEs, as it helps reduce costs, improve product quality, and develop new business models [9, 10].

The process-oriented approach to TPP becomes an essential part of digital transformation, as it allows for integrating digital technologies into management and production processes, thereby improving efficiency and adaptability to external changes [11]. It is also important to note that digital technologies enable precise monitoring and control of production processes, which is particularly crucial for parameters such as the quality of surface layers of parts, determining their wear resistance [12].

Despite the positive results of implementing digital technologies, certain challenges in the digitalization of TPP limit its full implementation. Specifically, insufficient integration of digital twins into traditional TPP methods is one of the main obstacles to optimizing production processes. Another key issue is the limited effectiveness of artificial intelligence (AI) and neural networks in predicting and optimizing the parameters of technological processing, which is caused by problems with the quality of training data and algorithm limitations [13, 14]. However, the use of advanced deep learning methods such as Convolutional Neural Networks (CNN) and Generative Adversarial Networks (GAN) significantly improves the accuracy and speed of defect detection in parts, which is essential for ensuring high-quality production in automated manufacturing environments [14, 15].

Another challenge is the lack of universal approaches to representing and utilizing 3D information in production processes. Modern 3D modeling technologies, which hold great potential for optimizing and automating production processes, have not yet achieved sufficient integration and adaptation to various stages of technological preparation for production. Existing modeling approaches often fail to account for the requirements of adaptive control in manufacturing systems, which is necessary to ensure high flexibility in the face of rapid market changes [11, 16, 17].

In this regard, studies indicate the need to develop new methodologies to assess digitalization's impact on SMEs' sustainable development and integrate digital twin methods and 3D information into comprehensive models of technological preparation for production. These methodologies should focus on optimizing design, processing, and quality control processes and ensuring strategic flexibility for businesses in the context of rapid market turbulence [10, 18].

The absence of a comprehensive model for optimizing the digitalization of TPP for SMEs is one of the leading scientific problems requiring further research. One promising strategy is the integration of digital twins into a discrete-mass production model, which allows for forecasting the state of a production system and adapting it to changes in market demand through digital scenario modeling. However, to date, there is no unified methodology for constructing such a model for SMEs, limiting its practical application [3, 19].

In conclusion, this article aims to overcome the limitations of current approaches in ensuring that engineering SMEs are stable, particularly in market fluctuations. By addressing the challenges of digital transformation, such as integrating digital twins, artificial intelligence, and 3D modeling technologies into production processes, the article highlights the potential for enhancing the adaptability and efficiency of SMEs. Developing new methodologies and models for optimizing technological preparation for production is crucial to helping these businesses remain competitive and flexible in a rapidly changing market environment.

3 Research Methodology

3.1 Definition of the 3D Concept and Its Role in Ensuring SMEs' Sustainability

The essence of the proposed concept lies in enhancing enterprise sustainability under unstable conditions by integrating digital data about the manufacturing object at 3D nano-, micro-, meso--, macro-, and global levels. This data is represented in the form of digital information, digital models, and digitized real-world data of physical objects, ensuring both vertical and horizontal 3D integration (Table 1).

Table 1. Structural and logical scheme of digital integration within the 3D concept

Level	Key Aspects of Digitalization and 3D Information	Integration with Other Levels
Nano-Level (Nanostructures)	- Influence of periodic 3D structures on the physico-mechanical properties of the material - Formation of surface effects enhancing operational characteristics. - Investigation of the relationship between nanorelief and functional parameters of the product	- Transmission of nanorelief characteristics to the micro-level for integration into digital processing models
Micro-Level (Surface and Structural Features)	- Analysis of material microstructure to predict behavior under load - Formation of multi-scale roughness for tribological and mechanical property control. - Influence of microscopic texture on aerodynamic and thermomechanical characteristics	- Receives data from the nano-level regarding microstructural features - Transmits information on roughness and texture to the meso-level for product geometry development

(*continued*)

Table 1. *(continued)*

Level	Key Aspects of Digitalization and 3D Information	Integration with Other Levels
Meso-Level (Geometry and Optimization)	- Optimization of product geometry considering acquired material characteristics - Digital models are used to test different design variations - Accounting for microstructural influence on durability and mechanical strength	- Integration with the macro-level to determine physical parameters and quality control methods - Receives micro-roughness data for incorporation into geometric solutions
Macro-Level (3D Models and Physical Information)	- Formation of an integrated digital product model that includes physical material properties - Analysis of dynamic characteristics and load prediction - Consideration of manufacturing process parameters in longevity calculations	- Interaction with the globo-level to ensure end-to-end digitalization of the product life cycle - Receives geometric solutions from the meso-level for stability and functionality testing
Globo-Level (Digitalization of the Product Life Cycle)	- Digital strategy for managing the product life cycle - Digital twins are integrated to predict operational characteristics - Influence on material selection and structural optimization based on collected digital data	- Transfers global optimization parameters downward to all levels - Receives data from the macro-level for creating comprehensive digital models

This multi-level 3D concept consolidates information on digital models of both manufacturing processes and physical objects, incorporating real physical characteristics (such as roughness, hardness, etc.) to monitor state changes in real-time. The approach involves collecting, analyzing, synthesizing, and integrating manufacturing object data from nano- to global levels for rapid and optimal decision-making. The concept aims to ensure sustainability and enhance enterprises' resilience to stress by integrating multi-level information on products and manufacturing processes at all production, realization, operation, and disposal stages (Table 1).

3.2 Integration of the 3D Concept into a Novel Technological Chain

We propose implementing 3D concept tools into a fundamentally new technological chain for creating high-intelligence engineering products such as assemblies and components in response to current market demands (Fig. 1).

The core of this strategy lies in maximizing the role of digitalization and 3D information and introducing a fundamentally new sequence of steps in the technological chain for manufacturing demanded products. This is especially critical at the initial stage of technological preparation for the production of complex, multi-component engineering assemblies. Given the rapidly changing market needs and the continuous emergence of new customer requirements, it has become necessary to prioritize modernization

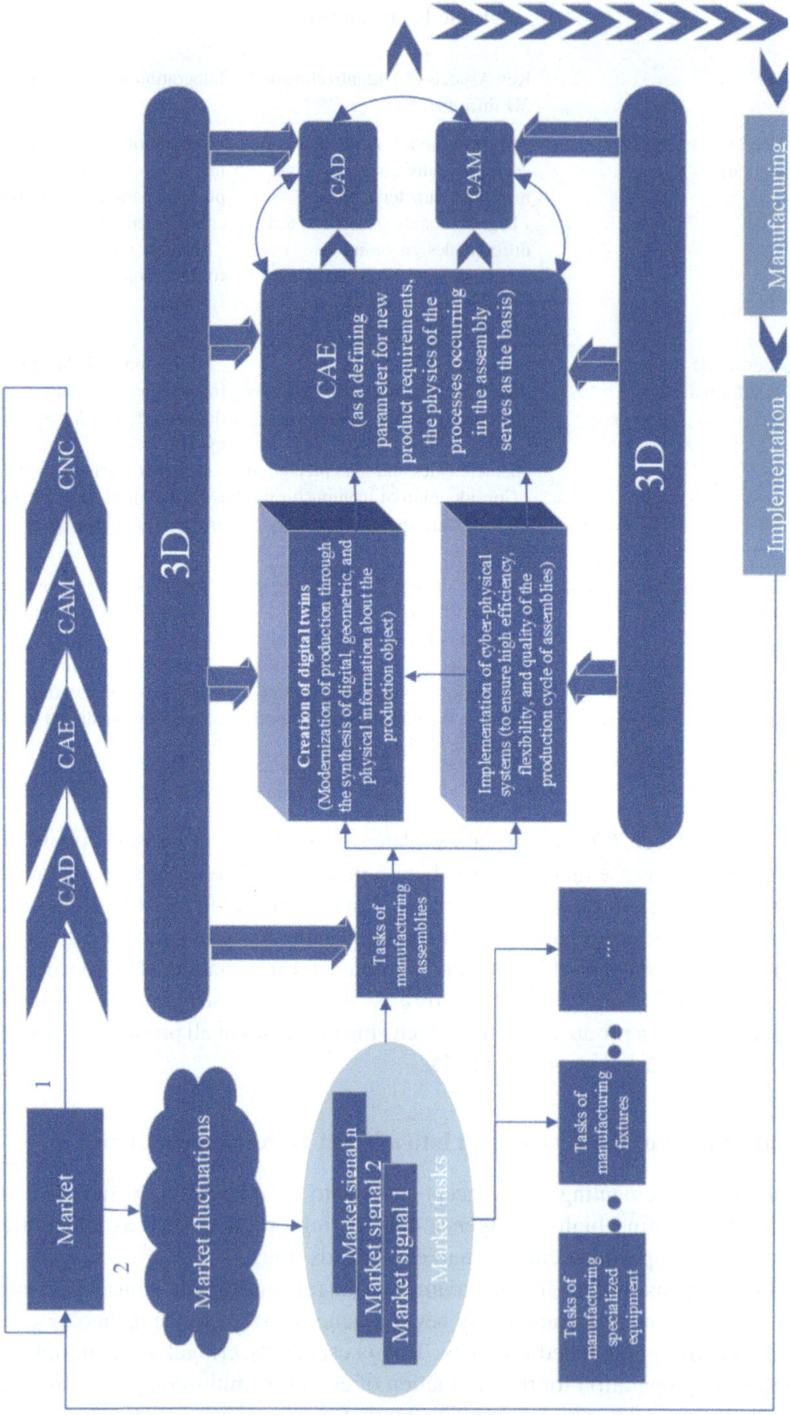

Fig. 1. Role of the 3D concept in the technological chain for creating high-intelligence engineering products: 1 – path for developing new products or individual parts; 2 – path for modernizing product characteristics in response to current market demand.

and enhancement methods for existing market products, which serve as prototypes in this case. Thus, a reliable verification and testing system must be established at the initial product development (modernization) stage to adapt the production of complex assemblies and components to new customer specifications.

3.3 Mathematical Justification of the Technological Chain

In the new sequence of the technological chain for creating an innovative product with a digital core, key process parameters are defined by updated requirements for the physical phenomena occurring in critical components of modernized systems.

This approach necessitates the optimization of energy flows, mechanical load distribution, heat exchange, and aerodynamics, which are essential for determining the final product's key characteristics: accuracy, stiffness, load resistance, and compliance with modern energy efficiency and environmental standards.

Let's define the optimization problem mathematically. Suppose the system's performance function is represented as:

$$F = \sum_{i=1}^{n} w_i f_i(x), \tag{1}$$

where $f_i(x)$ are the individual optimization criteria, w_i are weight coefficients, and x represents the design parameters. The constraints of the system can be formulated as:

$$g_j(x) \leq 0, j = 1, \ldots, m, \tag{2}$$

where $g_j(x)$ are the system constraints related to mechanical strength, thermal stability, and energy consumption.

The results obtained at the «engineering solutions» stage serve as the foundation for further optimal material selection, processing methods (CAM), and product design (CAD), ensuring rapid production of innovative (modernized) products. Thus, prioritizing the «engineering solutions» level in the technological chain for creating innovative products with a digital core facilitates the integration of physical components with digital twins. This enables product adaptation to changing operating conditions, minimizes losses, enhances reliability, and significantly improves efficiency.

We propose prioritizing the «engineering solutions» level (CAE) in the new technological chain to address this challenge of developing the demanded products. This approach is feasible when working with digital twins of products in combination with their cyber-physical systems to create innovative products with a digital core based on visual, digital, and digitized 3D information.

4 Results

Implementing the principles of the 3D concept will be considered using the example of the production of a turbocharger for internal combustion engines (ICE) by small and medium-sized manufacturing enterprises (SMEs). Turbochargers play a key role in increasing the efficiency and environmental sustainability of ICEs, aligning with global trends in sustainable development and the transition to alternative energy sources, such as hydrogen fuel. Due to the increasing demand for eco-friendly transportation, the modernization and ecological improvement of ICEs remain relevant challenges. Hydrogen has emerged as a promising energy carrier that can be used in ICEs in two primary ways: as an additive to traditional hydrocarbon fuels or as a standalone fuel source. These advancements necessitate the development of reliable verification and testing systems at the early stages of turbocharger design and modernization to meet evolving market demands.

A key aspect of turbocharger optimization is the design of turbine and compressor blades, which significantly influence efficiency and operational stability. A comprehensive 3D modeling approach has been applied to analyze both structural and gas-dynamic parameters of mixed-flow turbines at the meso- and macro-levels. This integrated use of 3D modeling in turbocharger component design has enabled the determination of optimal geometric parameters for compressor and turbine blades and the inlet and outlet sections, optimizing their shapes to enhance aerodynamic efficiency. It has been established that for 100% load conditions, the most efficient blade deflection angle is −35°, which ensures maximum output work and improved efficiency due to an increased contact surface between the blades and the working medium.

Based on surface modeling for turbine structure optimization, the study facilitated simplifying calculations for dynamic characteristics and flow interaction with blade surfaces. The application of 3D modeling and CFD analysis in evaluating the aerodynamic characteristics of the flow through the turbine impeller has provided precise data on pressure distribution, velocity, and turbulence, all of which critically affect turbocharger performance (Fig. 2, Fig. 3).

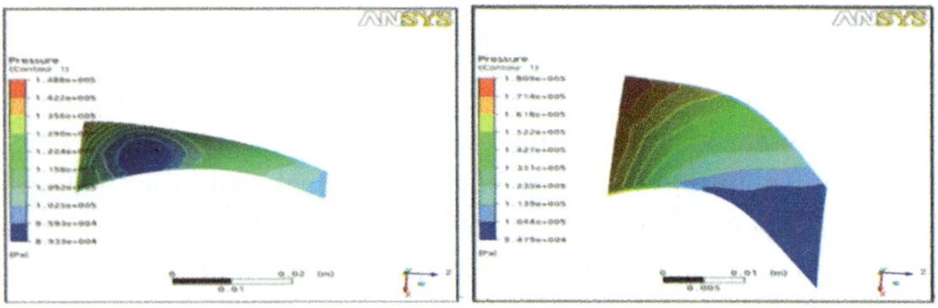

a b

Fig. 2. Static Pressure Contours: a - Near the Hub; b - Near the Shroud [20]

The mathematical model for blade geometry control, incorporating a Bézier polynomial for optimizing the blade's geometric shape, improved output performance by 2.16% and reduced rotor mass by 31.98%. Optimizing the blade diameter ratio also increased efficiency by 2.34%, although a slight increase in mass flow rate accompanied this. The temperature and pressure distribution analysis identified critical zones where pressure reduction could lead to blade erosion. The best results were achieved with a blade diameter ratio of 1.5–1.55.

The utilization of solid models in the digital simulation process enabled a detailed evaluation of structural efficiency and facilitated informed optimization decisions at

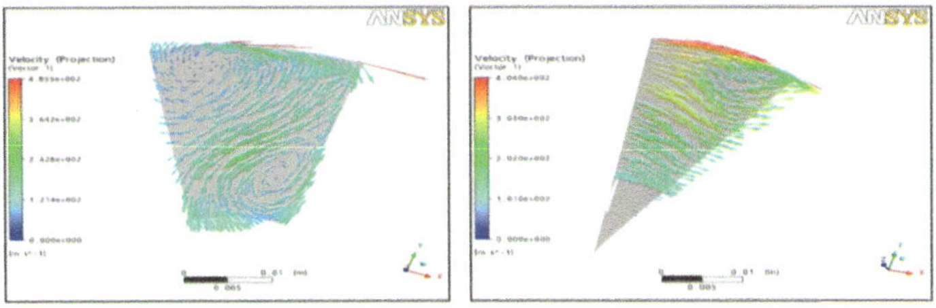

a b

Fig. 3. Field of Relative Tangential Velocity: (a) In the Rotor Inlet Zone; (b) In the Rotor Outlet Zone [20]

the early design stages. This approach significantly enhances the sustainability of enterprises in volatile market conditions. It was determined that the blade surface temperature remains uniformly high for the compressor wheel (Type B rotor). In contrast, for the turbine wheel (Type A rotor), the highest temperature is observed at the blade tips (Fig. 4, a, b).

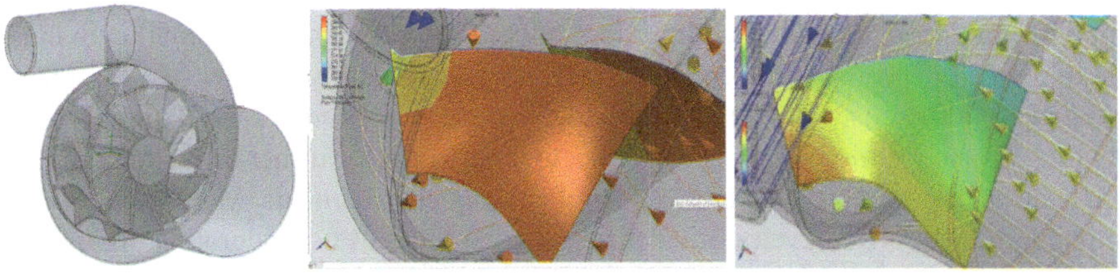

a b c

Fig. 4. The analysis of the temperature distribution across the blade surface is presented as follows: (a) the digital twin of the studied object, (b) for the compressor wheel (Type B rotor), and (c) for the turbine wheel (Type A rotor).

Within the proposed concept, the solid model serves as a crucial tool for adapting designs to market fluctuations, reducing experimental costs, and enhancing manufacturing efficiency by integrating digital twins.

Implementing digital twin and cyber-physical systems enabled the adaptation of the turbine design to real operating conditions, improving its efficiency and reducing energy losses. Integrating geometric and physical data helped identify the optimal blade exit angles, leading to a 2.83% increase in turbine efficiency and a 4.77% reduction in mass flow rate.

It was demonstrated that varying the blade deflection angle and selecting an optimal rotor diameter ratio significantly impact turbine performance. Specifically, it was established that optimizing geometric parameters results in a 2.16% increase in output power and a 3.52% reduction in mass flow rate, contributing to a 2.15% increase in isentropic efficiency. Simultaneously, this optimization reduces rotor mass by 31.98% without compromising structural integrity, achieving an optimal balance between weight and turbine efficiency.

Further advancements in manufacturing technologies require an in-depth study of the impact of digitalization at the micro- and nanoscale levels, where critical parameters of the processed surface are formed.

5 Conclusions

Using an integrated 3D information approach, real-time analysis of component wear and stress distribution can be conducted, enabling predictive maintenance and design modifications. The digital twin model allows simulation of various operational scenarios, optimizing airflow dynamics, mechanical load distribution, and thermal dissipation. The advanced technological chain developed based on the 3D concept meets current market requirements and establishes the foundation for the long-term development of innovative approaches in mechanical engineering. Integrating multi-level digital data within a vertical and horizontal framework optimizes the entire production process and minimizes response time to market demand. This supports the hypothesis that time is a key controlling parameter in ensuring the sustainability and flexibility of small and medium-sized enterprises (SMEs). It has been determined that the comprehensive use of surface and solid models at different phases of the product life cycle ensures the minimization of design time and product release in accordance with market demands. By adopting the proposed 3D concept, SMEs can significantly improve their production strategies, reduce costs, and enhance product innovation. Future research should focus on refining predictive models using AI and neural networks to further improve digital twin applications, particularly in adaptive manufacturing environments. Moreover, additional studies are required to develop standardized methodologies for integrating digitalization into various stages of the technological preparation of production in SMEs.

References

1. Tolmachov, D.: Digital transformation of enterprises. Problems and criteria of digitalization [Tsyfrova transformatsiia pidpryiemstv. Problemy ta kryterii tsyfrovizatsii]. Molodyi vchenyi **11**(111), 114–118 (2022). https://doi.org/10.32839/2304-5809/2022-11-111-24 (in Ukrainian)
2. Fadeev, V.A., Naboka, O.V., Mekhovich, K.S.: Technological reengineering of machine-building enterprises: Evolution, problems, solutions [Tekhnolohichnyi reinzhynirynh pidpryiemstv mashynobuduvannia: evoliutsiia, problemy, shliakhy vyrishennia]. In: Bulletin of the National Technical University "KhPI". Ser.: Techniques in a machine industry **2**(8), 114–121 (2023). (in Ukrainian)
3. Balushok, K.: Experience of designing and implementing an integrated system of technological preparation of production for manufacturing new aircraft engines. Mech. Adv. Technol. **7**(1), 24–35 (2023)
4. Bella, R.L.F., Leal Filho, W., Sigahi, T.F.A.C., et al.: Small- and medium-sized enterprises: trends and future perspectives for sustainability and digitalization in Germany. Sustainability **16**, 6900 (2024). https://doi.org/10.3390/su16166900
5. Spigarelli, F., Compagnucci, L., Lepore, D.: Blockchain unlocking collaborative opportunities for environmental sustainability through innovation intermediaries. J. Technol. Transf. (2024). https://doi.org/10.1007/s10961-024-10106-5

6. Sagala, G.H., Őri, D.: Toward SMEs digital transformation success: a systematic literature review. Inf. Syst. E-Bus. Manage. **22**, 667–719 (2024). https://doi.org/10.1007/s10257-024-00682-2
7. Le-Dain, M.A., Benhayoun, L., Matthews, J., et al.: Barriers and opportunities of digital servitization for SMEs: the effect of smart product-service system business models. Serv. Bus. **17**, 359–393 (2023). https://doi.org/10.1007/s11628-023-00520-4
8. Tripathi, A., Singh, A.: SMEs awareness and preparation for digital transformation: exploring business opportunities for entrepreneurs in Saudi Arabia's Ha'il region. Sustainability **16**, 3831 (2024). https://doi.org/10.3390/su16093831
9. Priyono, A., Moin, A., Putri, V.N.A.O.: Identifying digital transformation paths in the business model of SMEs during the COVID-19 pandemic. J. Open Innov. Technol. Market Complexity **6**(4), 104 (2020). https://doi.org/10.3390/joitmc6040104
10. Teece, D.J., Linden, G.: Business models, value capture, and the digital enterprise. J. Organ. Des. **6**, 8 (2017)
11. Treityak, V.V.: Process model of production technical preparation management in the environment of integrated automated systems [Protsesna model upravlinnia tekhnichnoiu pidhotovkoiu vyrobnytstva v seredovyshchi intehrovanykh avtomatyzovanykh system]. In: Automation and Control of Mechanical and Technological Systems and Complexes, **49**, 74–78 (2015) (in Ukrainian)
12. Tarelnyk, N.V., Dumanchuk, M.Y., Maifat, M.M., Dotsenko, A.O.: A systematic approach to the selection of technologies for controlling the quality parameters of the surface layers of machine parts working in conditions of hydroabrasive wear. In: Bull of Sumy National Agrarian University. The Series: Mechanization and Automation of Production Processes **1**(55), 85–89 (2024). https://doi.org/10.32782/msnau.2024.1.12
13. Liu, Q., Liu, M., Jonathan, Q.M., et al.: A real-time anchor-free defect detector with global and local feature enhancement for surface defect detection. Expert Syst. Appl. **246**, 123199 (2024). https://doi.org/10.1016/j.eswa.2024.123199
14. Zhang, W., Hu, Y., Shan, H., et al.: An online automatic carbide insert high-resolution surface defect detection system based on template-guided model. Expert Syst. Appl. **238**, 122089 (2024). https://doi.org/10.1016/j.eswa.2023.122089
15. Ameri, R., Hsu, C.-C., Band, S.S.: A systematic review of deep learning approaches for surface defect detection in industrial applications. Eng. Appl. Artif. Intell. **130**, 107717 (2024). https://doi.org/10.1016/j.engappai.2023.107717
16. Loginova, O.: Analysis of the impact of digitalization on the development of small and medium-sized business opportunities [Analiz vplyvu tsyfrovizatsii na rozvytok mozhlyvostei maloho ta serednoho biznesu]. Ekonomichni Horizonty 2–3(28), 259–265 (2024). https://doi.org/10.31499/2616-5236.3(28).2024.311521 (in Ukrainian)
17. Maksymenko, D., Osadchyi, Ye., Metelskyi, A.: Digital transformation of core activities of small and medium-sized enterprises: Tools and methods [Tsyfrova transformatsiia osnovnoi diialnosti malykh ta serednikh pidpryiemtsiv: instrumenty ta metody]. Ekonomichnyi Prostir **191**, 250–255 (2024). https://doi.org/10.32782/2224-6282/191-41 (in Ukrainian)
18. Ivzhenko, O.V., Antonova, H.V.: Project of technology for processing basic parts with high surface quality [Proekt tekhnolohii obrobky bazovykh detalei z vysokoiu yakistiu poverkhni]. In: Proceedings of Tavria State Agrotechnological University, vol. 21, no. 1, pp. 310–316 (2021) (in Ukrainian)
19. Demeter, K., Losonci, D., Szalavetz, A., Baksa, M.: Strategic drivers behind the digital transformation of subsidiaries: a longitudinal approach. Postcommunist Econ. **35**, 744–769 (2023)
20. Chelabi, M.A., et al.: The effect of blade angle deviation on mixed inflow turbine performances. Appl. Sci. **12**(8), 3781 (2022). https://doi.org/10.3390/app12083781

Comparative Study of Power Consumption of Overhead Traveling Crane Based on the Experiment

Valentyn Kovalenko[1], Maksym Strelkov[1], Vsevolod Stryzhak[1](✉), Magomediemin Gasanov[1], and Edl Milan[2]

[1] National Technical University «Kharkiv Polytechnic Institute», 2 Kyrpychova Street, Kharkiv 61002, Ukraine
{valentyn.kovalenko,Vsevolod.Stryzhak}@khpi.edu.ua
[2] University of West Bohemia, Plzen 30100, Czech Republic
edl@kpv.zcu.cz

Abstract. The study aims to determine experimentally the dependence of the change in energy consumption on the operating time of cranes equipped with different types of drives - based on a motor with a phase rotor and variable frequency motor. The aim is achieved by solving the following problems: selection of research objects and measuring equipment, determination of the methodology of measurements, ordering of experimental data, and determination of the dependence of the average power consumption on the operating time. In the process of solving the set tasks, it was found that in the long-term study of power consumption of an overhead traveling crane, the average value is reduced to 50%. At the same time, it was found that the crane equipped with a variable frequency drive at the same lifting capacity has better technical and economic indicators, the main of which is the weight of the crane and metal consumption. The most significant result is to obtain the final expressions for determining the dependence of electricity consumption of the crane and the reduction of electricity consumption when switching to an energy-efficient variable frequency drive. The significance of the obtained results is that the obtained dependencies can be applied in justifying the feasibility of modernization of cranes.

Keywords: Overhead travelling crane · Drive · Variable frequency drive · Motor power · Mechanism

1 Introduction

Overhead cranes play a key role in the industry, lifting and moving heavy loads. Improving the energy efficiency of the control systems of these cranes is an important task to reduce energy consumption, improve environmental safety, and reduce operating costs. A current trend is the use of variable frequency drives for crane mechanisms, which is known from numerous sources to reduce dynamic loads and energy consumption. This

D. D. Cioboată and J. Machado (Eds.): ICoRSE 2025, LNNS 1592, pp. 240–251, 2025.
https://doi.org/10.1007/978-3-032-02508-1_21

trend has existed for a long time, and during this time, experience has been gained in creating cranes designed and built to power drives from a frequency converter. This makes it possible to trace how the overall parameters of such cranes have changed compared to those of cranes with a traditional motor drive with a phase rotor, which are still in use.

2 Research Overview

The energy consumption of cranes and the reduction of this consumption are of great interest to researchers around the world, which indicates the undeniable relevance of the topic. An essential area of research into energy consumption reduction is the optimization of the supporting structure of cranes. Paper [1] proposes an approach to scheduling the operation of gantry cranes that takes into account the balance between efficiency and energy consumption, transforming this task into a vehicle routing problem. An integrated simulation optimization method is developed that uses a genetic algorithm for global search and a particle swarm method for local search, with computational experiments to verify its effectiveness.

In multi-crane systems, collisions between cranes often occur, which leads to significant energy consumption to avoid them. Article [2] proposes an approach to energy-efficient planning of multi-crane systems and selection of the required number of cranes, which allows for accurate assessment of the energy consumption of cranes in different phases of operation, modeling a multi-crane system in virtual space, identifying collisions, and proposing methods to avoid them.

The paper [3] analyzes the energy consumption of different engines of a non-tracked gantry crane. An experimental study was carried out in the port of Felixstowe, where data was collected during eight days of normal operation. The data was analyzed for operating and idling conditions of the machinery, as well as the energy consumption of the different engines. The analysis showed that, on average, about half of the energy consumed could potentially be recovered. This is expected to save fuel and reduce CO_2 emissions.

Article [4] addresses the problem of high energy consumption due to the suboptimal design of crane steel structures. An energy-saving design method is proposed to solve the problem of significant weight and material consumption. Two optimization models are proposed to minimize the weight of the structure. A multi-criteria optimization model for energy-saving design is developed to minimize waste during production.

The article [5] discusses the problems of designing crane systems, focusing in particular on energy costs and consumption. The complexity lies in fuzzy determinants such as costs, loads, and energy consumption, which is exacerbated by the need to balance the tension between multi-objective requirements in real-world conditions. The authors have developed a mathematical model to minimize costs and energy consumption, taking into account factors such as crane load, recovery period, and budget constraints.

Article [6] discusses the energy consumption resulting from the deflection of the main beam of an overhead crane. In this case, the trolley is on an inclined surface, which leads to the need to overcome additional resistance during movement. Dependencies are proposed that take into account the influence of welded joints and optimize deflection, but the conclusions regarding energy savings are based on general principles of mechanics without taking into account the design of the trolley or drives.

In [7], the problem of optimal beam deflection is solved by taking into account the probabilistic distribution of the value of the load lifted by the crane over its service life instead of the fixed or nominal loads that are traditionally used. This is done by analyzing the density functions of the load distribution and, accordingly, the energy consumption distribution. The assumption of lifetime load distribution is based on the formulas of international standards that regulate crane operating modes. However, this study also estimates the reduction in energy consumption based on the general principles of mechanics to reduce the vertical movement of the load and trolley.

Paper [8] presents a numerical study of the energy consumption of an overhead crane. The analysis is based on a dynamic model of the crane, consisting of models of the bridge movement mechanisms, trolley, and lifting mechanism. The experimentally measured power consumption of each control unit is also presented, and thus each numerical model was experimentally validated on a real crane. The study focuses on analyzing the energy consumption of an overhead crane during the movement, lifting, and lowering of a load. The possibility of energy recovery when lowering the load is considered. The possibility of recovering kinetic energy during the braking of a trolley or crane is not considered, which remains an unexplored part of the task of reducing energy consumption by crane mechanisms.

A number of studies draw conclusions about the efficiency of energy consumption and suggestions for improving energy efficiency, considering the crane as a whole.

Article [9] considers the entire energy flow of crane mechanisms. Energy consumption and losses are investigated using an equivalent induction motor circuit and a power balance equation consisting of many links. It is noted that the regulated drive has an advantage over the drive that regulates the speed by changing the rotor resistance of the electric motor. At the same time, the conclusions and recommendations for improving energy consumption modes are more focused on improving the efficiency of transmission links and mechanisms, and no assessment of consumption reduction for different types of drives is provided.

In [10], the energy consumption of an overhead crane is assessed in the 'full life cycle'. The model for assessing the energy efficiency of an overhead crane is based on the principles of fuzzy logic. For this purpose, ten overhead cranes of different designs were used, and their advantages and disadvantages in terms of energy consumption were analysed.

The article [11] draws attention to the fact that excessive energy consumption occurs due to the significant weight of overhead cranes. Therefore, to solve this problem, it is proposed to optimize this parameter using the multi-split reflection algorithm (M-SRA), which combines the specular reflection algorithm (SRA) with traditional swarm intelligence optimization. The authors note that the main parameter that affects the weight of the crane is the cross-section of the main beam in the middle of the span. Therefore, the optimization is aimed at determining the dimensions of the cross-sectional elements that give a reduced weight while meeting the constraints (in this case, static strength). The study considers only the nominal forces from the weight of the beam itself and the trolley with the load. The influence of dynamic forces and the possibility of their reduction using different types of actuators are not taken into account. Consideration of this possibility requires further study, as it contains additional reserves for reducing the weight of the crane and, accordingly, wider opportunities for reducing energy consumption that are not directly related to the electrical characteristics of drives and motors.

The optimization of the main beam cross-section to reduce weight and improve energy efficiency is discussed in [12]. As in the previous study, simplified dependencies are used to determine the stresses, but the stiffness of the structure is taken into account.

A number of studies consider the lifting mechanism as a whole. Article [13] considers the lifting mechanism of overhead and gantry cranes. In accordance with graph theory, a linkage model was created to study energy consumption in the lifting mechanism. The study proposes a method for estimating the loss parameters of each element, which ultimately leads to a conclusion about the impact on the energy efficiency of each of them in the form of influence coefficients. As in a number of other studies, the integrated approach is focused mainly on mechanical parts: gearboxes, couplings, and intermediate shafts. Among the electrical components, the electric motor is considered, but not the drive system as a whole. Taking into account the electric drive system as a whole can significantly change the component of the electrical part in the overall picture of the mechanism's energy efficiency and requires further study. Article [14] considers the lifting mechanism as a whole. The main focus is on the links that transmit motion, namely the gearbox and the system of ropes and blocks. Thus, the mechanical power and its losses are considered. However, it does not take into account the nature of the change in torque generated by the electric motor, which significantly affects the oscillation processes and, consequently, the choice of clearances and elastic deformations in the mechanical components of the mechanism. The shape of the force change during the duty cycle depends on the drive system, which is not taken into account in this study, limited to the statics formulas.

The problem of reducing energy consumption can also be solved by using special types of laws for changing the speed of movement of mechanisms. In [15], the optimization problem of simultaneously starting several mechanisms of a tower crane is formulated and solved. To obtain the laws of motion, a system of differential equations is used that takes into account the simultaneous movement of the mechanisms. The task of improving energy efficiency is solved by minimizing the RMS value of the total power. As a result, motion laws are obtained that minimize the power consumption of drive mechanisms and reduce dynamic loads.

Optimal control studies also address the problems of load sway, or the study of pendulum motion, as in [16]. This is because load oscillations create additional obstacles to positioning and, accordingly, lead to energy consumption.

The low-frequency oscillations of the load are studied in [17], where they are given renewed attention. In this case, a tower crane and a slewing mechanism are considered.

As noted in a number of papers, crane beams are optimized in terms of material consumption, taking into account static loads. However, the endurance conditions may be violated. The residual service life is studied in [20]. The approaches considered in these papers allow us to assess the effects of load perception.

The issue of studying the residual life and damage to the crane during operation is considered in [21]. The issue of studying effective stresses is also covered in [21]. However, this approach only allows us to assess the effects of load perception. To reduce fatigue damage, it is necessary to study the dynamic processes created by electric drives that affect the rate of damage accumulation.

When considering control, mathematical models describing movement over bumps, as in [22], or movement along the boom of a tower crane [23], can be considered. Optimization of movement parameters leads to a reduction in dynamic loads and energy consumption.

Increasing the efficiency of machines and mechanisms in design and operation according to the criteria of effort and energy savings is a general trend and research topic in the fields of robotics [21].

The papers reviewed reflect current trends in energy saving: route planning, improving metal structure design, and optimizing machine control. The issue of energy saving of existing cranes has not been sufficiently addressed. The issue of comparing the energy consumption of cranes operating under the same conditions, for example, with the same average reloading cycle, remains insufficiently covered. This article is dedicated to addressing this issue.

The problem of reducing energy consumption is relevant in terms of enterprise sustainability and improving logistics processes (reducing costs and saving resources). The implementation of solutions is in line with the concept of Industry 4.0, which makes the company modern and competitive [22, 23].

Thus, this review of current research shows that the energy efficiency of cranes is investigated by different approaches: structure design, drives, and controls. However, there is a lack of a comprehensive experimental approach that would allow, based on real measurements in production conditions, taking into account the complexity of the listed factors and random crane operating cycles. Such an approach would provide scientifically sound empirical regularities for practical use to assess the reduction of energy costs and would create certainty for technical decision-making in crane modernization and design.

3 Aim and Objectives

Aim of the study is to experimentally determine the dependence of changes in energy consumption on the operating time of cranes equipped with different types of drives - based on a motor with a phase rotor and a variable frequency drive.

4 Research Results

Reducing energy consumption for one working cycle of the crane mechanism is possible at all stages, even under normal control. The most effective means on this path is the choice of a rational type of drive and its operating modes. Therefore, it is relevant to assess the energy efficiency of a frequency-controlled drive in the mechanisms of lifting machines and compare it with a traditional asynchronous drive with a phase rotor. A comparison of energy consumption during the working cycle is shown in Fig. 1.

As a comparative experiment, we studied the amount of energy consumed during the operation of two cranes operating in the same production shop in one span in Kharkiv at the production facilities of PSE LLC (Fig. 2). The cranes operate in the area with mechanical welding operations for the manufacture of metal structures.

The drives of the crane marked with the letter 'a' in Fig. 1 are equipped with motors with a phase-mouth. This drive has been widely used for lifting machines due to a number of advantages, the main ones being lower cost compared to other drives and simplicity of construction and ease of maintenance.

In existing cranes, this drive remains in operation, as in this example. The crane marked with the letter 'b' has a modern design and is equipped with a more advanced drive - a variable frequency drive.

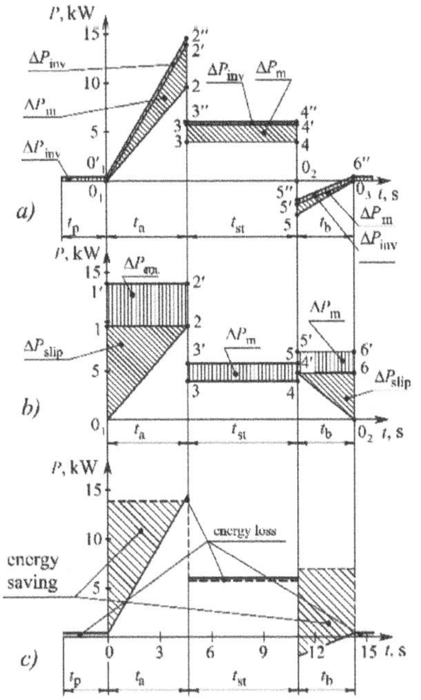

Fig. 1. Scheme of changing the power consumed from the network and recovered in the bridge crane mechanism. a – energy consumption and recovery scheme for a variable frequency drive; b – theoretical energy consumption for an asynchronous drive with a phase rotor and braking by switching to lower rheostat stages; c – overlay of power diagrams of both types of drives (shaded area - excess energy consumption in an asynchronous motor with a phase rotor; shaded area - excess energy consumption in a variable frequency drive)

Since the cranes are installed in the same span and are involved in the same process cycle of the enterprise, it is assumed that their average handling cycle is the same. That is, both cranes move a load of the same average weight over the same average distance. This assumption is all the more valid as the observation and measurement of energy consumption take place over a period of months and any deviations are averaged. Table 1 shows the parameters of the cranes.

The table shows that with the same lifting capacity, lifting range, and span, a crane with a variable frequency drive has lower gear speeds and, as a result, lower electric motor power.

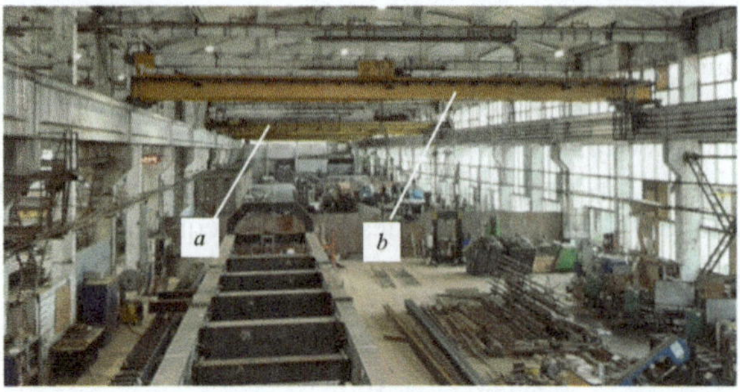

Fig. 2. Cranes in the production facility are used for a comparative study. a– crane with drives based on motors with a phase rotor; b - crane with a variable frequency drive

Table 1. Comparison of the parameters of the studied cranes.

Name	Figure 1, a	Figure 1, b
Crane lifting capacity, tonnes	10,0	10,0
Span, m	22,5	22,5
Lifting range, m	12,0	12,0
Weight of the crane, t	17,8	12,2
Lifting speed, m/min	7,0	5,0
Travelling speed of the trolley, m/min	36,0	25,0
Crane travel speed, m/min	50,0	33,0
Motor type and power of the lifting mechanism	MTN-411-8, Pn = 15 kW	DRS132M4BE11, Pn = 7.5 kW
Type and power of the trolley motor	MT-11–6, Pn = 2 × 2.2 kW	DR2S71M4, Pn = 2 × 0.5 kW
Type and power of the engine of the crane movement mechanism	MTV-311-8, Pn = 2 × 7.5 kW	DRS90M4 Pn = 2 × 1.5 kW
Current type, frequency, voltage	3f, 380 V, 50 Hz	3f, 380 V, 50 Hz
Environmental conditions	general industrial	general industrial
Control system	Relay contactor with resistor output in the phase rotor	Based on frequency converters
Control location	radio remote control	radio remote control

However, in our case, it can be assumed that this circumstance will not affect the final result of the experiment due to the fact that we measure the amount of energy consumed over a long period of time and take into account the entire handling cycle. Therefore, assuming that the crane operating conditions are the same and the average handling cycle is the same, it follows that a mechanism with lower speeds must operate longer to perform the same mechanical work. This means that less instantaneous power will be consumed for a longer time. An important parameter is the weight of the crane. In the case of a variable frequency drive, it should be noted that a crane of the same lifting capacity and span has a lower weight. This indicates an additional advantage that requires further study.

The meters shown in Fig. 3 were used to measure the consumed electricity.

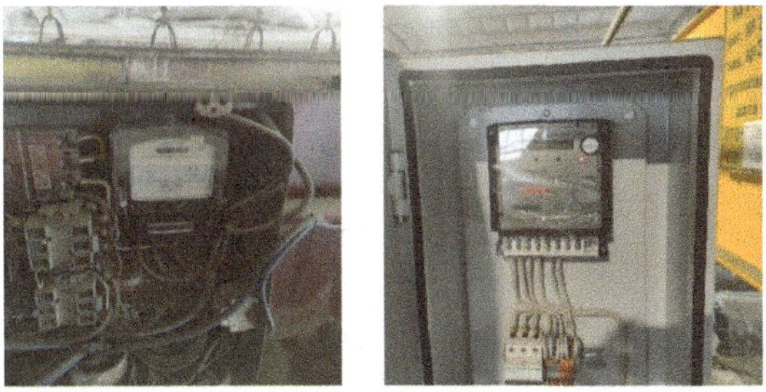

Fig. 3. Measuring devices used in the comparative study. a - crane electricity meter with drives based on motors with a phase rotor; b - crane electricity meter with a variable frequency drive

Each crane is equipped with a separate industrial-strength meter with a corresponding seal. These meters were used to measure the amount of energy consumed by the crane as a whole once a month. The measurement results are shown in Table 2.

Table 2. Results of measuring the amount of energy consumed by months by cranes with phase-locked motors (PSM) and variable frequency drives (VFD), kW.

Month	faz, kW· h	freq, kW· h
May 2022	29	18
June 2022	60	38
July 2022	65	40
August 2022	65	35
September 2022	60	40
October 2022	84	61
November 2022	85	63
December 2022	83	55
January 2023	75	45
February 2023	35	19,5
March 2023	20	16
Σ	661	430,5

Let's present the above measurement results in the form of a graph, where the horizontal axis shows the serial numbers of months, and the vertical axis shows the amount of energy consumed for each month (Fig. 4). The figure shows that cranes with phase-rotor motors have consistently higher energy consumption (the blue dots on the graph are located higher). For a quantitative assessment, we will find approximating functions of linear form that average the results and also place them on the graph. The approximating functions are found by the least squares method, using the built-in Excel functions.

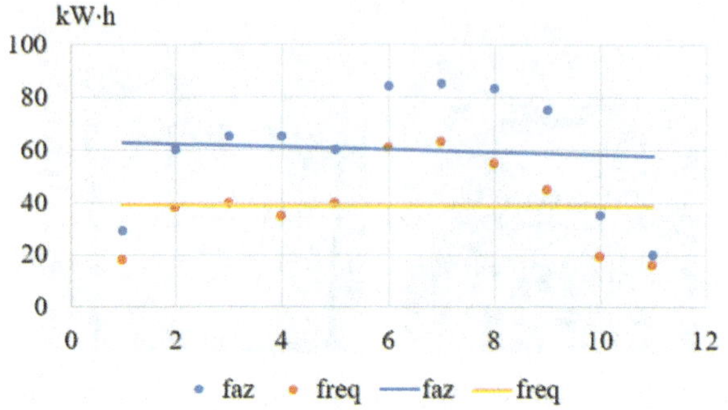

Fig. 4. Comparative analysis of electricity consumption measurements. Faz - the crane is equipped with motors with a phase rotor, freq - the crane is equipped with a variable frequency drive

The results of the comparative analysis show significant energy savings when using a variable frequency drive. The average reduction in energy consumption is 50%.

5 Discussion

The significant relevance of the topic of reducing energy consumption by cranes is confirmed both by the number of publications in this area and by the variety of approaches that are proposed: optimization of the movement of mechanisms, increasing the efficiency of mechanical components, reducing material consumption, and improving the geometry of the path of moving trolleys. However, the issue of an integral consideration of the crane as a whole, or mechanisms as a whole, their operating modes, and features of drive systems is not sufficiently reflected, and an in-depth study of drive systems is more common in studies of electrical equipment. The closest study is the work [8]. The authors of this work also study the overload cycle of the crane as a whole. However, this work does not take into account different types of drive systems but mainly considers the ability of the drive to recover energy. In addition, the research is more focused on the calculation and laboratory conditions of the overload cycle. The study [9] is also devoted to a comprehensive consideration of the crane, noting the advantage of a variable-speed drive over drive systems based on phase-rotor motors, which are traditional and the cheapest for crane electric drives. However, the main attention is focused on the mechanical part. This study provides a quantitative assessment of the advantage of a variable-speed drive of crane mechanisms compared to a drive based on a phase-rotor motor. In this case, not a separate mechanism is considered, but the lifting machine as a whole. Another advantage of the study is that the method of measuring the power consumption allows you to take into account both mechanical elements and all components of electrical circuits that take part in supplying current to the actuators, which is usually not taken into account. Another important advantage is that the data is obtained from a real production cycle over a long period. This allows you to minimize probability factors that could distort the assessment results. Thus, the results of this study are easily transferred to real production conditions. Another important conclusion of this study, which follows from the analysis of crane parameters, is that new cranes designed for adjustable drive have a lower mass. This issue requires further study in order to assess the impact of dynamic loads on the change in the safety factor and the possibility of optimizing the crane cross-section with this factor in mind. The main significance of this study is that the data provided provide objective grounds for substantiating crane modernization projects, making it possible to quantitatively assess energy savings.

6 Conclusions

The article solves the problem of experimentally determining the dependence of the change in energy consumption on the operating time of cranes equipped with different types of drives - based on a motor with a phase rotor and a variable-frequency motor. To achieve this goal, real cranes with the same lifting capacity and operating conditions were selected. The assumption was made that the average reloading cycle was the same. Measuring devices were selected, and energy consumption data were collected. The experimental data were processed, and the reduction of energy consumption by a crane with a variable frequency drive was estimated. It was found that an overhead crane with this type of drive consumes an average of 50% less electricity. In addition, the analysis of the parameters of existing cranes has shown that a crane with a variable frequency drive can have a lower weight, ceteris paribus (lifting capacity, span, average reloading cycle), which may be the subject of further study.

References

1. He, Y.J., Huang, Y., Yan, W.: Yard crane scheduling in a container terminal for the trade-off between efficiency and energy consumption. Adv. Eng. Inform. **29**(1), 59–75 (2015). ISSN 1474-0346. https://doi.org/10.1016/j.aei.2014.09.003
2. Zhao, N., Fu, Z., Sun, Y., Pu, X., Luo, L.: Digital-twin driven energy-efficient multi-crane scheduling and crane number selection in workshops, J. Clean. Prod. **336**, 130175 (2022). https://doi.org/10.1016/j.jclepro.2021.130175
3. Papaioannou, V., Pietrosanti, S., Holderbaum, W., Becerra, V.M., Mayer, R.: Analysis of energy usage for RTG cranes. Energy **125**, 337–344 (2017). https://doi.org/10.1016/j.energy.2017.02.122
4. Xu, H., Xu, G., Dong, Q., Xin, Y.: Comprehensive research on energy-saving green design scheme of crane structure based on computational intelligence. AIP Adv. **11**(7) (2021). https://doi.org/10.1063/5.0050653
5. Su, T.-S., Wu, C.-C., Yang, H.-R.: An analysis of energy consumption and cost-effectiveness for overhead crane drive systems by using fuzzy multi-objective linear programming 6241–6253 (2018)

6. Yifei, T., Lijin, L., Guomin, S., Dongbo, L., Xiangdong, L.: Camber deformation estimation of bridge crane and energy-consumption analysis. Proc. Inst. Mech. Eng. Part B: J. Eng. Manuf. **230**(2), 313–318 (2016). https://doi.org/10.1177/0954405414548468
7. Yifei, T., Zhaohui, T., Song, M., Guomin, S., Feng, G.: Research on energy-saving design of overhead travelling crane camber based on probability load distribution. Math. Probl. Eng. (2014). https://doi.org/10.1155/2014/484635
8. Kosucki, A., Stawiński, Ł., Malenta, P., Zaczyński, J., Skowrońska, J.: Energy consumption and energy efficiency improvement of overhead crane's mechanisms. Eksploatacja i Niezawodnosc **22**(2) (2020). https://doi.org/10.17531/ein.2020.2.15
9. Fei, S., Zhihao, G., Yifei, T., Xiangdong, L.: Characteristics analyses of energy consumption for bridge crane based on the energy flow. IOP Conf. Ser. Mater. Sci. Eng. **758**(1) (2020). https://doi.org/10.1088/1757-899X/758/1/012002
10. Ye, W., Tong, Y., Li, D., Li, X.: The bridge crane's energy efficiency fuzzy evaluation based on the rough set theory. Harbin Gongcheng Daxue Xuebao/J. Harbin Eng. Univ. **35**(8) (2014). https://doi.org/10.3969/j.issn.1006-7043.201305027
11. Qi, Q., Yu, Y., Dong, Q., Xu, G., Xin, Y.: Lightweight and green design of general bridge crane structure based on multi- specular reflection algorithm. Adv. Mech. Eng. **13**(10) (2021). https://doi.org/10.1177/16878140211051220
12. Tian, G., Zhang, S., Sun, S.: The optimization design of overhead traveling crane's box girder. Adv. Mat. Res. 538–541 (2012). https://doi.org/10.4028/www.scientific.net/AMR.538-541.2850
13. Hua, Z.: Energy consumption model of hoisting mechanism of bridge and gantry crane based on bond graph theory. Model. Simul. **09**(04) (2020). https://doi.org/10.12677/mos.2020.94053
14. Ge, Z., Tong, Y., Zhong, M., Li, X.: Study on the measurment of hoisting mechanism efficiency for bridge crane. J. Phys. Conf. Ser. **1345**(6) (2019). https://doi.org/10.1088/1742-6596/1345/6/062007
15. Perig, A.V., Stadnik, A.N., Deriglazov, A.I.: Spherical pendulum small oscillations for slewing crane motion. Sci. World J. **2014**, 451804 (2014). https://doi.org/10.1155/2014/451804
16. Loveikin, V., Romasevich, Y., Spodoba, O., Loveykin, A., Pochka, K.: Optimization of the mode of movement of the boom system of the loader crane. Strength Mater. Theory Struct. (111) (2023). https://doi.org/10.32347/2410-2547.2023.111.223-236
17. Gubskyi, S., et al.: Integrated approach to determine operational integrity of crane metal structure. Periodica Polytechnica Mech. Eng. **63**(4), 319–325 (2019). https://doi.org/10.3311/PPme.14064
18. Sergienko, N., Hubskyi, S., Pavlova, N., Turchyn, O., Hasiuk, O., Židek, K.: Obstacle-resistant wireless strain gauge complex for automated monitoring of the steel structures condition. In: Balog, M., Iakovets, A., Hrehova, S. (eds.) EAI International Conference on Automation and Control in Theory and Practice. EAI ARTEP 2023. EAI/Springer Innovations in Communication and Computing. Springer, Cham (2023). https://doi.org/10.1007/978-3-031-31967-9_2
19. Martin, I.A., Irani, R.A.: A generalized approach to anti-sway control for shipboard cranes. Mech. Syst. Signal Process. **148**. https://doi.org/10.1016/j.ymssp.2020.107168
20. Loveikin, V.S., Romasevich, Y.O., Stekhno, O.V.: Optimization of the movement mode of the mechanism for changing the departure of the tower crane cargo with a horizontal boom. Mech. Eng. (20) (2017). https://doi.org/10.32347/2410-2547.2023.111.223-236
21. Turek, J., Miskarik, L., Vojtesek, J., Kopecek, L., Svacinova, L., & Mizera, A. (2025). Soft Robotics in Industrial Automation: Adaptive Industrial Gripper Design and Evaluation. *EAI Endorsed Transactions on Digital Transformation of Industrial Processes*, *1*(1)

22. Basova, Y., Dobrotvorskiy, S., Balog, M., Iakovets, A., Chelabi, M.A., Zinchenko, A.: Increasing SME supply chain resilience in the face of rapidly changing demand with 3D model visualization. Int. J. Mechatronics Appl. Mech. **14**, 35–47 (2023). https://doi.org/10.17683/ijomam/issue14.5
23. Basova, Y., Dobrotvorskiy, S., Yakovenko, I.: Investigation of the impact of force majeure circumstances as a market instability factor on the flexibility and sustainability of engineering SMEs. EAI Endorsed Transactions on Digital Transformation of Industrial Processes **1**(1) (2025). https://publications.eai.eu/index.php/dtip/article/view/8951

Ways to Improve the Quality of Foundry Moulds

Mikhailo Vorobyov[1], Olga Ponomarenko[1], Tetiana Berlizieva[1], Nataliia Yevtushenko[1](✉), Pavlo Shelepko[1], and Andreea Stanciu[2]

[1] National Technical University "Kharkiv Polytechnic Institute", 2 Kirpicheva Street, Kharkiv 61002, Ukraine
natalya_0899@ukr.net

[2] National Institute of Research and Development in Mechatronics and Measurement Technique INCDMTM, 6-8 Pantelimon, 2Nd District, 021631 Bucharest, Romania

Abstract. The quality of the final product depends on the quality of castings, their accuracy and cost–effectiveness. Given the operating conditions of the casting mold, it must have certain mechanical, chemical and technological properties that ensure the production of high–quality castings. The development and application of cold–hardening mixtures for casting molds while preserving the indicators of their basic physical, mechanical and technological properties, and the development of their preparation technology are urgent tasks for foundry production.

The aim of the research is to improve the quality of the casting mold by improving the physical and mechanical properties of molding mixtures using the Alpha–set process. The parameters such as compressive strength, tensile strength, bending strength, crumbleness and vitality were studied, which were determined by standard methods. The article presents the results of research and optimization of technological and physical and mechanical properties of mixtures.

The regularities of the interaction of the resin with the hardener and quartz sand were established. Based on the planned experiment, mathematical models of the properties of mixtures on quartz sand with phenol–formaldehyde resin were developed. The compositions of cold – hardening resin mixtures on quartz sands were optimized, which allows obtaining high–quality forms with high strength properties. Industrial studies have shown that the optimal composition of the mixture is the amount of resin from 0.8 to 2.0%, the amount of hardener - from 0.4 to 1.0%.

Keywords: Foundry Mold · Quality · Cold–hardening Mixture · Physical and Mechanical Properties

1 Introduction

Today, the foundry industry and the metallurgical complex, which are the basis of the machine–building and defense industries, have found themselves in very difficult conditions and have lost a large share of their capacity as a result of military aggression [1]. Therefore, in In this critical situation, the burden of producing responsible castings in single and small–batch production conditions has mainly fallen on small foundry enterprises that are still operating.

D. D. Cioboată and J. Machado (Eds.): ICoRSE 2025, LNNS 1592, pp. 252–263, 2025.
https://doi.org/10.1007/978-3-032-02508-1_22

Castings and products constitute a significant share in terms of mass and labor intensity of manufacturing any type of mechanical engineering product. The quality and competitiveness of the final products – engines, machine tools, cars in [2, 3].

Increasing complexity, precision and reduction of wall thickness of cast parts, along with the requirements of minimizing labor costs and effective environmental protection, have significantly affected the development of casting production technologies. This fully applies to technology for manufacturing casting molds and cores [4].

The growing requirements for the quality of casting molds in the modern world and the efficiency of their manufacture lead to the fact that the requirements for the properties of molding and core mixtures, especially in recent years, are continuously increasing.

Currently, there are many foundry enterprises where the mold manufacturing process is not yet mechanized. In the conditions of individual and small–scale production, the requirements for the properties of molding mixtures are constantly increasing [5].

The quality of the casting has a decisive influence on the quality of the mold preparation, therefore, special requirements are imposed on the molds. The mold must: provide the necessary configuration and specified accuracy of the casting; withstand thermal effects and force loads of the metal and the casting; ensure the solidification of the casting at the optimal speed; create conditions that prevent the formation of external and internal defects of the castings; create conditions for obtaining the necessary cleanliness of the surface of the castings; ensure a directed process of metal crystallization in the castings; intensify the process of cooling and solidification of the castings in the mold in order to reduce their production cycle [6].

The strength of the mold should be such that the configuration and dimensions of the mold do not change during the process of filling with the melt, during the crystallization of the casting and for some time during its subsequent cooling, until the poured material acquires sufficient strength to preserve the configuration of the casting [7, 8].

2 Literature Review

The properties of the casting mold determine the quality of the castings. In this regard, the study of materials used for the preparation of molding and core mixtures, as well as understanding the theory of the formation of their properties, is the basis for the correct design of the casting technology and ensuring high quality of cast parts [8, 9].

Since mixtures have a complex and diverse effect on the entire technological process of casting production, they must have many properties: a certain granulometric structure, plasticity, strength in the wet and dry states, ductility, which is knocked out of the castings after the latter have hardened, a certain heat–accumulating ability, impact.

In recent years, the production of molds has been developing taking into account cold–hardening mixtures (CHM). At the same time, a number of advantages have been revealed that radically change the organization of foundry production: the possibility of selecting mixtures for molds and cores from the same components; reduction of production areas, since there is no need for mixture preparation departments; reduction by 3…4 times of the volume of molding mixtures and, accordingly, the volume of intra–shop transport operations; there is no need for molding machines with bunker trestles and extended belt conveyor systems; upper and lower molds can be manufactured on one installation [10, 11]. The transition to the manufacture of molds and cores from CHS sharply increases the accuracy of castings, production culture, minimizes losses from defects due to molds and cores, and reduces the labor intensity of production.

Today, in foundry production, more and more attention is paid to cold–hardening mixtures with synthetic resins. In recent years, cold–hardening mixtures have become the main ones in the manufacture of rods and molds in single and serial production in industrialized countries. The volume of use of synthetic resins in foundry production is very large. This is explained by the high strength of the mixture with a small consumption of resin, the ability to adjust the hardening speed of the mixture in a wide range, the absence of a drying operation and the need for drying equipment, easy knocking out of mixtures from the internal cavities of castings and castings from molds, low energy consumption of the process, due to which the cycle is significantly simplified and shortened [12, 13].

Mixtures contain a filler, a binder and a hardening material. The basis of the mixture is the filler. This component gives the mixture the necessary compressive strength, fire resistance, and certain thermophysical characteristics. Quartz sand is most often used as a filler for molding and core mixtures. The shape, size, and uniformity of its grain composition have a decisive influence on the properties of the mixtures [14, 15].

The Alpha–set process is one of the methods for manufacturing molds and rods from cold–setting mixtures. The process is universal, i.e. it can be used with the same binder material in the production of cast iron, steel and non–ferrous castings. It is especially suitable for use in the manufacture of medium and large steel castings, including alloy steels [16, 17]. Alpha – set process is based on the use of the same type of binder – liquid phenol–formaldehyde oligomer, converted by excess alkali (caustic potash or caustic soda) into the form of polyphenolate, which is a stable form, not prone to autocondensation during prolonged exposure. The mass fraction of free phenol is usually absorbed in 1%, free formaldehyde – less than 0.1% [18].

Hardeners are esters of the general structure RCOOR. The ester is hydrolyzed in an alkaline medium according to the scheme:

$$\mathrm{RCOOR} + \mathrm{HOH} \rightarrow \underset{\text{alcohol}}{\mathrm{ROH}} + \underset{\text{acid}}{\mathrm{RCOOR}}$$

The released acid (acetic, carbonic or oxybutyric) is neutralized by the potassium ion in the phenolate groups (OP), converting them to hydroxyl (OH). In this case, the phenol–formaldehyde oligomer passes from the phenolate to the normal form and then, due to the formation of additional methylene bonds, three–dimensional crosslinking occurs with the formation of a strong polymer structure [19].

Depending on the composition of the hardener (the ratio of slow and fast–acting reagents in it), the vitality and, accordingly, the rate of rejection can be regulated within the range from 3…5 to 30…40 or more minutes [20, 21].

The mixture is prepared in continuous or batch mixers. Depending on the quality of the sand (granulometry, grain shape, content of clay component and impurities), the resin content is usually 1.6…2.0% of the mixture mass, the amount of hardener is 30% of the resin mass. Any fillers can be used – quartz, zircon, distensillimanite, magnesite, since the properties of the mixture do not depend on alkalinity and depend little on the humidity of the sand [22].

The properties of molding and core mixtures depend on the uniform distribution of binding materials throughout the volume of the mixture. Uniform distribution of components and creation of binding shells around sand grains is achieved during the preparation of the mixture.

Rods or forms are made by compacting the mixture on vibrating tables or manually. The time to reach manipulator strength is 2…4 times longer than the vitality of the mixture. CHM using the Alpha – set process successfully compete with cold-hardening mixtures based on furan resins.

Advantages of the Alpha – set binding system: rapid volumetric solidification, which allows filling molds with metal 2…3 h after preparing the mixture and forming; low levels of free phenol, formaldehyde and methanol vapor emissions (compared to acid–hardened phenol–formaldehyde and furan resins) in the core and forming areas; practical absence of emissions of sulfur dioxide and organosulfur compounds in the areas of pouring, cooling and knocking out of molds, typical for CHM with furan resins hardened with aromatic sulfonic acids; reduced tendency to carburization of the surface of steel castings, elimination of saturation of the surface of castings with gray, absence of nitrogen and associated gas porosity in castings; a combination of high heat resistance and a certain flexibility of molds and rods during high–temperature heating, which reduces (although does not completely eliminate) the likelihood of casting defects such as undercuts and hot cracks; less sensitivity to moisture compared to CHM on furan resins and especially cold-hardening mixtures on phenol isocyanate binders; versatility in terms of alloys for casting – it is possible to produce both cast iron and steel castings (without any restrictions); in addition, there is the possibility of using sands of different chemical nature [23].

The disadvantages include: slightly lower strength, especially the strength of the edges of rods and molds, compared to CHM based on furan resins of medium and higher categories (with a furfuryl alcohol content in the resin of 60% or more) and cold-hardening mixtures based on a phenol–isocyanate binder with a higher binder consumption (1.5 wt. Parts of CHM, the total binder consumption is 1.2… 1.6 wt. Parts); a more complex process of regeneration of spent mixtures, which requires, in addition to traditional methods of mechanical processing, its chemical neutralization to get rid of increased alkalinity with subsequent heat treatment at a temperature of about 700 °C [24, 25].

Currently, there are many processes for manufacturing molds and cores, but for each enterprise it is necessary to develop its own mathematical model, the use of which allows obtaining high–quality molds with high physical, mechanical and technological properties.

Therefore, the development and application of cold–hardening mixtures for casting molds while preserving the indicators of their basic physical, mechanical and technological properties, and the development of their preparation technology are urgent tasks for foundry production.

The aim of the research is to improve the quality of the casting mold by improving the physical and mechanical properties of the molding mixtures using the Alpha–set process.

3 Research Methodology

JF –103 M brand, and a hardener of the HQG 20 brand were used as materials.

The mixture is prepared as follows: for 100 parts of quartz sand, add 0.5… 1.5 parts of hardener. The mixture is thoroughly mixed for 60 s, then 1.0… 3 parts of phenol–formaldehyde resin is added to this mixture and mixed thoroughly again for 120 s.

The work investigated properties such as compressive strength, tensile strength, bending strength, crumbleness strength and vitality.

Molding materials must have a number of properties that correspond to a specific molding or core mixture. These include the conditions of interaction of the mold with the liquid metal during mold filling, solidification and cooling of the casting; core preparation technology.

Mechanical properties determine the strength characteristics of the casting mold during its manufacture, pouring it with alloy, and solidification of the casting.

Compressive strength tests of mixtures were carried out in accordance with standards. For this purpose, a device for determining the ultimate strength of molding and core mixtures mod. 04116U with a device for compressing dry samples up to 15 MPa was used.

Technological properties characterize the conditions for obtaining high–quality molds and cores, as well as the conditions for manufacturing castings with the least labor intensity and high surface quality (without cracks and blockages).

The vitality of self–hardening mixtures is the main parameter limiting the production cycle of rods and foundry castings. The vitality indicator of cold –hardening mixtures was usually the time (in minutes) after which its strength value decreased by 30% of the maximum.

The crumbleness of the molding and core mixtures was determined on standard cylindrical samples with a diameter and height of 50 mm. The samples were tested in a dry state. The crumbleness of the mixture was assessed by the mass loss of a standard sample placed in a rotating mesh drum, with the crumbleness value expressed in percent.

4 Results

Recently, a computational and analytical method based on a planned experiment has been widely used to solve the problem of controlling the properties of molding mixtures and their stabilization. This approach opens up new opportunities for controlling the properties of molding mixtures due to the efficiency of process regulation when changing the properties of the starting materials.

To establish the patterns of interaction between the binder based on phenol–formaldehyde resin of the JF–103 M brand, the hardener HQG 20, and quartz sand of the PK–070–3 brand, an active planned experiment was conducted.

The intervals of variation of factors and their values at the main, upper and lower levels are given in Table 1.

Table 1. Conditions for conducting experiments for molding and core mixtures with the addition of phenol–formaldehyde resin and hardener.

Factors	Amount of resin, %	Amount of hardener, %
Code	x 1	x2
Basic level	2	1.0
Variation interval	1	0.5
Upper level	3	1.5
Lower level	1	0.5

The planning matrix for conducting Experiment 2^{3-1} is given in Table 2.

Table 2. Planning matrix and output parameter

No. Research	Amount of resin (x_1), %	Catalyst quantity (x_2), %	x_0	x_1	x_2	x_1 x_2	Compressive strength y_1, MPa	Tensile strength y_2, MPa	Bending strength y_3, MPa	Crumbleness y_4, %	Vitality y_5, min
1	3	1.5	+	+	+	+	1.92	1.4	1.55	0.075	13
2	1	1.5	+	–	+	–	0.66	0.17	0.50	0.48	21
3	3	0.5	+	+	–	–	1.55	0.43	1.25	0.19	12
4	1	0.5	+	–	–	+	0.76	0.2	0.54	1.15	15

Typically, in the Alpha–set process, the resin content is 1.6...2.0% of the mixture mass, the hardener amount is 30% of the resin mass.

During this experiment, the resin content varied from 1% to 3%, and hardener – from 0.5% to 1.5%.

The variable factors were: the amount of resin (x_1), and hardener (x_2). The main physical and mechanical indicators of the properties of the molding mixture were selected as the optimization parameter (y): compressive strength after 3 h (y_1), tensile strength (y_2), bending strength (y_3), crumbleness (y_4) and vitality (y_5).

As a result of processing the obtained data, the following system of equations was obtained in a coded scale:

$$\text{for } y_1 = 1,22 + 0,51x_1 + 0,12x_1x_2, \text{MPa} \quad (1)$$

$$y_2 = 0,55 + 0,37x_1 + 0,24x_2 + 0,25x_1x_2, \text{MPa} \quad (2)$$

$$y_3 = 0,96 + 0,44x_1 + 0,085x_1x_2, \text{ MPa} \tag{3}$$

$$y_4 = 0,21 - 0,09x_1 - 0,05x_2, \ \% \tag{4}$$

$$y_5 = 15,3 - 2,8x_1 + 1,8x_2 - 1,3x_1x_2, \text{ min.} \tag{5}$$

In these equations, we substitute the values of the factors in the coded scale, which are calculated from the corresponding values in the natural scale according to the formulas:

$$\mathrm{x}_1 = \frac{\mathrm{X}_1 - 2}{1} \quad \mathrm{x}_2 = \frac{\mathrm{X}_2 - 1}{0,5} \tag{6}$$

Checking the obtained mathematical models for the significance of the coefficient in and adequacy was performed using the Student's test and Fisher's exact test.

With the help of Figs. 1, 2, 3, 4 and 5 can be traced to the effect of the amount of resin and hardener on compressive strength, tensile strength, bending strength, crumbleness and vitality, respectively.

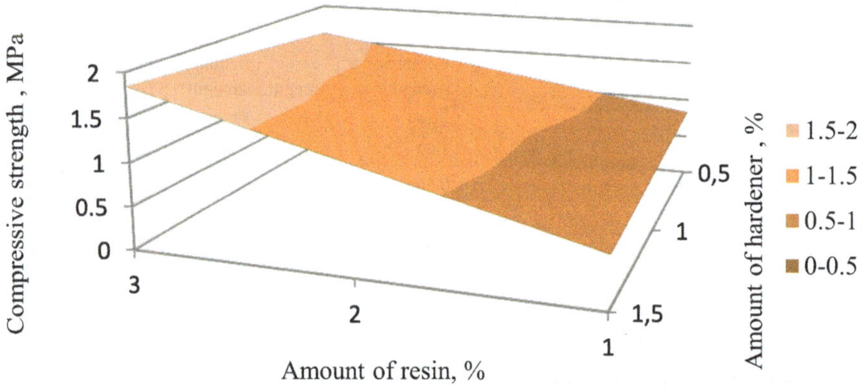

Fig. 1. Effect of resin and hardener amounts on the compressive strength of the molding compound

When analyzing the obtained data, the following can be noted: the influence of variable factors on the optimization parameters corresponds to theoretical ideas about the formation of the properties of the mixture during its preparation.

Compressive, tensile and bending strength increases with increasing resin content and increasing hardener content.

The crumbleness of the mixtures decreases with increasing resin and hardener content. The vitality of the mixture decreases with increasing resin content and increases with increasing hardener content.

Based on the developed mathematical models, a nomogram was created that describes the dependence between the parameters of the technology and the properties of the molding mixtures. They are presented in Fig. 6, on their basis it is possible to adjust the parameters of the sample preparation process by changing the amount of the

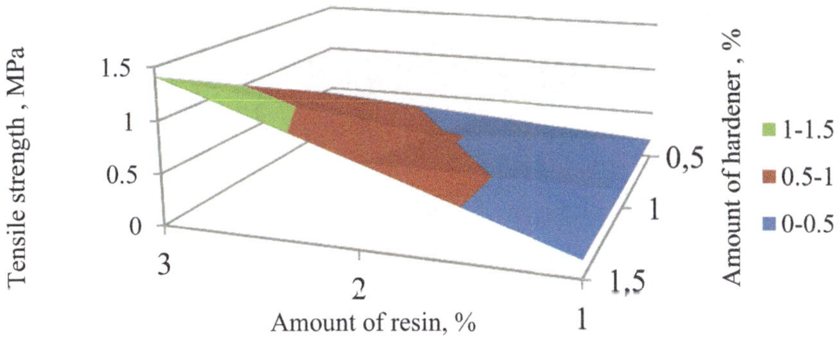

Fig. 2. Effect of resin and hardener amounts on the tensile strength of the molding compound

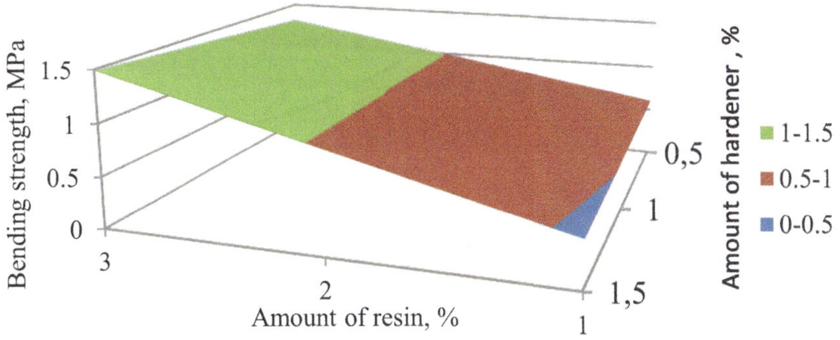

Fig. 3. Effect of resin and hardener amounts on bending strength of molding compound

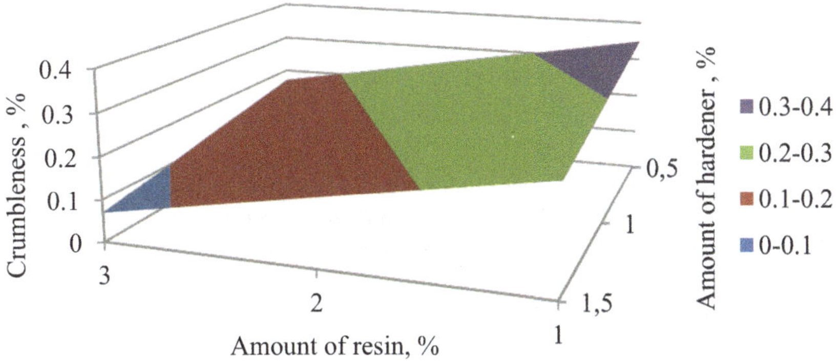

Fig. 4. Effect of the amount of resin and hardener on the crumbleness of the molding mixture

binder in the mixture, which can be used for operational control of the properties of the mixture. Nomograms are a means of graphically solving this problem.

The dependence is built on the basis of the obtained mathematical models for compressive strength, tensile strength, bending strength, bending resistance, and vitality. We

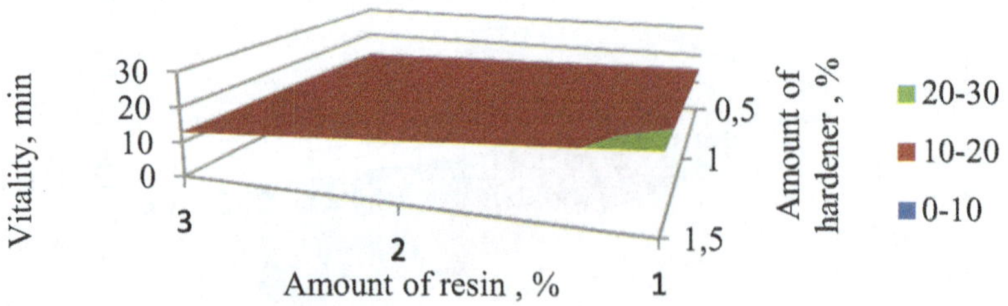

Fig. 5. Effect of resin and hardener amounts on the vitality of the molding compound

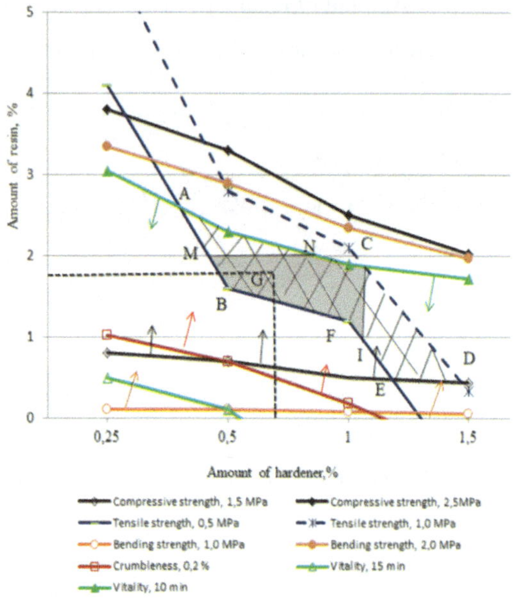

Fig. 6. Nomogram for determining the optimal composition of the mixture based on and resin and hardener

believe that we are satisfied with the compressive strength, which is within 1.5…2.5 MPa, tensile strength – 0.5 … 1.0 MPa, bending strength, which is within 1.0 … 2.0 MPa, crumbleness – below 0.2%, vitality is within 10 to 15 min.

The area of the polygon ABCDEF is the zone of optimal values, each point of which meets the requirements of the technological process of preparing the mixture. The arrows show the direction towards increasing the compressive strength, bending strength and vitality of the mixture and reducing the crumbleness.

For example, point *G* on the nomogram shows that if the resin is used in the mixture in an amount of 1.7 wt. % and hardener – 0.7 wt. %, the mixture will meet the requirements for the quality of the mixture. Point M shows the minimum amount of components that must be taken to obtain a mixture of the required quality.

Industrial studies have shown that the optimal compositions of the mixture are the amount of resin from 0.8 to 2.0%, the amount of hardener – from 0.4 to 1.0%. The area of the polygon BMNCIF is the zone of optimal values of the mixture.

5 Conclusions

Currently, cold–hardening mixtures with synthetic resins have become more widespread for the production of casting molds. This is explained primarily by their high strength of the mixture, low binder consumption (1…2%), and the ability to adjust the strength and hardening speed over a wide range.

In the course and development of the technological process for the manufacture of high–quality casting molds using the Alpha–set process was carried out in a planned experiment. A mathematical model was built, which is a system of equations that relate the compressive strength, tensile strength, bending strength of the mixture, crumbleness and vitality with controlled variables of a technological nature, which are considered appropriate: the amount of resin, as well as the amount of hardener.

It was found that the compressive, tensile and bending strengths increase with increasing resin content and increasing hardener content. The crumbleness of the mixtures decreases with increasing resin and hardener content. The vitality of the mixture decreases with increasing resin content and increases with increasing hardener content.

Based on the developed mathematical models, a nomogram was created that describes the dependence between the parameters of the technology and the properties of the molding mixtures. The dependence is built on the basis of the obtained mathematical models for compressive strength, tensile strength, bending strength, crumbleness and vitality. We assume that the compressive strength is within 1.5…2.5 MPa, tensile strength – 0.5…1.0 MPa, bending strength, which is within 1.0…2.0 MPa, crumbleness – below 0.2%, vitality is within 10 to 15 min.

The zone of optimal values is the polygon BMNCIF, each point of which meets the requirements of the technological process of preparing the mixture. Industrial studies have shown that the optimal composition of the mixture is the amount of resin from 0.8 to 2.0%, the amount of hardener - from 0.4 to 1.0%.

References

1. Manjunath Patel, G.C., Chate, G.R., Parappagoudar, M.B.: Modelling and optimization of alpha–set sand moulding system using statistical design of experiments and evolutionary algorithms. In: Optimization of Manufacturing Processes, Springer, Cham, 1–28 (2020). https://doi.org/10.1007/978-3-030-19638-7_1
2. Chate, G.R., Patel, G.C.M., Bhushan, S.N.B., et al.: Comprehensive modelling, analysis and optimization of furan resin–based moulding sand system with sawdust as an additive. J. Braz. Soc. Mech. Sci. Eng. **41**, 183 (2019). https://doi.org/10.1007/s40430-019-1684-0
3. Zhang, Y., Sappinen, T., Korkiala-Tanttu, L., Vilenius, M., Juuti, E.: Investigations into stabilized waste foundry sand for applications in pavement structures. Resour. Conserv. Recycl. **170**, 105585 (2021). https://doi.org/10.1016/j.resconrec.2021.105585

4. Ponomarenko, O., Yevtushenko,N., Berlizieva, T., Shelepko, P., Yevtushenko, Y.: Controlling the gas mode of a mold for producing thin-wall castings. In: Advances in Design, Simulation and Manufacturing VII. Lecture Notes in Mechanical Engineering, pp. 433–441. Springer, Cham (2024). https://doi.org/10.1007/978-3-031-63720-9_38
5. Drossel, W.G., Ihlemann, J., Landgraf, R., Oelsch, E., Schmidt, M.: Media for dimensional stabilization of rubber compounds during additive manufacturing and vulcanization. Materials **14**(6), 1337 (2021). https://doi.org/10.3390/ma14061337
6. Subramanian, S., Khan, Q., Ku, T.: Strength development and prediction of calcium sulfoaluminate treated sand with optimized gypsum for replacing OPC in ground improvement. Constr. Build. Mater. **202**, 308–318 (2019). https://doi.org/10.1016/j.conbuildmat.2018.12.121
7. Simatupang, M., et al.: The mechanical properties of fly–ash–stabilized sands. Geosciences **10**(4), 132. (2020). https://doi.org/10.3390/geosciences10040132
8. Berladir, K., Trojanowska, J., Ivanov, V., Pavlenko, I.: Materials selection in product development: challenges and quality management tools. In: Hamrol, A., Grabowska, M., Maletič, D. (eds) Advances in Manufacturing III. Manufacturing 2022. Lecture Notes in Mechanical Engineering, pp. 72–86. Springer, Cham (2022). https://doi.org/10.1007/978-3-031-00218-2_7
9. Lutiy, R.V., Guriya, I.M.: Forming materials. KPI named after Igor Sikorsky, Kyiv (2020)
10. Dragobetsky, V.V., Zahirnyak, O.D.: Technological foundations of mechanical engineering. "Tochka" Publishing House, Kharkiv (2019)
11. Khrychikov, V.E., Meniailo, H.V., Semenov, O.D., Aftandiliants, Y.G., Gnyloskureno, S.V.: Graphitizing modification of the axial zone of cast iron rolling rolls in the liquidus-solidus temperature range. Naukovyi Visnyk Natsionalnoho Hirnychoho Universytetu. Dnipro, Ukraine **1**, 67–73 (2023). https://doi.org/10.33271/nvngu/2023-1/067
12. Kostyk, K., Chen, X., Kostyk, V., Akimov, O., Shyrokyi, Y.: New complex treatment to ensure the operational properties of the surface layers of machine parts. Lect. Notes Mech. Eng., 284–293(2023)
13. Khoroshylov, O.: Influence of technological parameters of the continuous casting process on the process of accumulation of damage in the billet. In: Cioboată, D.D. (eds) International Conference on Reliable Systems Engineering (ICoRSE) - 2023. ICoRSE 2023. Lecture Notes in Networks and Systems, Springer Cham, vol. 762. (2023). https://doi.org/10.1007/978-3-031-40628-7_19
14. Ponomarenko, O.I., Yevtushenko, S.D., Yevtushenko, N.S., Berlizieva, T.V., Vorobiov, M.M.: Robust methods for controlling casting processes and the quality of castings. IOP Conf. Ser. Earth Environ. Sci. **1254**(1), 012007 (2023). https://doi.org/10.1088/1755-1315/1254/1/012007
15. Demin, D., Frolova, L.: Construction of a logical–probabilistic model of casting quality formation for managing technological operations in foundry production. EUREKA: Phys. Eng. **6**, 104–118 (2024). https://doi.org/10.21303/2461-4262.2024.003518
16. Aftandiliants, Y., Gnyloskurenko, S., Meniailo, H., Khrychikov, V.: Influence of melt properties on the dendritic structure of steel castings. Arch. Foundry Eng. Katowice Branch Polish Acad Sci. **24**(1), 5–13 (2024). https://doi.org/10.24425/afe.2024.149245
17. Sheikh, A.K., Khan, M.A.A.: Mold design optimization and quality assessment of steel castings through integrated simulations and experiments. J. Mech. Sci. Technol. **34**, 2975–2983 (2020). https://doi.org/10.1007/s12206-020-0629-y
18. Rodríguez-González, P., Fernández-Abia, A.I., Castro-Sastre, M.A., Barreiro, J.: Heat treatments for improved quality binder jetted molds for casting aluminum alloys. Addit. Manuf. **36**, 101524 (2020). https://doi.org/10.1016/j.addma.2020.101524
19. Suthar, J., Persis, J., Gupta, R.: Analytical modeling of quality parameters in casting process–learning–based approach. Int. J. Quality Reliab. Manage. **40**(8), 1821–1858 (2023). https://doi.org/10.1108/IJQRM-03-2022-0093

20. Liutyi, R., Petryk, I., Mogylatenko, V., Popovych, V., Shatska, H.: Research thermal fields in the crystallization process of steel cast parts. Adv. Mater. Sci. Eng. **7331866**, 9 (2022). https://doi.org/10.1155/2022/7331866
21. Al–Saraireh, F. M.: Cold–hardening mixtures based on biopolymer lignin complex for casting production in single and small–series conditions. AIMS Mater. Sci. **10**(5), 876–890 (2023). https://doi.org/10.3934/matersci.2023047
22. Dobrotvorskiy, S., Aleksenko, B.A., Basova, Y., Gnilitskyi, I.M., Kościński, M., Machado, J.: Light beam scattering from the metal surface with a complex mono- and two-periodic microstructure formed with femtosecond laser radiation. Appl. Sci. **14**(19), 8662 (2024) https://doi.org/10.3390/app14198662
23. Johnston, R.D., Gaul, R.T.: Lally C an investigation into the critical role of fiber orientation in the ultimate tensile strength and stiffness of human carotid plaque caps. Acta Biomater. **124**, 291–300 (2021). https://doi.org/10.1016/j.actbio.2021.02.008
24. Shilpa, M., Prakash, G.S., Shivakumar, M.R.: A combinatorial approach to optimize the properties of green sand used in casting mold. Mater Today Proc **39**, 1509–1514 (2021). https://doi.org/10.1016/j.matpr.2020.05.465
25. Amin, M., Agwa, I.S., Mashaan, N., et al.: Investigation of the physical mechanical properties and vitality of sustainable ultra–highperformance concrete with recycled waste glass. Sustainability **15**, 3085 (2023). https://doi.org/10.3390/su15043085
26. Hong, R., Zhang, Z., Pan, S.: Construction of PVA–lignosulfonate hydrogels for improved mechanical performances and all–in–one flexible supercapacitors. Int. J. Biol. Macromol.Macromol. **225**, 1494–1504 (2023). https://doi.org/10.1016/j.ijbiomac.2022.11.206

‘Additive Manufacturing of Compliant Mechanisms Designed for Small Displacements of Optical Components

Ciprian-Ion Rizescu, Robert Ciobanu(✉), Maria-Alexandra Mocanu, Dana Rizescu, and George Arhip

National Science and Technology University POLITEHNICA Bucharest, Splaiul Independentei 313, District 6, Bucharest, Romania
robert.ciobanu@upb.ro

Abstract. This paper primarily investigates the influence of two geometric parameters on the mechanical behavior of a cross-axis flexural pivot (CAFP), which enables motion along two axes. The objective of this study is to explore the potential for optimizing material consumption and reducing the mass of compliant mechanisms that incorporate CAFP structures. To achieve this, a design strategy was proposed in which each blade is replaced by two parallel blades of equal thickness, thereby minimizing material usage without compromising functionality. For reference, structures with whole blades were also studied. The varied parameters are the distance between the outer walls of a bar and, respectively, the distance between the inner walls of the blade. Due to the complexity of the structure, the most suitable manufacturing methods are 3D printing or injection. A Finite Element Analysis (FEA) was used to observe behavior under static loads in the plane of motion of the CAFP and to determine the modal frequencies. The results suggest that blade separation increases the resistance of CAFP against forces that would cause a rotational movement but decreases the resistance against forces that would cause compression. It was also observed that the volume of material used largely determines the modal frequencies of these CAFP structures with separated blades but has a more limited impact on those with whole blades.

Keywords: Compliant Mechanisms · 3D Printing · Finite Element Analysis · Optical Components · Multi-objective Optimization

1 Introduction

Compliant mechanisms are monolithic systems that use their deformation to transmit motion, force, and energy [1]. Their use reduces the number of classic kinematic couplings, reducing the risk of backlash and wear. As a result, these technologies are employed in fields such as medicine, aerospace, and precision engineering, with applications including biological cell handling, micro-vibration suppression and optical guides [2, 3]. They can be manufactured by various methods, such as 3D printing, injection molding, electrical discharge machining, laser cutting, or waterjet cutting [4, 5].

D. D. Cioboată and J. Machado (Eds.): ICoRSE 2025, LNNS 1592, pp. 264–275, 2025.
https://doi.org/10.1007/978-3-032-02508-1_23

Compliant mechanisms have numerous applications as alternatives to classic kinematic couplings [6, 7]. However, a challenge in the design of compliant kinematic couplings is the nonlinearity of the elasticity curve [3, 8]. The objective of this study is to investigate the potential for optimizing material consumption and minimizing the mass of compliant mechanisms incorporating cross-axis flexural pivots (CAFPs) [2, 10, 11], fabricated using 3D printing technologies, specifically fused deposition modeling (FDM), while simultaneously increasing the precision of micro-displacements through structural stiffening. The specimens were made from polylactic acid (PLA), a standard material used in FDM 3D printing [12]. Due to their versatility and efficiency, 3D printing technologies are widely used in diverse fields, ranging from mechanical engineering [13–15], such as the production of gears [16–19] and compliant mechanisms [20, 21], to biomedical applications [22, 23], aerospace components, and customized consumer products [24].

The compliant mechanism was designed for applications requiring high-precision positioning of optical elements [25]. In this case, a mirror can be attached to the mobile plate to enable micrometric adjustments for accurately directing a laser beam. To this end, a special structure was proposed, obtained by separating each flexure blade into two parallel blades of equal thickness, separated by an air gap (Fig. 1). This configuration is referred to hereafter as a bi-lamellar structure.

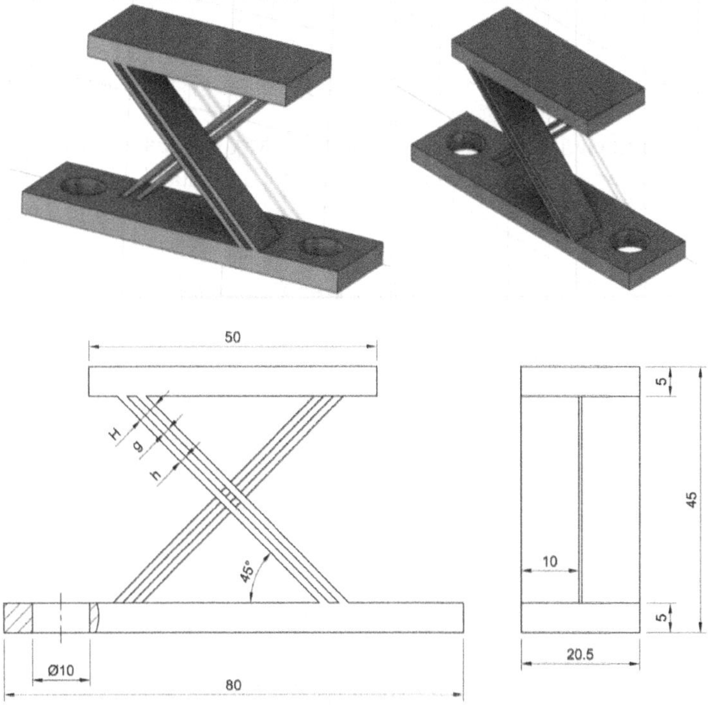

Fig. 1. The compliant structure studied

For reference, a structure with whole blades was evaluated. The study was carried out by varying two parameters (according to Table 2): the distance between the outer walls of a blade(H) and, respectively, the distance between the inner walls of the blade (g).

To evaluate the compliant mechanism, finite element analysis was conducted in Fusion 360, considering PMMA as the material. The corresponding material properties are presented in Table 1. Both effects of static loads in the plane of motion of the CAFP and modal frequencies were studied. In addition to the simulations, experimental determinations were also carried out for the most relevant specimens, which were manufactured using FDM 3D printing technology. These tests were performed using an Imada DSV-500N digital force gauge, allowing precise measurement of the force-displacement relationship under controlled conditions. The Hans-Schmidt HV manual vertical stand was used [26]. The digital force gauge has the capacity to test forces up to 500 N and is intended for various applications such as: compression and tensile load testing of different mechanical elements, machine parts or various mechanisms, mechanical strength testing of some electronic components and many other applications. A chisel stand accessory was used as a force/action element [26]. The experimental data were then compared with the simulation results to validate the FEM model and assess the influence of geometry on the specimens' performance.

2 Materials and Test Methods

The materials used in this study are polymethyl methacrylate (PMMA) for the finite element method (FEM) simulation and polylactic acid (PLA) for the experimental tests, with their physical parameters listed in Table 1. PMMA was selected for the simulation due to its similar properties to PLA and its availability in the standard Fusion 360 material library.

Table 1. PMMA and PLA properties

Material	PMMA	PLA
Young's modulus	2.74 GPa	2.758 GPa
Poisson's ratio	0.355	0.417
Yield stress	48.9 MPa	54.4 MPa
Tear stress	79.8 MPa	55.1 MPa

As mentioned before, the simulations were carried out using Fusion 360 software. The inner surfaces of the fastening holes were defined as fixed constraints, and the gravitational force was omitted. Three scenarios were analyzed: a vertical static load of 1 N uniformly distributed over the top surface of the mobile plate (Fig. 2.a); a frontal static load of 10 N uniformly distributed over the front surface of the mobile plate (Fig. 2.b); and a modal analysis to identify the natural frequencies of the system.

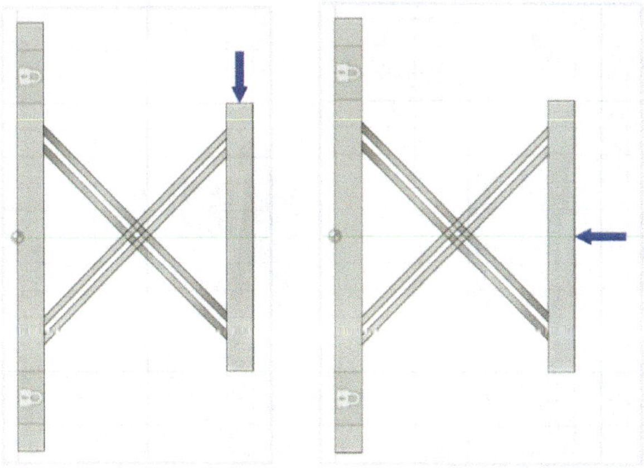

a) Vertical load b) Frontal load

Fig. 2. The static loads studied

A subset of the specimens analyzed through simulation was tested experimentally. The specimens were 3D printed using fused deposition modeling (FDM) technology. To avoid the need for support structures during printing, the mechanism was divided into two halves, which were subsequently assembled by connecting the two halves of the plates (Fig. 3).

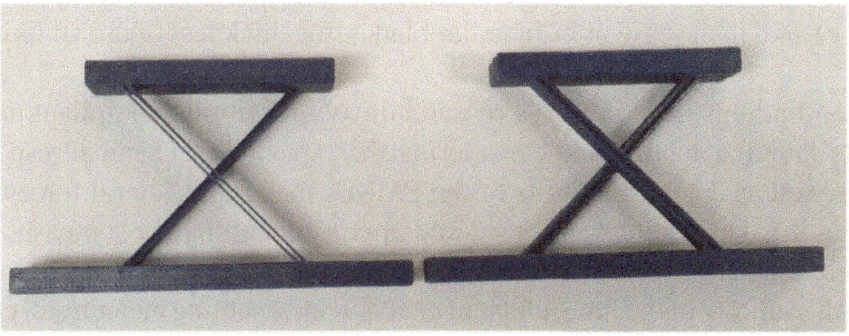

Fig. 3. The 3D printed compliant mechanism

Experimental measurements were performed using an Imada DSV-500N digital force gauge. The fixed plate was secured with two clamps, while a probe equipped with the force sensor was used to displace the mobile plate. The objective was to measure the force required to produce a set of predefined displacements. Two experimental cases were investigated: a vertical static load, with measurements corresponding to displacements of the mobile plate along the vertical axis within the 0–2 mm range (Fig. 4.a); and a frontal static load, with force measurements taken for displacements along the frontal axis within the 0–0.5 mm range (Fig. 4.b.).

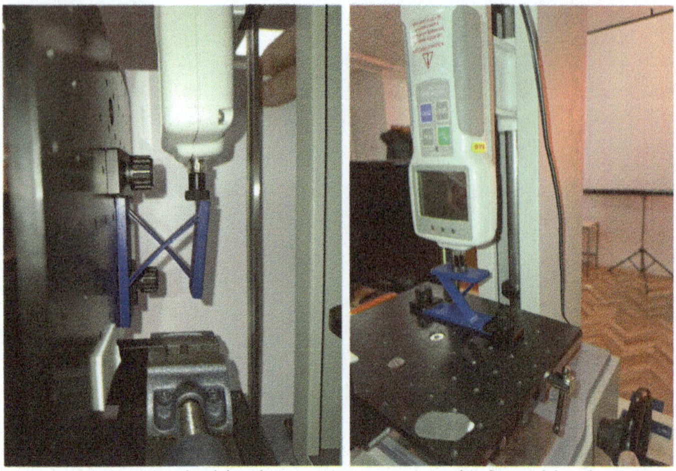

a) vertical load; b) frontal load

Fig. 4. The experimental stand for measuring the force necessary for a given displacement

3 Results

3.1 FEM Analysis

The parameters of the specimens evaluated through finite element method analysis are presented in Table 2. In addition, the table includes the thickness of the material used for a blade (H-g) and, in the case of bi-lamellar blades, the thickness of one of its component blades (h).

Based on findings from previous research involving similar compliant mechanisms subjected to lateral actuation and considering the expected mechanical response of the structure as well as the elastic limits of the PMMA material, a lateral force of 1 N was selected (Fig. 5). This value was considered appropriate to induce a measurable deformation while ensuring that the structure remained within the linear elastic regime, thus preserving the validity of the finite element analysis and enabling meaningful comparison with experimental results (Fig. 6).

Results indicate that, for the same amount of material (constant H − g), the maximum stress decreases as the distance between the flexure blades increases, suggesting that increasing the spacing between flexure elements can lead to improved stress distribution and reduced peak stress levels, which may contribute to enhanced durability and mechanical performance. Moreover, this effect is amplified as the sensitivity of the blades increases.

Given the substantially higher structural resistance under frontal loading, a force of 10 N was selected for this case (Fig. 7).

As with the previous loading case, it can be noted that the maximum stress is inversely proportional to both the blade width and the material thickness.

From Fig. 8 it is obvious that both the maximum stress and the total displacements are inversely proportional to the width of the blade and the thickness of the material.

Table 2. Specimens' parameters

Sample number	H [mm]	g [mm]	H-g = 2h [mm]	h [mm]
1	0.8	0	0.8	- [0.4]
2	1.2	0	1.2	- [0.6]
3	1.6	0	1.6	- [0.8]
4	2	0	2	- [1]
5	2.4	0	2.4	- [1.2]
6	2.8	0	2.8	- [1.4]
7	3.2	0	3.2	- [1.6]
8	3.6	0	3.6	- [1.8]
9	1.2	0.4	0.8	0.4
10	2	0.4	1.6	0.8
11	2.8	0.4	2.4	1.2
12	1.6	0.8	0.8	0.4
13	2.4	0.8	1.6	0.8
14	3.2	0.8	2.4	1.2
15	2	1.2	0.8	0.4
16	2.8	1.2	1.6	0.8
17	3.6	1.2	2.4	1.2

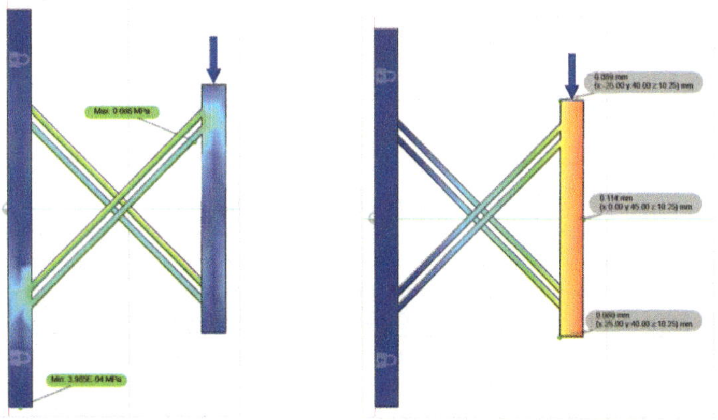

a) Maximum and minimum stress; b) Vertical displacement

Fig. 5. Results for specimen17

However, the geometric structure featuring separated flexure blades results in a higher maximum stress for frontal load.

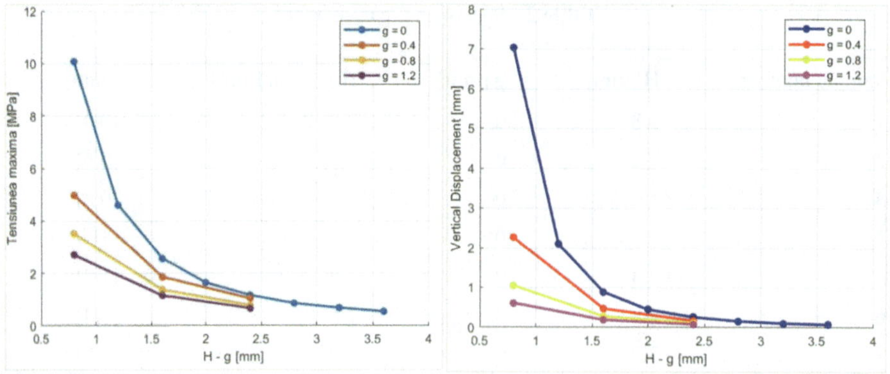

Fig. 6. The variation of maximum stress and vertical displacement for a H-g range of 0.8–3.6 mm, grouped by g

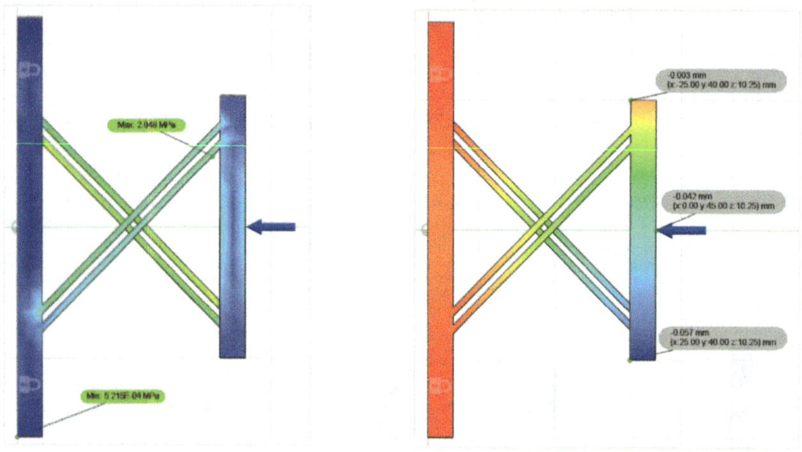

a) Maximum and minimum stress; b) Frontal displacement

Fig. 7. Results for specimen17

For a fully documented conclusion, the compliant mechanisms were evaluated in terms of their modal frequencies (Fig. 9). Modal analysis provides essential information about the dynamic behaviour of the structure by identifying the natural frequencies at which it tends to vibrate in the absence of continuous external excitation. In particular, the first modal frequency (the fundamental frequency) is of special importance, as it is directly correlated with the overall flexibility of the mechanism. A low value of this frequency indicates a more flexible structure, while a high value suggests a stiffer configuration. Therefore, the first modal frequency can serve as a key indicator in optimizing the balance between flexibility and stability.

From the perspective of both the order of magnitude and the function describing the modal frequencies, a clear separation can be made between mechanisms featuring

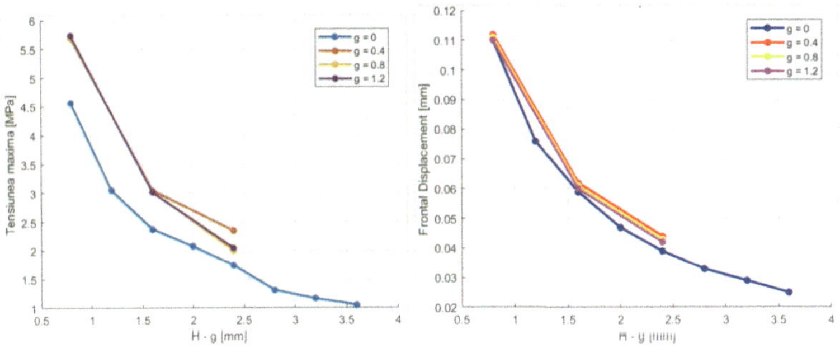

Fig. 8. The variation of maximum stress and frontal displacement for a H-g range of 0.8–3.6 mm, grouped by g

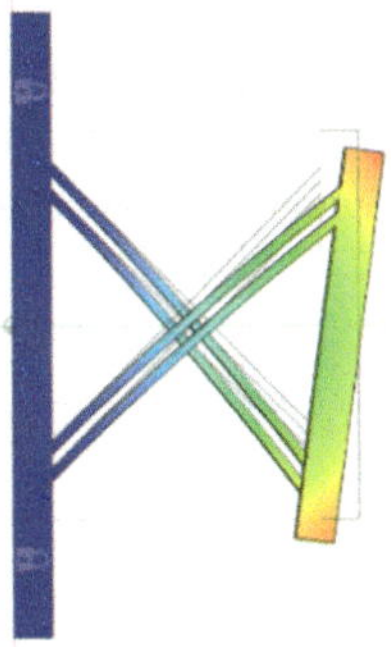

Fig. 9. Result for specimen 17 at the first modal frequency

solid flexure elements and those with bi-lamellar configurations. Figure 10 shows the evolution of the specimens with material thicknesses of 0.8, 1.6 and 2.4 mm (the H-g values of the bi-lamellar blade specimens studied).

The increase in modal frequencies is observed with increasing blade width and material thickness. The specimens with whole blades have an irregular variation, especially after the first 3 modes. In contrast, the specimens with bi-lamellar blades have a plateau in the range of modes 4–7. It is noted that the frequency values are very similar for the bi-lamellar blade specimens with the same material thickness.

3.2 Experimental Results

Specimens 3, 12, and 13 were evaluated experimentally, allowing for a comparison between the behavior of the specimens and the results obtained through finite element method (FEM) simulation. For this purpose, the force required to produce a given displacement was first determined experimentally, and then the effect of this load was simulated numerically to obtain the corresponding simulated displacement.

For the frontal static load, corresponding to displacements along the frontal axis in the range of 0–0.5 mm, the experimental results and those obtained from simulations are

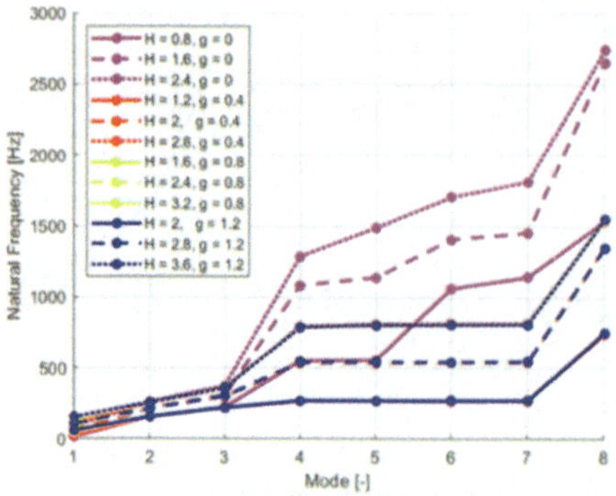

Fig. 10. Variation of modal frequency values depending on the mode of occurrence, grouped by the specimen studied

presented in Fig. 11. The experimental results exhibit nonlinear behavior, characterized by variations in stiffness depending on the geometry and the amplitude of displacement, leading to a noticeable discrepancy when compared to the simulation data.

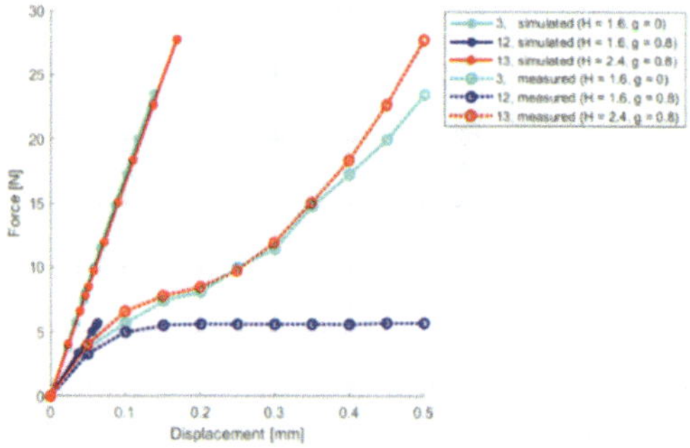

Fig. 11. The variation of frontal force for a displacement range of 0–0.5 mm

For the lateral loading case, displacements along the force axis were monitored within the range of 0–2 mm (Fig. 12). From a simulation perspective, the force–displacement relationship exhibits linear behavior for all specimens. Experimental results show similar linearity only for the monolayer configuration with H = 1.6 mm and g = 0 mm. In contrast, bi-lamellar structures exhibit a nonlinear response, characterized by a force threshold that marks the point at which the two blades come into contact as a result of the displacement. Beyond this point, the mechanical behavior becomes complex, involving both the elastic properties of the individual blades and the frictional interaction and action–reaction forces at the contact zones.

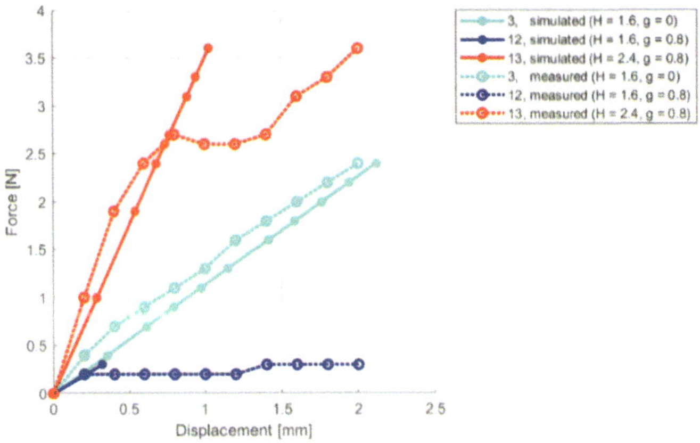

Fig. 12. The variation of vertical force for a displacement range of 0–2 mm

4 Conclusions

The results suggest that the separation of the blades increases the resistance of the CAFP against forces that would cause a rotational movement but decreases the resistance against forces that would cause compression. However, the much higher compressive strength of the structure offers the opportunity to use the bi-lamellar blades to improve rotational resistance without compromising compressive performance. This increase in rotational resistance also contributes to enhanced positioning precision, which is essential for applications involving micro-displacements of optical components.

Compliant mechanisms featuring a bi-lamellar configuration exhibit a higher first modal frequency compared to those with a whole blade structure, indicating increased stiffness.

In the case of bi-lamellar blade structures material thickness predominantly influences the modal frequency values.

From the perspective of lateral loading, the experimental data reveal a nonlinear force–displacement relationship, which differs from the results obtained through simulations.

For the pivoting motion, the bi-lamellar structure exhibits increased stiffness, with a linear behavior characteristic of each gap distance g between the bars, up to a critical point. This limit point corresponds to the force at which the bars come into contact due to displacement, causing the mechanism to operate in a regime influenced by multiple factors: the spring constant, the friction between the blades, and the non-uniform load distribution across the flexural elements.

Based on the results obtained, the optimal solution for achieving a pivoting micro-displacement of 0.5 mm is the bi-lamellar compliant mechanism with H = 2.4 mm and g = 0.8 mm. Compared to the single-blade configuration with H = 1.6 mm and g = 0 mm, which uses an equivalent amount of material, the proposed bi-lamellar structure exhibits significantly higher rotational stiffness. This increase in stiffness directly contributes to improved positioning precision and enhanced mechanical stability, making the structure particularly suitable for applications involving the micrometric adjustment of optical components. Such performance is essential in systems that require repeatable and precise laser beam steering, where even minimal deflections must be accurately controlled without introducing mechanical play or instability.

References

1. Howell, L.L.: Compliant Mechanisms. In: McCarthy, J. (ed.) 21st Century Kinematics, pp. 189–216. Springer, London (2013)
2. Domerg, M., Ostré, B., Joliff, Y., Grunevald, Y.-H., Garcia, A.D.: The development of a 3D-printed compliant system for the orientation of payloads on small satellites: material characterization and finite element analysis of 3D-printed polyetherketoneketone (PEKK). Aerospace **11**, 294 (2024). https://doi.org/10.3390/aerospace11040294
3. Arredondo-Soto, M., Cuan-Urquizo, E., Gómez-Espinosa, A.: A review on tailoring stiffness in compliant systems, via removing material: cellular materials and topology optimization. Appl. Sci. **11**, 3538 (2021). https://doi.org/10.3390/app11083538
4. Carollo, G., Ingrassia, T., Pantano, A.: Design of a low cost 3D printable single-component compliant mechanism for FWMAV's wing actuation. In: Proceedings of the Second International Conference on Design Tools and Methods in Industrial Engineering, Rome, Italy, pp. 39–49 (2021)
5. Ames, D.C., Smith, G.L., Nathan, L., Howell, L.L., Magleby, S.P.: Laser forming of compliant mechanisms. ASME Open J. Eng. (2023)
6. Wiesław, N.P., Aleksy, F., Michał, K.: Experimental studies on 3D printed compliant mechanisms in gripper applications. Adv. Sci. Technol. Res. J. Soc. Polish Mech. Eng. Tech. (SIMP) **17**(6), 228–244 (2023). https://doi.org/10.12913/22998624/174109
7. Zolfagharian, A., Mohammad, L., Ranjbar, S., Tadesse, Y., Bodaghi, M.: 3D printing non-assembly compliant joints for soft robotics. Results Eng. **15**, 100558 (2022)
8. Merriam, E.G., Howell, L.L.: Lattice Flexures: geometries for stiffness reduction of blade flexures. Precis. Eng. **45**, 160–167 (2016). ISSN 0141–6359, https://doi.org/10.1016/j.precisioneng.2016.02.007
9. Ma, W., Liu, X., Qiu, X., Zhou, Y., Liu, Y., Fu, W., et al.: Comparative folding/unfolding performance of notch-type compliant joints. Case Stud. Construct. Mater. **18**, e01760 (2023)
10. Haringx, J.A.: The cross spring pivot as a constructional element. Appl. Sci. Res. **1**(1), 313–332 (1949)
11. Jensen, B.D., Howell, L.L.: The modeling of cross-axis flexural pivots. Mech. Mach. Theory **37**, 461–476 (2002)
12. Simion, I., Arion F.A.: Dimensioning rules for 3D printed parts using additive technologies (FDM), U.P.B. Sci. Bull., Ser. D **78**(2) (2016). ISSN 1454–2358;
13. Ciobanu, R., Donţu, O., Besnea, D., Cioboata, D., Nicolae, C.: 3D CAD/CAM and rapid prototyping applied for malta cross mechanism fabrication. Rom. Rev. Precis. Mech. Optics Mechatron., 46 (2014)
14. Perşinaru, Ş., Popescu, D., Desuraune, V.: 3D-printed adapter for a robot gripper: decisions and failures, U.P.B. Sci. Bull., Ser. D **83**(2) (2021), ISSN 1454–2358
15. Cartal, L.A.: Mechanical amplification system for displacements produced by piezoelectric actuators, U.P.B. Sci. Bull., Series D **73**(2), 239–248 (2011), ISSN 1454–2358
16. Ciobanu, R., Rizescu, C.I., Rizescu, D., Gramescu, B.: Surface durability of 3D-printed polymer gears. Appl. Sci. **14**, 2531 (2024). https://doi.org/10.3390/app14062531
17. Rizescu, C.I., Ciobanu, R.R., Rizescu, D.: Double flank gear roll inspection of spur gears manufactured using fused deposition modeling. IOP Conf. Ser. Mater. Sci. Eng. **514**, 012031 (2019). https://doi.org/10.1088/1757-899X/514/1/012031, SCOPUS;

18. Bezzini, R., Bassani, G., Avizzano, C.A., Filippeschi, A.: Design and experimental evaluation of multiple 3D-printed reduction gearboxes for wearable exoskeletons. Robotics **13**, 168 (2024). https://doi.org/10.3390/robotics13110168
19. He, B., Tian, H., Yu, G., et al.: A review of composite gears. Mech. Compos. Mater. **60**, 1133–1154 (2025). https://doi.org/10.1007/s11029-025-10250-5
20. Dearden, J., Grames, C., Orr, J., Jensen, B.D., Magleby, S.P., Howell, L.L.: Cylindrical cross-axis flexural pivots. Precis. Eng. **51**, 604–613 (2018)
21. Gonçalves Junior, L.A., Theska, R., Lepikson, H.A., Ribeiro Junior, A.S., Linß, S., Gräser, P.: Theoretical and experimental investigation of performance characteristics and design aspects of cross-spring pivots, Int. J. Solids Struct. 185–186, 240–256 (2020). ISSN 0020–7683, https://doi.org/10.1016/j.ijsolstr.2019.08.023
22. Moraru, E., Dontu, O., Besnea, D., Constantin, V.: Study and realization of prosthetic dental models by additive technologies. IOP Conf. Ser. Mater. Sci. Eng. **444**, 042017 (2018). IOP Publishing
23. Bashir, K.S., Syed, I., Zhengjun, Z., Yuan, W.: Redefining medical applications with safe and sustainable 3D printing. ACS Appl. Bio Mater. (2025). https://doi.org/10.1021/acsabm.4c01923
24. Calusaru, I.M., Baran, N., Patulea, A.: The influence of the constructive solution of fine bubble generators on the concentration of oxygen dissolved in water. Adv. Mater. Trans Tech Publications, Switzerland, 538–541, 2304–2310 (2012). https://doi.org/10.4028/www.scientific.net/AMR.538-541.2304
25. Ciobanu, R., Pascu, A.T., Rizescu, D.: Application of image processing methods for contact lens evaluation. In: Proceedings of SPIE 12170, Advances in 3OM: Opto-Mechatronics, Opto-Mechanics, and Optical Metrology, 121700N (2022). https://doi.org/10.1117/12.2621598
26. Test Stand HV-500N II. https://www.hans-schmidt.com/en/produkt-details/test-stand-hv-500n/

Cognitive-Aware Retention Algorithms – a Case Study Analysis

Andrei Scurtu(✉)

Transilvania University, Blvd. Eroilor No. 29, Brasov, Romania
andrei-bc.scurtu@student.unitbv.ro

Abstract. Retention algorithms are designed to keep users engaged on a digital platform or service. These mechanisms are used by companies to increase the frequency of user interactions and generate more revenue through ads or subscriptions, collecting and analyzing data on user behavior to personalize the experience and encourage continuous interaction. While this brings significant benefits, excessive use of technology can generate negative effects, especially on the mental and behavioral health of users. The objectives of this study are: (1) to understand the significance and the ethical implications of the problem, (2) to identify methods to measure user's state and build a model that simulates the brain to keep track of its state while scrolling, (3) to build and optimize a recommendation system that keeps user state within safe thresholds, (4) to test and compare the built model to the actual Instagram model.

Keywords: Logical Regression · Neurobiological Optimization · User State Quantification · Neurobiological Indexation

1 The Ethical Problem: Algorithmic Design and Cognitive Health

1.1 Social Media Retention

Recent years have seen a categorical increase in social media, especially in the years following 2010, which marked the founding of the renowned social media platform Instagram. In 2022, roughly 2 billion people used the platform [1]. Of these users, Pew Researchers have found in 2023 that 73% engage in passive interaction with the platform, while only 21% post weekly, and the remaining 6% rarely open the app. Of the passive interactions recorded by Instagram, Meta's Q3 Earnings Report has estimated that reels make an ever-growing percentage, which was evaluated in 2024 to be 50% of total time spent on the app [2]. It is estimated that 17.6 million hours are spent watching Reels per day [3].

1.2 Algorithmic Addiction Mechanisms Exploited by Instagram

As a consequence of the evident significance of the platform in people's lives, numerous studies have been conducted to analyze the effects it may have on the human mind.

D. D. Cioboată and J. Machado (Eds.): ICoRSE 2025, LNNS 1592, pp. 276–285, 2025.
https://doi.org/10.1007/978-3-032-02508-1_24

In the context of a growing male loneliness epidemic, and social tensions rising high, the question of whether this can be attributed to social media has become increasingly important. Numerous studies have concluded that passive social media use is directly correlated to negative effects. Key studies in this area include 2015's "Passive Facebook usage undermines affective well-being: Experimental and longitudinal evidence", which has found that passive use, such as scrolling and routinely checking social media feeds for new posts, is directly linked to increased envy, loneliness, and depression [4]. Similarly, in 2018, a study – Hunt (2018) "No More FOMO: Limiting Social Media Decreases Loneliness and Depression" [5]- confirmed earlier findings by conducting a randomized controlled trial where participants limited social media use to 30 min per day. The results clearly indicated reduced loneliness, symptoms, and depression, effects visible particularly in heavy users. Passive scrolling was identified as being the biggest predictor of negative outcomes. Another study, suggestively entitled "They've got more likes than you: towards a better understanding of the association between social media use and well-being" arrived at similar conclusions [6]. Another pivotal study, Kross et al. (2021) in "Social Media and Well-Being: Pitfalls, Progress, and Next Steps" [7] identified significant negative effects of overall social media use on well-being, emphasizing the detrimental impact of passive use. Additionally, Kross arrives to the conclusion that there is a negative relationship between using social network sites and subjective well-being. These studies, among a broader body of research, consistently highlight adverse mental health consequences of passive social media use, which underscores the issue of the toxic design of algorithms employed by such platforms.

The question that naturally emerges as a result is whether algorithms can be engineered in a way that actively promotes and even protects users' brain health. The current paper puts forward a few conceptual ideas and proposes an implementation of how that might be done.

2 Measuring User State and Brain Simulation via Code

2.1 User State Measurement by NBIs

The most logical and unsophisticated way is to introduce the concept of neurobiological indexes (NBIs). A first set is used to measure the user's state, and a second set allows the algorithm to accurately compute the user's state. This latter set's role is twofold: (1) provide accurate computational data for projecting users' state after a reel has been viewed, (2) function as labels for reels in the selection pool of the algorithm.

Regarding the plausibility of collecting data that would accurately assess the user's mental state, a plethora of scientific literature exists on the topic. Studies such as Harari et al. 2016 [8] suggest that behavioral data can be retrieved unobtrusively, *in situ*, from onboard sensors and phone logs. Smartphones enable large-scale, real-time data collection in natural settings, providing multimodal data (GPS, voice recordings) that permit the user's behavior to be accurately monitored. Evidently, clear consent and data security are critical, however, the collection of such data, their processing and displaying the resulting data could be done in app, where the user would have complete control over what the app would have access to. A different study [9], has conducted a comprehensive scoping review of 35 studies to understand how passive sensing and electronic usage data (social media use, device activity etc.) collected via smartphone devices can be used to detect and predict depression and anxiety. Using machine learning algorithms, the study outlines the following general behavioral trends: reduced mobility and routine disruption measured via lower entropy of GPS coordinates (less movement variety) and increased time spent at home are associated with higher depression severity. Sleep irregularities caused by late-night smartphone use and disrupted accelerometer-derived sleep patterns are anxiety predictors. For the paper's current scope, there are 4 key metrics that will be computed and updated in real time to assess the user's state.

Therefore, the user's mental state (uNBI – user neurobiological indexes) has been simplified to four key variables: (1) Attention Resource, (2) Emotional State, (3) Fatigue Levels, (4) Reward System Activation (AR, ES, FL, RSA).

To achieve accurate measurements that sufficiently assess the user's state, the uNBI will be affected by NBIs derived from physiological signals indicative of mental state, measurable via standard smartphone sensors. The following have been pinpointed and are broken down into two further categories: (a) NBIs derived by smartphone sensors (sNBIs), (b) NBIs that indicate long-term cognitive or emotional strain obtained from precomputed values stored on the phones, from the health app (hNBIs). The metrics below are normalized to a range of [0, 1].

The quality of sNBI data depends on sensor accuracy, environmental conditions, and computational processing. Reaction time data from touchscreens is highly reliable, with latency errors typically below 10 ms [10]. Blink rate detection via front-facing cameras achieves accuracy of 80–90% due to robust facial recognition algorithms such as DLib and MediaPipe. Pupil dilation measurement is sensitive to lighting variations, requiring software normalization, yielding cca. 80% accuracy in stable conditions. Typing speed variability is less precise due to user habits and keyboard differences, with a signal-to-noise ratio of cca. 0.7, but remains a useful proxy for attention [11].

The following Tables 1 and 2 outline the importance of each sensory data in relation to the user's state. Though all sensory data is used to predict AR, FL, ES, and RSA levels, these are majorly impacted by specific metrics: AR is majorly impacted by reaction time and typing speed variability, while ES levels are affected by heart rate variability and pupil dilation. Heart rate variability is considered essential for RSA predictions, and FL are best projected by computing reaction time, blink rate, pupil dilation and typing speed variability.

Table 1. NBIs derived from smartphone sensors (sNBIs)

sNBI	Sensor/Method	State Relevant
Reaction Time	Touchscreen (latency)	AR, FL
Blink Rate	Front-Facing Camera (video analysis)	FL
Heart Rate Variability	Camera-based photoplethysmography	ES, RSA
Pupil Dilation	Front-Facing Camera (video analysis)	ES, FL
Typing Speed Variability/Accuracy	Keyboard Input (comment analysis)	AR, FL

Table 2. NBIs calculated by default in the Health app (hNBIs)

hNBI	State Relevant
Daily Screen Time	FL, RSA
Rate of Time Spent on Entertainment/Education	ES, RSA
Usage Patterns	ES, FL
App Switching Frequency	AR, RSA
Task Interruption	AR, FL, RSA

The metrics above are used to compute initial conditions of the user's state and updated periodically at a rate r (15 s). Initial measurements can be modeled via the following equations:

$$AR = w_1 \cdot (1 - RT) + w_2 \cdot (1 - BR) + w_3 \cdot TS$$

$$ES = w_4 \cdot PD + w_5 \cdot (1 - HRV)$$

$$FL = w_6 \cdot RT + w_7 \cdot BR + w_8 \cdot (1 - HRV) + w_9 \cdot (1 - TS)$$

$$RSA = w_{10}PD + w_{11}$$

The constants w_i are weights that are experimentally obtained and checked for further accuracy by a machine learning algorithm and tests conducted on individuals. While scrolling, the algorithm keeps track of the metrics above and ensures that all states are kept within certain critical points. These are computed for each user, and the critical variance usually stabilizes around 0.30 – 0.45.

2.2 A Simple Model of the Brain

The objective of the brain model is to monitor and predict the dimensions assigned to it – ES, FL, AR, RSA – during prolonged social media engagement. The following approach combines Hierarchical Temporal Memory (HTM) and Spiking Neural Networks (SNNs). A comparison between alternative brain-inspired architectures is included at the end of the subchapter.

The brain model, as previously outlined, is defined by uNBIs – AR, ES, FL and RSA. Its aim is to predict the user's state quantitively from a scale of 0 to 1, with high, medium, low and balanced levels of AR, ES, FL and RSA respectively being associated with optimal conditions of user state. Thresholds of critical values depends on the users but for the scope of this simulation we define them as 0.7, 0.6, 0.2, 0.5. The model receives two streams of input: (A) reel data, characterized by the following NBIs: (1) salience (how captivating the reel is), (2) emotional value, (3) reward value (quantified as projected dopamine value aroused), and (4) attention demand; and (B) physiological metrics, sNBIs, updated every 15 s. The employed brain model is a hybrid algorithm that combines HTM for neocortical processes, implemented via the Python library htm.core, and SNNs, using BindsNET, for subcortical dynamics. This setup ensures sequential processing and continuous state tracking and predictions.

HTM encodes inputs received through the pipelines into Sparse Distributed Representations (SDRs), binary vectors with 2–5% active bits. For example, reel salience of 0.8 and pupil dilation of 0.7 are encoded into 2048-bit SDRs with 40 active bits, which are then processed by the Spatial Pooler. The Spatial Pooler identifies patterns, inspired by cortical columns it uses Hebbian learning to strengthen connections. However, it is important to keep a global inhibition ensuring 2% active mini-columns, crucial to mimic attention filtering. If a reel's attention demand exceeds AR, it is skipped, mimicking prefrontal cognitive control, and complete shutdown of the brain simulation is realized when AR converges towards 0. The brain model is configured such that RSA can override low AR, it can tolerate low AR as long as RSA, simulating an addictive mechanism. Although, the brain model can send a shutoff signal when AR reaches critical low levels, it is instructed to delay it to maximize RSA closing value after the decay following the spike. Temporal Memory then tracks the sequences, predicting state changes likes AR decay (logarithmic) or ES variance (oscillations) over time. The model receives reels – data that describes real Instagram reels – scrolls at a computed time. High oscillations in ES often results in full watch-time of the reel, which means that the data of the next reel is received only after that amount of time, while high attention demanding and low salience reel results in an instant scroll. Limitations of the brain simulation include not being able to replicate user want to rewatch the same reel because the model thinks it should search for the next dopamine high in the next reel. On the second pipeline, sensory data is transmitted every 15 s. SNNs model the subcortical processes, using spike-timing-dependent plasticity to simulate dopamine-driven RSA spikes and emotional arousal, which ensures greater accuracy in predicting RSA and ES. A high-reward reel results in the SNNs computing momentary RSA surges (+0.1 on average), with an immediate decay (–0.12 on average) post-spike. Updated states serve as the starting point for the next cycle/exposure to new data. Running the simulation shows that Temporal Memory predicts AR decay and FL accumulation over time, with outputs like low AR (0.2) and high FL (0.8) being obtained after 2 h of scrolling (480 cycles/reel data sets).

There is evidence in the scientific literature that HTM + SNN models can offer biological plausibility for cognitive state prediction, though alternatives like Transformers and RL show increased accuracy. LSTMs (long short-term memory) predict cognitive state with high accuracy (99.5%), but require large, labelled data [12]. Transformers predict brain states excelling in long-term sequences but demand computational resources, but struggle with direct state prediction without defined actions [13]. In this case, HTM + SNN is a favoured approach as it offers real-time adaptability, most importantly, it is able learn continuously without retraining it.

3 Algorithmic Re-Optimization for Cognitive Well-Being

3.1 Full Stack Analysis of Instagram Algorithm

Instagram's Reels' recommendation system is a highly sophisticated, multi-staged pipeline designed to keep users maximally engaged. It is easily a triumph of Machine Learning, Big Data Management, and Handling. Though the exact structure of the system isn't open source, there are many rough estimations as to its architecture.

Instagram's Reels Recommendation system operates through a multi-stage funnel. However, the evidence isn't indicative of the whole architecture; rather, only parts of the algorithm are discussed in various technical blogs and articles.

The system's key process is Candidate Generation. The process has four key stages: (1) retrieval, (2) first-stage ranking, (3) second-stage ranking, and (4) final reranking [14].

(1) Retrieval: Its objective is to narrow the billions of reels to only a few thousand candidates. This is done with the Two-Tower Neural Network and is described in a Meta Engineering Blog [15]. It consists of two separate dynamic neural networks (DNNs), covering user features and reel features. Embeddings are derived from the towers, and then the algorithm checks for overlaps and similarities to find the most suitable reels.

(2) First-Stage Ranking: This part reduces the candidate pool even more to around 150 reels using a lightweight version of the Two-Tower neural network, described as lightweight due to its cache ability property, focusing on filtering out lower candidates [15].

(3) Second-Stage Ranking: refines to around 50 candidates with heavy models. As noted here [15], the step uses extensions of the Two-Tower model and incorporates contextual features, such as time of day, device type.

(4) Final Reranking: Selects the top 10–20 reels for display. By leveraging an MTML neural network to optimize multiple objectives, such as watch time, shares, and diversity, it obtains the final candidates that are then displayed to the user.

3.2 Tweaking the Recommendation System

Changing the objective of the Recommendation Algorithm requires first replicating a simplified system locally. The key 4-step process still has the same objectives, however, with a number of 1000 reels, mock data generated by a Two-Tower Neural Network. The reels have features such as hashtags, likes, comments, shares, hook measurement (a weighted sum of the time to like, comment, or save), watch time, topics, and rNBIs.

The recommendation algorithm has the same stages and roughly the same algorithmic models. Below is a detailed account of the employed models:

(1) Candidate Generation (Retrieval)

Model used: Two-Tower Model (User Tower + Reel Tower) with TensorFlow Recommenders.

Goal: Efficiently retrieve 150 candidates from the pool of 1000 reels. The model is trained on implicit feedback (calculated via the equation mentioned previously), and the embeddings are exported to a vector database.

(2) First-Stage Ranking (Lightweight)
Model used: Logistic Regression
Goal: Narrows down the candidates quickly – it was faster than Gradient-Boosted Trees. Deployed with ONNX (Oriented Neural Networks).
(3) Second-Stage Ranking (Precise Ranking)
Model used: Neural Network (DNN)
Goal: Combining the precomputed embedding with sNBIs. It optimizes for balanced AR, ES, RSA, and FL states using TFRS.
(4) Final Reranking
Model used: simple sorting algorithm
Goal: The resulting list of arrays is of length 7, and, thus, the last part orders the reels optimizing at different times for different uNBIs that depend on the user's current state.

The ideal is to find the optimal spot of maximizing retention while minimizing addiction risk. If the current working state of the Instagram algorithm keeps the user hooked for hours, this algorithm's aim is to cut that to a healthy 30 min while providing the maximum excitement, or dopamine arousal.

Analyzing NBI Injection Points – Experimentation

In order to find the optimal point of integrating rNBIs, I have conducted experiments by injecting them at different stages and comparing results in the uNBIs. The table below contains the user's state variables after the exposure of the brain simulation to 50 reels.

Table 3. uNBIs after injecting rNBIs at different points on the same Brain Simulation

Experiment	AR	ES Variance	FL Increase	RSA
Baseline	0.4	0.35	0.25	0.65
rNBI at Retrieval	0.6	0.14	0.10	0.3
rNBI at First Rank	0.57	0.17	0.18	0.37
rNBI at Second Rank	0.5	0.23	0.13	0.48
rNBI at Final Rank	0.41	0.27	0.24	0.60
rNBI at All Stages	0.63	0.08	0.08	0.3

The data in Table 3 clearly indicates that the Retrieval Stage is the best for minimizing the proxy measurements associated with negative mental variables, however, the reward is, expectedly, minimal. The data in the following section has been computed for integrating rNBIs at Retrieval. The experiment reveals that placing the rNBIs at retrieval minimizes negative effects but is less likely to keep the user scrolling.

4 Setup Analysis for Neuroprotective Algorithms Testing

To compare the simulated user metrics (the brain simulation) to the actual Instagram algorithm, a machine learning algorithm has been trained to identify key characteristics of the reels. I've automated a system that firstly downloads the Instagram reel, analyzes it, generates labels, tags, and other metrics, and then feeds it to the brain simulation. This process takes around 15–20 s, which gives a latency of 25–33%, considering the average reel length to be 60 s.

The following experiment was conducted: 8 brain simulations with defining characteristics of AR, ES, FL, and RSA were 'exposed' firstly to the Instagram algorithm, and then to the reoptimized algorithm. To simulate different users, the 'brains' are divided into 4 groups of 2, with one metric depleting or increasing at different rates than the others. Group 1 (Sims 1 and 2) has a slower fatigue simulation, with $r_{11} = 0.5$ and $r_{12} = 0.3$, group 2 (Sims 3 and 4) exhibit faster attention decay, with $r_{21} = 2$, $r_{22} = 3$, group 3 (Sims 5 and 6) exhibit higher dopaminergic sensitivity, with the RSA variance being modified by $r_{31} = 2$, $r_{32} = 3$, and, finally, group 4 (Sims 7 and 8) had greater emotional volatility, with the ES variance being affected by $r_{41} = 1.5$, $r_{42} = 2$.

The brain simulations were tested against 200 Instagram and mock reels. The way the user's NBIs are affected is twofold: (1) on one hand, the user's AR are considered to be continually depleting as an inversed natural logarithmic function with spikes that correlate to RSA, that are lowered after the spike has occurred; ES was designed as an oscillating sine function with spikes that either increased or decreased the base level or that shortly modified the amplitude; RSA was modeled similarly to the dopamine levels in the brain: a base level that spikes for reels with high Salience and Reward Value, only to decrease the base level after spike; FL was modeled as a logarithmic natural function with increases when high variance in RSA and ES are detected. As mentioned previously, the brain model receives two sets of data: (1) the reel's data and (2) the sNBIs. The hNBIs and sNBIs are used to compute the user's initial state, which is averaged with an initial 10-s screening test via the set of equations mentioned above in part 2.1. Sensory data has been generated across 3 levels and 3 categories, which is fed to the brain sim as a function of the overlapping of the brain sim's preferences, which are the same. The sensory indexes reflect no excitement, moderate excitement, and high excitement in terms of emotional positive and negative responses and dopamine arousal (a proxy for laughter). When the overlap between user preferences and reel topic is high at a certain moment, then an algorithm sends an appropriate response of sensory metrics, in which there are variations depending on the degree of similarity in the user's preferences and the reel's topics. Then, rNBIs are taken into account by the brain model, which computes spike projections for the uNBIs.

5 Simulated System Results and Conclusion

The tables below contain the projected data across the 8 simulations. Hardware limitations reduced extensive testing possibilities, with the average reel exposure time being 30 min. One experimental simulation was compiled and then executed for two hours, which seemed to indicate correct AR and FL levels, however ES and RSA predictions were inaccurate, with ES and RSA spikes having a delay of around pi/2 (high ES and RSA levels should synchronize graphically, however, using python's matplotlib to display the oscillations in these levels there is phase angle difference), caused by hardware failure to process inputted data at correct times. The ES and RSA signals initially spiked in synchrony, but over time a growing phase angle developed between them, suggesting desynchronization likely due to hardware drift and timing inconsistencies (Tables 4 and 5).

Table 4. Simulations exposed to the reoptimized algorithm

Simulation	AR	FL Final Variance	ES Final Variance	RSA
Sim. 1	0.6	0.05	0.14	0.3
Sim. 2	0.6	0.03	0.14	0.3
Sim. 3	0.3	0.1	0.14	0.3
Sim. 4	0.2	0.1	0.14	0.3
Sim. 5	0.6	0.1	0.14	0.6
Sim. 6	0.6	0.1	0.14	0.9
Sim. 7	0.6	0.1	0.21	0.3
Sim. 8	0.6	0.1	0.25	0.3

Table 5. Simulations exposed to the Instagram algorithm

Experiment	AR	FL Final Variance	ES Final Variance	RSA
Sim. 1	0.4	0.15	0.21	0.6
Sim. 2	0.4	0.09	0.21	0.6
Sim. 3	0.1*	0.3	0.21	0.6
Sim. 4	0.066*	0.3	0.21	0.6
Sim. 5	0.4	0.3	0.21	1.2*
Sim. 6	0.4	0.3	0.21	1.8*
Sim. 7	0.4	0.3	0.315	0.6
Sim. 8	0.4	0.3	0.375*	0.6

* Numbers exceed safe/threshold values

Obtained data indicates that prolonged social media use impacts AR, ES, FL, and RSA levels, which would be supported by longitudinal studies. A 2023 study [16] found that habitual social media checking increases brain sensitivity to social rewards in regions such as the amygdala, aligning with RSA variance predictions. Another independent study [17] reports that high social media use relates to poor attention and high distractibility, supporting AR depletion.

It is important to note that hardware limitations severely restricted the scaling of the simulations. Future directions would include incorporating Raspberry Pi technology into the system for additional computational power. Simulations would be refined for individuals, with tests measuring initial and final user state via a series of tests to refine data predictions.

In this paper, I have shown that it is both feasible and beneficial to re-architect content-recommendation systems, at least for the user. As the results indicate, passive use time would decrease, which would result in a categorical benefit for the user, as the studies cited in chapter one revealed. However, it should be mentioned that, as users don't engage as much with the platforms, Instagram would take a hit in terms of ad revenue, which is why such an approach hasn't been researched before: it isn't profitable. Thus, the idea of protecting the young generation from early negative neurological effects is worth pursuing, as it might yield a positive impact on incoming generations in terms of mental health, social relations, and work motivation. The goal of the paper is to start a discussion of how neuroprotective algorithms can be engineered and show that a retrieval-stage injection of neurobiological information does result in a reduction in the negative effects inflicted.

References

1. statista.com, "Statista". https://www.statista.com/statistics/183585/instagram-number-of-global-users/. Accessed 17 Apr 2025
2. Zuckerberg, M.: "hypebot.com". https://www.hypebot.com/hypebot/2024/04/threads-has-150m-monthly-users-50-of-instagram-time-spent-is-on-reels.html. Accessed 17 Apr 2025
3. Wall Street Journal. https://www.wsj.com/articles/instagram-reels-tiktok-meta-facebook-documents-11662991777. Accessed 19 Apr 2025
4. Verduyn, P., et al.: Passive usage undermines affective well-being: experimental and longitudinal evidence. J. Exp. Psychol. Gen. **144**(2), 480–488 (2015)
5. Hunt, M.G., Marx, R., Lipson, C., Young, J.: No More FOMO: limiting social media decreases loneliness and depression. J. Soc. Clin. Psychol. **37**(10) (2018)
6. Faelens, L.: They've got more likes than you: towards a better understanding of the association between social media use and well-being. University of Gent, Gent (2018)
7. Kross, E., Verduyn, P., Sheppes, G., Costello, C.K., Jonides, J., Ybarra, O.: Social media and well-being: pitfalls, progress, and next steps. Trends Cogn. Sci. **25**(1), 55–56 (2021)
8. Harari, G.M., Lane, N.D., Wang, R., Crossier, B.S., Campbell, A.T., Gosling, S.D.: Using smartphones to collect behavioral data in psychological science: opportunities, practical considerations, and challenges. Perspect. Psychol. Sci. **11**(6), 838–854 (2016)
9. Beames J.R., et al.: Use of smartphone sensor data in detecting and predicting depression and anxiety in young people (12–25 years): a scoping review. Heliyon **30**(10) (2024)
10. Raisamo, R., Rakkolainen, I., Majaranta, P., Salminen, K., Rantala, J., Farooq, A.: Human augmentation: past, present and future. Int. J. Hum. Comput. Stud. **131**, 131–143 (2019)
11. Barry, C., de Souza, J., Xuan, Y., Granholm, E., Holden, J., Wang, E.J.: At-Home Pupillometry using Smartphone Facial Identification Cameras. ACM Digital Library (2022)
12. Kim, H., Hong, S., Yoo, G.: Prediction of cognitive load from electroencephalography signals using long short-term memory network. Bioengineering **10** (2023)
13. Sunn, Y., et al.: Predicting Human Brain States with Transformer. [Preprint] arxiv.org (2024)
14. "aman.ai". https://aman.ai/recsys/ranking/. Accessed 19 Apr 2025
15. Shugaepov, I., Vorotilov, V.: https://engineering.fb.com/2023/08/09/ml-applications/scaling-instagram-explore-recommendations-system/. Accessed 19 Apr 2025
16. Maza, M.T., et al.: Association of habitual checking behaviors on social media with longitudinal functional brain development. JAMA Pediatr. **177**, 160–167 (2023)
17. Alaparthi, K.: Technology and digital media's impact on attention span in teenagers and young adults. SSRN Soc. Sci. Res. Network. https://ssrn.com/abstract=4872178 (2024)

Theoretical Justification for Increasing the Stability of Trailer Links of a Tractor-Transport Train

Ivan Koliesnik[1(✉)], Evgeny Kalinin[1], Juliana Koliesnik[1], Oksana Pankova[2], and Vadim Raputa[3]

[1] National University of Life and Environmental Sciences of Ukraine, Kyiv 03041, Ukraine
ivankolesnik@nubip.edu.ua
[2] Kharkiv National Automobile & Highway University, Kharkiv 61002, Ukraine
[3] Sumy National Agrarian University, Sumy 40021, Ukraine

Abstract. The paper presents an analysis of studies on the movement of multi-link transport trains, which showed that the last links are considered the most unstable. They are especially unstable on turns, because they are most susceptible to the action of not only active disturbing loads, but also in the presence of random influences from the road due to a small number of links they exhibit extreme instability of movement, this leads to an increase in oscillatory processes, which negatively affects traction and coupling properties, controllability, stability, etc. Therefore, reducing dynamic loads and increasing the efficiency of road trains is an urgent task.

It must be said that the conditions under which overturning is possible are also influenced by the stiffness of the suspension, but the need to take this factor into account also leads to the further possibility of increasing the stability of the train as a whole. Consequently, and increasing its basic characteristics by changing the properties of the connections between its links.

Thus, during rectilinear movement, the train links are acted upon by forces only in the longitudinal direction, and transverse vibrations arise as a result of the effects of road irregularities and the elastic interaction of the wheels with the road. It is also desirable to get rid of this phenomenon, and there are already a number of proposals and specific measures. It should be noted that in a turn, a whole group of forces act in the transverse direction, the influence of which cannot be neglected. Unfortunately, this does not help to increase the stability of the train during a turn.

Keywords: Tractor-technological unit · Dynamics · Traction and coupling device · Trailer · Stability

1 Introduction

A tractor train can be considered as a mechanical system of moving solid bodies, with masses concentrated in the centers of gravity, which have certain connections between their links: dynamic, kinematic, control signals (or the law of rotation of the steered wheels).

D. D. Cioboată and J. Machado (Eds.): ICoRSE 2025, LNNS 1592, pp. 286–300, 2025.
https://doi.org/10.1007/978-3-032-02508-1_25

Until now, a special difficulty in building a model is the theoretically unsolved issue of modeling rolling when turning an elastic tire [1–4]. At the same time, many researchers have solved the issue too simply, which led to obvious limitations in solving general problems. One of the tasks of studying the features of the movement of a multi-link train is to determine its characteristics when moving on an uneven road (surface). In the idealized case (in the first approximation), the surface remains for all elements located relative to a flat axis. However, during modeling, longitudinal resistance is applied to all links with a small sequential time delay relative to the link in front.

Analysis of studies on the movement of multi-link transport trains has shown that the last links are considered the most unstable. They are especially unstable on turns, because they are most susceptible to the action not only of active disturbing loads, but also in the presence of random influences from the road due to the small number of connections they exhibit extreme instability of movement, which is dangerous by overturning, or in poor road conditions by skidding with subsequent overturning [5, 6].

2 Analysis of Literary Sources

In this paper [7], a hierarchical control system for trajectory tracking and lateral stability is proposed by combining distributed electric drive technology. Firstly, a linear three-degree-of-freedom dynamics reference model is established for obtaining the ideal values of vehicle state quantities. Secondly, the upper trajectory tracking controller is designed based on model predictive control (MPC) and PID control theory for accurately tracking the reference trajectory.

In paper [8], the articulated heavy vehicles considered exhibit low lateral stability, which can lead to unstable driving modes, such as trailer swaying and bumping, which cause serious accidents. Changing the relevant vehicle parameters improves the static stability, but worsens the dynamic one. Previous studies have focused only on static or dynamic stability.

Aiming at the problems of folding tail-flicking and rollover of the semi-trailer train in the dangerous section of the mountain highway in the cold area, taking a domestic semi-trailer train as the research object, the dynamic mathematics of the six-degree-of-freedom semi-trailer train is established based on the vehicle dynamics theory model. Using Truck-sim to complete the establishment of the semitrailer train vehicle model and dangerous road section model, according to the principle of semi-trailer train rollover and yaw instability, determine the judgment conditions for semi-trailer train rollover and yaw instability. Taking the difference between the ideal state parameters and the actual state parameters of the semi-trailer train as the control target a control strategy for the anti-rollover and yaw instability of the semitrailer train based on the tinear quadratic regutator (LQR) control algorithm and differential braking is established with the help of Simulink [9].

Existing stability analyses for a tractor semi-trailer cannot fully explain the instability phenomenon during combined steering and braking conditions, in which the lateral tire forces are affected by the longitudinal tire forces through the adhesion force ellipse. In response to the problem posed in this paper [10], the coupled bifurcation and stability zone of a tractor semi-trailer at different deceleration and yaw angles are obtained.

The braking response delay of articulated heavy vehicles, braking coordination between the tractor and the semi-trailer, braking shock, etc. affect the driving stability. To solve the above problems, in article [11], a configuration scheme is developed and a model of a pneumatic electronic control braking system is established, and a joint braking control strategy for the tractor and the semi-trailer is proposed.

This paper [12] proposes a control method for a Rear Wheel Steering (RWS) system aimed at improving both the sway stability and maneuverability of vehicles equipped with trailers.

In this work [13], a stability-guaranteed computationally efficient model predictive path tracking framework is devised. In particular, a comprehensive tractor semi-trailer stability constraint (TSSC) consisting of a three-dimensional (3D) yaw stability envelope and rollover limitation is proposed to guarantee the yaw and roll stability.

3 Description of the Mathematical Model of the Tractor

It must be said that the conditions under which a rollover is possible are also influenced by the stiffness of the suspension, but the need to take this factor into account also leads to the further possibility of increasing the stability of the train as a whole. Consequently, and increasing its main characteristics by changing the properties of the connections between its links.

Figure 1 shows a flat diagram of the forces acting on the train when making a turn. This diagram shows that when creating an additional dynamic force between the links P_{cil}, forces P'_{kr1i} or (and) P'_{kr2i} are created that counteract the rollover.

So, during rectilinear movement, forces act on the train links only in the longitudinal direction, and transverse vibrations arise due to the effects of road irregularities and the elastic interaction of the wheels with the road. It is also desirable to get rid of this phenomenon, and there are already a number of proposals and specific measures. Unfortunately, this does not help to increase the stability of the train during a turn.

It should be noted that in a turn in the transverse direction, a whole group of forces acts, the influence of which cannot be neglected. These include the lateral forces of the resistance of the drawdown (P_δ). These forces ensure the stability of the trailer when turning and are a non-holonomic connection of the kinematic link with the surface [3, 4, 14, 15]. The value of these forces should not exceed the value of the force of adhesion of the wheel to the surface. With a small value of the centrifugal force of inertia (F_{in}), a fairly safe and stable movement is ensured, with a large value - movement occurs in a critical mode, or a rollover may occur.

The total moment of the overturning forces has a counteraction in the form of moments of several forces:

- gravity (mg);
- dynamic force in the suspension elements (applies only to active suspensions) (P_{susp});
- reactions in the connections between the links (P_{kr}).

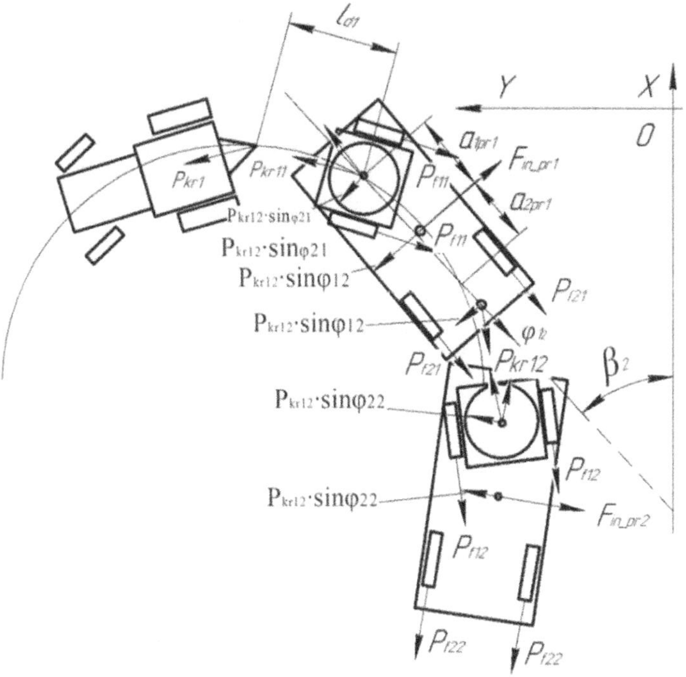

Fig. 1. Planar dynamic diagram of the movement of a multi-link tractor train and the forces acting on the trailer

Thus, it was decided to calculate the required force P'_{kr1i} in the traction and coupling device of the i-th trailer, to ensure the required level of the coefficient of stability against overturning according to the formula (1):

$$M_{stab\_pr\_i} = P'_{kr1i} h_{pr2i} + P'_{kr2i} h_{pr1i} + mg \cdot \frac{B_{pr}}{2} \tag{1}$$

$$M_{res\_\text{э}k\_i} = F_{in\_pr2i} \cdot h_{pr2i\_in} \tag{2}$$

$$F_{in\_pr2i} = \frac{m_{pr2i} \left(\upsilon_{pr2i} \right)^2}{R_{pov}} \tag{3}$$

where h_{pr_i} - height of connection of the trailer and the towing device;

$h_{pr\_in\_i}$ - height of the central point of the trailer;

P'_{kr1i} - tractive effort on the turning wheel of the trailer, which counteracts overturning, m;

P'_{kr2i} - tractive effort on the trailer hook in the presence of the following trailer, which counteracts overturning, m;

The value of the efforts P'_{kr1i}, P'_{kr2i} is equal to

$$P'_{kr1i} = P_{kr1i} \cdot \sin \phi_{2i}$$

$$P^{/}_{kr2i} = P_{kr2i} \cdot \sin \phi_{1i+1} \quad (4)$$

where P_{kr2i} - total tractive effort on the trailer's steering wheel, m;

φ_{2i} - angle between the longitudinal axis lines of the first kinematic link and the second kinematic link of the trailer;

φ_{1i+1} - angle between the longitudinal axis lines of the first kinematic link of the next trailer and another kinematic link of the first trailer.

Thus, it is possible to determine the magnitude of the total force on the hook to provide a certain value of μ, the value of which can be determined in advance to ensure safety against overturning, based on the conditions of movement, maneuver and road profile, load mass, etc. [15, 16].

We obtain:

$$P^{/}_{kr1i} + P^{/}_{kr2i} = \frac{F_{in\_pr\_i} \cdot h_{pr\_in\_i}}{h_{pr\_i}(1 - \mu_{caps})} - m_{pr\_i}g \cdot \frac{B_{pr\_i}}{2h_{pr\_i}} \quad (5)$$

$$P^{/}_{kr1i} = \frac{\left(\frac{F_{in\_pr\_i} \cdot h_{pr\_in\_i}}{1-\mu_{caps}} - mg \cdot \frac{B_{pr\_i}}{2}\right)}{h_{pr\_i}} - P^{/}_{kr2i} \quad (6)$$

$$P^{/}_{kr1i} = \frac{\left(\frac{F_{in\_pr\_i} \cdot h_{pr\_in\_i}}{1-\mu_{caps}} - m_{pr\_i}g \cdot \frac{B_{pr\_i}}{2}\right)}{h_{pr\_i}} \quad (7)$$

$$P^{/}_{kr1i} = P_{kr1i} \cdot \sin \phi_{2i};\ P_{kr1i} = P_{kr} + P_{cil}; \quad (8)$$

Based on the following formulas (5) – (8), you can determine the necessary force on the hooks of the tractor hitch; due to the obviousness of one reason, you can determine the necessary traction of the tractor (7). With due respect to those who, during transport operations, the amount of traction supports is small, as is obvious, and the hook on the hook, to remove the required hook, you can use an additional special device to close the additional hook (by looking at the additional hydraulic cylinder). Therefore, for an active traction-traction device, it is possible to calculate, and the values necessary for expansion of the area of the cylinder cross-section, its stroke and other parameters. True, many parameters from this complex value can be calculated numerically only after the development of a system of differential levels that describes the collapse of the entire train, and adequately evaluate all the officials who contribute to the disruption [17].

The vicoristic residual formulas (7) and (8) were determined by the values of P_{kr1i} and $P_{cil.}$, at $\mu = 0.7$ depending on the inertia force, which in its turn was calculated for changes in the speed of the steering wheel, the turning radius, and the weight of the head (with vantage).

In the diagrams below, you can determine the necessary force in the traction device P_{kr1i} to ensure the lateral stability of the traction strips [18–23], ensuring that on the traction devices even with the addition of traction forces supports (supports the roller and the barrel of the train tracks) already have a traction force P_{kr}. In the diagrams, in order to fully demonstrate the mutual interaction, the inertia force Fin is shown, which acts on the connecting strips, the force P_{kr1i}, which must be added to the traction device to increase the traction force. Improved lateral stability. The value of the additional value

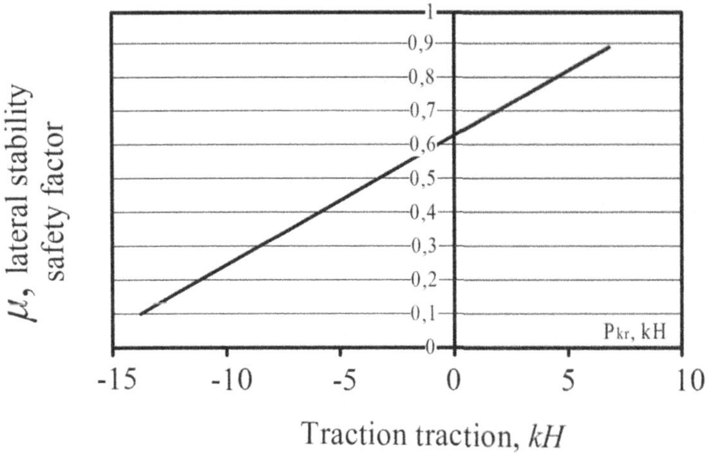

Fig. 2. Tack strength as a factor to the lateral strength reserve

P_{cil} is shown, which is necessary for assessing what is required for the design of the device.

Analysis of Fig. 2 allows us to theoretically estimate the values of the traction force to ensure the necessary margin of lateral stability. Further, taking the cob value $\mu = 0.7$, there are significant dynamic forces that are generated (P_{kr}, F_{in}, P_{kr1i}, P_{cil}) for the rudder of a train with mixed-mass vehicles with speeds of 15, 18, 25 km/h at turns with a radius of 10 m (Figs. 3, 4 and 5). Here you can clearly note that with the increase in speed, the subcenter force of inertia significantly increases, leading to a significant increase in the necessary force.

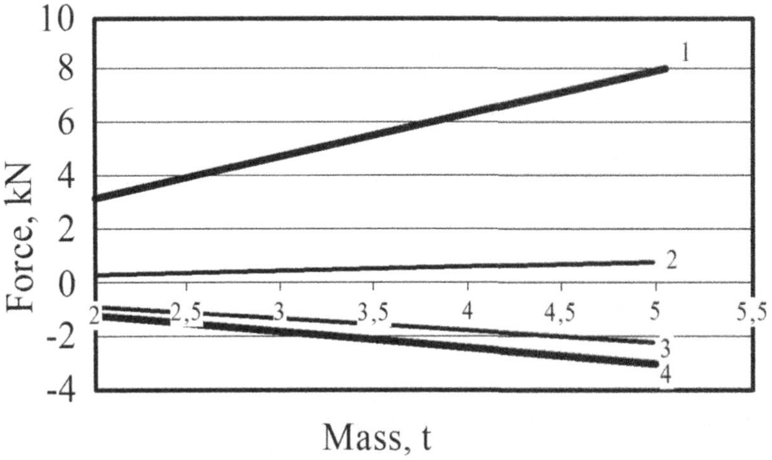

Fig. 3. Deposit of the main forces in the mass of rakes at a speed of 15 km/h, turning radius of 10 m 1-*Fin*, 2-*P*kr, 3-*Pkr1i*, 4-*Pcil*

An analysis of Fig. 1 gives a similar picture. Figures 6 and 7 shows the strength of dynamic forces against the fluidity of the steering wheel at different turning radii [24].

Analyzing Figs. 2, 3, 4, 5, 6, 7 and 8, and the most productive steering speed when turning (by adjusting the distribution of the value of turning radii [1, 15, 25]) it means that with turning radii from 10 to 20 m and speeds consistent with the mode axle from 10 to 25 km/h, the value of the additional force of the traction and coupling device is required in the range of 1–5 kN.

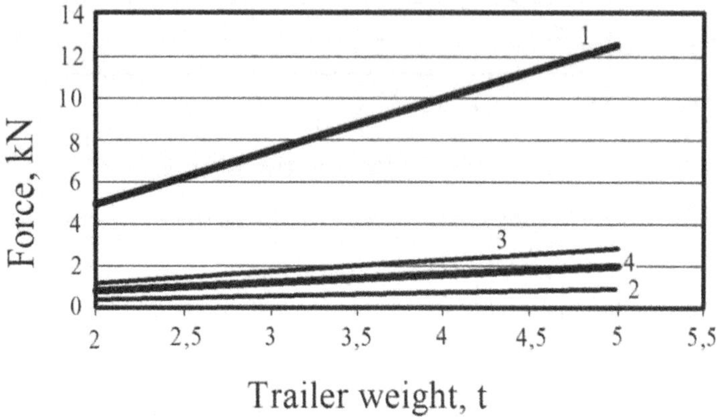

Fig. 4. Deposit of the hulls in the mass of the trailer at a speed of 18 km/h 10m 1-*Fin*, 2-*Pkr*, 3-*Pkr1i*, 4-*Pcil*

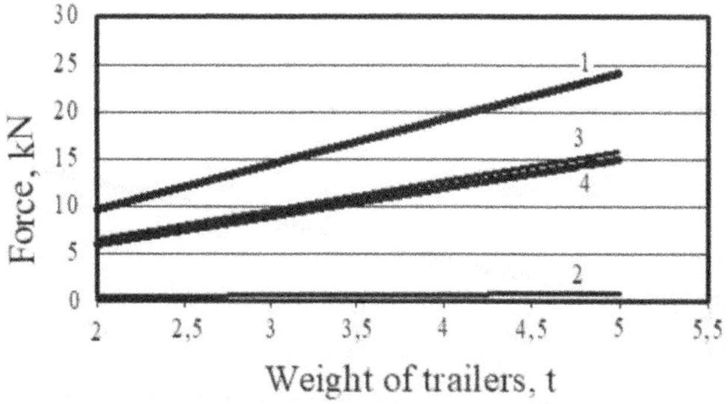

Fig. 5. Deposit of hydraulic forces in the mass of rakes at a speed of 25 km/h, turning radius 10 m 1-*Fin*, 2-*Pkr*, 3-*Pkr1i*, 4-*Pcil*

The result is that the increase in the lateral stability of a multi-lank tractor train is determined by the alignment of all the main factors of the building in relation to the change in the inflow of the sub-center force inertia, including for the rate of movement of the zusilla on the hook by a small amount in the concentration of the minds of the rukh.

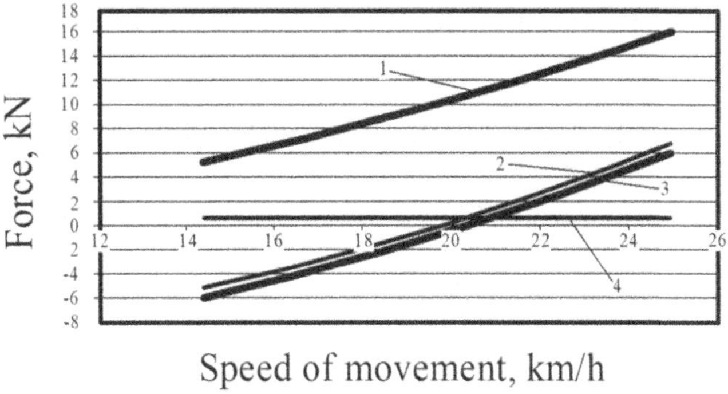

Fig. 6. Loading of the girders with the smoothness of the handle, weight 5000 kg and turning radius 15 m 1-F_{in}, 2-P_{kr}, 3-P_{kr1i}, 4-P_{cil}

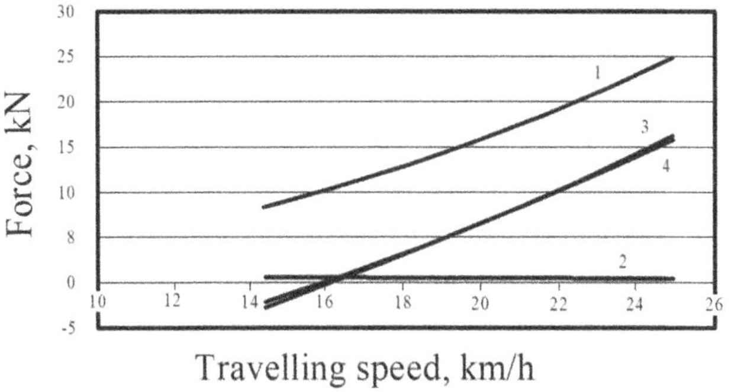

Fig. 7. Loading of the gears with fluidity of the handle, weight 5 t and turning radius 10 m 1-*Fin*, 2-*Pkr*, 3-*Pkr1i*, 4-*Pcil*

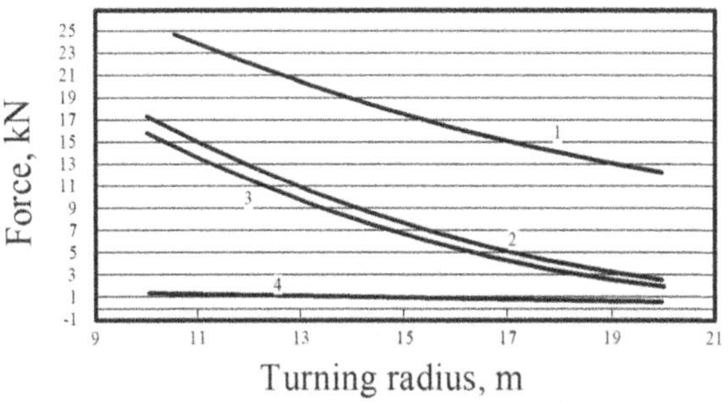

Fig. 8. Depth of hauling forces depending on the turning radius of the train, train weight 5 tons, speed 25 km/h 1-*Fin*, 2-*Pkr*, 3-*Pkr1i*, 4-*Pcil*

4 Mathematical Model of the Rover of a Multi-Lane Tractor Train on a Turn

The use of the decomposition method allows the initial methods of mathematical analysis to be used to complete folding systems through obvious kinematic connections (non-holonomic connections between the tractor along the plane and between the tractor and the driving force Lanka).

Systems of differential alignments, which describe a tractor train with two traction drives, can be represented in the formal form of alignments (13, 14–19) [26–30].

Such is the power of P_{kr} is also strong, as it depends on the type of structure used for the different structure of the formula. In practice, this describes the operation of a new traction-drive device with an active drive to reduce the negative influx of inertial forces.

$$\begin{aligned}&\frac{m\upsilon_{tr}}{L}\left[b\left(\frac{\Omega_K}{\cos^2\theta}-\delta_1\right)-a\delta_2\right]=K_{y2}\delta_2+K_{y1}\delta_1\cdot\cos\theta-\\&-tg\theta\cdot K_{y1}\cos\theta-P_{f1}\sin\theta-\frac{m\upsilon_{tr}^2}{L}-\\&-(tg\theta+\delta_2+\delta_1)-\frac{m\upsilon_{tr}}{L}(b(tg\theta-b_1)-b_2)-P_y\end{aligned}\tag{9}$$

$$\begin{aligned}&\frac{J_{31}}{L}\left(\frac{\Omega_k}{\cos^2\theta}+\dot{\delta}_2-\dot{\delta}_1\right)=K_{y1}a_1\delta_1\cos\theta-K_{y2}a_2\delta_2\sin\theta-\\&-\frac{J_{31}\dot{\upsilon}_\upsilon}{L}(tg\theta+\delta_2-\delta_1)-M_z\end{aligned}\tag{10}$$

For a flat rozhunkovy model, the tractor's axle movement on spring tires is carried out in two levels, without changing the coordinates of the rigid system. To the material point representing the center of gravity of the tractor, the already determined force that acts in the transverse direction on the tractor is added, and the sum of the moments of these forces is equal to the center of gravity of the tractor [31]. All external forces and moments of these forces are represented by two levels (9) and (10):

where

$$Э_y = P_{kr}\cdot\sin\phi_1 \tag{11}$$

$$M_z = P_{kr}\cdot\sin\phi_1(b+c) \tag{12}$$

For a clear representation of these deposits, the corner deformation of the spring tires and the removal of the single-arm type of the zagal system of the belts, transforms the data of the belts to the canonical form of the differential belts (13).

5 Description of the Mathematical Model of the Tractor

The final view of the system is equal to the tractor's ruhu. Views from the system:

$$\begin{cases} m_{tr}\cdot\dot{\upsilon}_x=\left(P_{k2}-P_{f2}\right)\cos\delta_2-P_{f1}\cos(\theta-\delta_1)-P_{\delta1}\sin(\theta-\delta_1)-P_{\delta2}\sin\delta_2-P_{kr} \\ \dot{\delta}_1=\frac{\upsilon_x}{L}(\theta+\delta_2-\delta_1)-\frac{P_{\delta1}}{\upsilon_x}\left(\frac{1}{m_{tr}}+\frac{a_1^2}{J_{э1}}\right)-\frac{P_{\delta1}}{\upsilon_x}\left(\frac{1}{m_{tr}}-\frac{a_1a_2}{J_{э1}}\right)+ \\ +\frac{P_y}{\upsilon_xm_{tr}}+\frac{M_za_1}{\upsilon_xJ_{э1}}+\Omega_K \\ \dot{\delta}_2=\frac{\upsilon_x}{L}(\theta+\delta_2-\delta_1)-\frac{P_{\delta1}}{\upsilon_x}\left(\frac{1}{m_{tr}}-\frac{a_1a_2}{J_{э1}}\right)-\frac{P_{\delta2}}{\upsilon_x}\left(\frac{1}{m_{tr}}-\frac{a_1^2}{J_{э1}}\right)+ \\ +\frac{P_y}{\upsilon_xm_{tr}}+\frac{M_za_1}{\upsilon_xJ_{э1}} \\ \dot{x}_{tr}=\upsilon_x\left[\cos\beta_{tr}-\frac{a_2(\theta-\delta_1)-a_1\delta_2}{L}\sin\beta_{tr}\right] \\ \dot{y}_{tr}=\upsilon_x\left[\sin\beta_{tr}-\frac{a_2(\theta-\delta_1)-a_1\delta_2}{L}\cos\beta_{tr}\right] \\ \dot{\beta}_{tr}=\omega_{tr}=\frac{\upsilon_x}{L}(tg\theta+\delta_2-\delta_1) \end{cases} \tag{13}$$

In this case, it is necessary to note that the steering of the tractor and the entire system as a whole is modeled by additional input, which is a change in the direction of rotation of the tractor wheels. The way to turn the wheels of the tractor, if necessary, is modeled by the ordering of the model of the actions of the tractor driver in different minds and during various maneuvers depending on the nature of the maneuvers (rearrangement, overtaking, etc.) [32–35].

In the simplest version, which we accept to simplify the layout, where the rotation changes according to the linear law and looks like:

$$\begin{cases} 0<t\le T_1^{\top}\Omega_K=0_{\max}/T_1 \\ T_1<t\le T_2^{\top}\Omega_K=0 \\ T_2<t\le T_3^{\top}\Omega_K=0_{\max}/T_3-T_2 \end{cases} \tag{14}$$

As was described earlier, the model of the tractor's axle is based on two coordinate systems, one of which is non-rotary and connected to the surface of the earth *OXYZ*, and the other is the axle coordinate system $C_nX_nY_nZ_n$, in which we denote the axle of all elements of the tractor (wheels, axles, spring parts, etc.) [15].

The system (11 – 13) has the following expressions for determining the meaning of the components that go before it:

Support force for one axle of the tractor:

$$P_{f1,2}=fm_{tr}g/2, \tag{15}$$

de t – coefficient of support for the rigidity.

Support force for the side tire outlet of one tractor axle:

$$P_{\delta 1,2} = \delta_{1,2} k_{\delta 1,2}, \tag{15}$$

de $\delta_{1,2}$ - kut of the tractor wheels 1 and 2 axles in line, degrees.

$k_{\delta 1,2}$ - coefficient of support for the side wheel output, kN/deg.

The force Ru and moment Mz are calculated in ranks (13) by ranks (11 and 12).

One of the important moments is the crushing force of the train - the additional force P_{k2} - which is seen in the system in a very simplified way, when wheel slipping, deformations in the kinematic transmission system to the torque that is turned on, are turned on. Transmission, deformation in the coupling. We can see only the robot engine, the torque on the shaft of which is in position in the direction of rotation (meaning mediated by the rotation frequency of the tractor wheels), described by formula (16) [36, 37]:

$$P_{dv} = M_{dv\_\max}\left(1 + \frac{\omega_{dv}}{\omega_{M\max}} - \left[\frac{\omega_{dv}}{\omega_{M\max}}^{\frac{M_{dv\_\max}}{M_{dv\_nom}}}\right]\right) \tag{17}$$

$$P_{k2} = \frac{M_{dv} i_{tr} \eta_{tr}}{r_k} \tag{18}$$

$$\omega_{dv} = \upsilon_{tr} i_{tr} / r_d, \tag{19}$$

where M_{dv_max} – the maximum motor torque beyond the control characteristic, Nm;

ω_{dv} – the rotation frequency of the motor's rotary shaft (meant by formula (18), rad/s;

r_d - dynamic radius of the wheel, m;

υ_{tr} - the steering speed of the tractor, determined by the differential alignment adjustment.

ω_{dv_Mmax} - motor turning frequency with the maximum turning torque behind the control characteristic, rad/s.

One of the tasks of studying the peculiarities of the ruk of a rich-lank train is the importance of its characteristics on an uneven road (surface). In the idealized form (at the first closest) the surface is forfeited for all elements moved along a completely flat axis. However, when modeling, on all the lanks with small subsequent deposits, an hour before the lanks, a later support is applied, which is indicated by the virus:

$$R_{add} = R_k \cos\left(\frac{\pi t + \sum_i x_i}{n}\right) \tag{20}$$

de R_k – the average value of the additional support for a given road distance, N;

x_i – hour of washing with the program for the i-th skin area, s;

n – frequency factor.

6 Description of the Mathematical Model of the Principle

Like a tractor (tractor), the side arms collapse under the influx of applied forces, and the influx from the tractor is replaced by the traction force on the hook. On all the lanks of the cream there is a strong support for rigidity, rotation and others [38]. Since these straps may be very large, then in the indicated forces and description of the design parameters we take the meaning of the i-th principle with two kinematic straps (the first kinematic strap is the drawer, the other is the primer).

System of alignment of the 1st kinematic lanyard of the prichep:

$$\begin{cases} m_{pr1i}\dot{\upsilon}_{pr1i} = P_{kr1i} - R_{1i} - P_{kr2i}\cos\phi_{pr1i} \\ \omega_{pr1i} = \dfrac{\left[\upsilon_{1i}\sin\phi_{pr1i} - \omega_{tr}(a_{2i}+c)\cos\phi_{pr1i1}\right]}{l_{dt}} \\ \dot{x}_{pr1i} = \upsilon_{pr1i}\cos\beta_{pr1i} \\ \dot{y}_{pr1i} = \upsilon_{1i}\sin\beta_{pr1i} \\ \dot{\beta}_{pr1i} = \omega_{pr1i} \\ \dot{\phi}_{pr1i} = \omega_{tr} - \omega_{pr1i} \end{cases} \tag{21}$$

Rivalry to the handle of another kinematic lantern:

$$\begin{cases} m_{pr2i}\dot{\upsilon}_{2i} = P_{kr2i} - R_{2i} - P_{kr1i(i+l)}\cos\phi_{pr2i} \\ \omega_{pr1i} = \dfrac{\upsilon_{pr2i}\sin\phi_{pr2i}}{l_{pr\_i}} \\ \dot{x}_{pr2i} = \upsilon_{pr2i}\cos\beta_{pr2i} \\ \dot{y}_{pr2i} = \upsilon_{pr2i}\sin\beta_{pr2i} \\ \dot{\beta}_{pr2i} = \omega_{pr2i} \\ \dot{\phi}_{pr2i} = \omega_{pr1i} - \omega_{pr2i} \end{cases} \tag{22}$$

It is significant that the support for the side tire output of the traction strips is modeled more simply, lower for the tractor wheels, the fragments of the skin-related kinematic strap (one or the other) have less than one support for the whole [5, 39, 40], and of similar importance where the lateral outlet is located in the kinematic dozhniy lanka, for the purpose of the drainage value of the dozhni l_{di} - changes.

$$\delta_{pr\_1i} = \frac{\omega_{pr1i} l_{dt}}{\upsilon_{pr1i}} \tag{23}$$

$$\delta_{pr\_2i} = \frac{\omega_{pr2i} l_{pr\_i}}{\upsilon_{pr2i}} \tag{24}$$

Additional ligaments on the mind of the hand are superimposed with rivets, which describe the operation of the traction and coupling device. In this case, it should be noted that one of the tasks of this robot is the need to obtain equal results from the replacement of the serial and modernized traction-hauling device.

7 Conclusions

A mathematical model of the rudder of a tractor-transport train on a turn, the rotational forces that act on all the rails of the train, the characteristics of the engine, the characteristics of the wheels of the rails of the train, the thrust at the later straight line from the supports of the road surface:

- a methodology has been developed for calculating the reserve of lateral stability for kinematic slats with a modernized traction and coupling device for increasing and stabilizing the traction force;
- the optimal power of the traction-traction device was determined, which allows increasing the margin of lateral stability (damping coefficient and the value of the rigidity of the device, the value of the additional traction force);
- it is theoretically confirmed that the lateral strength margin is increased by 40–60% when using the modernized traction drive unit with optimal parameters.

References

1. Koliesnik, I., Kalinin, E., Shevchenko, I., Koliesnik, J., Stepanov, O.: Conceptual energy model of the dynamics of a transport-technological unit for the appearance of two drive axles. In: Lecture Notes in Networks and Systems. 4th International Conference on Reliable Systems Engineering, ICoRSE 2024, vol. 1129 LNNS, pp. 314–327 (2024). https://doi.org/10.1007/978-3-031-70670-7_27
2. Kozhushko, A., Tkachov, V., Mittsel, M., Sokolik, S.: Optimization of traction properties the electric tractor based on the simulation dlg-powermix test cycles. Int. J. Mechatron. Appl. Mechan. **15**, 70–78 (2024). https://doi.org/10.17683/ijomam/issue14.5
3. Yue, Y., Wenyao, Z., Zhengdi, Z., Qinsheng, B.: Analysis on the motion of nonlinear vibration with fractional order and time variable mass. Appl. Math. Lett. **124**, 987–993 (2022)
4. Valeh, B., Nail, M.: Dynamic analysis of the mechanical motion of a variable-mass composition rocket system. In: Proceedings of the International Astronautical Congress, IACTOM 2023-October2023 74th International Astronautical Congress, IAC 2023 Baku, pp. 197–203 (2023)
5. Yanmin, L., Tingting, Z., Fengxiang, M.: Stable variable mass mechanical systems constructed by using a gradient system with negative-definite matrix. Lixue Xuebao/Chinese J. Theor. Appl. Mech. **50**, 109–113 (2018)
6. Kayuk, Y.F., Denisenko, V.I.: Motion of a mechanical system with variable mass-inertia characteristics. Int. Appl. Mech. **40**, 814–820 (2020)
7. Deng, Z., Deng, S.: Automatic driving systems considering trajectory tracking and lateral stability control of distributed electric drive semi-trailer truck. Int. J. Veh. Perf. **11**(1), 1–25 (2025). https://doi.org/10.1504/IJVP.2025.10069055
8. Tarun, S., Yuping, H.: On trade-off relationship between static and dynamic lateral stabilities of articulated heavy vehicles. Designs **8**, 23 (2024). https://doi.org/10.3390/designs8050103
9. Du, X.-J., Ning, W., Jie, Z., Pei, Y.-L.: Control strategy of lateral stability of semi-trailer train in dangerous section of cold area. Jilin Daxue Xuebao (Gongxueban)/J. Jilin Univ. (Eng. Technol. Edition) **54**(4), 996–1006 (2024). https://doi.org/10.13229/j.cnki.jdxbgxb.20221059
10. Zhao, J., Wu, X., Song, Z., Sun, L., Wang, X.: Research on coupled bifurcation and stability of tractor semi-trailer under combined braking and steering conditions. In: Proceedings of the Institution of Mechanical Engineers, Part D (2024). https://doi.org/10.1177/09544070241265395

11. Wang, X., Fang, B., Zhao, Y., Wang, J., Feng, P.: Research on cooperative control strategy for braking system of articulated heavy vehicles. Int. J. Heavy Veh. Syst. **31**(4), 443–470 (2024). https://doi.org/10.1504/IJHVS.2024.139689
12. Wanki, C., Youngsik, Y., Seung-Han, Y.: A study on sway stability control methods for tractor-trailer system using rear wheel steering. IEEE Access. **12**, 66952–66963 (2024). https://doi.org/10.1109/ACCESS.2024.3398715
13. Li, E., Yu, H., Xi, J., Ju, Z.: Stability-guaranteed model predictive path tracking of autonomous tractor semi-trailers under extreme conditions. IEEE Trans. Intell. Veh., 1–14 (2024). https://doi.org/10.1109/TIV.2024.3376676
14. Cveticanin, L., Zukovic, M., Cveticanin, D.: Oscillator with variable mass excited with non-ideal source. Nonlinear Dyn. **92**, 673–682 (2018)
15. Koliesnik, I., Kalinin, E., Koliesnik, J., Pankova, O., Shutko, V.: Dynamics of braking of the transport and technological train. Int. J. Mechatron. Appl. Mechan. (IJOMAM) **18**, 148–156 (2024). https://doi.org/10.17683/ijomam/issue18.17
16. Luo, S.: Relativistic variation principles and equation of motion for variable mass controllable mechanical system. Appl. Math. Mech. (English Edition) **17**, 683–692 (2020)
17. Djordje, M.: General energy change law for systems with variable mass. Eur. J. Mech. A/Solids **18**, 719–730 (2021)
18. Zeković, D.N.: Dynamics of mechanical systems with nonlinear nonholonomic constraints – II Differential equations of motion. ZAMM Zeitschrift fur Angewandte Mathematik und Mechanik **91**, 899–922 (2011)
19. Bojan, J., Radoslav, R., Aleksandar, O.: Analysis of the brachistochronic motion of a variable mass nonholonomic mechanical system. Theor. Appl. Mech. **43**, 19–32 (2016)
20. El-Nabulsi, Ahmad, R.: Free variable mass nonlocal systems, jerks, and snaps, and their implications in rotating fluids in rockets. Acta Mechanica **232**, 89–109 (2021)
21. Pesce, C.P., Orsino, R.M.M., Silva, L.S.P.: Nonlinear dynamics of variable mass oscillators. In: Castilho Piqueira, J.R., Nigro Mazzilli, C.E., Pesce, C.P., Franzini, G.R. (eds) Lectures on Nonlinear Dynamics. Understanding Complex Systems. Part F1825, pp. 217–252. Springer, Cham (2024). https://doi.org/10.1007/978-3-031-45101-0_8
22. Yi, Z., Chuanjing, S., Xianghua, Z.: Generalized gauss principle of least compulsion for variable-acceleration dynamical systems. Lixue Xuebao/Chin. J. Theor. Appl. Mech. **55**. pp. 1174–1180. (2023)
23. Pradhan, N.C., et al.: ANN-PID based automatic braking control system for small agricultural tractors. J. Field Rob. **41**(8), 2805–2831 (2024). https://doi.org/10.1002/rob.22393
24. Zhang, L.-B., Feng, S.-Y., Shan, H.-Y., Wang, G.-R.: Braking timing detection of tractor-trailer-train based on tire slip rate and monocular vision. J. Jilin Univ. (Eng. Technol. Edition) **53**(12), 3415–3423 (2023). https://doi.org/10.13229/j.cnki.jdxbgxb.20220092
25. Radzajewski, P., Guzek, M.: Assessment of the impact of selected parameters of tractor-semitrailer set on the braking safety indicators. Appl. Sci. **13**, 5336 (2023). https://doi.org/10.3390/app13095336
26. Borawski, A.: Effects of initial speed and load of universal semitrailer on braking system's friction pair heating. Heat Transfer Res. **51**(15), 1417–1428 (2020)
27. Samorodov, V., Kozhushko, A., Pelipenko, E.: Formation of a rational change in controlling continuously variable transmission at the stages of a tractor's acceleration and braking. Eastern-Europ. J. Enterprise Technol. **4**(7–82), 37–44 (2016)
28. Shiwei, X., Xiaopeng, Z., Yuan, J., Lulu, W., Jingjing, H., Xinyu, Z.: Research on the multi-mode composite braking control strategy of electric wheel-drive multi-axle heavy-duty vehicles. World Electr. Veh. J. **15**(3), 83 (2024)
29. Zhen, W., Feng, G., Jing, D., Xuelei, W.: Dynamics model and braking performance analysis of multi-axle vehicle. Zhongguo Jixie Gongcheng/China Mechanical Eng. **19**(3), 365–369 (2008)

30. Xinping, W., Rongnian, H., Han, G., Mengyu, C.: Electronic mechanical braking system executive mechanism design, calculation, and modeling based on dynamic control. Front. Mech. Eng. **10** (2024). https://doi.org/10.3389/fmech.2024.1368683 (2024)
31. Sadono, A., Mahendra, O., Saputra, R., Rijanto, E.: Dynamics and control simulation of DC series motor for electric vehicle in regenerative braking (2024). https://doi.org/10.1063/5.0205710
32. Davidyan, G., Bortman, J., Kenett, R.: Development of an operational digital twin of a freight car braking system for fault diagnosis. Adv. Theory Simul. 7 (2024). https://doi.org/10.1002/adts.202301257
33. Megantara, T.R., Supian, S., Chaerani, D.: Mathematical modeling on integrated vehicle assignment and rebalancing in ride-hailing system with uncertainty using fuzzy linear programming. J. Adv. Res. Appl. Sci. Eng. Technol. **42**(2), 133–144 (2024). https://doi.org/10.37934/araset.42.2.133144
34. Wang, Q., Wang, Z., Mo, J., Wang, R., Wang, K.: Effect and mechanism of wheel–rail adhesion on torsional vibration and stability of the braking system. Tribol. Int. **179**, 108182 (2022). https://doi.org/10.1016/j.triboint.2022.108182
35. Lebedev, A., et al.: Determining conditions for providing maximum traction efficiency of tractor as part of a soil tillage unit. Eastern-Eur. J. Enterprise Technol. **1**(127), 6–14 (2024). https://doi.org/10.15587/1729-4061.2024.297902
36. Cole, C., Spiryagin, M., Wu, Q., Sun, Y.Q.: Modelling, simulation and applications of longitudinal train dynamics. Veh. Syst. Dyn. **55**(10), 1498–1571 (2017)
37. Megantara, Y.T., Supian, S., Chaerani, D.: Mathematical modeling on integrated vehicle assignment and rebalancing in ride-hailing system with uncertainty using fuzzy linear programming. J. Adv. Res. Appl. Sci. Eng. Technol. **42**, 133–144 (2024). https://doi.org/10.37934/araset.42.2.133144
38. Kalinin, Y., Medvediev, Y., Lebedev, S., Kuchuk, H., Kuchuk, N.: Solving problems of dynamics of systems with elastic elements and variable masses. In: Cioboată, D.D. (eds) International Conference on Reliable Systems Engineering (ICoRSE) - 2024. ICoRSE 2024. Lecture Notes in Networks and Systems, vol. 1129. Springer, Cham (2024). https://doi.org/10.1007/978-3-031-70670-7_15
39. Kozhushko, A., Pelypenko, Y., Kravchenko, S., Danylenko, V.: Improving the procedure for modeling low frequency oscillations of the free surface liquid in a tractor tank. Eastern-Eur. J. Enterprise Technol. **122**(7), 61–68 (2023). https://doi.org/10.15587/1729-4061.2023.277254
40. Ovsyannikov, S., Kalinin, E., Koliesnik, I.: Oscillation process of multi-support machines when driving over irregularities. Int. Sci. Conf. Energy Manage. Municipal Facil. Sustain. Energy Technol. **1**, 307–317 (2020)

Creative Solution for Flood Response Assessment. A Challenge-Based Learning Implementation

Andreea Dana Alionte[1], Carlos Paredes[2], Cristian Gabriel Alionte[1](✉), Elcin Tan[3], Miguel Llorrente[4], Rogelio de la Vega[2], Luis Igelsias[2], Maria Sonia Medina[5], Marta Ortiz[5], and Adrian Rosario[4]

[1] National University of Science and Technology POLITEHNICA Bucharest, Splaiul Independenţei 313, 060042 Bucharest, Romania
cristian.alionte@upb.ro

[2] Department of Geological and Mining Engineering, School of Mines and Energy, Universidad Politecnica de Madrid, Ríos Rosas 21, 28003 Madrid, Spain

[3] Department of Climate Science and Meteorological Engineering, Technical University of Istanbul, İTÜ Ayazaga Campus, 34469 Maslak/Istanbul, Turkey

[4] Instituto Geológico y Minero de España (CN IGME - CSIC), Ríos Rosas 23, 28003 Madrid, Spain

[5] Department Economía de La Empresa (ADO), Economía Aplicada II y Fundamentos Análisis Económico, Fac. CC. Economía y Empresa, Universidad Rey Juan Carlos, P. Artilleros S/N, 28032 Madrid, Spain

Abstract. To enhance the evaluation and education of geohydrological risk management, creative activities can be utilized through the deployment of challenge-based learning, which is based on collaborative GIS/hydrodynamical modeling and serious gaming. To answer the demand for efficient flood mitigation techniques in the context of increasing climate-driven threats, this interacts advanced simulation with practical problem-solving. Throughout the process of modeling crisis scenarios, participants create and evaluate response plans in an environment that is within control. The activity is designed to be accomplished in a collaborative manner, and it is assisted by professionals, which helps to enhance the practical application of interdisciplinary scientific and technological knowledge. Therefore, engineers can be prepared to deal with catastrophes from a variety of perspectives. These perspectives are not limited to flood assessment; rather, they must be able to manage the full extent of the hazard. This includes the exploration of floods, the modeling of hydrodynamics, the promotion of resilience to disasters, and the preparation of plans for prevention, mitigation, and evacuation. Consequently, we anticipate that participants will actively learn to apply theoretical concepts in practical situations. Participants will accomplish this by incorporating novel approaches, empathizing, sharing, and focusing on current socio-economic and environmental needs. Additionally, we expect participants to project an interdisciplinary perspective over a practical challenge in a real geohydrological menace case.

Keywords: Academic training · Challenge-based learning · Natural geohydrological hazard · Preparedness · Resilience

D. D. Cioboată and J. Machado (Eds.): ICoRSE 2025, LNNS 1592, pp. 301–315, 2025.
https://doi.org/10.1007/978-3-032-02508-1_26

1 Introduction

Challenge-based learning emphasizes collaboration and practical engagement, requiring students to interact with peers, educators, and community experts to enhance their understanding of academic subjects, address challenges, contribute positively to their communities, and disseminate their findings globally. The interdisciplinary, collaborative, and practical characteristics of challenge-based learning render it suitable for students across all age groups. Students engage in collaborative groups and utilize technology to address real-world issues within the context of their school, family, or local community. Teachers are tasked with collaborating with students to integrate multidisciplinary, standards-based content; relate it to contemporary global issues; and facilitate experiences that enable students to impact their community positively. Achieving this objective requires providing students with structure, support, checkpoints, and appropriate tools to complete their work effectively, while also granting them sufficient autonomy to be self-directed, creative, and inspired [1]. Throughout the challenge, the students, both in teams and as individuals, can create a variety of products, including a challenge proposal video, a set of guiding questions, research plans and results, solutions with beta testing plans and evaluation parameters, a solution video, student journals, and individual reflection videos. Once the challenge has been determined, students generate sets of questions that will guide their search for a solution. The question sets should be extensive and represent the needs of each group member. The students create this product to organize and document their efforts to research and answer the guiding questions. A well-prepared plan will ensure a thorough and organized learning experience as students search for a solution. The plan also provides a way to manage the learning experience. The teacher serves as a guide or mentor to ensure the timeline and milestones are appropriate for the time allotted for the challenge.

1.1 The Framework of "Challenges in Geohydrological Risk Management: As a First Response to Flood Exploration, Modelling and Resilience in Action"

The event titled "Challenges in Geohydrological Risk Management: A First Response to Flood Exploration, Modelling, and Resilience in Action" was organized by Universidad Politécnica de Madrid, in collaboration with the Technical University of Istanbul and the National University for Science and Technology Politehnica Bucharest. The course aimed to equip postgraduate engineering students with competencies in controlling geohydrological hazards, focusing on initial flood exploration, hydrodynamic modeling, and enhancing resilience to disasters. Participants are anticipated to implement theoretical principles in practical contexts through challenge-based learning and will employ hydrodynamic models to explore flood dynamics across several scenarios. They will possess fundamental knowledge of resilience, sensors, and robotics applicable in natural disaster scenarios. The suggested activity substantially addresses social concerns and promotes the United Nations Sustainable Development Goals (SDGs), notably SDG 11: Sustainable Cities and Communities and SDG 13: Climate Action. The event had three key objectives: promoting knowledge exchange, enhancing awareness and education, and advancing policy and decision-making. The event is centered on the geo-description of an area and the hazard situation (for three challenge locations: Bisbal, Fluvia, and

Sureda), as well as the identification of issues to be addressed. Clear answers to each of the specific questions/objectives are provided as product components. An academic synthesis of the teaching-learning process includes what and how students learned, the problems encountered and how they were overcome, and how the team arrived at the proposed solution. A graphic and documented description of the suggested answer to the challenge outlines what is missing to complete the outcome. The challenge-based solution begins at stage one (Fig. 2a), where the big idea (significant on a global scale, enabling students to collaborate to gain deep transdisciplinary content knowledge and understanding), essential questions (should be answerable through research, assisting students in focusing their efforts and providing a framework), and the challenge (the task of participants for creating a local solution to a global problem) can be defined. In stage two (Fig. 2b), the student teams can use guidance (to identify the expertise required to design a solution to the situation), leading exercises (to help them acquire the knowledge needed to answer the guiding questions and develop a creative, insightful, practical solution), resources (that must relate to their inquiries and activities), and workflow (a graphical representation of the process) (Fig. 1).

a)

b)

c)

d)

Fig. 1. Different stages of the challenge-based solution: a) stage 1: Big Idea, b) stage 2: Brainstorming, c) stage 3: student teams can find one solution, c) stage 5: Examples of posters.

At stage three (Fig. 2c), the student teams can find one solution through prototyping, experimenting, or another method and can determine how to implement their plan. In stage four, the students can implement the remedy, evaluate the results, and reflect on what worked and what didn't by tracking the progress of the task through meetings. Stage five is dedicated to the results and products which are presented in an exhibition using posters and videos from the representative spokesperson. The products (Fig. 2d) must follow this structure: big idea, team information, essential questions, significance or context, an example, and the challenge. The teams must demonstrate their ability to manage the workflow of the process, find a solution, assess implementation, and reflect on the lessons learned. Because the primary goal of the event is learning, and was the first event of this type, it does not make any sense to do a performance analysis of flood models built by students. But the learning results can be compared with the traditional methods where it can be seen that in 30 h of dedicated learning the results are similar to the traditional methods (48 h, 2 h per week).

2 Modeling the Floods

2.1 Conceptual Modeling

The conceptual modeling is used when the least understood and removed from the modeling process; it is considered as an art due to the lack of defined methods and procedures and is performed by the modeler that has a thorough understanding of the operations of the system being modeled. It is the most critical part of the simulation modeling process. The main elements of conceptual modeling are problems and objectives (purpose of the model and the simulation project), assumptions (uncertainties or beliefs about the real world being modeled), simplifications (ways of reducing the complexity of the model), and inputs (elements that can be altered/measured to create an improvement).

The structure of a model is made of inputs, outputs (results of the simulation model), and content (components in the model and their interactions). Usually, the models describe structure models in terms of entities, relationships, and constraints or describe behavior or functional models in terms of states, transitions among states, and actions performed in states and transitions or describe interactions and user interfaces in terms of messages sent and received and information exchanged. A conceptual model can be shown as a list of system parts (which includes the components like entities, attributes, or activities), a process flow diagram (which maps how the entities move through processes), or a logic flow diagram (which maps the logical choices made during the process flow). However, conceptual models tend to be simplified and therefore are not always perfectly accurate because sometimes, the real world is more complicated than conceptual models allow.

Modeling can be done by self-organized criticality that explains phenomena with power-law distributions and combines self-organization (the ability of certain non-equilibrium systems in the absence of control or manipulation by an external agent to develop specific structures and patterns) and criticality (emphasizing the similarity with phase transitions, the system stays at the border of stability and chaos) to describe complexity. This kind of modeling does not require external tuning with control; it organizes itself into critical behavior and is constantly driven by some random energy input, evolving into a critical state that is maintained. The quasi-stationary state (quasi, not absolute, because the system is critical) is established solely because of the dynamical interactions (mainly by shear forces); the state is self-organized.

Collecting spatial data can occur manually through techniques such as surveying, geodesy, or Global Navigation Satellite Systems (GNSS), or through remote methods like GPS, GLONASS, BEIDOU, remote sensing, photogrammetry, or LIDAR (Light Detection and Ranging - Laser Imaging Detection and Ranging). The spatial data can be stored and analyzed using cartography or Geographic Information Systems (GIS) [2]. Mapping is a critical tool in geohydrological risk management because it can facilitate the identification of risk zones (through GIS analysis, it identifies most vulnerable hazard locations effectively), data integration and visualization (it combines diverse information sources for comprehensive hazard mapping), risk assessment and prioritization (mapping aids decision-makers in prioritizing high-risk areas intervention), early warning systems (mapping is essential for timely Collaborative mapping is a method in which numerous people work together to create or amend a digital map by gathering and sharing geographic data via a collaborative platform. It is used in urban planning, environmental monitoring, disaster response, and community building.

Collaborative mapping has several advantages, including improved accuracy (crowd-sourced data is frequently updated and refined), cost-efficiency (reduces the need for expensive surveying, particularly in inaccessible areas), community engagement (involving local populations directly in mapping their surroundings), and rapid data collection (useful for immediate needs such as disaster relief or event planning). However, collaborative mapping has some drawbacks and limitations, including data quality control (ensuring reliability and accuracy), privacy concerns (personal data or sensitive locations may be inadvertently mapped), technical challenges (requiring technical skills and access to digital tools), and data standardization.

2.2 Flow Modeling and Flood Hazard

Prior to doing any flood hazard assessment, it is advisable to inspect the research area. The primary objective of visiting the study location is to comprehend the issue we intend to analyze and to assess the extent of a potential inundation. Field campaigns are crucial for obtaining fundamental data (e.g., topography) and for identifying potential bottleneck effects that may influence the numerical modeling of flood events, including alterations in land use and the existence of structures (bridges, culverts, etc.), among others. The study area should align with the objectives of the primary research; however, it cannot be confined exclusively to the zones occupied by the targets. Sufficient distance must be ensured to prevent interference with flow propagation caused by the boundaries of the study area, which primarily include the road traversing the entire area, additional facilities located upstream of the campsites, and the bridge abutments situated just downstream of the campsite.

Topographical data from geoportals is essential for conducting the inundation study, since it delineates the geometry of the study region, accurately depicting features such as the riverbed, floodplains, and levees. Hydraulic numerical modeling methods facilitate the characterization of flow hydrodynamics, including the evolution of water depth, velocity, flood extent, and associated hazards. The hydrodynamic parameters of the flood, such as depth and velocity, are frequently employed to characterize the inundation hazard, either as standalone metrics or through the integration of many factors (added, multiplied, etc.). The danger criteria may differ based on the legislation of each nation. Numerical models vary in efficacy based on spatial discretization, numerical schemes, and other factors.

Modeling facilitates forecasting through flow modeling and flood hazard assessment, design, evaluation, asset management, floodplain zoning to direct suitable development, flood mapping to enhance community awareness, and various other components of flood risk management.

Various methods can be employed to ascertain water flow by computational predictions of water levels, flow rates, flood depth, inundation areas, or by utilizing model approximations of actual physics. Numerical methods employ several methodologies, including one-dimensional, two-dimensional, and three-dimensional dimensions; partial differential equations; and specific objectives. Fluid dynamics constitutes a three-dimensional (3D) issue characterized by the Navier-Stokes equations, which, in general, lack an analytical solution for this system of partial differential equations (PDEs). Numerical approaches facilitate the approximation of solutions by positing certain hypotheses (e.g., incompressible fluid) and discretizing the study region into clearly defined control points where the equations are resolved.

Depending on the desired precision (both geographically and temporally), dozens or even millions of calculations may be performed to derive the solution. Consequently, computational tools are essential for efficiently resolving the issue within a suitable timeframe and with minimal exertion. The integration of numerical and computational approaches is referred to as Computational Fluid Dynamics (CFD) when utilized for fluids, as they facilitate the estimation of fluid dynamics. CFD delineates the progression of pressure, velocity, density, temperature, and other variables as functions of spatial and temporal dimensions. In flood hazard assessment studies, one-dimensional (1D) or two-dimensional (2D) spatial methodologies are frequently employed to discretize the spatial domain, owing to the extent of the river reach under analysis. The depth of free surface water and its average velocity at each control element may be contingent upon the modeler's expertise. In the 1D model and 2D model, the flow proceeds down the axis, with hydraulic variables defined in spatial dimensions, a singular value for water level and velocity, and no requirement to predefine flow paths. Two-dimensional models address the Shallow Water Eqs. (2D-SWE), which comprise the mass and momentum equations derived from the Navier-Stokes equations for incompressible fluids (constant density—div v = 0), with no temperature fluctuations and constant viscosity (Newtonian).

2.3 Mathematical Model

Iber software [3] uses two-dimensional shallow-water Eqs. (2D-SWE) via an integrated hydrological-hydraulic modeling methodology. Iber incorporates precipitation and infiltration dynamics into the shallow-water equations, enabling the concurrent simulation of rainfall-runoff conversion and river flooding phenomena. The model employs an unstructured finite-volume solver for the 2D shallow water equations, enabling the user to customize the mesh according to basin morphology and to specify varied mesh sizes in various sections of the research area [4].

The model starts with the hydrodynamic equations:

$$\frac{\partial h}{\partial t}+\frac{\partial q_x}{\partial x}+\frac{\partial q_y}{\partial y}=r-f \tag{1}$$

$$\frac{\partial q_x}{\partial t}+\frac{\partial}{\partial x}\left(\frac{q_x^2}{h}+\frac{gh^2}{2}\right)+\frac{\partial}{\partial y}\left(\frac{q_x q_y}{h}\right)=-\mathrm{gh}\frac{\partial z_b}{\partial x}-ghI_x \tag{2}$$

$$\frac{\partial q_y}{\partial t}+\frac{\partial}{\partial x}\left(\frac{q_x q_y}{h}\right)+\frac{\partial}{\partial y}\left(\frac{q_y^2}{h}+\frac{gh^2}{2}\right)=-\mathrm{gh}\frac{\partial z_b}{\partial y}-ghI_y \tag{3}$$

where x and y are the two horizontal directions, t is the time, z_b is the bed elevation,

h is the water depth, q_x and q_y are the two components of the unit discharge in the two horizontal directions, I_x and I_y are the two components of the bed friction slope, g is the gravity acceleration, r is the rainfall rate, and f is the infiltration rate. The bed friction slope is computed using Manning empirical formulation.

The temporal and spatial evolution of the concentration for each sediment class is computed from the following depth-averaged scalar transport equation:

$$\frac{\partial hC_k}{\partial t} + \frac{\partial q_x C_k}{\partial x} + \frac{\partial q_y C_k}{\partial y} = D_{rdd,k} + D_{rdrd,k} + D_{fdd,k} + D_{fdrd,k} + D_{dep,k} \quad (4)$$

where C_k is the depth-averaged suspended sediment concentration of class k, $D_{rdd,k}$ and $D_{fdd,k}$ are the rainfall-driven and flow-driven detachment rates of sediment class k from the original soil layer, $D_{rdrd,k}$ and $D_{fdrd,k}$ are the rainfall-driven and flow-driven re-detachment rates of sediment class k from the loose sediment layer $D_{dep,k}$ is the deposition rate of sediment class from the water column into the loose sediment layer.

The mass conservation equation is solved to compute the time evolution of each particle class in the loose sediment layer:

$$\frac{\partial M_{s,k}}{\partial t} = -\left(D_{rdd,k} + D_{rdrd,k} + D_{fdd,k} + D_{fdrd,k} + D_{dep,k}\right) - \rho_S\left(\frac{\partial q_{bx,k}}{\partial x} + \frac{\partial q_{by,k}}{\partial y}\right) \quad (5)$$

where $M_{s,k}$ is the mass of each particle, ρ_s is the density of each particle, and $q_{bx,k}$ is the bed load transport capacity for each particle class k, calculated using any empirical formulation (Meyer Peter-Müller, Wong-Parker, Einstein-Brown, van Rijn, Engelund-Hansen, etc.).

The evolution of the bed elevation is computed from the mass conservation equation, which includes all the terms implying transfers of sediment particles from or to the loose sediment layer and the original soil layer:

$$\frac{\partial z_b}{\partial t} = -\sum_{k=1}^{N_s}\left(\frac{1}{(1-\Phi)\rho_S}\frac{\partial M_{s,k}}{\partial t}\right) \quad (6)$$

where Φ is the porosity of the soil layer.

The mathematical model defined by 2D-SWE (Eqs. 1–3), sediment transport (Eqs. 4–6), are solved with a finite volume solver for unstructured grids, and the numerical details of the finite volume method applied are well known in literature [4]. At each time step the new bed elevation is updated in the 2D-SWE to ensure an appropriate coupling between the overland flow and soil erosion.

At the river reach size, the primary influences are the intake discharges of water and sediment at the upstream boundary. The user must supply the inlet hydrograph and sedigraph (or the depth-averaged sediment concentration) at the upstream boundary. At the outlet boundary, only the water depth needs to be specified. This can be accomplished either as a user-specified water level or as a supercritical flow condition.

2.4 Modeling Data

Hydrological parameters, such as the roughness of the terrain, are defined by the land use concept. We can depict land uses in as much detail as we want; however, this definition must be aligned with the goals of the study. In general, we can define 5 land uses, corresponding to rivers, single buildings, infrastructure (roads), other facilities, and sparse vegetation (flood plains). Iber is compatible with several GIS and CAD formats and even allows for importing calculation meshes directly. Geometric entities are points, lines, or surfaces. A geometric entity can be created using the "Create/lines" menu option: first create a closed polygon of the whole study area and for each part of the study area (e.g., river, road, floodplains, etc.). Then complete the geometry of the model by generating all surfaces using "Create/NURBS surface" based on the lines of each closed polygon. Iber's intuitive interface and methodology allow for the incorporation of a backdrop image utilizing an orthophoto. For flow modeling and flood hazard assessment, two boundary conditions must be established: a hydrograph as the inlet boundary condition (red line) and an unrestricted exit boundary condition (white line).

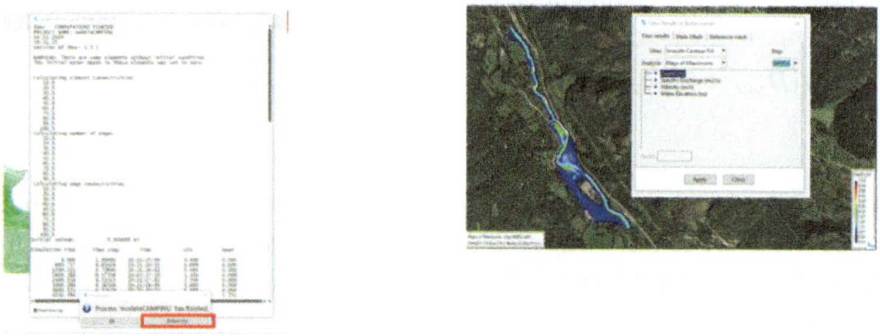

a) b)

Fig. 2. Examples of the results: a) tabular results, b) graphical results

During the pre-process phase (hydraulic conditions), we can modify the hydrodynamic parameters specified in "Data/Hydrodynamic" and adjust the boundary conditions via the "Border Conditions" menu. A file containing the measured discharge values can be utilized for the intake, while the exit can be designated according to the flow state of "Supercritical/Critical". The terrain roughness can now be delineated by land uses, employing the identical geometric definition utilized by the model via the option "Land uses" in the menu "Data/Roughness/Land use.". Iber supplies a catalog of land uses together with a roughness value that correlates to the Manning coefficient for each land use. We may apply roughness to each surface. Select each land use, modify the Manning's value as necessary, and allocate it to the appropriate surface (river – 0.035, buildings – 0.070, road – 0.020, industrial – 0.100, floodplains – 0.080). Ultimately, to conclude the pre-processing phase, we must create the mesh that discretizes the domain into distinct calculation points, or elements, and incorporate the topographical data. The mesh must be delineated in accordance with the requirements of the study and the Iber software requires a mesh size and can utilize it as a default parameter.

The final step is to revise this mesh with the topographical, and you need to navigate to the menu "Tools/Mesh/Edit/Set the elevation from the file" to utilize the Digital Terrain Model files, specifically "DTM_l.txt" and "DTM_2.txt," sequentially.

The time parameters for the calculation can be defined in the tool "Data/Problem Data/Time Parameter." Finally, the simulation is calculated using the menu "Calculate/Calculate.". This will run the model in normal mode, and we can check some simulation parameters by going to "Calculate/View.". In Fig. 2, the results can be seen where both campsites are flooded, but the road is also affected.

2.5 Flood Hazard Assessment

The assessment of flood hazards is a concept extensively utilized in civil engineering to ascertain the level of danger posed by a flood. It is typically characterized as a function of several hydraulic variables (e.g., water depth and velocity), but it may also be linked to probabilistic scenarios of precipitation or flooding events. The danger criteria vary by country and even by regions or towns within the same nation. Despite the campgrounds being located in Spain, we will establish a specific danger criterion for this study based on the hydraulic parameter of water depth. Each criterion value represents the upper threshold of the criterion itself; for instance, "low" danger is defined as water depths exceeding 0.2 m and reaching up to 0.5 m. Reopen Iber and load the previously computed model. In the Custom Hazard tab of the "Data / Problem Data" menu, we must establish the prior hazard criteria.

Constructing floodwalls to safeguard the town center is a cost-effective and efficient solution, enabling the prevention of floods in a densely populated region. However, the river's inability to overflow within the hamlet has resulted in an increase in its depth, leading to overflowing downstream, where the primary agricultural lands are situated. The user may opt to establish a flood control zone in the river. This method, referred to as Sustainable Drainage Systems (SUDs), is predicated on the principle of honoring the river's natural cycles by redirecting flooding to a designated location. The primary purpose of these structures is to provide a hydraulic function: during heavy rainfall, they act as rainwater retention basins, mitigating the risk of downstream flooding in the flood control region. In addition to its hydraulic role, it enriches and safeguards the river's flora and fauna and can be structured to provide leisure and recreational opportunities during dry periods. In summary, it is a multidisciplinary solution that fosters sustainability and resilience in the region of its construction. This method mitigates the risk of river overflow while honoring the river's natural characteristics.

3 Resilience

Resilience denotes a material's capacity to restore its original form following deformation by bending, stretching, or compression. Moreover, resilience is the capacity to regain happiness, prosperity, and other positive states following adversity or negative experiences. Resilience is the capacity to swiftly revert to a prior favorable state following adversity. In flood risk management, resilience denotes the capacity of populations, infrastructure, and ecosystems to foresee, endure, adjust to, and recuperate from flood occurrences while preserving vital functions and mitigating long-term effects. This method transcends just flood response; it includes proactive planning, adaptive infrastructure, and community readiness to mitigate susceptibility. Flood resilience encompasses various dimensions of preparedness (enhancing awareness, implementing early warning systems, and providing public training to equip communities for potential floods), adaptation (designing flexible urban planning and infrastructure capable of enduring diverse flood intensities), and recovery (facilitating swift recovery processes to restore normalcy promptly after a flood, thereby minimizing economic and social disruption). Resilience in flood risk management seeks to revolutionize community responses to floods by shifting from reactive strategies to a proactive approach that incorporates long-term adaptation, infrastructure robustness, and community readiness. This method diminishes the incidence and intensity of flood-related damages and improves the ability for rapid recovery. Future directions and innovations in flood resilience include emerging trends in flood resilience include green infrastructure, adaptive urban planning, and sophisticated hydrological modeling; the significance of technology and data science in bolstering resilience; the potential of machine learning and artificial intelligence to enhance predictive accuracy and response times during flood events.

Resilience in disaster management encompasses various forms: personal resilience (the capacity to remain composed and make sound judgments), organizational resilience (the capability to strategize and coordinate among entities), and community resilience (the ability to mobilize and provide support within the community). The significance of catastrophe management lies in mitigating adverse effects, expediting recovery, and enhancing the resilience triad (individual, organizational, and societal). Personal resilience encompasses calmness and effective decision-making, a positive disposition, adaptability, and problem-solving capabilities derived from individual resources such as skills (first aid training, stress management), support networks (family, friends, colleagues), physical and mental well-being (maintaining good health), and soft skills including emotional intelligence (emotion regulation), adaptability (adjusting to new circumstances), effective communication (articulating and listening), and problem-solving (innovative solutions).

Organizational resilience is characterized by anticipation, preparation, response, and adaptation to change, alongside strategic planning, a robust organizational culture, and effective communication. It relies on resources such as infrastructure and equipment (secure facilities, emergency apparatus), human capital (trained personnel and rapid response teams), organizational capabilities, foresight (planning and procedures: contingency plans, emergency protocols), trust, resilient leadership, and strategies for collective response, as well as a strong organizational culture (shared values), effective communication (clear channels), adaptation, innovation, and flexibility (open-mindedness).

Social resilience within a community refers to its ability to withstand, adapt to, and recuperate from challenges while preserving social cohesion and functionality. This resilience encompasses the capacity to respond to adverse events and to reorganize and enhance itself through resources such as education, health, infrastructure, communication, and management planning.

The mechanisms for cultivating resilience include fostering a positive mindset and practicing gratitude, establishing connections, engaging in self-care (such as seeking professional assistance), setting realistic objectives and embracing change, enhancing stress management and problem-solving abilities, promoting self-efficacy, strengthening social ties, and improving access to resources and services through the development of resilient infrastructure, raising awareness of flood impacts, elucidating the interrelationship between natural phenomena and the cultural and social advancement of communities, and encouraging collaboration and leadership.

The objectives of disaster management encompass minimizing adverse effects, expediting recovery, enhancing resilience, ensuring preparedness (in terms of damage and response speed), training emergency personnel, establishing evacuation centers, fortifying networks, conducting regular drills, implementing response measures (to safeguard lives and mitigate damage), mobilizing emergency teams, coordinating evacuations, establishing emergency communication systems, delivering essential services, facilitating recovery infrastructure, rehabilitating affected areas, and supporting critical restoration efforts.

Leadership positively influences resilience and is enhanced through coordinated efforts, effective communication, and community support and mobilization. Conversely, it can be diminished by poorly defined objectives, lack of commitment, power imbalances, absence of legitimacy, and concealed agendas (political interests). Resilience can be enhanced by a collective response involving stakeholder mediation and negotiation, fostering discourse, transparency, communication, trust-building, resource prioritization, and the establishment of clear and equitable criteria.

4 Technology Used in Natural Disasters: Robot Technology

Robotics, a rapidly expanding field of study, will have an impact on individuals from all socioeconomic backgrounds within the next few years. These robots will use their systems and detection techniques to assist rescuers in precisely locating victims and saving lives. Robots must do several tasks either with human control or autonomously. Here, we present a limited list of tasks that robots can perform:

- Reconnaissance. We must carry out this crucial task in the area affected to assess the situation and ensure the successful deployment of search and rescue teams. Unmanned aerial aircraft may prove highly beneficial in this task. [5]
- Search: The search function is a critical task. Victims must be located initially to facilitate their rescue. Small, unmanned ground vehicles can navigate rubble more effectively than humans, hence facilitating quicker victim recovery by rescuers. Robots can potentially utilize their ability to navigate debris for structural inspection.
- Rubble removal and victim extraction: Operations for rubble removal in search and rescue scenarios significantly differ from those in construction or demolition activities. Rescue robots and exoskeletons may assist rescuers in this endeavor.
- Telepresence: The term refers to a robotic application that employs a rescue robot as the "ears and eyes" for a rescue crew operating outside the debris. Consequently, the team was able to interact more swiftly. Furthermore, telemedicine may be highly beneficial.

Wheeled robots represent one of the most prevalent categories of mobile robots, distinguished by their simplicity and efficiency in locomotion. Wheeled robots are advancing due to improvements in sensor technology, artificial intelligence, and materials, enhancing their versatility and capabilities. Wheeled robots are simpler to design, construct, and program than legged robots. Their simple mechanics render them suitable for several applications. Wheels facilitate rapid and efficient locomotion on level surfaces, rendering these robots appropriate for inside settings such as residences, workplaces, and industrial facilities. Wheels necessitate reduced energy for movement, rendering them ideal for prolonged operations; they also facilitate rapid transit on level terrain, advantageous for time-critical applications. However, wheeled robots encounter difficulties on rocky, uneven, or steep surfaces and are unable to surmount obstacles when compared to legged robots. Categories of wheeled robots: two-wheeled robots (often require balancing mechanisms to stay upright), three-wheeled robots (these have two powered wheels and one free-rotating wheel for balance), four-wheeled robots (these are the most stable and are commonly used in applications requiring robust and reliable movement), omnidirectional robots (equipped with special wheels that allow movement in any direction without changing orientation).

Legged robots can navigate rugged and irregular terrain, ascend stairs, and surmount obstacles; they are applicable in diverse locations, ranging from metropolitan areas to natural landscapes; and they can modify their stride and posture, enabling them to adapt to various surfaces and situations. Designing and managing legged robots is more intricate than that of wheeled robots due to the necessity for balance and coordination; they frequently demand greater power to function, particularly during dynamic motions. Ultimately, the sophisticated technology and materials required can render their development and production costly. Types of legged robots: bipedal robots (these robots have two legs and are designed to walk similarly to humans), quadrupedal robots: with four legs, these robots resemble animals like dogs or cats), hexapod robots (these robots have six legs, providing excellent stability, and are often inspired by insects), multi-legged robots (robots with more than six legs, such as octopods, which can offer even greater stability and redundancy).

Aerial robots, referred to as drones or UAVs (unmanned aerial vehicles), represent a swiftly evolving domain within robotics. They have distinctive capabilities owing to their capacity for flight and access to inaccessible locations. Aerial robots can traverse obstacles and access locations inaccessible to terrestrial robots, rendering them suitable for diverse applications. They can be outfitted with various sensors and instruments, enabling them to execute a diverse array of tasks. Numerous aerial robots can function independently, employing sophisticated algorithms and artificial intelligence to navigate and execute tasks without human oversight. Drones can execute jobs more economically and swiftly than conventional techniques, such as piloted aircraft or terrestrial inspections. Issues with drones include restrictive airspace laws, limited flight duration due to battery constraints, and concerns about potential crashes and privacy violations. Types of aerial robots: multicopters(these include quadcopters, hexacopters, and octocopters, which use multiple rotors for lift and stability), fixed-wing drones (these drones have wings like an airplane and are more efficient for long-distance flights), hybrid drones (combining features of multicopters and fixed-wing drones, these can take off and land vertically but fly efficiently like fixed-wing drones).

Mobile mimetic robots, or biomimetic robots, are engineered to replicate the movements and behaviors of real animals. These robots get inspiration from nature to address intricate engineering challenges and improve their functionality. These robots emulate the physical and behavioral characteristics of animals or people, enabling them to traverse surroundings in manners that conventional robots cannot. Imitating natural movements can enhance navigation efficiency and task execution effectiveness. These robots can be built for diverse settings and purposes. Biomimetic designs frequently yield creative solutions unattainable by conventional engineering methods. The challenges associated with these robots include the complexity of designing and controlling biomimetic robots, which necessitate advanced algorithms and meticulous engineering; the high costs of development and production due to the intricate technology involved; and the significant challenge of ensuring these robots can endure harsh environments and extended usage. Types of mobile mimetic robots: insect-inspired robots (these robots mimic the movement of insects like ants, beetles, or bees; they are often small, agile, and capable of navigating tight spaces), animal-inspired robots (these robots replicate the locomotion of animals such as dogs, cats, or snakes), human-inspired robots (also known as humanoid robots, these robots mimic human movements and behaviors).

In the literature some research demonstrates that the sensor fusion algorithm will improve the efficacy of identifying potential victims. Therefore, we will provide examples of onboard sensors employed in mobile robots applicable to natural catastrophe scenarios. The study conducted by M. Yang [5] presents an automatic laser scan registration in urban situations. Initially, two separate laser scans with low overlaps can be acquired from significantly different viewpoints (wide baseline). Secondly, artificial edifices commonly display several structures with similar appearances. This can lead to significant aliasing in the matching process. We suggest a new way to enhance current 2D feature methods to help automatically align two scans taken from far apart by combining laser data and camera images. The primary objective is to utilize the obtained 3D geometry to improve the efficacy of 2D image features for wide baseline matching subsequent to the extraction of predominant planar structures from 3D point clouds. The paper by Donghoon [6] introduces Radio Tomographic Imaging (RTI), a novel method for localizing physical objects within a geographical area supported by wireless networks [7]. The concepts of radio tomographic methods have been employed to create spatial loss fields (SLFs), which are maps that measure the attenuation of electromagnetic waves in radio frequency (RF) bands at each spatial location. The study by Kang [8] introduces a multithreaded RGB-D SLAM framework that integrates a precise terrestrial Light Detection and Ranging (LiDAR) point cloud, thereby reducing cumulative errors and enhancing system robustness in extensive and complex environments. The study by Jauhar [9] presents a Ubiquitous Sensor Network designed to monitor environmental conditions, specifically on bridges. An accelerometer identifies vibrations. Data will be conveyed between sensors and servers using the ZigBee communication protocol. Elliptic curve encryption and key agreements will be employed for security protocols. Menezes-Qu-Vanstone. "Ubiquitous" originates from the Latin term "ubique," signifying "everywhere." The USN has enabled the collection of data in real-time, regardless of location or time. The platform is a useful tool for checking, fixing, and monitoring, capable of finding, locating, identifying, and measuring different types of damage in various materials all at the same time.

A useful technology used fo sensor fusion is the AI that presents considerable potential in numerous domains but artificial intelligence in natural disasters exhibits numerous constraints:

- Data availability and quality: The training and validation of AI systems require significant quantities of high-quality data and collecting datasets from various natural disaster events poses considerable difficulties.
- The quality and consistency of data may fluctuate due to sensor limitations, noise interference, and data deficiencies. Data analysis assists AI systems in understanding operational mechanisms; nonetheless, seismologists have superior experience and knowledge in this field.
- Constraints of generalization: AI may struggle to produce innovative replies when trained on intricate datasets. The interaction of AI with various natural disaster scenarios, including different regimes, locations, and geographical regions, requires the utilization of many data sets.
- Inadequate reliability and interpretation: Deep learning models may provide irrational outcomes. Developing an algorithm for prediction without understanding sensor signals leads to intrinsic uncertainty in the results.
- Significant computational resources are essential for tackling multifaceted issues in natural disasters. Strong processing capabilities are essential for creating effective models that amalgamate diverse data sources, including pictures and time series.
- Ethical and social dilemmas arise when certain AI models fail to discern the root causes of false alarms in early warning systems, which may result from assaults or human malfeasance [1].

5 Conclusions

The activity has demonstrated a comprehensive perspective on the challenges of effective flood risk management, encompassing prevention, mitigation, and post-disaster strategies for resilient recovery and reconstruction. A teaching-learning process has been developed from the student's perspective to facilitate experiential learning, utilizing numerical simulation and role-playing games within a challenge-based learning environment, supported by a multidisciplinary team of instructors and trainers from UPM, UPB, ITU, IGME, and URJC. The success is attributed to the variety of products developed for real-world scenarios by participants organized into teams, including a comprehensive strategy for flood management, prevention, mitigation, and disaster response during extreme events. The products were showcased through a poster, a collaborative presentation, and a brief movie.

Acknowledgements. This communication has been developed in the frame of project EELISA, co-funded by the European Union through the 3rd EELISA Joint Call, and the ATHENS International Network (Advanced Technology Higher Education Network/SOCRATES) program.

References

1. Gârdan, I.P., et al.: Adopting AI in education: optimizing human resource management considering teacher perceptions. Front. Educ. **10** (2025). ISSN=2504–284X, https://doi.org/10.3389/feduc.2025.1488147
2. Vales-Bravo, J.J., et al.: LiDAR-photogrammetry coverage, bathymetry and infrastructures survey to support flood risk analysis in the Genil River sub-basin. Revista De Teledetección **63**, 79–97 (2024). https://doi.org/10.4995/raet.2024.20481
3. Bladé, E., et al.: Iber: herramienta de simulación numérica del flujo en ríos. Rev. Int. Metod. Numer. **30**, 1 (2014). https://doi.org/10.1016/j.rimni.2012.07.004
4. Cea, L., García-Feal, O., Nord, G., Piton, G., Legoût, C.: Implementation of a GPU-enhanced multiclass soil erosion model based on the 2D shallow water equations in the software Iber. Environ. Modelling Software **179**, 106098 (2024). ISSN 1364–8152
5. Yang, M.Y., Cao, Y., McDonald, J.: Fusion of camera images and laser scans for wide baseline 3D scene alignment in urban environments. ISPRS J. Photogramm. Remote Sens. **66**, S52–S61 (2011)
6. Lee, D., Berberidis, D., Giannakis, G.B.: Adaptive Bayesian Radio Tomography. IEEE Trans. Signal Process. (SUBMITTED, APRIL 4, 2018)
7. Thanu Thavasi, P., Suriyakala, C.D.: Sensors and tracking methods used in wireless sensor network based unmanned search and rescue system- a review. Procedia Eng. **38**, 1935–1945 (2012)
8. Kang, X., Li, J., Fan, X., Wan, W.: Real-Time RGB-D simultaneous localization and mapping guided by terrestrial LiDAR point cloud for indoor 3-D reconstruction and camera pose estimation. Appl. Sci. **9**, 3264 (2019). https://doi.org/10.3390/app9163264
9. Jauhar, I., Sudarsono, A., Yuliana, M.: Secure ubiquitous sensor network based on elliptic curve MenezesQu Vanstoneas status data supply of environment in disaster management. EMITTER Int. J. Eng. Technol. **1**(1) (2013). ISSN: 2443–1168

Optimal Use of Low-Power Solar Panels for an Autonomous IoT Soil Monitoring Station

Valentina-Daniela Băjenaru[1,2(✉)], Constantin Anghel[1], and Simona-Elena Istrițeanu[1,2]

[1] National Institute of Research and Development in Mechatronics and Measurement Technique, 021631 Bucharest, Romania
valentina.bajenaru@incdmtm.ro

[2] National University of Science and Technology POLITEHNICA Bucharest, Bucharest, Romania

Abstract. Photovoltaic (PV) systems, leveraging the photoelectric conversion of solar radiation into electrical energy, have undergone substantial technological advancements in recent decades. These advancements have yielded increased conversion efficiencies and an expansion of large-scale deployments, ranging from utility-scale power plants to in-situ distributed micro-systems. This paper presents a technical analysis of the efficiency of terrestrial photovoltaic (PV) panels. The study evaluates their power performance, focusing on the variable factors that introduce uncertainty in predicting incident solar radiation and its subsequent conversion into electrical energy. Consequently, to optimize the operating point of a photovoltaic (PV) system intended for powering an in-situ Internet of Things (IoT) station, a focused laboratory investigation is necessary. This study aims to determine the optimal utilization of a specific solar panel for powering a soil characteristic monitoring station. Furthermore, this paper will explore the potential and evolutionary trends of IoT applications, encompassing communication technologies and air-to-ground sensors, in enhancing the operating point of in-situ mounted PV systems. The results demonstrate the suitability of the proposed PV system for the autonomous powering of the soil monitoring station. This presents a promising solution for realizing an autonomous IoT pilot station, integrating a solar photovoltaic system for intelligent remote and real-time monitoring of soil parameters.

Keywords: Photovoltaic panel · IoT · Autonomous monitoring stations

1 Introduction

Within the global imperative of escalating energy consumption and the critical need to mitigate greenhouse gas emissions, the shift towards renewable energy sources has become paramount. Solar energy, owing to its widespread availability and inherent sustainability, is emerging as a compelling solution for electricity generation. Photovoltaic (PV) systems, which harness the photoelectric conversion of solar radiation, have undergone substantial technological advancements in recent decades. These advancements are

D. D. Cioboată and J. Machado (Eds.): ICoRSE 2025, LNNS 1592, pp. 316–330, 2025.
https://doi.org/10.1007/978-3-032-02508-1_27

characterized by increasing conversion efficiencies and a broadening scope of large-scale deployment, spanning from centralized utility-scale solar power plants to decentralized in-situ micro-generation units.

Given the escalating integration of solar energy into the global energy infrastructure, the optimized deployment of photovoltaic (PV) systems is now critical. Ground-mounted and rooftop PV systems represent two prevalent solutions, each presenting unique advantages and limitations. Ground-mounted systems facilitate the installation of larger PV arrays and offer design flexibility in terms of orientation and tilt angles, thereby maximizing solar energy capture. Conversely, rooftop systems leverage existing urban infrastructure, minimizing land use requirements. Looking ahead, the incorporation of Internet of Things (IoT) applications will be instrumental in enhancing the efficiency of both system types. This will be achieved through real-time performance monitoring and control, predictive fault analysis, and the dynamic optimization of energy generation based on prevailing weather patterns and consumption profiles.

This research endeavors to investigate the potential and developmental trends of Internet of Things (IoT) applications, encompassing both communication technologies and air-to-ground sensors, in optimizing the efficiency and sustainability of in-situ mounted photovoltaic (PV) systems designed to power ground monitoring stations. Initially, the impact of installation variables on PV system performance will be examined, with a focus on analyzing factors such as tilt angle, azimuthal orientation, and shading losses. Subsequently, the study will analyze the operational characteristics of photovoltaic systems in in-situ applications, alongside methodologies for determining key photovoltaic cell parameters.

2 IoT Technologies Specific to Autonomous Monitoring Stations

The Internet of Things (IoT), when applied to autonomous monitoring stations, constitutes a network of physical devices—including electronic objects embedded with sensors, communication protocols, and other enabling technologies. This infrastructure allows these devices to interconnect via a platform utilizing the Wide Area Network (WAN) to facilitate data exchange. Specifically, these sensors are designed to acquire environmental data from both terrestrial and subsurface domains.

By integrating sensor technology with telecommunications infrastructure, Internet of Things (IoT) devices can enable value-added services that would otherwise be unattainable [1]. These services encompass remote monitoring, automation, control functionalities, and advanced data analytics. Consequently, IoT finds applications across a diverse spectrum of domains, including smart homes, smart cities, agriculture, healthcare, industrial processes, and transportation [2]. In the context of this research, the focus is on leveraging these capabilities to determine the aridification trends of the specific geographical areas under observation.

2.1 Sensors Specific to Soil Monitoring

Sensors for monitoring soil parameters are foundational elements of the IoT monitoring station, responsible for acquiring real-world data from pilot ground installations

[3]. These sensors are capable of measuring a diverse array of parameters, including but not limited to temperature, humidity, motion, pressure, and light, as illustrated in Fig. 1. The integration of these sensing capabilities enables IoT devices to monitor and respond dynamically to environmental variations [4]. This integration is a key driver of innovation, facilitating the development of novel IoT applications and services, enhancing efficiency and productivity, enabling the automation of processes, and optimizing real-time data acquisition operations in field deployments [5].

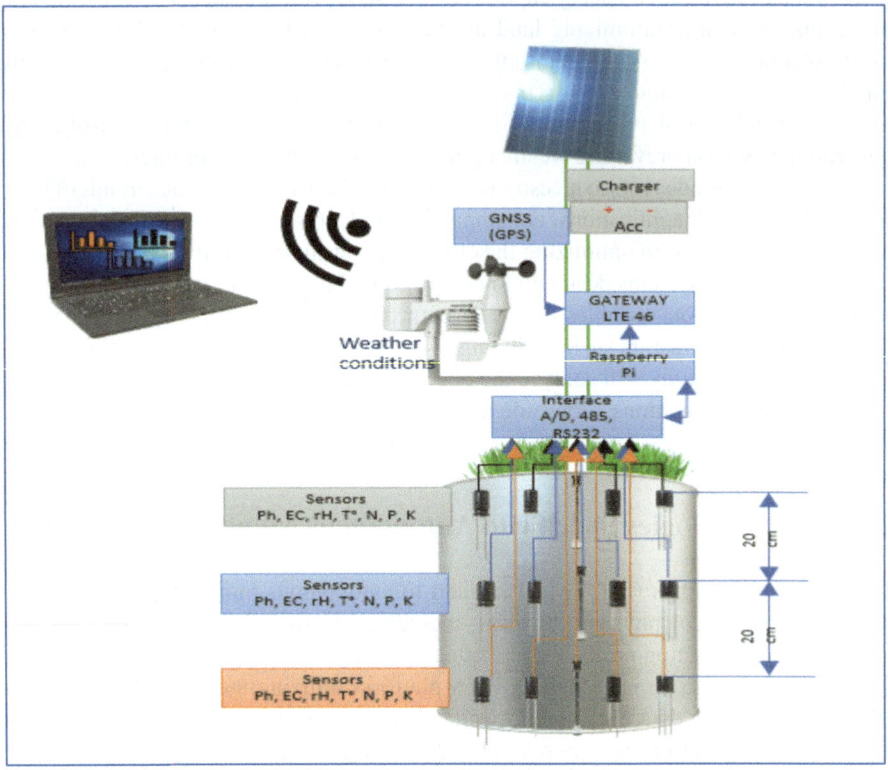

Fig. 1. In situ monitoring station using IoT technology

Cloud computing infrastructures offer critical solutions for managing the substantial data volumes generated by smart meters and other intelligent grid components, providing efficient storage, energy resource management capabilities, and virtual power plant functionalities. While the Cloud paradigm presents numerous advantages for processing IoT-derived data, the inherent latency due to the geographical separation between edge devices and centralized data centers restricts its applicability in scenarios demanding immediate responsiveness, such as real-time streaming analytics.

2.2 Communication Technologies

Data acquired from both ground-based and aerial sensors can be transmitted across a network utilizing a range of IoT connectivity technologies. These include Wi-Fi, Bluetooth, Zigbee, LTE (4G or 5G), and LPWAN (low-power wide-area wireless networks), facilitating communication between devices and with cloud-based platforms [6].

The Cloud platform serves as a central infrastructure for storing, processing, and analyzing the data streams generated by IoT devices. Furthermore, it offers a suite of tools and services for comprehensive device management, insightful data visualization, and streamlined application development.

Technologies specific to communication protocols:

- LPWAN (Low-Power Wide-Area Network): Low-power, long-range wireless communication technologies such as Lora WAN and NB-IoT, which are ideal for IoT applications with low data consumption and long battery life.
- Edge computing: Processing data at the edge of the network, closer to IoT devices, to reduce latency and improve performance.
- Artificial Intelligence (AI) and Machine Learning (ML): Using AI and ML algorithms to analyze IoT data, identify patterns, and make intelligent decisions.
- Blockchain: Using blockchain technology to ensure the security and integrity of IoT data.
- Flash/SD memory for local data storage in case of connection loss.

The Internet of Things (IoT) is characterized by continuous and rapid evolution, with new technologies and applications emerging consistently. Consequently, IoT can be effectively implemented in isolated microgrid systems, particularly in scenarios demanding uninterrupted power supply. Within such systems, all networked devices possess the capability for mutual interaction. In the context of our specific application, we have selected LTE 4G technology for its communication capabilities.

3 Laboratory Characterization of Photovoltaic Panel Performance

Figure 2 illustrates a simplified block diagram, outlining the interconnection pathway of components that enable the utilization of energy generated by a solar panel within the experimental monitoring station under investigation.

The performance of solar panels is significantly influenced by their azimuth and elevation angles, which directly impact the incident solar irradiation and consequently, the electrical energy output. Optimizing these parameters is crucial for maximizing system efficiency [7]. Conversely, deviations from the optimal values can lead to a substantial degradation in performance, potentially reaching 25–30% under unfavorable conditions. As demonstrated in studies [8] such as:

- Achieving optimal orientation for solar panels can enhance annual energy production by as much as 15–20% compared to suboptimal configurations. Furthermore, each degree of deviation from the ideal tilt angle can result in an approximate 0.5% reduction in efficiency.

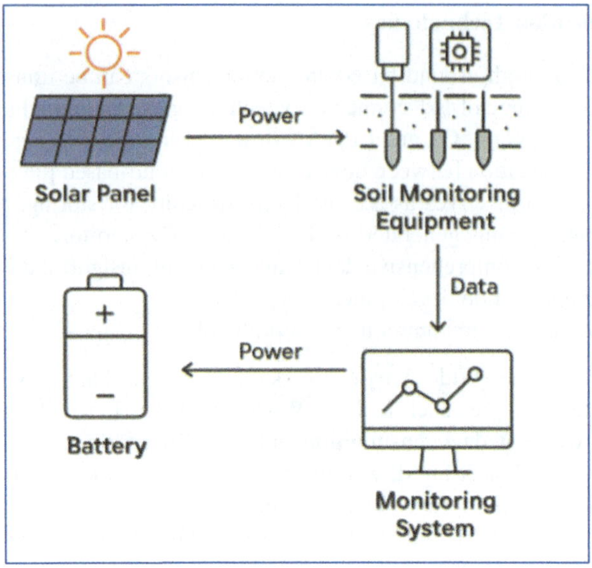

Fig. 2. Block diagram of the PV system for powering the soil monitoring station

- Properly oriented panels can yield up to 30% more energy during the winter months, a period often characterized by higher energy consumption.

To thoroughly characterize the electrical performance, efficiency, and behavior under varying conditions of the 12V and 60W solar panel in a laboratory setting, a series of targeted tests can be conducted.

Figure 3 illustrates the integration of the ESP32-S3 microcontroller with the LTE 4G SIM7670G module, highlighting the interconnection and key components involved:

- SIM7670G requires 3.8V–4.2V, and consumption can reach 2A at peak.
- An LDO or a 12V to 4V/3A buck converter is required (e.g. MP2315).
- Low ESR capacitors are needed close to the module to prevent voltage drops.

Therefore, to effectively utilize a PV panel with a nominal power of 60W and an output voltage of 20V, it necessitates the selection of a charge controller compatible with the $LiFePO_4$ battery. For comprehensive energy management, a system comprising the following was implemented:

1 MPPT controller from 20V to 12V that ensures efficient battery charging. Two models were tested:
  - EPEVER MPPT 10A (Tracer1210AN)
  - Victron Smart Solar MPPT 75/10
2. Over discharge protection – To prevent battery damage, an integrated or separate BMS (Battery Management System).
3. DC-DC converter
  - 5V (for microcontroller and sensors)
  - 3.3V (if you have sensors that require this voltage)

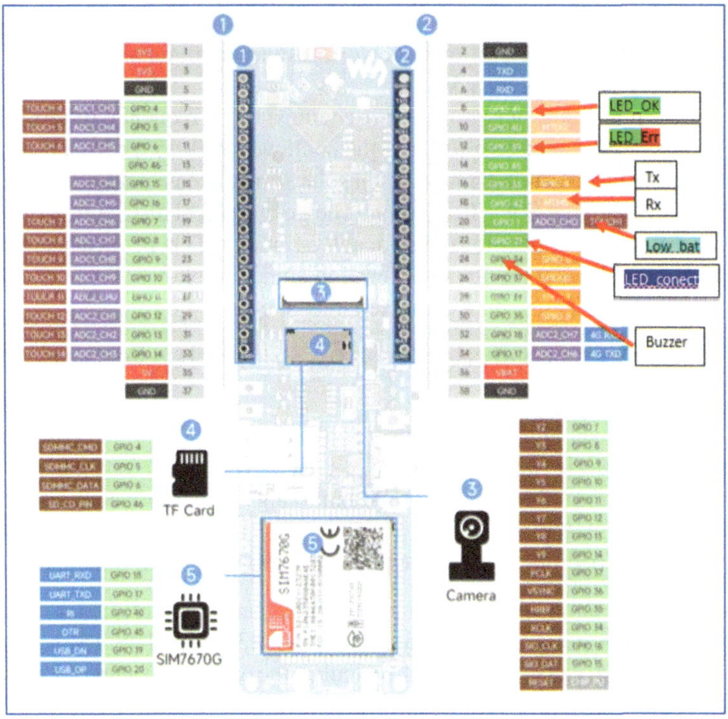

Fig. 3. Wiring diagram for integrating ESP32-S3 with SIM7670G

4. Voltage and current monitoring – An INA219 or similar module to track battery status and system consumption.

Communication with the LTE module is established via UART (Universal Asynchronous Receiver-Transmitter) protocol for the transmission of AT commands, with an optimal baud rate configured at 115200 bits per second. The communication sequence is as follows:

- The ESP32-S3 GPIOs are on 3.3V, and the SIM7670G works with 1.8V/2.8V logic level converter (ex: TXB0108).
- Deep sleep mode has been enabled for ESP32-S3 and puts SIM7670G in low power mode when not transmitting.
- Load data at regular intervals to save energy (e.g. every 30 min.).

3.1 Calculating the Azimuth Angle

Within the geographical context of Romania, situated in the northern hemisphere, orienting photovoltaic panels towards the south to optimize azimuth is a fundamental strategy for maximizing incident solar radiation. This southward alignment ensures the most effective capture of the daily photonic flux. However, the following considerations are crucial [9]:

- Minor Azimuth Deviations: Deviations of up to 20° from the ideal southern azimuth, whether towards the southeast or southwest, result in a minimal reduction in energy production, typically less than 5%.
- Regional Optimization: In regions experiencing a high frequency of morning fog, a slight deviation in orientation towards the southwest (10–15°) can optimize energy capture during the afternoon hours, when solar irradiance is more consistent.
- Variability of Regional Solar Irradiation: The geographical distribution of solar irradiation across Romania exhibits significant variations. This necessitates the careful adaptation of both the azimuth and elevation angles of photovoltaic panels to align with the specific solar resource characteristics of each region.

Determining the optimal tilt angle for photovoltaic panels in Romania is influenced by several key factors:

- Geographic Latitude of the Location;
- Seasonal Variations in the Sun's Path;
- Specific Objectives of the Installation (Maximizing Annual Production Vs. Optimizing for the Cold Season);
- Local Climatic Conditions (E.G., Frequency of Snowfall in Mountainous Areas).

Calculating the optimal tilt angle for photovoltaic panels in three geographical areas.

Given Romania's geographical location between 43°37' and 48°15' north latitude, several methodologies can be employed to calculate the optimal tilt angle for photovoltaic panels across different regions:

Simplified method: Tilt angle = Location latitude.

For Bucharest (44°26' N), this would mean approximately 44°.

Seasonal method: Summer: Latitude - 15° (approximately 29° for Bucharest).

Winter: Latitude + 15° (approximately 59° for Bucharest).

$$\sphericalangle Opt = 3.7 + 0.69\ \mathrm{L} \tag{1}$$

where:

$\sphericalangle Opt$ is the optimal angle and L is the latitude.

Applying this formula for Bucharest, located at approximately 44.43° N latitude, yields an optimal tilt angle of approximately 44.43° − 10° = 34.43°, which aligns with the value you provided.

Optimizing both the azimuth and elevation angles of photovoltaic panels not only maximizes the efficiency of solar energy conversion into electricity but also generates a substantial economic impact, primarily manifested by:

- Reduction of the investment payback period: Under conditions of optimal solar irradiation achieved through correct panel orientation, the time required to recoup the initial investment can be significantly shortened.
- Increase of economic efficiency: By maximizing energy generation, the operational costs associated with electricity consumption are reduced, leading to enhanced economic efficiency.

Furthermore, the implementation of photovoltaic systems with high energy efficiency contributes to extending the operational lifespan and autonomy of in-situ pilot ground stations (Fig.4).

Precise determination of the azimuth and elevation angles for photovoltaic panels is a critical design and implementation consideration for solar energy systems. A thorough analysis and accurate execution of these parameters can distinguish between a system with average performance and one that delivers maximum long-term economic and energy benefits. To calculate the optimal tilt angle of the photovoltaic panel for three regions of interest concerning the stage of soil aridification, MATLAB was utilized, and the resulting values are presented in Table 1:

% Usage examples:

Latitude Bucharest = 44.43; % Latitude Bucharest.

Latitude_Dolj = 43.43; % Latitude_Dolj.

Latitude of Giurgiu = 43.53; % Latitude of Giurgiu with results:

Table 1. The optimal annual angle for 3 locations in Romania

Angle	*Bucharest*	*Dolj*	***Giurgiu***
Optimal annual angle	44.43°	43.43°	43.53°
Optimal winter angle	68.987°	68.087°	68.177°
Optimal summer angle	16.487°	15.587°	15.677°

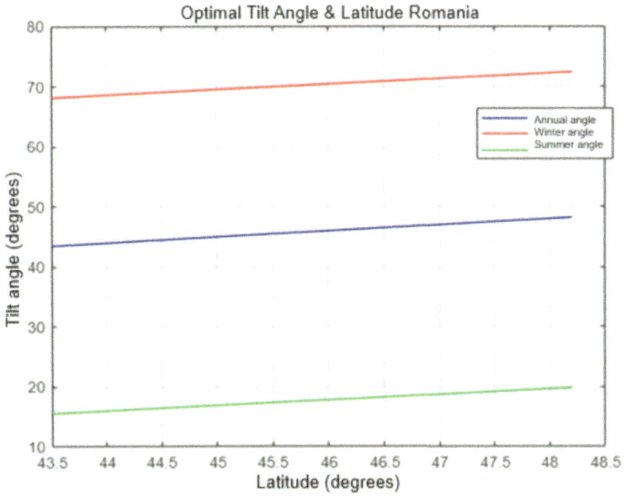

Fig. 4. Graphical representation of the optimal PV tilt angle in Romania

Research conducted within the geographical context of Romania has identified the following key correlations regarding the impact of the elevation (tilt) angle of photovoltaic panels on energy efficiency [10]:

- Minor deviations: Deviations of up to 5° from the optimal elevation angle cause a negligible reduction in annual energy production, below 1%.
- Moderate deviations: An underestimation of the elevation angle by 10° induces a decrease in annual energy production of between 1.5% and 3%.
- Horizontal installation: Photovoltaic panels installed at an elevation angle of 0° (horizontal) can suffer a loss of up to 15% of their production potential, compared to those installed at the optimal elevation angle.

For the majority of residential and commercial photovoltaic installations across Romania, a fixed tilt angle within the range of 30° to 40° is generally recommended. This range provides a practical compromise, ensuring reasonably good energy production levels across both summer and winter seasons, taking into account the seasonal variations in solar irradiance at this latitude [11].

3.2 Solar Panel Evaluation

Solar irradiance, representing the power density of solar radiation incident on a surface, is a critical parameter for photovoltaic system performance [12]. The average extraterrestrial irradiance at the top of Earth's atmosphere is approximately 1.36 $\frac{\mathrm{kW}}{\mathrm{m}^2}$. Due to the Earth's elliptical orbit around the Sun, this value experiences a variation of ± 3.4% throughout the year. The total solar power continuously incident on the Earth's atmosphere is estimated to be around 1.75 × 10^5TW. Considering an atmospheric transmission rate of 60%, approximately 1.05 × 10^5TW of solar power continuously reaches the Earth's surface in Bucharest [13].

From a solar energy resource perspective, Romania is favorably situated, experiencing approximately 210 sunny days annually. The country receives a total solar irradiance ranging from 1,000 to 1,300 kWh/m2/year, with an estimated technically exploitable potential of 600 to 800 kWh/m2/year [14]. This indicates a significant capacity for solar energy generation within the region of Bucharest. [15].

The regions of Northern Dobrogea and Oltenia represent the areas with the highest solar energy potential in Romania, exhibiting an average annual solar irradiance of 1,600 kWh/m2/year. This highlights these regions as particularly favorable for solar energy harvesting [16].

To determine the forecasted production, for each individual consumption point, the calculation relationship can be applied:

$$W_{month} = A \cdot n_{days} \cdot \eta_n \cdot I_m \cdot C_p \quad [KWh_{mouth}] \tag{2}$$

were:

- *A[m2] is the total area of the PV installation,*
- *η_n is the nominal efficiency of the PV panels,*
- *Im [kWhm^{2}days] is the monthly average daily irradiance, determined using solar radiation sensors*
- *n_{days} is the number of days in the month,*
- *Cp is the coefficient of performance of the PV system.*

The coefficient of performance (COP) serves as a crucial qualitative metric for photovoltaic (PV) systems, providing an indication of the system's overall performance, independent of specific installation parameters such as orientation, tilt angle, or incident irradiance. The COP inherently accounts for all power and energy losses within the PV system [17], with the most significant contributors being:

- Inverter losses (ΔPinvertor) – between 4% and 10%,
- Photovoltaic cell losses due to temperature (ΔPtemp) – between 5% and 20%,
- Shading losses (ΔPumbră) – between 0% and 80%, specific to each geographical area, taking into account the degree of sunny days,
- Losses due to inverter operation at low load (ΔPmin, inverter) – between 3% and 7%,
- Losses due to dust, snow, atmospheric pollution, or other climatic conditions that can lead to the coating of photovoltaic cells with solid particles of materials (ΔP) – approx. 2%,
- Other power and energy losses, unquantifiable.

To ascertain the necessary energy production for the operation of the IoT-SOL station, the nominal power rating of the proposed photovoltaic panels will be considered under standard test conditions (STC): an average irradiance of 1 kW/m2 and an average ambient temperature of 20 °C. For the sake of simplicity in this initial analysis, the dependence of the panels' efficiency on variations in average ambient temperature has been neglected. It's worth noting that for many photovoltaic panels, this temperature coefficient typically has a value of approximately −0.4% per degree Celsius deviation from the STC temperature.

3.3 Methods for Determining Photovoltaic Cell Parameters

Accurate determination of photovoltaic cell parameters is essential for several key reasons: achieving high-efficiency photovoltaic panels, maximizing their operational lifespan, and enabling precise estimations of the energy generated by these panels under real-world operating conditions. Understanding these parameters allows for optimized system design and performance prediction.

Photovoltaic panel manufacturers typically determine critical performance parameters and their associated temperature coefficients under Standard Test Conditions (STC). A secondary standard, Nominal Operating Cell Temperature (NOCT), is also recognized, particularly in research settings, although it sees less widespread adoption in industrial applications. The specific conditions defined for these standards are outlined below:

- STC (Standard Test Conditions) – the measurement conditions are: irradiance I = 1000W/m2, temperature T = 250C, and air mass AM = 1.5;
- NOCT (Nominal Operating Cell Temperature) – the measurement conditions are: irradiance IT = 800 W/m2, temperature T = 200C, wind speed is 1m/s, and the inclination angle of the photovoltaic cell or panel is 45°.

Table 2 presents a selection of representative photovoltaic cell types that have been developed primarily within research laboratories, rather than in large-scale industrial production environments. These examples often showcase cutting-edge materials and

Table 2. Efficiency of the most representative types of photovoltaic cells under standard measurement conditions

Cell type	Efficiency [%]	Producer
Monocrystalline silicon	25.8	FhG-ISE
Multi crystalline silicon	22	Jinko solar
GaAs – thin film	28.8	Alta Devices
CIGS	21.7	Solar Frontier
CdTe	21	Toshiba
Multi-junctionInGaP/GaAs/InGaAs	37.9	Sharp
Perovskite	20.9	KRICT

innovative designs aimed at pushing the boundaries of solar energy conversion efficiency [18].

Figure 5 provides a schematic representation of the battery charging circuit integrated within the controller, which is powered by the photovoltaic panel. This diagram illustrates the flow of energy from the PV panel to the battery and the control mechanisms involved in the charging process.

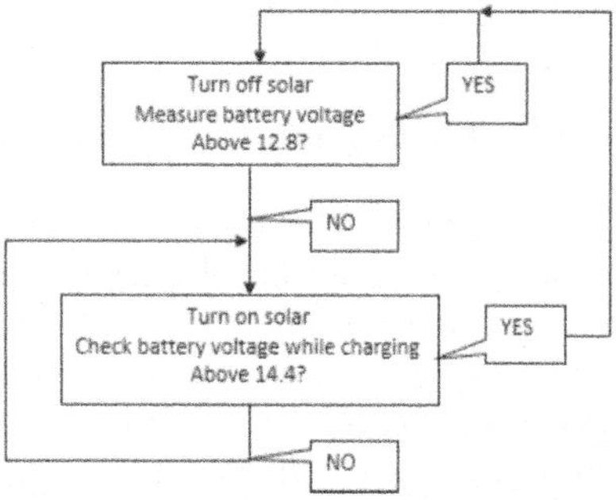

Fig. 5. Solar panel charger flow diagram

To experimentally characterize the power output of the solar panel in a laboratory setting, the following experimental setup was constructed, as depicted in Fig. 6. This setup allows for controlled measurements of the panel's voltage and current under varying load conditions, enabling the determination of its power characteristics.

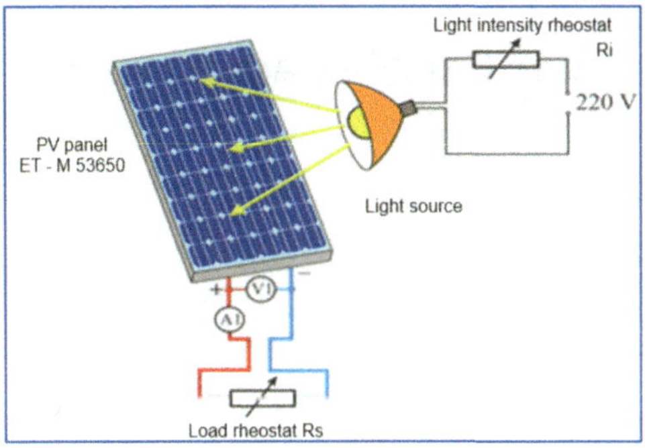

Fig. 6. Generic schematic of laboratory experiment setup

4 Results

Table 3 presents the data obtained from the measurements conducted using the experimental setup illustrated in Fig. 7. This table likely includes corresponding values of voltage, current, and potentially power output of the solar panel under the tested conditions.

Table 3. Experimental results for various loads (R load).

Rs [Ω]	U out [V]	Iout [A]	Power [W]
60.00	21.60	0.00	0.00
30.00	20.50	1.06	21.75
20.00	19.20	1.50	28.75
15.00	18.10	1.89	34.17
13.00	17.50	2.11	36.86
11.00	16.80	2.39	40.10
9.00	15.80	2.75	43.38
7.00	14.20	3.17	44.97
6.00	13.20	3.43	45.30
0.00	0.00	5.62	0.00

Figure 7 visually and clearly indicates the maximum power point (MPP) of the solar panel under the tested conditions, revealing a peak power output of 45.3 W. This graphical representation aids in understanding the optimal operating point for maximizing energy extraction from the panel.

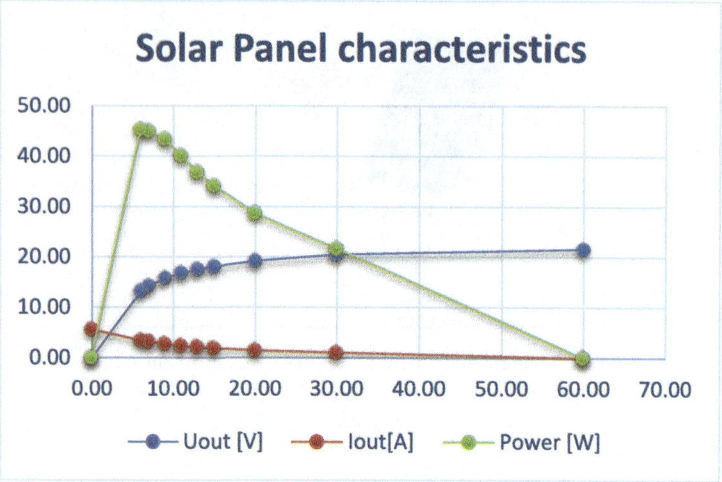

Fig. 7. Power chart representation function of load Rs

5 Conclusions

Based on the experimentally determined values under a reduced irradiance of 500 W/m2, the photovoltaic device exhibits the following key characteristics: an open-circuit voltage (Voc) of approximately 21.6 V and a short-circuit current (Isc) of roughly 3.6 A. The maximum power point (MPP) was observed at a voltage (Vmp) of 13.2 V and a current (Imp) of 2.20 A, yielding a maximum power output (Pmax) of 29.04 W.

Considering the reduced irradiance conditions (500 W/m2, half of the Standard Test Conditions - STC of 1000 W/m2) and an assumed device area of 0.46 m2, the calculated efficiency (η) is approximately 12.6%. This value is consistent with the expectation of a lower power output and consequently lower efficiency under reduced irradiance.

The provided note suggests that under STC (1000 W/m2), an idealized efficiency of around 25% could be theoretically achieved. However, it is crucial to acknowledge that real-world efficiencies under STC are typically lower, estimated to be in the range of 17–20%. This discrepancy arises due to various factors including, but not limited to, temperature coefficients, series and shunt resistances within the device, and optical losses that are not fully accounted for in idealized calculations.

Therefore, while the experimental results at 500 W/m2 provide a baseline understanding of the device's performance, extrapolating to STC requires careful consideration of these non-ideal factors. The estimated real-world efficiency range of 17–20% under STC offers a more realistic expectation of the device's performance under standard operating conditions. Further investigation and characterization under STC would be necessary to precisely determine the device's efficiency and validate these estimations.

Results extracted from experimentally determined values:

- Voc $\approx$ 21.6V
- Isc $\approx$ 3.6A
- Vmp $\approx$ 13.2V

- Imp ≈ 2.20A
- Pmax ≈ 29.04W

(Note: values are below specification due to laboratory testing at 500 W/m2, i.e. half of the STC irradiance, i.e. power is ~ 50% lower).

The efficiency (yield) calculated at 500 W/m2 is:

$\eta = 29.04W500{\cdot}0.46 = 29.04230 \approx 12.6\%$ (3).

At 1000 W/m2 it is estimated that around 25% can be achieved, but this is idealized (in reality, the real efficiency is 17–20%).

Acknowledgements. This work is supported by the Ministry of Education and Research - National Research Authority (Romania), CCDI – UEFISCDI, Project number: PN 23 43 03 01 - 17N/2023, and PN 23 43 05 02 - 17N/2023, within PNCDI IV - NUCLEU Program.

References

1. https://nswits.com.au/iot-in-telecom-industry/. Accessed 20 Apr 2025
2. https://www.ibm.com/think/topics/internet-of-things. Accessed 21 Apr 2025
3. Afridi, W.A.K., Vitoria, I., Mukhopadhyay, S.C.: Development of IoT soil sensor node: installation and modeling. In: Pradhan, B., Mukhopadhyay, S. (eds) IoT Sensors, ML, AI and XAI: Empowering A Smarter World. Smart Sensors, Measurement and Instrumenta-tion, vol 50. Springer, Cham (2024). https://doi.org/10.1007/978-3-031-68602-3_5
4. Elisabeta, N., Dimitrescu, A., Alecusan, A.M., Babis, C., Chivu, O., Dascalu, L.: Structural – Functional Analysis of the Soil Level Machine, pp.94–99 (2018), Reliability Journal, Tg.Jiu, Romania, Durability - Friability & Durability No 2/ 2018 "Academica Brâncuşi" Publishing House, Târgu Jiu, ISSN 2344 – 3669, ISSN-L 1844 – 640X
5. Landaluce, H., Arjona, L., Perallos, A., Falcone, F., Angulo, I., Muralter, F.: A review of IoT sensing applications and challenges using RFID and wireless sensor networks. Sensors **20**, 2495 (2020). https://doi.org/10.3390/s20092495
6. Eleonora, T., Eugenia, T., Maria, G.M.: Internet of Things in energy: technologies, applications, challenges and future trends. Rom. J. Inf. Technol. Autom. Control **31**(2), 45–58 (2021). https://doi.org/10.33436/v31i2y202104, ISSN 1220–1758
7. Rotaru, A., Luciana D.: Reaction Forces in parallel mechanism pairs. Rom. J. Tech. Sci. Appl. Mech. **61**(3), 261–270 (2016)
8. Ioan, V.I.B.: Optimization solutions of photovoltaic energy conversion into electricity. Bacau (2011). https://doi.org/10.13140/RG.2.1.3856.8160
9. Boncătă, A.E., Nechita, R.M., Ulerich, O., Cristoiu, C.: Transforming libraries. The role of autonomous robots in sorting, inventorying and book search. Int. J. Mechatron. Appl. Mech. (17), 87–95 (2024). https://doi.org/10.17683/ijomam/issue17.10
10. Silaghi, D.: Theoretical and experimental research on the possibilities of harnessing solar energy as a non-polluting source. Thesis for the scientific title of Doctor of Engineering at the "Politehnica" University of Timişoara in the field of Mechanical Engineering (2008)
11. https://www.electricon.ro. Accessed 04 Apr 2025
12. Usman, Z., Tah, J., Abanda, H., Nche, C.: A critical appraisal of PV-systems' performance. Buildings **10**, 192 (2020). https://doi.org/10.3390/buildings10110192
13. http://geomodelsolar.eu. Accessed 06 Apr 2025
14. https://www.rri.ro/en/features-and-reports/green-planet/solar-energy-in-romania-id622498.html. Accessed 08 Apr 2025

15. resentation Memorandum for the "SIMIN" project, chrome-extension://efaidnbmnnnibpcajpcglclefindmkaj/https://www.meteoromania.ro/wp-content/uploads/comunicate/Memoriu-Prezentare.pdf
16. Irena, R.M., Bianca, M., Bianca, M.: Consequences of setting up photovoltaic parks - Related land use/land cover changes in Giurgiu County rural area (Romania). Carpathian J. Earth Environ. Sci. **10**, 211–218 (2015)
17. Popescu, V.S., Popescu, I.M.: Impact of UV post processing on simple 3D printed parts using masked stereolithography. Int. J. Mechatron. Appl. Mech. **2024**(18), 86–92 (2024). https://doi.org/10.17683/ijomam/issue18.10
18. Green, M.A., Hishikawa, Y., Dunlop, E.D., Levi, D.H., Hohl-Ebinger, J, Ho-Baillie, A.W.Y.: Solar cell efficiency tables (version 52). Progress Photovoltaics: Res. Appl. **26**, 427–436 (2018)

Analysis of Motors Equipped on an Omnidirectional Platform from the Perspective of the Commutation Signal

Daniel Balc[1], Aurel Mihail Titu[2,3](✉), Emanuel Balc[2], and Ligia-Luminița Petrescu[4]

[1] National University of Science and Technology POLITEHNICA Bucharest, 313 Splaiul Indpendenței, București, Romania

[2] Lucian Blaga University of Sibiu, 10 Victoriei Street, Sibiu, Romania
mihail.titu@ulbsibiu.ro

[3] Academy of Romanian Scientists, 3 Ilfov Street, Bucharest, Romania

[4] National Institute of Research - Development for Mechatronics and Measurement Technique - INCDMTM, Bucharest, Romania

Abstract. This paper presents an analysis conducted on direct current (DC) electric motors used in an omnidirectional mobile platform, with a focus on the commutation signal as a primary indicator of their dynamic behavior and functional condition. Omnidirectional platforms provide significant advantages in mobile robotics applications due to their ability to move in any direction without requiring a change in orientation, a capability made possible by the use of four individually controlled DC motors.

The study emphasizes the commutation signal as a diagnostic tool for evaluating motor performance, as it directly reflects the interaction between the motor's mechanical and electrical components. The analysis aims to identify key variations in current, partial short circuits, and non-uniformity in the commutation process between the brushes and the commutator segments. Furthermore, the behavior of multiple motors mounted on the same platform is compared in order to assess operational uniformity and its influence on the platform's omnidirectional motion.

Keywords: Motor · Commutation waveform analysis · Signal acquisition · Brush-commutator interaction · DC motor diagnostics

1 Introduction

In the context of the accelerated development of autonomous robotic systems, omnidirectional mobile platforms are becoming increasingly prominent in applications that demand enhanced mobility, high maneuverability, and operational flexibility in complex environments or confined spaces [1]. Unlike traditional vehicles equipped with two or four fixed wheels, omnidirectional platforms can move in any direction without requiring a change in orientation. This feature offers significant advantages in applications such as internal logistics, collaborative robots (cobots), autonomous vehicles for industrial environments, mobile medical assistance systems, as well as in research and education [2].

D. D. Cioboată and J. Machado (Eds.): ICoRSE 2025, LNNS 1592, pp. 331–339, 2025.
https://doi.org/10.1007/978-3-032-02508-1_28

The operation of such a platform involves a number of critical technical requirements, including precise motion coordination, wheel synchronization, stability during rapid maneuvers, and, importantly, the reliability of the electric motors that provide propulsion. Typically, these platforms use four direct current (DC) electric motors, each connected to a mecanum or omni wheel, enabling lateral, diagonal, or in-place rotational movement [3]. Each motor operates independently, yet must be fully integrated into a control system that ensures overall synchronization of the platform.

In this context, careful monitoring of each motor's behavior becomes essential, along with a detailed analysis of parameters that can provide insight into the motor's operational condition and potential faults over time. One such relevant parameter is the commutation signal, generated within the DC motor during operation [4]. The commutation signal is generated by the interaction between the brushes and the commutator and can reveal important details about the quality of electrical contact, the mechanical state of internal components such as brush wear or rotor imbalance and the motor's overall dynamic behavior during operation. Analyzing the commutation signal provides a practical and non-invasive way to assess motor performance, allowing valuable insights to be gained without the need for disassembly or intrusive inspection methods [5, 6]. In this study, an oscilloscope was used to acquire the commutation signal in real time, allowing for waveform shape, amplitude, and frequency analysis.

In addition, consistent performance across all motors is essential in omnidirectional platforms, as even small differences can lead to unwanted effects such as deviations from the intended trajectory, increased vibrations, or instability during movement [7]. This paper aims to highlight the potential of using the commutation signal as an analysis method for DC motors integrated into an omnidirectional platform [8, 9]. The theoretical background of commutation signal generation is presented, along with an examination of how it is influenced by both mechanical and electrical parameters of the motor.

2 Structural and Functional Overview of the Analyzed System

The system analyzed in this study is a prototype of an omnidirectional mobile platform, specifically developed for applications that require precise and flexible movement in multiple directions. The platform relies on Mecanum wheels, which are widely used in mobile robotics due to their ability to enable holonomic motion without the need to change orientation. However, unlike conventional wheels, Mecanum wheels involve complex dynamics that are still not fully understood [10, 11]. The configuration consists of four direct current (DC) electric motors, each driving a mecanum wheel, thereby enabling movement in any direction without altering the platform's orientation. The system layout is illustrated in Fig. 1.

The omnidirectional mobile platform examined in this study incorporates a straightforward yet effective electric drive system, specifically designed to enable both individual and synchronized control of each mecanum wheel. This configuration is built upon a hardware setup composed of standard electronic components, carefully selected and arranged to support multidirectional mobility as well as Bluetooth command capabilities. The primary purpose of this system architecture is to facilitate experimental testing of DC motor behavior particularly from the perspective of commutation signal analysis under controlled and reproducible conditions.

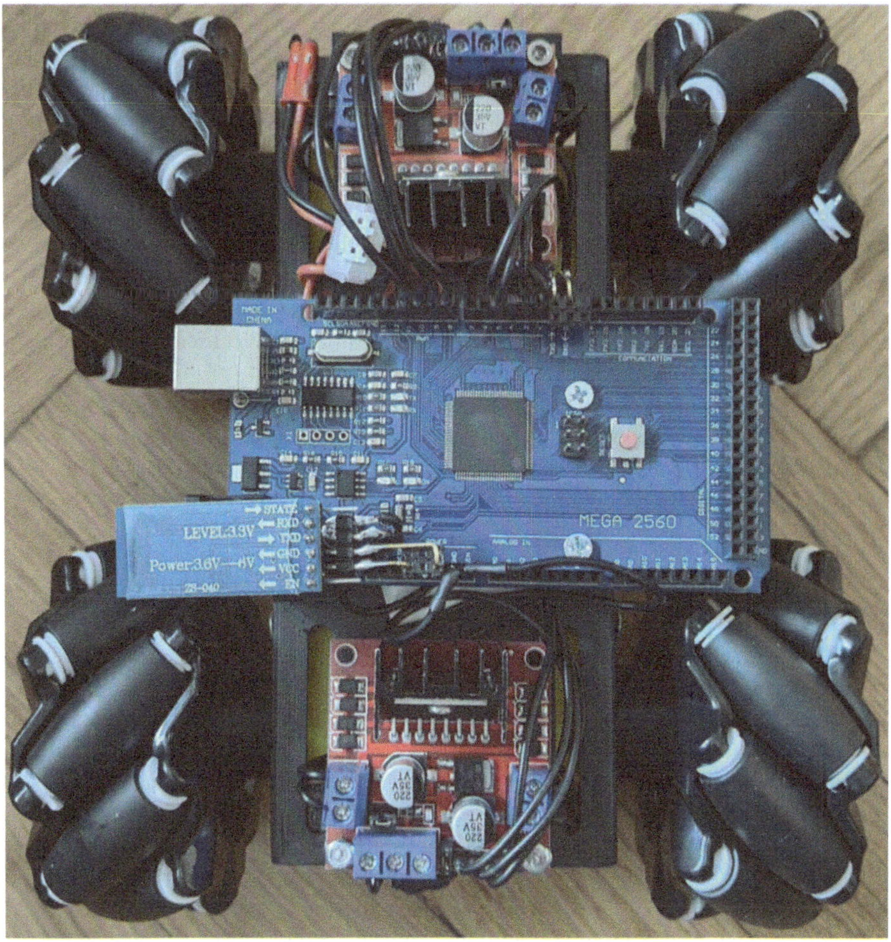

Fig. 1. Omnidirectional platform

At the heart of the control system lies the Arduino Mega 2560 development board, a versatile microcontroller equipped with a sufficient number of digital and analog I/O pins to manage the four DC motors in parallel, while also handling external communication tasks. The Arduino Mega is tasked with generating PWM control signals, receiving user input via Bluetooth, and transmitting appropriate commands to the motor driver modules.

Motor control is achieved through two L298N dual H-bridge driver modules, each capable of controlling two DC motors independently. These drivers enable bidirectional rotation and speed modulation via PWM input, allowing precise manipulation of motor behavior. With two modules, the system can control all four motors with minimal hardware overhead, eliminating the need for more complex switching arrangements or additional electronic resources.

On the communication side, the platform integrates a HC-05 Bluetooth module, connected to the Arduino through a UART interface. This module enables wireless command transmission from a mobile device or computer, thereby allowing real-time remote control of the platform. During experimental procedures, the wireless link proves especially useful by removing the constraints of physical cabling and allowing dynamic testing across various operating conditions, including while the platform is in motion.

The four 6V DC motors mounted on the chassis are mechanically coupled to mecanum wheels, which provide the unique capability for omnidirectional movement including lateral, diagonal, and rotational motion depending on the speed and direction commands assigned to each wheel. The torque output of the motors is sufficient to drive the unloaded platform, making it suitable for evaluating the electrical and mechanical behavior of each actuator during testing.

This modular and scalable architecture supports not only individual motor control but also the analysis of synchronized behavior among all actuators. Such an arrangement is particularly beneficial when assessing the consistency of commutation characteristics and the influence of motor coordination on the overall performance of the platform.

3 Evaluation Methodology of the Motors from the Perspective of the Commutation Signal

For data acquisition, an MDO4054C oscilloscope was used, which enabled the recording of signals from each of the four motors in a manner that allowed direct comparison in terms of commutation behavior.

The MDO4054C oscilloscope is an advanced measurement instrument designed for multifunctional analysis in the field of electronics and real-time systems. It is part of the Tektronix Mixed Domain Oscilloscope (MDO) series and is distinguished by its ability to integrate time and frequency domain measurements into a single device. The model features 4 analog channels, a bandwidth of up to 500 MHz, and a sampling rate of 2.5 GS/s per channel, providing high fidelity in capturing fast or noisy electrical signals.

One of the main advantages of this oscilloscope is its frequency domain analysis capability, as it is equipped with a built-in spectrum analyzer an essential feature for diagnosing complex signals and identifying interference. To perform the commutation signal measurement, the oscilloscope probe was connected to the positive terminal of the motor, with the probe ground connected to the negative terminal as can be seen in the Fig. 2.

To ensure a fair and consistent evaluation of motor behavior, each DC motor was individually removed from the platform and tested under no-load conditions. This approach was chosen to eliminate external mechanical influences such as drivetrain friction or uneven load distribution resulting from the platform's geometry. By testing each motor separately, it was possible to observe its electrical behavior particularly the commutation signal with minimal external interference.

The motors were powered by a regulated DC power supply, configured to deliver a constant voltage throughout the entire duration of each test. Maintaining stable power supply conditions was essential to ensure that any variations observed in the recorded

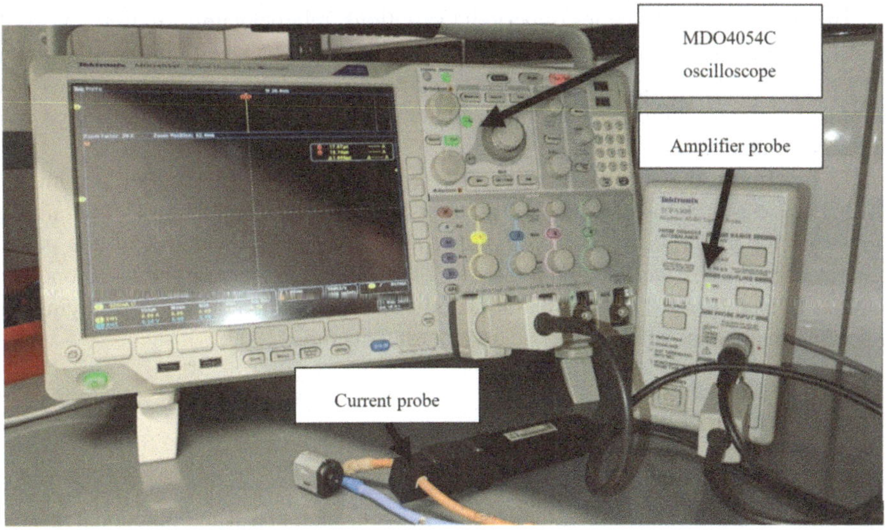

Fig. 2. Testing setup

signal reflected only the internal dynamics of the motor and not fluctuations from the power source.

All waveforms captured during testing were saved for subsequent analysis. Performing the tests under identical conditions for all four motors made it possible to draw meaningful comparisons between their behaviors. These results provided a foundation for drawing preliminary conclusions and for defining future directions of investigation related to commutation signal quality and motor consistency [12, 13].

In the context of commutation signal evaluation, the measurements produced a characteristic waveform, as illustrated in the Fig. 3.

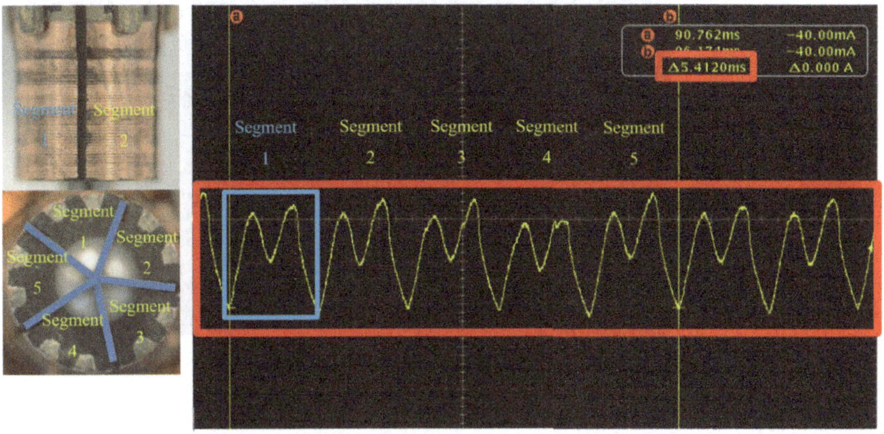

Fig. 3. Commutation signal correlation with rotor segments

The signal displays a series of periodic variations, relatively uniform in shape, each corresponding to a commutation event that is, the moment when current transitions from one commutator segment to the next as the motor shaft rotates. Each voltage pulse identified in the waveform reflects the switching action between two adjacent segments of the commutator [14].

To determine the motor's rotational speed, expressed in revolutions per minute (RPM), it is essential to know the total number of commutator segments. This parameter allows for the identification of a complete rotor revolution by counting all the pulses generated by successive commutations [15]. Once the time corresponding to one full rotation is established, it can be extrapolated to a one-minute interval by applying the following mathematical relationship: 60/recorded ms time*1000.

4 Results and Discussion

As a first step following the acquisition of oscilloscope measurements, each DC motor was carefully disassembled in order to perform a thorough visual and functional inspection of the internal components. This stage served a dual purpose. On one hand, it allowed for the precise identification of the number of commutator segments, a critical parameter required for accurate waveform interpretation and the calculation of rotational speed based on the number of commutation events per revolution.

On the other hand, the inspection focused on assessing the physical condition of key internal elements, including the commutator surface, the contact area between the brushes and the commutator, and the stator magnets. These components play a fundamental role in motor performance, and their wear or degradation can directly affect the quality and stability of the commutation signal observed during operation (Fig. 4).

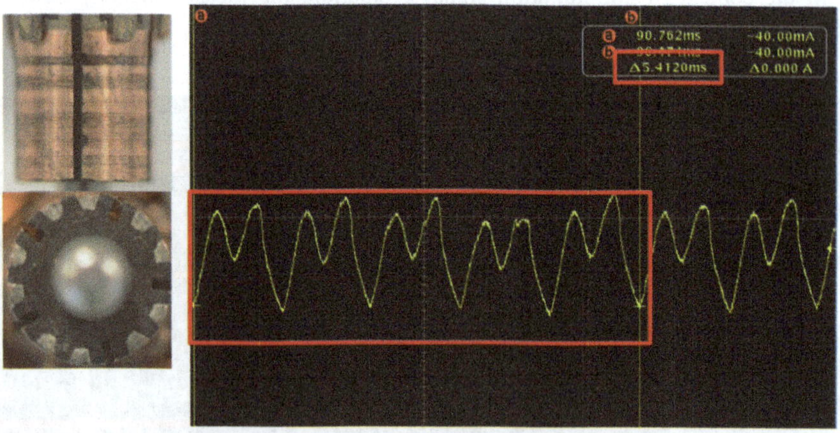

Fig. 4. Commutation signal for Motor 1 (11086 rpm)

Based on the recorded results, it can be inferred that Motor 1 is equipped with a commutator featuring five segments, which implies that a full mechanical rotation corresponds to five distinct commutation peaks observable in the waveform captured by the oscilloscope. These peaks highlighted with a red frame in the Fig. 5 represent the electrical transitions that occur as the brushes sequentially contact each segment of the commutator during rotation.

By measuring the time interval required to complete one full rotation determined by identifying the duration encompassing five successive commutation peaks it was possible to calculate the motor's rotational speed. Using this time value (5.4120 ms), the rotational speed of Motor 1 was determined to be 60/5.4120*1000 = 11086 RPM.

In the following sections, the results corresponding to the remaining three motors are presented, following the same structure used for Motor 1 (Figs. 6 and 7).

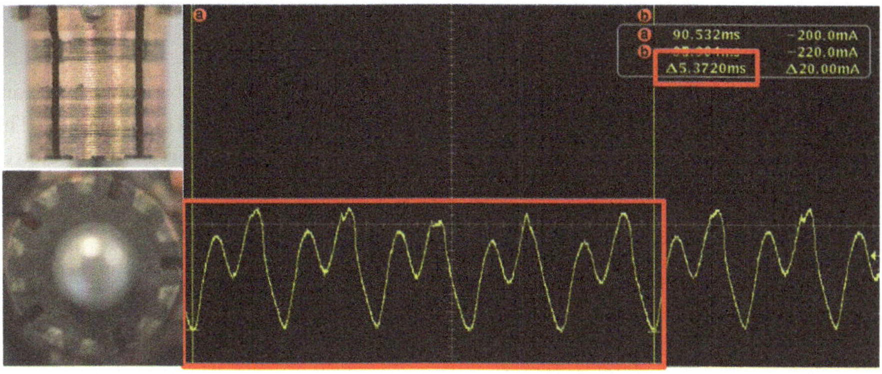

Fig. 5. Commutation signal for Motor 2 (11169 rpm)

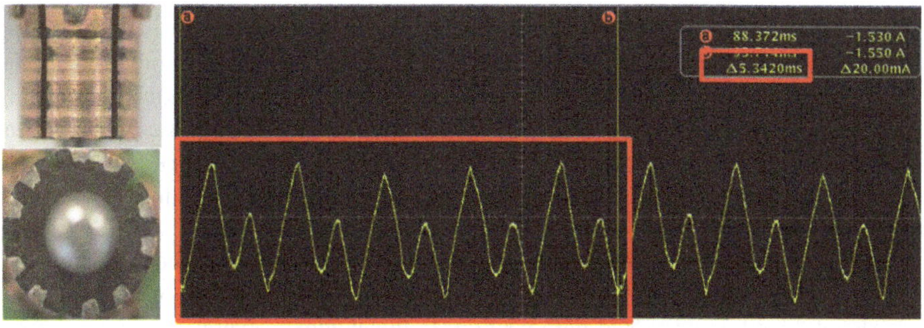

Fig. 6. Commutation signal for Motor 3 (11231 rpm)

Based on the results obtained in this study, a promising direction for future research involves the development of a compact, embedded data acquisition system capable of recording commutation signals in real time, directly onboard the omnidirectional platform. This system would enable continuous monitoring of each motor during actual operation, allowing the commutation behavior to be analyzed under real-world dynamic conditions, including load variations, directional changes, and varying surface interactions. Such an implementation would eliminate the limitations of static, lab-based

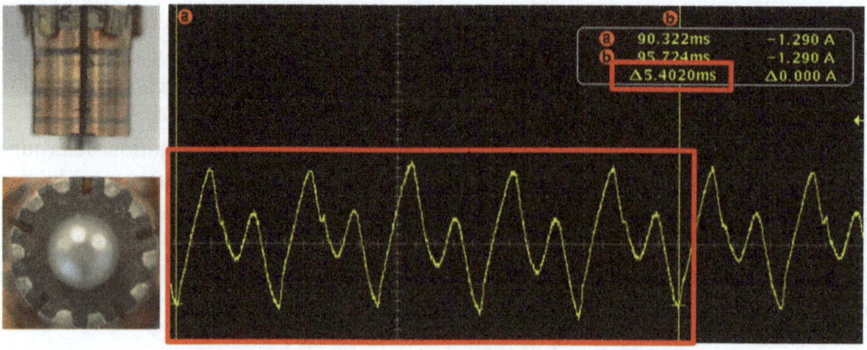

Fig. 7. Commutation signal for Motor 4 (11106 rpm)

measurements and provide a much more accurate picture of how electrical and mechanical factors influence the performance of DC motors in their functional environment. The collected data could also serve as input for adaptive control algorithms and predictive maintenance strategies, significantly improving the reliability and efficiency of mobile robotic platforms.

Additionally, this future work could include a comparative analysis using identical, factory-new DC motors to better quantify the influence of component wear on commutation quality. Combined, these efforts would contribute to the development of a comprehensive diagnostic framework for small-scale electromechanical systems.

5 Conclusions

This study has highlighted the relevance of commutation signal analysis in evaluating the dynamic behavior of direct current (DC) electric motors used in an omnidirectional platform. By employing a high-performance oscilloscope and a non-invasive testing methodology, both the rotational uniformity of the motors and irregularities in the current draw manifested as periodic drops were observed. These discontinuities can negatively affect the synchronization of the platform, especially under load conditions or during rapid maneuvers.

However, considering that the tested motors are not new and have been subjected to mechanical wear, the conclusions regarding the absolute performance of the commutation signal must be interpreted with caution. In this context, several key directions for future research emerge.

A first direction involves performing comparative testing between worn and new motors, in order to assess the impact of wear on the commutation signal and the stability of the current consumption. Such a comparison would enable the identification of early diagnostic indicators based solely on electrical signal analysis, without requiring mechanical disassembly.

Secondly, it is proposed to integrate a data acquisition system directly onto the mobile platform, with the objective of monitoring the commutation signal in real time under actual operating conditions. This approach allows for the direct correlation between signal variations and the kinematic behavior of the platform, providing valuable insights for control optimization and the development of predictive maintenance algorithms.

Finally, extending the analysis into the frequency domain and employing advanced signal processing techniques such as the Fourier transform could offer a deeper understanding of how deviations from ideal commutation occur. These investigations will contribute not only to enhancing platform performance but also to the development of robust diagnostic methods for a wide range of systems equipped with DC motors in intelligent mobile applications.

References

1. Bayar, G., Ozturk, S.: Investigation of the effects of contact forces acting on rollers of a mecanum wheeled robot. Mechatron. Sci. Intell. Mach. **72**(102467), 102467 (2020)
2. Wang, C., Wang, Q., Dong, R.-F., Cai, Y.-P.: Dynamic self-assembly of micro-nanomotor. Inorgan. Chem. Commun. **91**, 8–15 (2018)
3. Agha-Mohammadi, A.-A., Chakravorty, S., Amato, N.M.: FIRM: feedback controller-based Information-state Roadmap - a framework for motion planning under uncertainty. In: 2011 IEEE/RSJ International Conference on Intelligent Robots and Systems (2011)
4. van Meer, M., Witvoet, G., Oomen, T.: Optimal commutation for switched reluctance motors using Gaussian process regression. IFAC-PapersOnLine **55**(37), 302–307 (2022)
5. Bolderman, M., Lazar, M., Butler, H.: Structured physics-guided neural networks for electromagnetic commutation applied to industrial linear motors. Mechatronics **106**, 103291 (2025)
6. Senthilnathan, A., Palanivel, P.: A new approach for commutation torque ripple reduction of FPGA based brushless DC motor with outgoing phase current control. Microprocess. Microsyst. **75**, 103043 (2020)
7. Jaroszewicz, K., Walendziuk, W., Trochimczuk, R., Wolniakowski, A., Tomaszewski, R., Slanina, Z.: Mecanum wheel mobile platform for rapid prototyping of the autonomous control algorithms. IFAC-PapersOnLine **58**(9), 183–188 (2024)
8. Jaafari, A., Mohsenzade, S., Razi-Kazemi, A.A.: An adaptive current commutation for hybrid DC circuit breaker. Electr. Power Syst. Res. **247**, 111828 (2025)
9. Zhang, Q., Wen, B., He, Y.: Rotational speed monitoring of brushed DC motor via current signal. Measurement **184**, 109890 (2021)
10. Xiao, H., Yu, D., Chen, C.L.P.: Self-triggered-organized Mecanum-wheeled robots consensus system using model predictive based protocol. Inf. Sci. **590**, 45–59 (2022)
11. Manzl, P., Sereinig, M., Gerstmayr, J.: A Mecanum wheel model based on orthotropic friction with experimental validation. Mech. Mach. Theory **193**(105548), 105548 (2024)
12. Ben Abdeljawed, H., El Amraoui, L.: Simulation and rapid control prototyping of DC powered universal motors speed control: towards an efficient operation in future DC homes. Eng. Sci. Technol. Int. J. **34**, 101092 (2022)
13. Al-Amyal, F., Hamouda, M., Számel, L.: Performance improvement based on adaptive commutation strategy for switched reluctance motors using direct torque control. Alexandria Eng. J. **61**(11), 9219–9233 (2022)
14. Lynch, K.M., Marchuk, N., Elwin, M.L.: Brushed permanent magnet DC motors. Embed. Comput. Mechatron. PIC32, 399–425 (2016)
15. Fakih, B., Dienwiebel, M.: The structure of tribolayers at the commutator and brush interface: a case study of failed and non-failed DC motors. Tribol. Int. **92**, 21–28 (2015)

The Influence of Soft Value Dimensions on Consumers' Purchase Intentions for Electric Vehicles

Anita Subotić[1] and Nina Antičić[2](✉)

[1] Coca-Cola HBC Hrvatska, Milana Sachsa 1, Zagreb 10000, Republic of Croatia
[2] RIT Croatia, D. T. Gavrana 15, Zagreb 10000, Republic of Croatia
Nina.anticic@croatia.rit.edu

Abstract. The electric vehicle (EV) market has expanded significantly in recent years. As adoption accelerates, understanding the underlying motivations behind EV purchases becomes increasingly important. Functional, economic, and environmental factors have traditionally driven EV adoption. However, recent research highlights the growing importance of soft value dimensions—such as status, symbolism, prestige, exclusivity, and recognition. This research explores how EV ownership aligns with the buyer's personal identity, social standing, and emotional appeal, and how soft value dimensions influence purchase intentions for EVs. The collected empirical data were processed and analyzed using multivariate analysis methods and structural equation modeling with the partial least squares method (PLS-SEM). The findings extend current knowledge of EV adoption by highlighting the role of soft value dimensions in shaping consumer behavior. A key managerial implication of the research findings is that soft value dimensions may accelerate adoption if effectively integrated into marketing strategies.

Keywords: Electric vehicles · Purchase intentions · Perceived value

1 Introduction

The electric vehicle (EV) market expanded rapidly over the past decade, with global sales reaching around 13.8 million units in the year 2023. Electric cars represented around 18% of all cars sold in the year 2023, in comparison to the previous year, 2022, which accounted for 14% of EVs sold (IEA, 2024). In the United States, new electric car registrations accounted for 1.4 million in 2023, increasing by more than 40% compared to the previous year, 2022. Relative annual growth in 2023 was slower than in the preceding two years, but the demand for electric cars and absolute growth remained strong (IEA, 2024). On the other hand, Norway continues to be an industry leader in the adoption of EVs. In 2023, EVs represented almost 93% of new vehicle purchases, due to the underlying impact of government incentives and supportive policies for consumers (Norwegian EV Association, 2024).

As the market continues to grow rapidly, it is crucial to understand the motivations that drive the purchase of EVs. EV purchase decisions are influenced by purchasing

D. D. Cioboată and J. Machado (Eds.): ICoRSE 2025, LNNS 1592, pp. 340–354, 2025.
https://doi.org/10.1007/978-3-032-02508-1_29

power, word of mouth, environmental awareness, technological openness, perceived benefits, and ease of use, while major disruptions to usage patterns have little impact (Tu and Yang, 2019). Hence, inertia in usage patterns serves as an obstacle to market development.

There are conflicting findings in the literature regarding the influence of social norms on purchase intentions. While family and friends' opinions do not significantly influence consumers' actual purchase intentions (Tu and Yang, 2019), emotional and social values, for instance, shape the consumers' perceptions toward EVs by enhancing the perceptions of sustainability, innovation, and social status. Earlier studies, designed about ten to fifteen years ago, indicate that economic and environmental benefits are crucial to the adoption of EVs. However, more recent research suggests that social dimensions related to the aspect of prestige, exclusivity and recognition tend to be just as important. Around 30% of buyers who purchase luxury vehicles prefer to own unique vehicles, supporting the role of status in the adoption of EVs and aligning with the notion that EVs pose as status-inducing products (Bertoncello et al., 2022). As technology advances, more marketers are positioning their vehicles as premiums and as models of symbols, innovation, and prestige, appealing to those customers who view their vehicles as an extension of their social identity. How these values translate to purchase intentions, we investigate in this paper.

2 Theoretical Background

Given the various motives behind consumers' adoption and use of EVs, their decisions appear to be shaped by a broad set of perceived benefits and the accompanying monetary and non-monetary costs. This perspective is consistent with the concept of perceived value. Not only does the perceived value drive the competitive advantage, but it also helps marketers predict consumer behavior (Wang et al., 2024). Perceived value refers to consumers' evaluation of a product and its excellence and superiority to other products. It differs from objective quality and is more of an abstract evaluation rather than an evaluation based on the product's actual attributes (Zeithaml, 1988). Since customer perceived value strongly influences purchase intentions, one can argue that the purchase intentions are the outcomes of the perceived value (Widing et al., 2003) and a strong indicator of actual buying behavior (Yu, 2023). In the context of EVs, the perceived value directly influences the likelihood of a purchase (Yu, 2023). Henceforth, the soft value dimensions believed to influence EV purchase intentions are further elaborated.

2.1 Status

Objects or traits can carry status value, which reflects self-esteem and honor through ownership and shaping decisions beyond price (Thye, 2000, as cited in Zabkar and Hosta, 2012). Status has also been described as a symbolic attribute, as in the statement, "An electric car gives me status" (Noppers et al., 2015; Schuitema et al., 2013, as cited in White et al., 2017). This relates to status consumption, defined as the motivation to enhance social standing through buying prestigious products (Eastman, Goldsmith, and Flynn, 1999, as cited in Goldsmith and Clark, 2011).

2.2 Symbolism

For a product to be symbolic, it must be socially recognized and have a clearly defined meaning within different social groups (Grubb, 1967). In the case of electric vehicles, symbolic attributes are connected to identity, allowing individuals to express self-image and gain social status (Rezvani et al., 2015; Burgess et al., 2013, as cited in White and Sintov, 2017). Symbolic value traditionally reflects an owner's social standing and desire for differentiation (Dalal and Aljarah, 2021; Gilai et al., 2018, as cited in Zhuoqun, Tat, and Keong, 2023). When a product aligns with environmental identity and personal values, consumers are more likely to create positive attitudes and purchase intentions (Schuitema et al., 2013). White and Sintov (2017) identified two dimensions of symbolism: environmentalist and social innovator, with environmentalist symbolism being the strongest predictor of EV adoption. Ultimately, symbolism shapes consumer attitudes toward EVs, influencing their adoption beyond just functionality.

2.3 Prestige

Prestige products are those described as goods that are not frequently purchased and require high-involvement decision making and a higher level of interest and knowledge (Vigneron and Johnson, 1999). Their worth comes from the superiority of their design, financial sacrifice, and how others view them (Vigneron and Johnson, 2004; Thomas, 2007, as cited in Femina and Santhi, 2024). The desirability to be recognized by others by owning prestige products boosts the owner's self-worth and status (Moore and Birtwistle, 2005, as cited in Femina and Santhi, 2024).

2.4 Exclusivity

The adoption of EVs and consumer preferences are significantly influenced by exclusivity. A McKinsey analysis from 2022 (Bertoncello et al., 2022) states that the market for luxury EVs is expanding because consumers' desire for exclusive experiences that create a feeling of social standing significantly increases. Liao and Molin (2016) provide more evidence in favor of the desire for exclusivity by pointing out that characteristics like brand prestige and distinction play a significant role in influencing consumers' preferences for EVs. Furthermore, a 2023 study by Li et al. highlights how positive interactions between buyers and sellers can raise the perceived value of EVs, making customer exclusivity a crucial demand driver.

2.5 Recognition

Brand recognition and strong brand image boost brand trust, increasing EV purchase intention (Belay, 2022, as cited in Lady and Angelino, 2024). Brand recognition influences consumer preferences, especially in advertising (Hauser, 2011, as cited in Tho-ma and Williams, 2023). While some consumers prefer popular brands (Deval et al., 2013; Steinhart et al., 2014, as cited in Wan, Wang, and Wu, 2018), others opt for more exclusive ones for self-expression.

2.6 Purchase Intentions

Purchase intentions reflect a buyer's commitment to a product or brand (Oliver, 1999; Widing et al., 2003). In the context of EVs, these intentions are shaped by factors like environmental concern, self-image, and functional congruity, consumers are more likely to repurchase when a brand aligns with their identity (Lee, Kang and Tang, 2012). Brand environmental reputation also plays a key role in shaping purchase intention (Zhao et al., 2024), as do positive emotional experiences, which increase the likelihood of repeat purchases (Sari and Burhanudin, 2023). Additionally, emotional experiences during the buying process impact future behavior positive emotions increase the likelihood of return purchases (Sari and Burhanudin, 2023).

Perceived value positively influences the purchase intentions, either mediated by customer satisfaction or directly. Specifically, economic, functional, and environmental factors are the main drivers that influence the intention to purchase an EV. However, current studies lack examination of how soft value dimensions like status, symbolism, exclusivity, prestige, and recognition influence intentions to purchase EVs. That considered, the objective of this research is to explore the role of these value dimensions in the formation of purchase intentions. A conceptual model that presents these influences is put forward in Fig. 1.

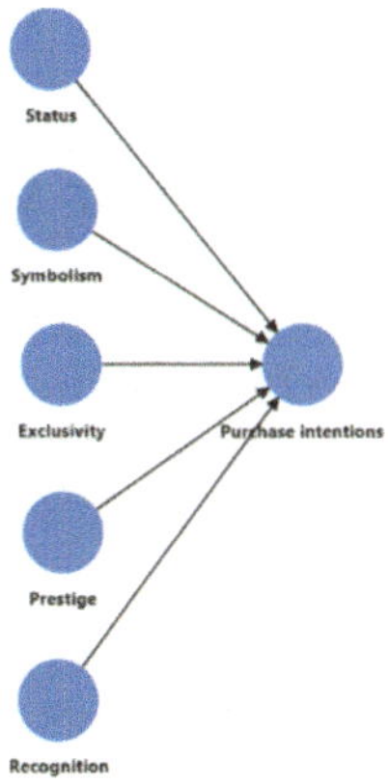

Fig. 1. Conceptual model. Source: Authors' research

3 Methodology, Sample, Procedures and Measures

The initial phase of the research elaborates theoretical concepts of consumer values related to EVs—status, prestige, exclusivity, symbolism, and recognition—based on a literature review to define and assess their influence on purchase intentions. Measurement scales were adapted from prior studies (Eastman et al., 1999; Klisheimer, 1993; Lynn and Harris, 1997; Han and Terpstra, 1988; Sweeney and Soutar, 2001; Escalas and Bettman, 2005; Widing et al., 2003), with some items developed by the researchers. A structured questionnaire was used to measure attitudes on a 5-point Likert scale. The survey began with demographic variables (gender and age), followed by 26 items: 20 assessing independent variables (status, prestige, exclusivity, symbolism, recognition) and 6 measuring the dependent variable (purchase intentions). Each construct was measured using 3–6 items. The study employed a survey distributed via social media posts to a random sample of potential EV consumers. A self-selection sampling method was used, where participants voluntarily completed the questionnaire based on their self-identified interest in EVs. Data collection occurred from mid-November to early December 2024.

After excluding incomplete or inconsistent responses (e.g., excessive neutral answers, identical ratings, or missing data), 76 valid responses remained. The gender split was 53.2% female and 46.8% male. Respondents were primarily aged 18–30 (45.5%), followed by 31–45 and 46–55 (each 23.4%), and 7.8% were over 55. The scales were analyzed using multivariate techniques. Reliability was assessed via Cronbach's Alpha ($\alpha \geq 0.70$). Exploratory factor analysis was conducted using principal component analysis (PCA) with Varimax rotation, applying a 0.50 loading threshold and the Kaiser criterion for factor extraction. Data suitability was confirmed by the Kaiser-Meyer-Olkin (KMO) measure (>0.6) and Bartlett's Test of Sphericity ($p < 0.05$). In the second phase, following factor extraction and validation, the conceptual model was revised and hypotheses formulated. Confirmatory factor analysis was performed using PLS-SEM. Internal consistency was assessed via Cronbach's Alpha, Rho_A, and composite reliability (all preferably > 0.7). Convergent validity was confirmed with AVE > 0.5 and outer loadings preferably > 0.7. The PLS algorithm used path weighting, 5,000 iterations. Discriminant validity was verified through cross-loadings, the Fornell-Larcker criterion ($\sqrt{AVE}$ > inter-construct correlations), and the HTMT ratio (<0.85). Hypotheses were tested via bootstrapping (5,000 resamples, percentile confidence intervals, two-tailed test, $\alpha = 0.05$) with parallel processing and a fixed seed for reproducibility.

Statistical data processing was performed in the statistical programs IBM SPSS Statistics 23 and SmartPLS 4 (Ringle, Wende and Becker, 2024). OpenAI's ChatGPT language model (April 10 - May 6, 2025) was used to improve English language wording and to refine academic language.

4 Results

The results of the analysis of measurement scales in Table 1 provide insights into their reliability.

Table 1. Reliability of measurement scales.

Scale	Status	Recognition	Symbolism	Prestige	Exclusivity	Purchase Intentions
Cronbach's Alpha	0,817 After removing STAT_2	0,820 After removing RECOG_4	0,738 after removing SYMB_2	0,812	0,804	0,894

Source: Authors' research

All scales met or exceeded the reliability threshold and demonstrate strong reliability. These results confirm the appropriateness of the items for subsequent exploratory factor analysis (EFA). The reduction of measurement scales and the removal of manifest variables that compromised their reliability (STAT_2, RECOG_4 and SYMB_2) yielded a twenty-three items model for further research which will be further evaluated through exploratory factor analysis.

Data suitability for factor analysis was confirmed (KMO = 0.880 > 0.6; Bartlett's $\chi^2 = 1212.724$, df = 253, $p < 0.001$), indicating high sampling adequacy and sufficient inter-variable correlations. Principal component analysis (PCA) was used for factor extraction, with the Kaiser criterion guiding factor retention. Initially, four factors emerged; SYMB_1 (no loading) and PREST_2 (cross-loading) were removed. A revised analysis yielded a stable three-factor solution (Table 2).

Table 2. Rotated Factor Solution – Principle Component Analysis.

Rotated Component Matrix[a]			
	Component		
	1	2	3
STAT_1			,883
STAT_3			,514
STAT_4			,747
STAT_5	,706		
RECOG_1	,788		
RECOG_2	,716		
RECOG_3	,640		
SYMB_3	,711		
PREST_1			,751
PREST_3		,764	
PREST_4		,702	
EXCLUS_1	,598		
EXCLUS_2	,834		
EXCLUS_3	,651		
EXCLUS_4	,754		
PURCHINT_1		,739	
PURCHINT_2		,805	
PURCHINT_3		,774	
PURCHINT_4		,754	

(*continued*)

Table 2. (*continued*)

Rotated Component Matrix[a]			
	Component		
	1	2	3
PURCHINT_5		,850	
PURCHINT_6		,780	

Extraction Method: Principal Component Analysis.
Rotation Method: Varimax with Kaiser Normalization.
[a]Rotation converged in 5 iterations.
Source: Authors' research

Component 1 groups items primarily related to recognition, exclusivity, and partly symbolism. High loadings were observed for RECOG_1 (0,788), RECOG_2 (0,716), RECOG_3 (0,640), SYMB_3 (0,711), EXCLUS_1 (0,598), EXCLUS_2 (0,834), EXCLUS_3 (0,651), and EXCLUS_4 (0,754).

Component 2 is defined mainly by items measuring purchase intentions and prestige. Items PREST_3 (0,764), PREST_4 (0,702), PURCHINT_1 (0,739), PURCHINT_2 (0,805), PURCHINT_3 (0,774), PURCHINT_4 (0,754), PURCHINT_5 (0,850), and PURCHINT_6 (0,780) loaded strongly on this component.

Component 3 primarily consists of status indicators, with high loadings from STAT_1 (0,883), STAT_3 (0,514), and STAT_4 (0,747). Additionally, PREST_1 (0,751) loaded here, suggesting a link between social status perception and prestige.

All factor loadings are above 0,50, indicating strong relationships between variables and their respective components. The clean separation of items across components supports the discriminant validity of the constructs.

Since Kaiser's criterion may overestimate factor count, eigenvalues and explained variance were reviewed. The three factors accounted for 65.829% of total variance, exceeding the 60% threshold recommended for social sciences (Fazlić and Đonlagić, 2016). Results are shown in Table 3.

Table 3. Total variance explained.

Component	Initial Eigenvalues			Rotation Sums of Squared Loadings		
	Total	% Variance	Cumulative %	Total	% Variance	Cumulative %
1	9,100	43,331	43,331	5,311	25,292	25,292
2	3,170	15,097	58,428	5,238	24,eees941	50,233
3	1,554	7,401	65,829	3,275	15,596	65,829

Extraction Method: Principal Component Analysis.
Source: Authors' research

After PCA, the following conclusions are put forward:

The measurement scales are reliable, as indicated by the values of the Cronbach Alpha coefficient (higher than 0,70).

Measurement scales are convergently valid because their items have high factor loadings on the corresponding factors (more than 0,5).

Measurement scales are discriminately valid because the corresponding items have a high factor load on the relevant factors, and lower cross-loadings on other factors.

The component matrix indicates that the PCA extracted three factors, hence, the initial conceptualization of the five dimensions of value influencing purchase intentions is not sustained.

Component 1 captures aspects of social image value, encompassing exclusivity, recognition, and partially symbolic elements. Based on the convergence of items reflecting a desire for exclusivity, social acknowledgment, and symbolic evaluation by others, this factor was named *social image value*. This label more accurately reflects the construct's conceptual focus on how EV ownership enhances the consumer's perceived identity, social differentiation, and social image in the eyes of others.

Component 2 is best interpreted as representing *purchase intentions*, as it is primarily composed of items that directly capture consumers' willingness, commitment, and support for buying an EV. Convergence of prestige items is theoretically justified due to the behavioral and evaluative wording of the items related to prestige (Lee, Kang and Tang, 2012; Zhao et al., 2024). This outcome supports existing findings that prestige and value congruence can act as direct motivational triggers (Yu, 2023). Finally, they are consistent with the idea that prestige, when framed through the lens of personal relevance and product quality, functions as a reinforcing factor within the broader construct of purchase intention.

Component 3 primarily consists of status value indicators and one element of social prestige. These indicators highlight perceptions of EVs as prestigious, high-status products, which are associated with wealth and personal achievement. The items go beyond being simply tools for recognition or self-expression; they reflect a deeper admiration for EVs as symbols of success, elevated social standing, and aspirational status. Given this mix of prestige, status, and aspirational success, the factor was named *social prestige value*. This label distinguishes it from social image value, which centers on external recognition and symbolic signaling, while social prestige captures deeper associations with societal rank, self-worth, and class elevation. This interpretation aligns with existing research on prestige consumption, which shows that consumers are drawn to products that communicate power, financial success, and personal accomplishment (Vigneron and Johnson, 1999; Moore and Birtwistle, 2005; Goldsmith and Clark, 2011).

These constructs will be further evaluated by confirmatory factor analysis. Since the model is simple, first-order, the measurement (outer) model and the structural (inner) model are presented simultaneously. The specification of the model is provided in Fig. 2.

Model quality indicators (Table 4) show Cronbach's Alpha, Rho_A, and composite reliability all above 0.7, confirming internal consistency. AVE values exceed 0.5, meeting model quality criteria.

Discriminant validity was confirmed using multiple criteria. Cross-loading analysis showed each indicator loaded highest on its intended construct, indicating clear construct separation. The Fornell-Larcker criterion supported this, as each construct's AVE

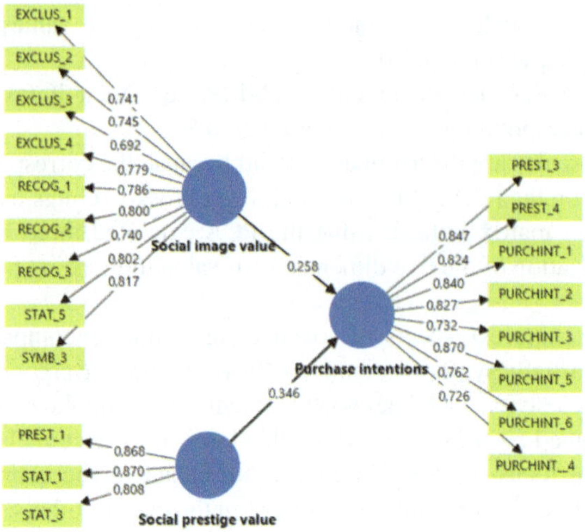

Fig. 2. Model of influence of social image value and social prestige value on purchase intentions of electric vehicles. Source: Authors' research

Table 4. Model quality.

Variable	Indicator	Outer loadings	Alpha	Rho_A	CR	AVE
Social image value	EXCLUS_1	0,740	0,913	0,918	0,928	0,589
	EXCLUS_2	0,746				
	EXCLUS_3	0,692				
	EXCLUS_4	0,779				
	RECOG_1	0,786				
	RECOG_2	0,799				
	RECOG_3	0,740				
	STAT_5	0,802				
	SYMB_3	0,816				
Social prestige value	PREST_1	0,846	0,744	0,782	0,842	0,579
	STAT_1	0,822				
	STAT_3	0,808				
Purchase intentions	PREST_3	0,844	0,923	0,938	0,936	0,649
	PREST_4	0,819				
	PURCHINT_1	0,839				
	PURCHINT_2	0,824				

(continued)

Table 4. (*continued*)

Variable	Indicator	Outer loadings	Alpha	Rho_A	CR	AVE
	PURCHINT_3	0,738				
	PURCHINT_4	0,729				
	PURCHINT_5	0,872				
	PURCHINT_6	0,766				

Source: Authors' research

square root exceeded its correlations with other constructs. All HTMT values were below 0.85, confirming sufficient construct distinctiveness. The model met all requirements for internal reliability, convergent validity, and discriminant validity.

4.1 Hypotheses

Based on the exploratory and confirmatory factor analysis results, the initial theoretical model of five value dimensions influencing EV purchase intentions was revised. The following hypotheses are proposed for empirical testing:

H1: Social image value positively influences purchase intentions.

H2: Social prestige value positively influences purchase intentions.

Both hypothesized relationships are supported. The path from social image value to purchase intentions is significant ($\beta = 0.258$, $p = 0.012$, $t = 2.500$; 95% CI [0.079, 0.482]), showing a positive, reliable effect. The path from social prestige value is also significant ($\beta = 0.346$, $p = 0.004$, $t = 2.891$; 95% CI [0.102, 0.570]). As neither confidence interval includes zero, both effects are confirmed, indicating that social image and prestige significantly influence EV purchase intentions. Results are shown in Table 5.

Table 5. Hypotheses testing.

Hypothesis	Direct effects	β, Path coefficient	p-value	t-value	Confidence interval [2.5%, 97.5%]
H1	Social image value > Purchase intentions	0,258	0,012	2,500	[0,079, 0,482]
H2	Social prestige value > Purchase intentions	0,346	0,004	2,891	[0,102, 0,570]

Source: Authors' research

5 Discussion of Results

In the model, brand exclusivity, symbolic expression, and peer recognition formed a unified social image value factor, while items related to personal prestige and social status loaded onto a distinct social prestige factor. This empirical structure aligns with theory, as all elements reflect EV owners' efforts to signal status and elevate social standing (Vigneron and Johnson, 1999; Moore and Birtwistle, 2005, as cited in Femina and Santhi, 2024).

While both variables relate to social perception, social image captures symbolic and recognition-based motivations, whereas social prestige reflects aspirational drivers like wealth and success. Purchase intentions emerged as an independent outcome construct, consistent with Widing et al. (2003), indicating their formation is shaped by perceived social value.

Social image value significantly influenced purchase intentions, supporting findings linking brand recognition with consumer preference (Hauser, 2011, as cited in Thoma and Williams, 2023). This effect is amplified in self-expressive products like EVs, where exclusivity enhances social image (Deval et al., 2013; Steinhart et al., 2014, as cited in Wan, Wang, and Wu, 2018), reinforcing perceived status (Bertoncello et al., 2022).

Similarly, social prestige value had a significant positive effect on purchase intentions, as EV ownership boosts both self-image and perceived social status (Moore and Birtwistle, 2005, as cited in Femina and Santhi, 2024).

6 Practical Implications

This study offers actionable insights for EV marketers, particularly in crafting campaigns that enhance product appeal through social value. Positioning EVs as symbols of identity, innovation, and prestige can elevate purchase intentions by strengthening the owner's social image.

Product design should highlight performance, futuristic styling, and visible eco-features that signal modernity and status. Personalization options—custom features, tech add-ons, accessories—can amplify self-expression. High-end or limited-edition models appeal to prestige-seeking consumers.

Pricing strategies should reinforce value and status. Premium pricing, justified by quality or branding, can boost perceived exclusivity (e.g., high-end EVs as status symbols). Mid-range models should remain accessible, supported by subsidies, trade-in incentives, and long-term value messaging (e.g., fuel and maintenance savings).

Place strategies should combine high-visibility retail (e.g., showrooms in trendy areas) with digital convenience. Online sales, home delivery, and a tech-forward retail experience align with modern, digital-first consumers.

Promotion should leverage innovative tools—virtual showrooms, VR test drives—to showcase advanced tech and appeal to tech-savvy buyers. Campaigns should link environmental and social benefits (e.g., low emissions, progress) to personal values and identity. Endorsements by influencers or public figures can enhance prestige, while exclusive owner communities and loyalty programs can foster brand attachment and peer influence.

In sum, by aligning the 4Ps with consumers' aspirations for recognition, identity, and status, EV marketers can design emotionally resonant, value-driven strategies to increase adoption in the competitive B2C market.

7 Limitations

This study emerged from a student research project within a consumer behavior course at RIT Croatia. Accordingly, the research design was tailored to fit undergraduate scope, time, and skill constraints—offering valuable learning opportunities but also introducing several limitations.

The sample was drawn primarily from the student researcher's social media network, limiting representativeness, cultural diversity, and generalizability to the broader EV market. The sample size was modest and lacked demographic variation.

The study used self-reported, cross-sectional data, which, while common in consumer research, is subject to social desirability bias and cannot capture changes over time. Additionally, the outcome variable was purchase intention rather than actual behavior. Though frequently used in consumer behavior studies, intentions may not fully reflect real-world decisions, especially for high-involvement products like EVs, where competing offers or negative feedback may deter purchases.

The literature review was intentionally narrow due to space constraints, limiting theoretical depth—particularly in constructs like recognition, which would benefit from stronger conceptual grounding. Some variables grouped differently than theoretically expected during factor analysis, likely due to the use of newly developed items that had not been previously validated. While the use of both exploratory and confirmatory factor analysis is a strength, more time for pre-testing and iterative refinement would have enhanced measurement reliability. Future studies should employ validated scales to improve construct clarity and consistency.

The convergence of exclusivity, recognition, and symbolism into a broader social image factor offers empirical insight but may obscure nuanced theoretical distinctions. Moreover, the study focused on "soft" value dimensions (e.g., status, prestige, symbolism), excluding others like emotional, hedonic, relational, or epistemic value. It also omitted "hard" value dimensions such as performance, utility, quality, range, infrastructure, and financial factors. Future research should incorporate both soft and functional dimensions for a more holistic view of EV purchase drivers.

8 Further Research

This study offers initial insights into the influence of symbolic and social values on EV adoption. Future research should expand the value framework to include functional and infrastructural attributes such as performance, reliability, and service quality. Comparative analyses of soft vs. hard values could clarify their relative influence on consumer attitudes and behavior.

Longitudinal studies are recommended to observe how symbolic drivers evolve and affect actual purchase behavior as EVs become mainstream. Investigating psychological mediators—such as identity congruence or satisfaction—could further illuminate value-driven decision-making.

Culturally and demographically segmented studies, including gender- and generation-specific analyses, would also provide actionable insights for targeted EV marketing across diverse consumer groups.

References

Beavers, A.S., Lounsbury, J.W., Richards, J.K., Huck, S.W., Skolits, G.J., Esquivel, S.L.: Practical considerations for using exploratory factor analysis in educational research. Pract. Assess. Res. Eval. **18**(6), n6 (2013)

Bertoncello, M., et al.: BEVolution of Luxury Vehicles: How Customer Preferences are Changing the Luxury Vehicle Market. McKinsey & Company (2022)

Bosshart, L., Macconi, I.: Media Entertainment. Centre for the Study of Communication and Culture (1998)

Chumpitaz Caceres, R., Paparoidamis, N.G.: Service quality, relationship satisfaction, trust, commitment and business-to-business loyalty. Eur. J. Mark. **41**(7/8), 836–867 (2007)

Chung, D.S., Yoo, C.Y.: Audience motivations for using interactive features: distinguishing use of different types of interactivity on an online newspaper. Mass Commun. Soc. **11**(4), 375–397 (2008)

Contzen, N., Perlaviciute, G., Sadat-Razavi, P., Steg, L.: Emotions toward Sustainable Innovations: a matter of value congruence. Front. Psychol. **12** (2021)

Damanpour, F.: Organizational innovation: a meta-analysis of effects of determinants and moderators. Acad. Manag. J. **34**, 555–590 (1991)

Davis, F.: Perceived usefulness, perceived ease of use, and user acceptance of information technology. MIS Q. **13**(3), 319–340 (1989)

de Winter, J.C., Dodou, D.I.M.I.T.R.A., Wieringa, P.A.: Exploratory factor analysis with small sample sizes. Multivar. Behav. Res. **44**(2), 147–181 (2009)

Du, X., Tseng, M.M.: Characterizing customer value for product customization. In: International Design Engineering Technical Conferences and Computers and Information in Engineering Conference, vol. 19746, pp. 315–325 (1999). American Society of Mechanical Engineers

Eastman, J.K., Eastman, K.L.: Perceptions of status consumption and the economy. J. Bus. Econ. Res. **9**(7), 9–19 (2011)

Edwards-Schachter, M.: The nature and variety of innovation. Int. J. Innov. Stud. **2**(2), 65–79 (2018)

Egbue, O., Long, S.: Barriers to widespread adoption of electric vehicles: An analysis of consumer attitudes and perceptions. Energy Policy **48**, 717–729 (2012)

Escalas, J.E., Bettman, J.R.: Self-Construal, reference groups, and brand meaning. J. Consum. Res. **32**(3), 378–389 (2005)

Fazlić, S., Đonlagić, S.: Primjena faktorske analize u identificiranju dimenzija kvalitete visokoobrazovne usluge. Poslovna izvrsnost **10**(2), 45–72 (2016)

Femina, E., Santhi, P.: Effects of perceived value dimensions on customer brand loyalty towards luxury cars. Rajagiri Manag. J. (2024)

Franke, N., Keinz, P., Steger, C.J.: Testing the value of customization: when do customers really prefer products tailored to their preferences? J. Mark. **73**(5), 103–121 (2009)

Freeburg, N.: Exploring our value of fun. Leadership Vision (2020)

Gault, F.: Defining and measuring innovation in all sectors of the economy. Res. Policy **47**(3), 617–622 (2018)

Goldsmith, R.E., Clark, R.A.: Materialism, status consumption, and consumer independence. J. Soc. Psychol. **152**(1), 43–60 (2011)

Grubb, E.L., Grathwohl, H.L.: Consumer self-concept, symbolism and market behavior: a theoretical approach. J. Mark. **31**(4), 22–27 (1967)

Guadagnoli, E., Velicer, W.F.: Relation of sample size to the stability of component patterns. Psychol. Bull. **103**(2), 265 (1988)

Hair Jr., J.F., Hult, G.T.M., Ringle, C., Sarstedt, M.: A Primer on Partial Least Squares Structural Equation Modeling (PLS-SEM). Sage Publications (2016)

Higueras-Castillo, E., Molinillo, S., Coca-Stefaniak, J.A., Liébana-Cabanillas, F.: Perceived value and customer adoption of electric and hybrid vehicles. Sustainability **11**(18), 4956 (2019)

Hur, Y., Ko, Y.J., Valacich, J.: A structural model of the relationships between sport website quality, e-satisfaction, and e-loyalty. J. Sport Manag. **25**(5), 458–473 (2011)

Ickin, S., Wac, K., Fiedler, M., Janowski, L., Hong, J.H., Dey, A.K.: Factors influencing quality of experience of commonly used mobile applications. IEEE Commun. Mag. **50**(4), 48–56 (2012)

International Energy Agency: Global EV Outlook 2024: Trends in electric cars (2024)

Lady, L., Angelino, K.: Analysis factors that influence purchase intention based on brand trust for electric cars. Int. J. Multi Discipline Sci. **7**(1), 83–95 (2024)

Lee, J.Y., Kang, J., Tang, L.: The role of store-image and functional image congruity in determining brand loyalty: cognitive, affective, and conative brand loyalty in the context of brand coffeehouses. Graduate Student Research Conference in Hospitality and Tourism, University of Massachusetts Amherst (2011)

Liu, J., McLeod, D.M.: Pathways to news commenting and the removal of the comment system on news websites. Journalism **20**(2), 1464884919849954 (2019)

Liu, R., et al.: The relationship between symbolic meanings and adoption intention of electric vehicles in China: the moderating effects of consumer self-identity and face consciousness. J. Clean. Prod. **288**, 125116 (2021)

MacCallum, R.C., Widaman, K.F., Zhang, S., Hong, S.: Sample size in factor analysis. Psychol. Methods **4**(1), 84 (1999)

Malhotra, N.K., Birks, D.F.: Marketing Research: An Applied Approach. Pearson Education Limited (2003)

Norwegian EV Association: Norwegian EV market. Elbil.no

Oliver, R.L.: Whence consumer loyalty? J. Mark. **63**, 33 (1999)

Parasuraman, A., Zeithaml, V.A., Berry, L.L.: A conceptual model of service quality and its implications for future research. J. Mark. **49**(4), 41–50 (1985)

Pecht, M. (ed.): Product Reliability, Maintainability, and Supportability Handbook. CRC Press (2009)

Qi, X., Cheng, Y.: Sports app user behavior analysis: participation motivation, satisfaction, and loyalty. Secur. Commun. Netw. (2022)

Rezvani, Z., Jansson, J., Bodin, J.: Advances in consumer electric vehicle adoption research: a review and research agenda. Transp. Res. Part D: Transp. Environ. **34**, 122–136 (2015)

Ringle, C.M., Wende, S., Becker, J.-M.: SmartPLS 4. Bönningstedt: SmartPLS GmbH (2024). https://www.smartpls.com

Riskos, K., Hatzithomas, L., Dekoulou, P., Tsourvakas, G.: The influence of entertainment, utility and pass time on consumer brand engagement for news media brands: a mediation model. J. Media Bus. Stud. **19**(1), 1–28 (2022)

Sari, S.L., Burhanudin.: Emotional experience and conative loyalty: the mediating role of revisit intention. Jurnal Manajemen dan Kewirausahaan **11**(1), 40–48 (2023)

Schuitema, G., Anable, J., Skippon, S., Kinnear, N.: The role of instrumental, hedonic and symbolic attributes in the intention to adopt electric vehicles. Transp. Res. Part A: Policy Pract. **48**, 39–49 (2013)

Sheth, J.N., Newman, B.I., Gross, B.L.: Consumption values and market choices: theory and applications. Cincinnati: South-Western Pub (1991a)
Simonson, I.: Determinants of customers' responses to customized offers: conceptual framework and research propositions. J. Mark. **69**(January), 32–45 (2005)
Squire, B., Readman, J., Brown, S., Bessant, J.: Mass customization: the key to customer value? Prod. Plan. Control **15**(4), 459–471 (2004)
Tu, J.C., Yang, C.: Key factors influencing consumers' purchase of electric vehicles. Sustainability **11**(14), 3863 (2019)
Turner, F., Merle, A., Gotteland, D.: Enhancing consumer value of the co-design experience in mass customization. J. Bus. Res. **117**, 473–483 (2020)
Tzeng, J.Y.: Perceived values and prospective users' acceptance of prospective technology: the case of a career eportfolio system. Comput. Educ. **56**(1), 157–165 (2011)
Valenzuela, A., Dhar, R., Zettelmeyer, F.: Contingent response to self-customization procedures: implications for decision satisfaction and choice. J. Mark. Res. **46**(6), 754–763 (2009)
Van De Ven, A.H.: Central problems in the management of innovation. Manag. Sci. **32**(5), 590–607 (1986)
Vigneron, F.a.L.W.J.: A review and a conceptual framework of prestige-seeking consumer behavior. Acad. Mark. Sci. Rev. **1999**, 1 (1999)
Vladica, F.: Understanding entertainment value: an investigation into the subjectivity of people who experience entertainment. Unpublished Doctoral Dissertation presented to Ryerson University, Toronto Metropolitan University (2012)
Wan, X., Wang, T., Wu, J.: I'll follow the minority: the effects of sales level on purchase intention of self-expressive products. Front. Psychol. **9** (2018b)
Wang, H.Y., Liao, C., Yang, L.H.: What affects mobile application use? The roles of consumption values. Int. J. Mark. Stud. **5**(2), 11 (2013)
Watson, J.C.: Establishing evidence for internal structure using exploratory factor analysis. Meas. Eval. Couns. Dev. **50**(4), 232–238 (2017)
White, L.V., Sintov, N.D.: You are what you drive: environmentalist and social innovator symbolism drives electric vehicle adoption intentions. Transp. Res. Part A: Policy Pract. **99**, 94–113 (2017)
Widing, R.E.:Customer Behaviour: Consumer Behaviour and Beyond. Thomson Learning (2003)
Yoshida, M., James, J.D., Cronin Jr., J.J.: Sport event innovativeness: conceptualization, measurement, and its impact on consumer behavior. Sport Manag. Rev. **16**(1), 68–84 (2013)
Yu, Q.: Research on factors affecting consumers' purchase intention. Adv. Econ. Manag. Polit. Sci. **37**(1), 61–66 (2023)
Zabkar, V., Hosta, M.: Willingness to act and environmentally conscious consumer behaviour: can prosocial status perceptions help overcome the gap? Int. J. Consum. Stud. **37**(3), 257–264 (2012)
Zeithaml, V.A.: Consumer perceptions of price, quality, and value: a means-end model and synthesis of evidence. J. Mark. **52**(3), 2 (1988)
Zhang, L., Zheng, L., Peng, T.Q.: Structurally embedded news consumption on mobile news applications. Inf. Process. Manag. **53**(5), 1242–1253 (2017)
Zhuoqun, L., Tat, H.H., Keong, C.S.: Exploring the impact of product symbolism on consumer loyalty: a perspective from consumption ritualization. Glob. Bus. Manag. Res. **15**(2), 1–24 (2023)
Zolkepli, I.A., Mukhiar, S.N.S., Tan, C.: Mobile consumer behaviour on apps usage: the effects of perceived values, rating, and cost. J. Mark. Commun. **27**(6), 571–593 (2021)

Method for Selecting a Design Scheme for a Continuously Variable Electromechanical Transmission Based on the Principle of an Energy Flow Adder

Andrii Kozhushko[1](✉), Ievgenii Veretennikov[1], Yevhen Pelypenko[1], Vadym Karpov[1], Vitalii Danylenko[1], and Daniela-Doina Cioboată[2]

[1] National Technical University "Kharkiv Polytechnic Institute", 2, Kyrpychova Street, 61002 Kharkiv, Ukraine
Andrii.Kozhushko@khpi.edu.ua

[2] National Institute of Research and Development in Mechatronics and Measurement Technique INCDMTM, 2nd District, 021631 Bucharest, Romania

Abstract. Recently, more and more transport and special vehicles have been equipped with hybrid transmissions. The main aim is to reduce fuel consumption and pollutant emissions. Today, hybrid transmissions are available in many design configurations. This state of affairs leads to the provision of recommendations regarding the structural structure of the hybrid transmission, the dimensions and characteristics of the power control components, which affect the efficiency of the hybrid transmission installation. The objective of this work is to search for the constructive structure of an electromechanical transmission based on the principle of a power flow adder, by analyzing the kinematic and power parameters of the transmission operation, taking into account the peculiarities of the operation of an electric motor and an internal combustion engine. The method used to solve the problem was based on the presentation of the principles of the structure of a continuously variable transmission and the mathematical description of the change in kinematic and force parameters during the operation of a continuously variable transmission. The result of the study is a scientifically based selection of two continuously variable electromechanical transmission schemes based on the principle of an energy flow adder.

Keywords: Vehicle · Continuously Variable Electromechanical Transmission · Planetary Gear · Angular Velocity · Torque · Methodology

1 Introduction

The continuous development of technologies in the field of car and tractor manufacturing is primarily aimed at reducing emissions of polluting gases, which leads to the introduction of hybrid technologies. And while these technologies have traditionally been used in car transmissions, work is now underway to introduce hybrid transmissions in special-purpose vehicles, such as agricultural tractors, engineering and construction equipment,

D. D. Cioboată and J. Machado (Eds.): ICoRSE 2025, LNNS 1592, pp. 355–366, 2025.
https://doi.org/10.1007/978-3-032-02508-1_30

etc. The design of hybrid transmissions in specialized vehicles involves the use of one or more electric motors - this involves the use of many variants of hybrid transmission design, from simpler ones such as parallel schemes to more complex series-parallel schemes.

The use of technology based on the analysis of the interaction of two power flows in hybrid transmissions is aimed at meeting the operational requirements of a specialized vehicle. Operational indicators primarily include speed and energy indicators. Therefore, when designing the transmission system, it is assumed that the speed and energy requirements of the vehicle will be met. The design of any continuously variable transmission (electromechanical or hydrovolumetric-mechanical) needs to ensure zero starting speed, which means balancing the power flows from the internal combustion engine and the auxiliary energy source.

2 Literature Review

A preliminary idea of the transmission operating modes is obtained by using an analytical approach to determine the specific nature of the change in power flows in a hybrid transmission. This will provide information on the advantages and disadvantages of the selected hybrid transmission system at the design stage. The use of a comprehensive evaluation of all possible circuit configurations of a hybrid transmission [5] provides an optimized evaluation of the transmission structure. In [6], the authors emphasize that for the parametric design of hybrid transmissions, it is necessary to use indicators of the entire life cycle of the vehicle in the economic assessment of the design of hybrid transmissions. However, solving such a problem requires that the operation of all control nodes is taken into account when implementing optimization modeling. The most common optimization method that allows the operation of all control units to be taken into account is the dynamic programming approach, which allows a global optimal solution to be found with some accuracy [7, 8]. In [9], a fast dynamic programming method for hybrid transmissions was developed to reduce the computational effort, which allows to recommend a certain transmission configuration by optimizing the distribution of power flows and the characteristics of the control units. A modernized dynamic programming with speed prediction approach to power flow distribution optimization is presented in [10]. In [11], a globally optimal method called hyperfast dynamic programming is proposed, which allows to significantly reducing the computation time by optimizing the control modes of the control nodes. In [12], the advantages of an innovative cascade approach to the optimization of hybrid transmission control parameters are evaluated. Here, dynamic programming is combined with a gradient-based optimization algorithm in a systematic and direct way, which increases the accuracy of the globally optimal result. In addition to the dynamic modeling approach, the Pontryagin minimum principle, which relies on dynamic modeling to solve an analytical problem, has also become widespread. In [13], an adaptive Pontryagin minimum principle was developed for a fuel cell/battery hybrid vehicle by estimating the joint state using predictive traffic information. In [14] the Pontryagin minimum principle is applied to the development of hybrid fuel cell vehicles. Hybrid transmissions are also designed using the strategy of minimizing equivalent consumption, which is based on converting electricity into equivalent fuel consumption and minimizing the equivalent fuel consumption using an equivalent coefficient. Thus, in [15], an analytical solution method based on quadratic programming was developed to determine the equivalent coefficient for the equivalent consumption minimization strategy. In [16], a hierarchical efficient real-time co-optimization strategy is developed for the automated operation of a plug-in hybrid vehicle, which is used to co-optimize vehicle speed and energy management in urban traffic conditions. In [17], numerous topics and research papers are reviewed that use machine learning methods in different categories to develop energy management strategies for a hybrid vehicle.

3 Research Methodology

3.1 Principles of Design of a Continuously Variable Electromechanical Transmission

It is necessary to formulate the principles of the scientific search for electromechanical transmission systems before proceeding to the calculation of mathematical equations. The placement of a planetary mechanism on the output link of the transmission, which plays the role of a power flow adder, is the basis for the creation of possible transmission schemes. As the transmission under consideration is a continuously variable transmission, one of the main criteria for its operation should be a zero start. The solution to this problem is to balance the power flows generated by the electric motor and the combustion engine. With this in mind, Fig. 1 shows a diagram of the electromechanical continuously variable transmission studied, which consists of a internal combustion engine (ICE), an electric drive and two planetary gears. Note that in Fig. 1 there are no cylindrical gears on the links that can increase or decrease the power flows, this requires separate study in further work.

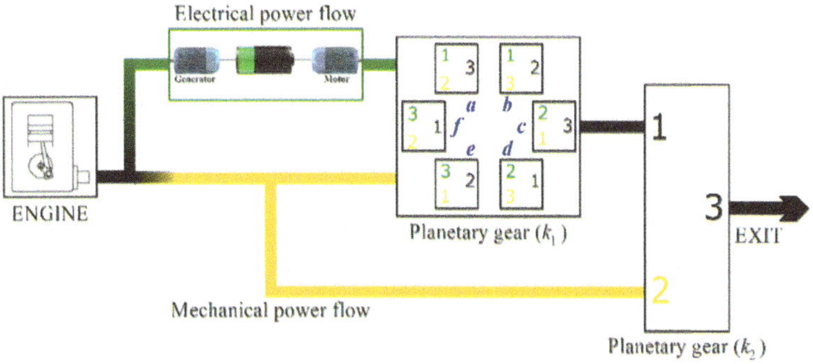

Fig. 1. Schematic of a continuously variable electromechanical transmission based on the power flow adder principle.

It should be noted that the search for possible structural schemes of electromechanical transmissions, formed according to the principle of adding up the power flows, will be carried out in relative quantities.

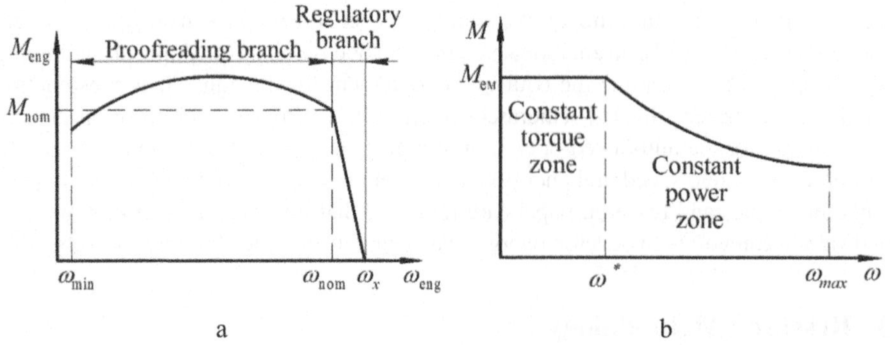

Fig. 2. Conditional external speed characteristic of ICE (a) and mechanical characteristic of an electric motor (b).

The characteristics (Fig. 2) of the internal combustion engine and the electric motor [18] must be taken into account.

Since the solution of the tasks is implemented by calculating the kinematic and power indicators of the transmission operation in relative quantities, we will assume the following initial kinematic dependence:

$$\omega_{\text{nom}} = \omega^* \cdot i_E = 1, \quad (1)$$

where ω_{nom} is a nominal angular speed of the crankshaft ICE; ω^* is a angular speed characterizing the transition from the constant torque zone to the constant power zone; i_E is a conditional indicator characterizing dependence $|\omega_{\text{nom}} / \omega^*|$. In the work, we assume $i_E \in [0; 2]$, where '0' is characterizes the zero angular velocity of the electric motor shaft ($\omega^* = 0$); '1' is characterizes the equivalence of angular velocities ($\omega_{\text{nom}} = \omega^*$); '2' is characterizes the dominance of the angular velocity of the electric motor over the angular velocity of ICE by two times ($\omega_{\text{nom}} = 2 \bullet \omega^*$).

As for the power dependencies, taking into account the mechanical characteristics of the electric machine (Fig. 2b), we propose the following dependency

$$M_{\text{nom}} - M_{\text{eM}} \cdot j_E = 0; \quad (2)$$

where M_{nom} is a nominal torque of the crankshaft ICE; M_{eM} is a constant torque of an electric machine; j_E is a conditional indicator that characterizes dependence $|M_{\text{nom}} / M_{\text{eM}}|$. In the work, we assume $j_E \in [0.1; 1]$, where '0.1' is indicates that the constant torque of the electric machine is equal to 10% of M_{nom} of ICE; '0.5' is indicates that the constant torque of the electric machine is equal to 50% of M_{nom}; '1' is indicates that the constant torque of the electric machine is equal to 100% of M_{nom}.

It should be noted that the paper makes an assumption regarding the operation of ICE. This assumption is that ICE operates exclusively in nominal mode. In the study of changes in the design parameters of planetary gear, it is customary to search within the limits of $k_1 \in [-1; -3]$ and $k_2 \in [-1; -3]$. The justification for the selection of such ranges of changes in design parameters is based on the structural characteristics of planetary gears used in the automotive and tractor industries.

3.2 Mathematical Description of Changes in Kinematic and Performance Parameters of the Continuously Variable Transmission

The structural diagram of the studied electromechanical transmission is formed in Fig. 3 based on the diagram in Fig. 1.

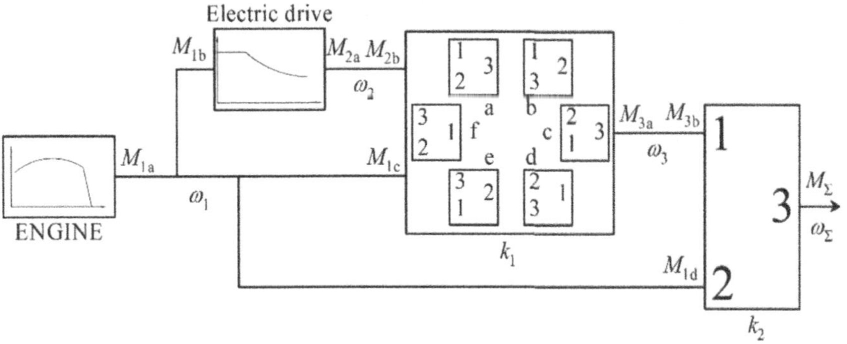

Fig. 3. Structural scheme of the electromechanical transmission that has been studied.

The description of the kinematic and power performance of the transmission is based on the calculation of fundamental equations in general form, i.e. under single initial conditions. The description of the parameters' kinematic properties is derived from a fundamental equation developed by Willis, which is provided below [4]:

$$\omega_1 - k\omega_2 + (k - 1)\omega_3 = 0; \tag{3}$$

where $\omega_{1,2,3}$ are angular velocities of the links of the planetary mechanism, namely the sun and epicyclical gears, the carrier; k is internal gear ratio of a planetary gear set is the ratio of the speeds of the different gears in motion relative to the gear that is being used to transmit power.

The power description of the transmission operation is based on the calculation of the following equations, which do not take into account torque losses due to friction in the planetary gear engagement [4]

$$M_1 + M_2 + M_3 = 0; \tag{4}$$

$$M_1 \cdot k + M_2 = 0; \tag{5}$$

where $M_{1,2,3}$ are torques on the links of the planetary mechanism (i.e. the sun and epicyclic gears) and the carrier must be calculated, without taking into account mechanical losses in the gearing.

The study adopted a simplification: mechanical losses of transmission components are not included. Assessing the efficiency of transmission means measuring its ability to transmit power. This is determined by the transmission power adder's various structural structures.

3.3 Search for Possible Schemes of Electromechanical Transmissions Built on the Principle of a Power Flow Adder

Carry out a kinematic and force analysis of the possible electromechanical transmission schemes shown in Fig. 3, determining the variation in angular velocity and torque at the output link of the transmission when the structure of the planetary mechanisms (k_1, k_2) is varied. For each possible scheme in Table 1, we will form mathematical equations describing the determination of the angular velocity (ω_Σ) and torque (M_Σ) of the output link of the transmission.

Table 1. Mathematical equations describing the determination of the initial angular velocity and torque.

№	Schematic combination of planetary gear (Fig. 1)	Output angular velocity (ω_Σ)	Output torque (M_Σ)
1	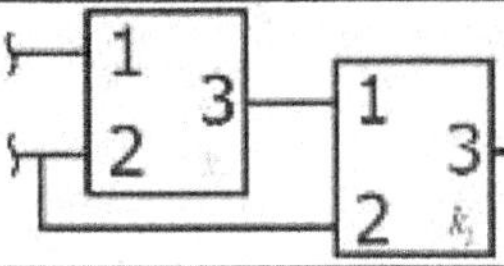	$\dfrac{k_2 - \dfrac{k_1 - i_E}{k_1 - 1}}{k_2 - 1}$	$j_E \cdot (k_1 - 1) \cdot (k_2 - 1)$
2	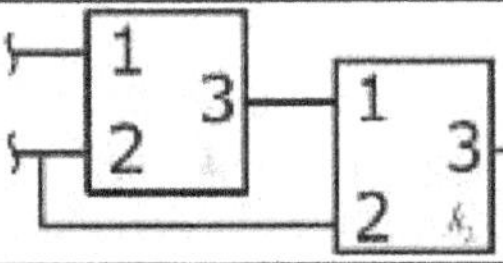	$\dfrac{k_2 - \dfrac{i_E + (k_1 - 1)}{k_1}}{k_2 - 1}$	$j_E \cdot k_1 \cdot (1 - k_2)$
3	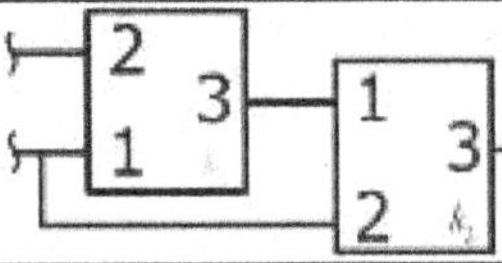	$\dfrac{k_2 - \dfrac{k_1 \cdot i_E - 1}{k_1 - 1}}{k_2 - 1}$	$j_E \cdot \left(\dfrac{1}{k_1} - 1\right) \cdot (k_2 - 1)$
4	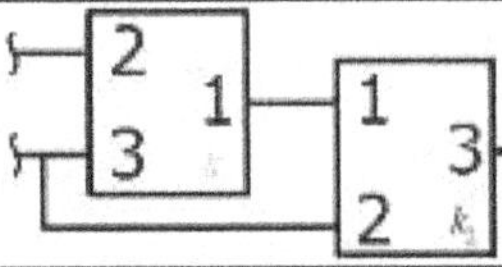	$\dfrac{k_2 - (k_1 \cdot i_E - (k_1 - 1))}{k_2 - 1}$	$\dfrac{j_E}{k_1} \cdot (1 - k_2)$
5	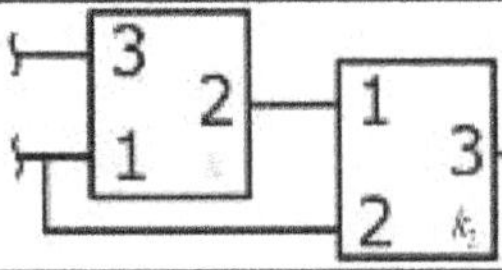	$\dfrac{k_2 - \dfrac{k_1 \cdot i_E - 1}{k_1 - 1}}{k_2 - 1}$	$-\dfrac{j_E \cdot k_1 \cdot (k_2 - 1)}{k_1 - 1}$
6	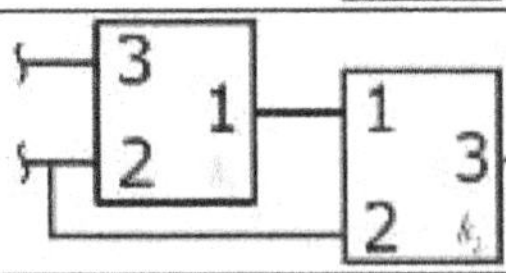	$\dfrac{k_2 - (k_1 - i_E \cdot (k_1 - 1))}{k_2 - 1}$	$\dfrac{j_E \cdot (k_2 - 1)}{k_1 - 1}$

4 Results

4.1 Kinematics Analysis of Transmission Operation

Note that regardless of the schematic structure of the electromechanical transmission (Fig. 3) and the combinations of planetary gear connections (k_1 and k_2), according to Table 1, the same angular velocity is ensured at the output link ω_Σ at $i_E = 1$, i.e. the condition $\omega_\Sigma = \omega_{\text{nom}} = \omega^* = 1$ must be fulfilled. The results of further studies of the change in the initial angular velocity ω_Σ as the indices i_E, k_1 and k_2 are varied according to the circuit structure from Table 1 are shown in Figs. 4 and 5 [19]

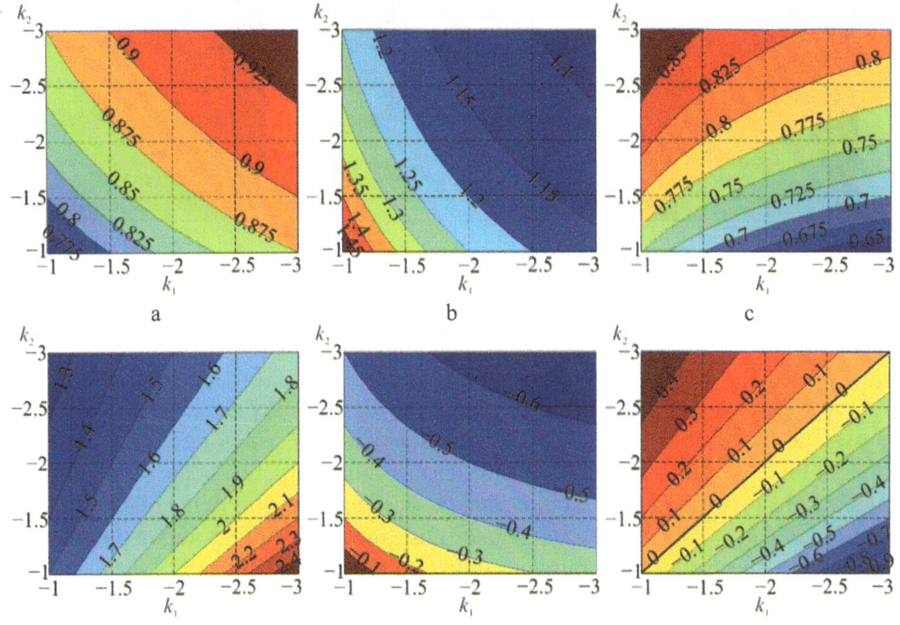

Fig. 4. The results of the study, depending on the structural scheme from Fig. 3(a–f) at $i_E = 0$

Analyzing the results of Figs. 4 and 5, i.e. the use of *Schemes 1, 3, 5 and 6*, it can be seen that as the angular velocity on the shaft of the electric motor increases, the output angular velocity increases (i.e. an acceleration process is observed). The use of *Schemes 2 and 4* shows that with increasing angular velocity on the shaft of the electric motor, the output angular velocity slows down (i.e. a braking process is observed). Note that using *Scheme 5* will increase the angular velocity in the opposite direction. Considering the conditions for the design of a continuously variable electromechanical transmission, i.e. ensuring zero starting speed when the shaft of the electric motor is stopped ($i_E =$ 0), based on the results of Fig. 4, two design schemes will be considered in the future: *Scheme 5* and *Scheme 6*.

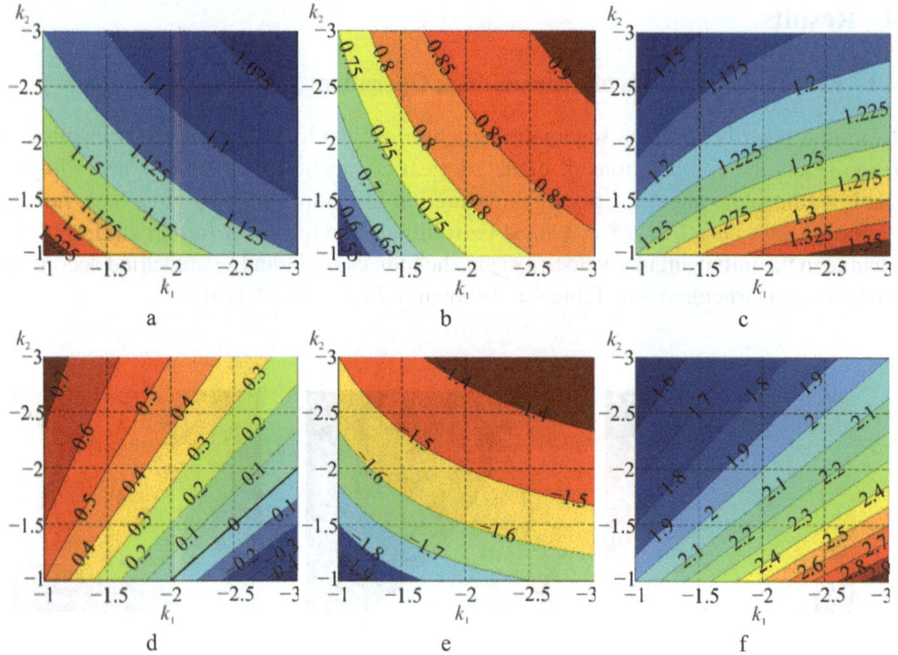

Fig. 5. The results of the study, depending on the structural scheme from Fig. 3(a–f) at $i_E = 2$

The use of *Scheme 5* requires the combination of planetary mechanisms with $k_1 = k_2 = -1$ to be observed, in addition to which a spur gear with a ratio of –1 must be added to the output link in order to change the direction of rotation of the output angular velocity. The use of *Scheme 6* requires that the design parameters of the planetary gears are equal, i.e. $k_1 = k_2$. *Schemes 1 – 4* can be implemented by incorporating them as part of the electromechanical transmission. Based on the above definition of the change in output torque, it is reduced to the calculation of only two schemes (*Schemes 5, 6*) from Fig. 3.

4.2 Forces Analysis of Transmission Operation

In the second stage, we will carry out a force analysis of the possible electromechanical transmission schemes presented in Fig. 3, determining the variation in torque (M_Σ) of the output link of the transmission when the structural structure of the planetary gears (k_1, k_2) is varied. Note that regardless of the schematic structure of the electromechanical transmission (Fig. 3) and the combinations of planetary gear connections (k_1 and k_2), according to Table 1, zero torque is guaranteed at the output link of M_Σ at $j_E = 0$, i.e. the condition $M_\Sigma = M_{em} = 0$ Nm at $M_{\text{nom}} = 1$ Nm must be met. The results of further studies of the variation of the output torque of the M_Σ when the indicators j_E, k_1 and k_2 are varied according to the circuit structure from Table 1 are shown in Figs. 6 and 7.

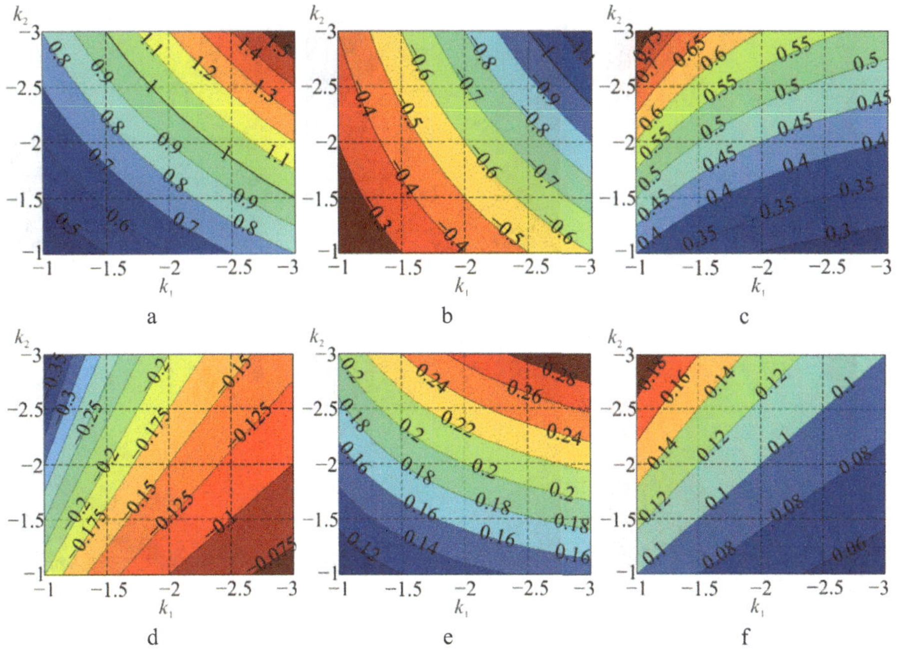

Fig. 6. The results of the study, depending on the structural scheme from Fig. 3(a–f) at $j_E = 0.1$

Analyzing the results from Figs. 6 and 7, i.e. using all the circuit solutions from Fig. 3, it can be seen that as the torque on the shaft of the electric motor increases, so does the output torque. The use of *Schemes 2, 4* shows that as the torque on the shaft of the electric motor increases, the output torque increases in the opposite direction.

4.3 Forces Analysis of Transmission Operation

Systematizing the results from Figs. 4, 5, 6 and 7, we find that given the outlined requirements for a continuously variable electromechanical transmission, *Scheme 5* (where $k_1 = k_2 = -1$) and *Scheme 6* (where $k_1 = k_2$) are the most appropriate. Therefore, in the next step we will analyze the dependencies of the output power, the conditional indicator j_E and i_E, the result is presented in Fig. 8.

As can be seen from Fig. 8, in order to achieve unit output power it is necessary to satisfy condition $\frac{i_E(i_E-5)}{j_E-2} = 4$, we note that the conditional values j_E and i_E must be either strictly positive or strictly negative. We also note that for *Scheme 5* in such a configuration it is necessary to implement a spur gear with a ratio $u_\Sigma = -1$ on the output link.

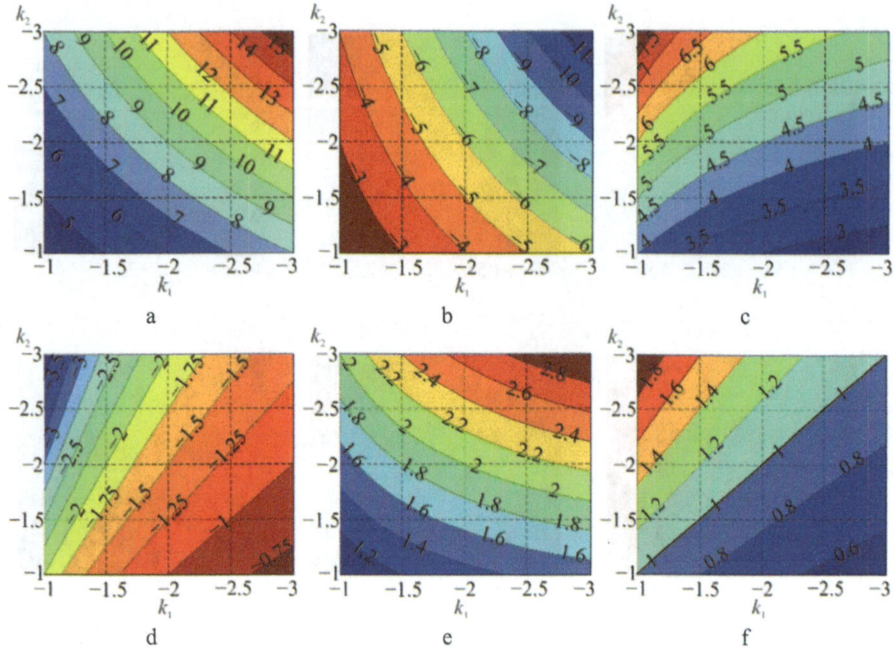

Fig. 7. The results of the study, depending on the structural scheme from Fig. 3(a–f) at $j_E = 1$

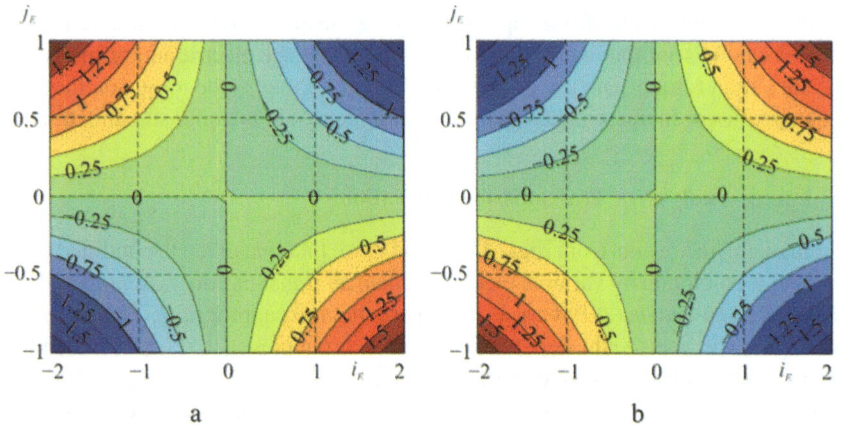

Fig. 8. Dependencies of the output power for *Scheme 5* (a) and *Scheme 6* (b) when changing the conditional indicators j_E and i_E under the condition $k_1 = k_2 = -1$.

5 Conclusions

A search was carried out for the structure of a continuously variable electromechanical transmission with power flow adders placed on the output link. An analysis of the kinematic and performance parameters of the transmission operation is presented, taking into account the peculiarities of the operation of the electric motor and the internal combustion engine. Two possible schemes are presented that provide a zero start, created by balancing the power flows from the electric motor and the combustion engine. In addition, the resulting design schemes ensure equality of input (from the ICE) and output power flowing into the continuously variable electromechanical transmission.

References

1. Samoilenko, D., Savchenko, A., Kravchenko, S.: Improving the performance of diesel engines fueled with water-fuel emulsion. Combustion Engines **194**(3), 116–122 (2023)
2. Dumenko, P., Kravchenko, S., Prokhorenko, A., Talanin, D.: Formation and study of static and dynamic characteristics of electronically controlled diesel engine. Latv. J. Phys. Tech. Sci. **56**(2), 12–23 (2019)
3. Gao, Y., Ehsani, M., Miller, J.M.: Hybrid electric vehicle: overview and state of the art. In: Proceedings of IEEE International Symposium on Industrial Electronics, Dubrovnik, Croatia, 20–23 June, pp. 307–315 (2005)
4. Kozhushko, A., et al.: Analysing the response of a dual-flow transmission (HMCVT) for wheeled tractors according to efficiency and productivity criteria. Int. J. Mechatron. Appl. Mech. **16**, 33–41 (2024)
5. Pei, H., Hu, X., Yang, Y., Tang, X., Hou, C., Cao, D.: Configuration optimization for improving fuel efficiency of power split hybrid powertrains with a single planetary gear. Appl. Energy **214**, 103–116 (2018)
6. Vora, A.P., Jin, X., Hoshing, V., Saha, T., Shaver, G., Varigonda, S.: Design-space exploration of series plug-in hybrid electric vehicles for medium-duty truck applications in a total cost-of-ownership framework. Appl. Energy **202**, 662–672 (2017)
7. Onori, S., Serrao, L., Rizzoni, G.: Hybrid electric vehicles: energy management strategies. Encycl Energy **277**(4), 197–213 (2016)
8. Jeong, J., Kim, N., Lim, W., Park, Y., Cha, S.W., Jang, M.E.: Optimization of power management among an engine, battery and ultra-capacitor for a series HEV: a dynamic programming application. J. Automot. Technol. **18**(5), 891–900 (2017)
9. Yang, Y., Pei, H., Hu, X., Liu, Y., Hou, C., Cao, D.: Fuel economy optimization of power split hybrid vehicles: arapid dynamic programming approach. Energy **166**, 929–938 (2019)
10. Li, G., Gorges, D.: Energy management strategy for parallel hybrid electric vehicles based on approximate dynamic programming and velocity forecast. J. Franklin Inst. **356**(16), 9502–9523 (2019)
11. Zou, Y., Yang, Y., Zhang, Y., Liu, C.: Computationally efficient assessment of fuel economy of multi-modes and multi-gears hybrid electric vehicles: a hyper rapid dynamic programming approach. Energy **313**, 133811 (2024)
12. Cipek, M., Kasać, J., Pavković, D., Zorc, D.: A novel cascade approach to control variables optimisation for advanced series-parallel hybrid electric vehicle power-train. Appl. Energy **276**, 115488 (2020)
13. Li, X., Wang, Y., Yang, D., Chen, Z.: Adaptive energy management strategy for fuel cell/battery hybrid vehicles using Pontryagin's Minimal Principle. J. Power. Sources **440**, 227105 (2019)

14. Song, K., Wang, X., Li, F., Sorrentino, M., Zheng, B.: Pontryagin's minimum principle-based real-time energy management strategy for fuel cell hybrid electric vehicle considering both fuel economy and power source durability. Energy **205**, 118064 (2020)
15. Yao, M., Zhu, B., Zhang, N.: Adaptive real-time optimal control for energy management strategy of extended range electric vehicle. Energy Convers. Manag. **234**, 113874 (2021)
16. Chen, Zh., Wu, S., Shen, Sh., Liu, Y., Guo, F., Zhang, Y.: Co-optimization of velocity planning and energy management for autonomous plug-in hybrid electric vehicles in urban driving scenarios. Energy **263**(F), 126060 (2023)
17. Jui, J., Ahmad, M., Molla, M., Rashid, M.: Optimal energy management strategies for hybrid electric vehicles: a recent survey of machine learning approaches. J. Eng. Res. **12**(3), 454–467 (2024)
18. Kozhushko, A., Tkachov, V., Mittsel, M., Sokolik, S.: Optimization of traction properties the electric tractor based on the simulation dlg-powermix test cycles. Int. J. Mechatron. Appl. Mech. **15**, 70–78 (2024)
19. Basova, Y., Dobrotvorskiy, S., Balog, M., Iakovets, A., Amine, C.M., Zinchenko, A.: Increasing sme supply chain resilience in the face of rapidly changing demand with 3D model visualization. Int. J. Mechatron. Appl. Mech. **14**, 35–47 (2023)

The Impact of Process Audits on Sustainability in Automotive Manufacturing

Ioan Danut Iuonas[1], Aurel Mihail Titu[2,3](✉), Camelia Cristina Dragomir-Pânzaru[3,4], Florin Claudiu Iuonas[1], and Florentina Badea[5]

[1] Faculty of Industrial Engineering and Robotics, National University of Science and Technology POLITEHNICA Bucharest, Splaiul Independentei No. 313, 6th District, Bucharest, Romania

[2] Faculty of Engineering, Industrial Engineering and Management Department, "Lucian Blaga" University of Sibiu, 10 Victoriei Street, 550024 Sibiu, Romania
mihail.titu@ulbsibiu.ro

[3] Academy of Romanian Scientists, 3 Ilfov Street, Bucharest, Romania

[4] Transilvania University of Brasov, 29 Eroilor Street, Brasov, Romania
camelia.dragomir@unitbv.ro

[5] National Institute of Research - Development for Mechatronics and Measurement Technique - INCDMTM Bucharest, Bucharest, Romania

Abstract. Sustainability has emerged as a strategic priority in the automotive industry, driven by increasing regulatory pressures, customer expectations, and the urgent need to minimize environmental impact. In this context, organizations are called to rethink their production processes and quality management approaches to align with sustainability objectives. Process audits, traditionally focused on compliance and operational efficiency, are now evolving to play a key role in promoting sustainable practices. By systematically evaluating production activities, process audits can identify inefficiencies, excessive resource consumption, and opportunities for waste reduction, energy saving, and improved environmental performance. This paper explores the impact of process audits on sustainability initiatives within the automotive manufacturing sector. It highlights how audits contribute not only to compliance with environmental standards but also to fostering a culture of continuous improvement oriented towards ecological responsibility. Key benefits such as optimized resource use, reduced emissions, and cost savings are discussed, along with challenges like the integration of sustainability metrics into traditional audit frameworks and resistance to change within organizations. Moreover, the paper examines emerging trends, including the use of digital tools and data analytics in enhancing the sustainability outcomes of audits. Through this analysis, the study emphasizes the critical role of process audits in driving sustainable transformation and offers insights into best practices and future directions for companies aiming to strengthen their environmental performance while maintaining high product quality standards.

Keywords: Process Audits · Sustainability · Automotive Industry · Environmental Performance · Quality Management

D. D. Cioboată and J. Machado (Eds.): ICoRSE 2025, LNNS 1592, pp. 367–375, 2025.
https://doi.org/10.1007/978-3-032-02508-1_31

1 Introduction

The automotive industry stands at the forefront of the global shift toward sustainability. Increasing regulatory demands, consumer expectations, and environmental concerns are pushing manufacturers to rethink their operations and adopt more sustainable practices (European Commission, 2020). Traditional quality management systems and production audits, historically focused on compliance and efficiency, are no longer sufficient to meet these new demands. Companies are now expected not only to deliver high-quality products but also to minimize their environmental footprint and promote social responsibility throughout their value chains (Elkington, 1997).

Process audits, long regarded as tools for verifying conformity to technical standards such as ISO 9001 and IATF 16949, are evolving into strategic mechanisms for driving sustainability performance (Hoyle, 2017). By expanding the scope of audits to include environmental, social, and governance (ESG) criteria, organizations can systematically identify opportunities for reducing resource consumption, minimizing waste, optimizing energy use, and promoting circular economy principles (Lee et al., 2019).

At the same time, the rapid digital transformation of the industry—through innovations such as the Internet of Things (IoT), blockchain, and big data analytics—enables more accurate, real-time monitoring of sustainability performance (Jabbour et al., 2020). Process audits are thus becoming dynamic, data-driven tools capable of supporting continuous improvement toward both operational excellence and ecological responsibility.

This paper explores the growing role of process audits as catalysts for sustainability in the automotive sector. It analyzes the main mechanisms through which audits contribute to sustainable development, presents recent data and best practices, and discusses future trends and recommendations for companies aiming to strengthen their environmental and social performance.

2 Process Audits as Drivers of Sustainability

Process audits have traditionally focused on verifying compliance with technical specifications, customer requirements, and regulatory standards (Hoyle, 2017). Their primary objective was to ensure product quality and operational efficiency. However, the emergence of sustainability as a critical business priority has expanded the scope and purpose of process audits.

Today, process audits serve not only to validate operational conformity but also to identify areas for improvement in environmental, social, and energy performance (Smith and Johnson, 2020). By systematically evaluating the resource inputs, energy usage, waste generation, and environmental impacts of production processes, audits can highlight inefficiencies and propose corrective actions that align with sustainability goals.

The table below illustrates the key differences between traditional process audits and sustainability-focused process audits (Table 1):

By integrating sustainability considerations into audits, companies can move from a purely reactive model—focused on identifying and correcting defects—to a proactive

Table 1. Comparison Between Traditional and Sustainability-Focused Process Audits

Aspect	Traditional process Audits	Sustainability-Focused Process Audits
Primary Objective	Compliance and operational efficiency	Environmental, social, and operational excellence
Scope of Evaluation	Processes, product specifications	Processes, resource use, environmental impact, social practices
Metrics	Conformity to standards (ISO 9001, IATF 16949)	Carbon footprint, energy efficiency, waste reduction, ESG compliance
Tools Used	Checklists, documentation reviews, interviews	IoT sensors, data analytics, sustainability KPIs
Output	Non-conformities, corrective actions	Sustainability reports, improvement roadmaps
Long-term Impact	Product quality assurance	Organizational resilience and sustainable growth

Sources: Hoyle, 2017*; Lee et al.,* 2019*; Jabbour et al.,* 2020

model that anticipates environmental risks, fosters innovation, and builds long-term resilience.

Additionally, sustainability-focused audits support compliance with emerging regulations such as the European Union's Corporate Sustainability Reporting Directive CSRD) and help organizations achieve recognized certifications such as ISO 14001 (Environmental Management Systems) and ISO 50001 (Energy Management Systems) (ISO, 2018).

Real-World Impact Example:

A 2023 industry survey (McKinsey and Company, 2023) found that automotive manufacturers who systematically integrated sustainability into their auditing practices reported a 12% average reduction in operational costs and a 20% improvement in environmental performance over three years.

The digital transformation of process audits has significantly enhanced their ability to support sustainability objectives. Modern technologies such as the Internet of Things (IoT), Artificial Intelligence (AI), and big data analytics enable real-time monitoring, predictive analysis, and more precise identification of sustainability-related risks (Jabbour et al., 2020).

IoT devices installed on production equipment can continuously track metrics such as energy consumption, emissions levels, and material usage. These data points are automatically collected and analyzed, allowing auditors to detect anomalies or inefficiencies without relying solely on periodic manual checks. Predictive maintenance systems, powered by AI algorithms, can forecast equipment failures, reducing downtime and optimizing energy usage.

Blockchain-based digital twins are gaining relevance as simulation tools for real-time monitoring and predictive sustainability assessment (Suhail et al., 2021).

Furthermore, blockchain technology is being explored for audit purposes to ensure the transparency and integrity of sustainability data across supply chains (Xu et al., 2021). Such technological integrations transform process audits from static compliance exercises into dynamic systems that drive real-time decision-making and continuous improvement.

A recent survey conducted by Deloitte (2023) among automotive manufacturers showed that companies leveraging digital technologies in their auditing processes achieved a 15% faster identification of sustainability improvement opportunities compared to companies using traditional audit methods.

The Fig. 1 illustrates the key benefits organizations have reported from implementing sustainability-focused process audits:

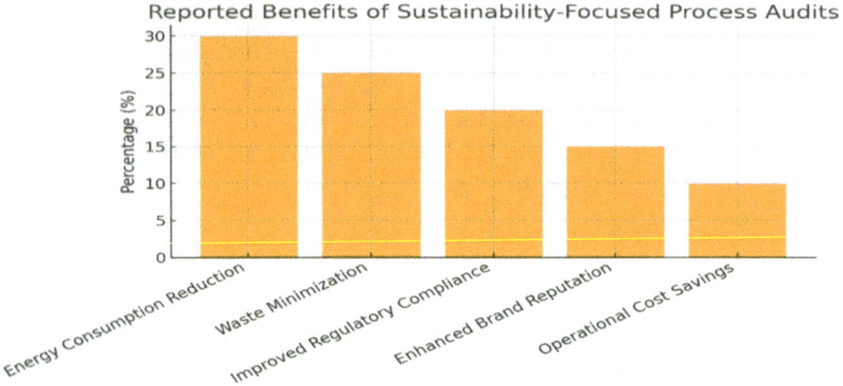

Fig. 1. Reported Benefits of Sustainability-Focused Process Audits

In conclusion, the role of process audits in driving sustainability within the automotive sector has become increasingly significant. Organizations that evolve their audit frameworks toward sustainability reap measurable benefits, including cost savings, improved compliance, and strengthened brand reputation.

The link between process audits and sustainability is thus essential for gaining strategic advantages, including brand differentiation, customer loyalty, operational cost reductions, and resilience against market volatility (Porter and Kramer, 2019).

A digital audit workflow is shown in Fig. 2.

3 Future Perspectives and Recommendations

The digitalization of process audits is expected to accelerate rapidly in the coming years. Technologies such as Artificial Intelligence (AI), Internet of Things (IoT), and blockchain are increasingly embedded into production and quality systems. AI can identify anomalies and forecast sustainability risks using pattern recognition on historical and real-time data (Xu et al., 2021). In smart factories, IoT sensors continuously track environmental parameters—such as emissions, energy consumption, and water usage—providing auditors with accurate, real-time data instead of retrospective reports.

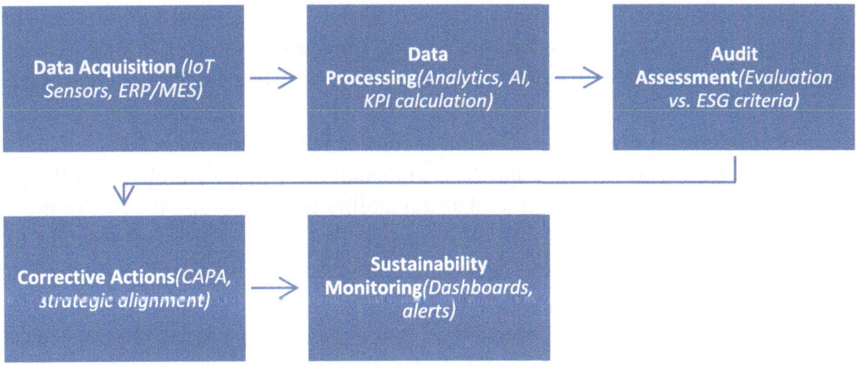

Fig. 2. Schematic representation of a modern sustainability-focused audit workflow, integrating digital data collection, KPI analysis, and continuous improvement mechanisms

Furthermore, blockchain is gaining momentum as a tool for ensuring transparency and traceability in sustainability reporting. For instance, blockchain-based systems are now deployed in logistics and warehouse applications to improve sustainability tracking across the supply chain (Fernandez-Carames et al., 2024).

Through immutable ledgers, companies can record sustainability-related transactions such as recycled material usage or carbon offsets, making audits more reliable and tamper-proof.

Recent developments also highlight how blockchain-based AI models are being integrated in industrial IoT environments to enhance traceability and autonomous decision-making (Rahman et al., 2024) (Fig. 3).

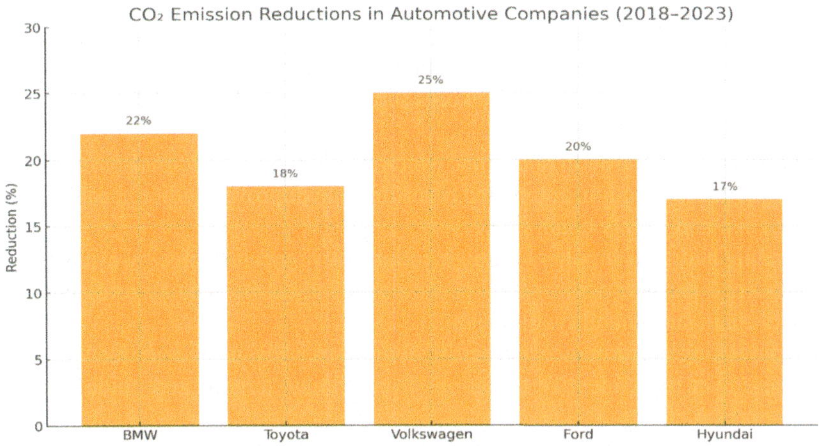

Fig. 3. Estimated CO_2 emission reductions achieved by selected automotive manufacturers between 2018 and 2023. Data is based on publicly available sustainability reports and reflects progress toward carbon neutrality goals.

Legislative pressure is intensifying across all sectors, especially in automotive, where environmental impact is closely scrutinized. The European Union's Corporate Sustainability Reporting Directive (CSRD), which came into effect in 2024, requires large companies to disclose not only financial but also environmental, social, and governance (ESG) performance in standardized formats (European Commission, 2020).

This regulation expands upon the Non-Financial Reporting Directive (NFRD) and mandates assurance-level verification of sustainability data—making the role of process audits crucial in supporting credible disclosures. Companies will need to audit their carbon emissions (Scope 1, 2, and eventually 3), water use, biodiversity impacts, and human rights policies across the value chain.

Additionally, the European Green Deal and Fit for 55 package impose concrete goals like carbon neutrality by 2050, which further emphasizes the importance of integrating sustainability into internal auditing mechanisms (European Commission, 2020). Future process audits must go beyond production processes and include areas such as procurement, logistics, and end-of-life product management.

Frameworks such as the Global Reporting Initiative (GRI, 2021), the Sustainability Accounting Standards Board (SASB, 2023), and the European Sustainability Reporting Standards (EFRAG, 2023) are being harmonized to provide consistent audit reference points.

To prepare for this evolving landscape, automotive companies must adopt a strategic approach to sustainability audits.

Based on recent research (e.g., Lee et al., 2019; Ghobakhloo and Fathi, 2021), the following strategic actions are recommended to enhance the effectiveness of sustainability-focused process audits in the automotive industry:

- Embedding ESG metrics into audit criteria, including energy consumption, greenhouse gas emissions, diversity and inclusion, and responsible sourcing practices, to reflect a comprehensive sustainability performance assessment.
- Training multidisciplinary audit teams, with competence not only in quality management, but also in environmental regulation compliance, data analytics, and standardized sustainability reporting frameworks such as GRI, SASB, and ESRS (GRI, 2021; SASB, 2023; EFRAG, 2023).
- Implementing digital audit platforms, which enable centralized cloud-based planning, real-time data collection, automated reporting, and improved audit traceability across production and supply chain processes.
- Extending audits to suppliers through collaborative mechanisms, by evaluating Tier 1 and Tier 2 partners against ESG performance indicators, thus ensuring end-to-end supply chain sustainability alignment.
- Benchmarking performance and setting long-term KPIs, using audit data to monitor progress over time, identify deviations from climate targets, and guide continuous improvement efforts.
- Promoting an internal culture of sustainability, through employee engagement, training programs, and integrating environmental responsibility into individual and departmental performance evaluations (Table 2).

Table 2. Key performance indicators (KPIs) commonly integrated into sustainability-oriented process audits in the automotive industry. These metrics help assess resource efficiency, environmental impact, and supply chain responsibility.

KPI	Unit of Measure	Relevance in Automotive
Energy Consumption	kWh/unit produced	Emissions, cost control
CO2 Emission	kg CO2/unit	Greenhouse gas impact
Water Usage	Liters/Unit	Resource efficiency
Material Waste	%Waste/recyclable	Circular economy
Supplier ESG Compliance Rate	% compliant suppliers	Supply chain sustainability

3.1 Audit Digital Maturity Levels

Another emerging direction in the audit transformation process is the evaluation of digital maturity within organizations. Audit frameworks can include a structured assessment of how integrated digital tools and sustainability metrics are across production, quality control, and supply chain management. Companies at higher maturity levels exhibit real-time dashboards, automated data collection, and predictive maintenance systems, which significantly increase the accuracy and timeliness of sustainability audits. A standardized maturity model for digital audits can help benchmark progress and prioritize future investments (Ghobakhloo and Fathi, 2021; Lee et al., 2019).

Nanotechnologies and virtual environments are also beginning to intersect with audit processes, particularly in simulation-based assessments and digital auditing platforms. As highlighted by Pirnau et al. (2024), the integration of nanotechnologies in virtual frameworks creates new opportunities for high-precision monitoring and analysis, supporting more advanced forms of sustainability evaluation.

Emerging approaches such as federated learning integrated with blockchain offer promising frameworks for secure, decentralized audit data management (Kang et al., 2022).

3.2 Lifecycle-Based Audit Approach

Future-oriented audits are also expected to adopt a full lifecycle perspective. Rather than focusing solely on the manufacturing stage, sustainability audits will evaluate environmental and social impacts across the product's entire lifecycle: from raw material sourcing and production to use phase, disassembly, and end-of-life treatment. This broader approach aligns with circular economy goals and EU sustainability regulations and enables more comprehensive sustainability reporting and decision-making (Kalmykova et al., 2018; European Commission, 2020).

Moreover, fostering a culture of sustainability requires not only technical systems but also educational and organizational development. As shown by Țîțu et al. (2013), creativity, quality education, and collaborative networks are key drivers in building innovative and resilient industrial clusters, which are increasingly relevant in the transition to green manufacturing ecosystems.

Process audits are increasingly becoming a foundation for obtaining and maintaining sustainability-related certifications. Standards such as ISO 14001, ISO 50001, and future ESG-related audit schemes require consistent internal auditing. Companies that embed sustainability in their audit cycles are better positioned to respond to customer demands, enter green supply chains, and access financing tied to ESG performance (ISO, 2018; Global Reporting Initiative, 2021). This trend reinforces the strategic value of audits not only for compliance but also for competitive positioning.

4 Conclusions

This paper has demonstrated that process audits play a growing and transformative role in driving sustainability within the automotive manufacturing industry. Traditionally focused on compliance and operational efficiency, process audits are now evolving into strategic tools that promote environmental and social responsibility, support innovation, and reinforce resilience across operations and supply chains.

Through the integration of digital technologies such as IoT, AI, and blockchain, audits are becoming increasingly data-driven, predictive, and capable of capturing real-time sustainability metrics. Moreover, the growing pressure from international regulations—including the Corporate Sustainability Reporting Directive (CSRD) and the European Green Deal—makes it essential for companies to adapt their audit methodologies to reflect both environmental performance and social impact.

The future of process audits lies in their ability to go beyond production lines and assess the full product lifecycle, integrate ESG criteria, and benchmark progress using standardized sustainability indicators. Organizations that embrace this evolution will be better positioned to meet regulatory demands, strengthen supply chain accountability, and enhance their competitive standing in a decarbonizing global economy.

Therefore, companies in the automotive sector must recognize the strategic value of sustainability-focused process audits—not merely as tools for risk control, but as levers for long-term transformation and alignment with global sustainability goals.

References

Elkington, J.: Cannibals with Forks: The Triple Bottom Line of 21st Century Business. Capstone Publishing, Oxford (1997)

European Commission: The European Green Deal. https://ec.europa.eu. Accessed 20 May 2024

EFRAG: European Sustainability Reporting Standards (ESRS). https://www.efrag.org. Accessed 20 May 2024

Fernandez-Carames, T.M., Fraga-Lamas, P.: Towards an autonomous industry 4.0 warehouse: a UAV and blockchain-based system for inventory and traceability applications in big data-driven supply chain management. arXiv preprint arXiv:2402.00709 (2024)

Ghobakhloo, M., Fathi, M.: Industry 4.0 and digital supply chain: enhancing digital audit maturity. J. Clean. Prod. **289**, 125155 (2021)

Global Reporting Initiative (GRI): GRI Standards. https://www.globalreporting.org. Accessed 20 May 2024

Hoyle, D.: ISO 9000 Quality Systems Handbook 7th edn. Routledge, London (2017)

ISO: ISO 14001:2015 Environmental Management Systems—Requirements with Guidance for Use. ISO, Geneva (2018)

Jabbour, C.J.C., de Sousa Jabbour, A.B.L., Sarkis, J., Filho, M.G.: Unlocking the circular economy through new business models based on large-scale data: an integrative framework and research agenda. Technol. Forecast. Soc. Change **144**, 1–14 (2020)

Kalmykova, Y., Sadagopan, M., Rosado, L.: Circular economy – from review of theories and practices to development of implementation tools. Resour. Conserv. Recycl. **135**, 190–201 (2018)

Kang, J., Yu, R., Xiong, Z., et al.: Blockchain-based federated learning for industrial metaverses: incentive scheme with optimal AoI. arXiv preprint arXiv:2206.07384 (2022)

Kotter, J.P.: Leading Change. Harvard Business Review Press, Boston (2012)

Lee, K., Park, J., Kim, H.: Integrating sustainability into quality management systems: challenges and opportunities J. Clean. Prod. **210**, 1234–1245 (2019)

Pirnau, M., Pirnau, C., Priescu, I., Ţîţu, A.M., Ghiculescu, L.D., Ghena, M.F.: The impact of nanotechnologies in the virtual world. Acta Tech. Napoc. Appl. Math. Mech. Eng. **67**(3) (2024)

Porter, M.E., Kramer, M.R.: Creating shared value. Harv. Bus. Rev. (2019)

Rahman, A., Shahin, M., Kabir, M.A.: Blockchain-based AI methods for managing industrial IoT: recent developments, integration challenges and opportunities. arXiv preprint arXiv:2405.12550 (2024)

SASB: Sustainability Accounting Standards for Automobiles. https://www.sasb.org. Accessed 20 May 2024

Smith, A., Johnson, B.: The role of process audits in enhancing sustainability in automotive manufacturing. Int. J. Automot. Technol. Manag. **189**(2), 150–165 (2020)

Suhail, S., Jabbar, S., Khan, M.: Blockchain-based digital twins: research trends, issues, and future challenges. arXiv preprint arXiv:2103.11585 (2021)

Ţîţu, M.A., Pirnau, C., Pirnau, M.: Creativity, education and quality for sustainable development, the real support for the innovative cluster's European network. In: 8th Research/Expert Confernce on "QUALITY 2013", pp. 19–24. Bosnia & Herzegovina (2013)

Xu, X., Sun, J., Wang, S.: A review of machine learning applications in sustainability assessment. J. Clean. Prod. **286**, 125465 (2021)

AI-Driven Supplier Quality Assurance: Enhancing Compliance and Traceability in Automotive Supply Chains

Aurel Mihail Titu[1,2](✉), Claudiu-Alexandru Covaci[3], Camelia Cristina Dragomir-Pânzaru[2,4], and Dănuţ-Iulian Stanciu[5]

[1] Faculty of Engineering, Industrial Engineering and Management Department, "Lucian Blaga" University of Sibiu, 10 Victoriei Street, 550024 Sibiu, Romania
mihail.titu@ulbsibiu.ro

[2] Academy of Romanian Scientists, 3 Ilfov Street, Bucharest, Romania
camelia.dragomir@unitbv.ro

[3] Faculty of Industrial Engineering and Robotics, National University of Science and Technology POLITEHNICA Bucharest, Splaiul Independentei No. 313, 6th District, Bucharest, Romania

[4] Transilvania University of Brasov, 29 Eroilor Street, Brasov, Romania

[5] National Institute of Research - Development for Mechatronics and Measurement Technique - INCDMTM Bucharest, Bucharest, Romania
danut.stanciu@incdmtm.ro

Abstract. This paper examines the transformative potential of integrated RFID-EPCIS-AI systems for quality assurance in automotive supply chains. The study demonstrates how combining RFID sensor networks with EPCIS standards and artificial intelligence enables real-time monitoring of critical quality parameters (humidity, shock, temperature) across the entire supply chain - from supplier warehouses to final assembly. Key findings reveal three core advantages of this technological integration: First, it establishes unprecedented supply chain transparency through automated data capture and standardized information exchange. Second, machine learning algorithms applied to RFID-EPCIS data streams enable predictive quality control, identifying potential defects before they impact production. Third, the system significantly improves operational efficiency by reducing quality incident response times and minimizing manual inspection requirements. The research highlights how AI-driven analysis of sensor data can detect subtle patterns correlating transportation conditions (e.g., vibration levels) with future failure risks. Case evidence shows these systems facilitate proactive interventions at critical control points, transforming traditional reactive quality management into a preventive approach. This study contributes to both academic literature and industry practice by providing a comprehensive framework for implementing smart quality assurance systems that combine physical sensor networks with digital traceability standards and advanced analytics. The results demonstrate measurable improvements in defect prevention, compliance assurance, and supply chain resilience in automotive manufacturing contexts.

Keywords: EPCIS · RFID · Artificial Intelligence · Machine Learning · Predictive quality control · Automotive

D. D. Cioboată and J. Machado (Eds.): ICoRSE 2025, LNNS 1592, pp. 376–384, 2025.
https://doi.org/10.1007/978-3-032-02508-1_32

1 Introduction

The automotive manufacturing sector depends critically on maintaining rigorous supplier quality standards, given its extensive reliance on complex supplier networks. The operational effectiveness, cost optimization, and market position of vehicle producers are fundamentally influenced by the caliber of supplied materials and components. This complex challenge encompasses multiple dimensions, from supply chain coordination to risk mitigation and quality assurance protocols.

The relationship between supplier quality management and manufacturing performance has been well-documented in recent studies. Academic findings demonstrate a strong positive association between superior supplier quality and key operational indicators including production adaptability, on-time delivery rates, and cost-effectiveness (Raissouni, 2023). Notably, supply chain resilience has been identified as a critical factor that sustains quality metrics during operational disruptions (Raissouni, 2023). Maintaining high standards in supplied components serves dual purposes: ensuring adherence to the automotive industry's exacting quality requirements while providing protection against potential operational breakdowns caused by substandard parts.

The strategic integration of supply chain operations serves as a critical driver for quality enhancement in supplier networks. By aligning various supply chain components - encompassing supplier partnerships, customer interfaces, and internal operational frameworks - organizations can create synergistic effects that elevate quality performance. Empirical evidence demonstrates that effective partnership models between automotive manufacturers and their suppliers form the foundation for sustained quality excellence, facilitated by transparent information exchange and collaborative quality improvement programs (Wiengarten, 2010). This cooperative paradigm aligns with established theories of supply chain agility, where organizational responsiveness to quality-related disruptions emerges as a key competitive advantage.

Digital transformation has significantly expanded the scalability of quality management systems in automotive supply chains. Artificial intelligence demonstrates particular value through its adaptive capabilities, enabling dynamic adjustments to quality assurance protocols based on fluctuating production requirements and diverse supplier competencies (Chowdhury, 2024). This adaptability aligns with Industry 4.0 principles, where interconnected smart systems combine IoT-enabled sensors with AI analytics to continuously monitor and verify quality metrics throughout manufacturing processes.

The digital toolkit extends to blockchain solutions, which provide immutable transaction records to strengthen supply chain traceability. When integrated with AI systems, blockchain creates a synergistic quality management ecosystem that ensures both data reliability and process transparency (Chowdhury, 2024). This technological combination empowers manufacturers to:

- Track supplier compliance in real-time
- Maintain comprehensive audit trails
- Prevent quality deviations before they impact production

2 Blockchain and AI in Automotive Material Sourcing

The automotive industry increasingly relies on digital technologies to enhance supplier quality control, with blockchain and artificial intelligence emerging as transformative solutions. Research by Surjandy (2020) demonstrates how blockchain technology establishes immutable records of component provenance, effectively addressing Indonesia's significant challenge with counterfeit automotive parts that impact both manufacturer reputations and consumer safety. Their proposed Blockchain-based Supply Chain Model (BlcSCM) integrates four critical dimensions of the Leavitt Diamond framework, automating compliance through smart contracts while ensuring adherence to local regulations such as Indonesia's Minister of Transportation Regulations No. 33/2018 (Surjandy, 2020).

Complementing these findings, Ada (2023) present empirical evidence of blockchain's operational benefits through a detailed case study implementation. Their research documents a 61.29% reduction in mean waiting time achieved through real-time data sharing between supply chain nodes, along with a 6% decrease in daily operational costs attributable to automated inventory tracking (Ada, 2023). The study particularly highlights blockchain's capacity to improve Inventory Quality Ratio (IQR) by enabling precise differentiation between active and obsolete stock.

Artificial intelligence applications further enhance these blockchain implementations by enabling predictive quality control. Machine learning algorithms analyze historical supplier performance data to identify potential quality deviations before they disrupt production schedules (Ada, 2023). This proactive approach to quality management complements blockchain's documentation capabilities, creating a comprehensive system for supplier evaluation and risk mitigation.

Implementation challenges identified in both studies include significant supplier onboarding costs, particularly for smaller vendors lacking digital infrastructure (Surjandy, 2020). Data standardization issues across different supply chain partners and varying regional compliance requirements present additional barriers to widespread adoption (Ada, 2023).

Building upon these blockchain and AI foundations, emerging applications of EPCIS (Electronic Product Code Information Services) standards demonstrate potential for further enhancing quality management. The integration of EPCIS with existing blockchain architectures could enable:

- Granular tracking of individual components through standardized event recording (what, where, when, why), as demonstrated in the traceability frameworks of both Surjandy's BlcSCM and Ada's Hyperledger implementation.
- Enhanced predictive analytics by combining EPCIS event data with AI systems, extending Ada's defect prediction model to incorporate real-time supply chain events.
- Automated quality alerts when components deviate from specified handling parameters, building upon the smart contract automation principles in Surjandy's research.

This potential integration aligns with studies' findings on the importance of:

- Standardized data formats
- Real-time monitoring capabilities
- Automated compliance checks

3 Integrating EPCIS Standards for Enhanced Predictive Quality Control

Integrating EPCIS (Electronic Product Code Information Services) standards into automotive supply chains has the potential to enhance predictive quality control by creating a unified, real-time data infrastructure to capture critical events across the supply chain. This integration enables manufacturers to benefit from high-resolution traceability of components and materials, thereby improving the predictive capabilities of quality control systems.

One of the primary advantages of incorporating EPCIS standards is the establishment of a standardized framework for capturing and sharing event data (e.g., production events, shipping data, and storage conditions) across various stakeholders. Such standardization facilitates seamless communication among diverse systems and organizations. When this standardized data is consolidated within a collaborative architecture, it provides a reliable foundation for advanced predictive analytics models. These models can assess historical and real-time data to identify patterns that may indicate compromised component quality or process deviations.

Moreover, leveraging EPCIS-based architectures enhances traceability, which is critical in the automotive sector where quality requirements are stringent and product recalls can be extremely costly. By enabling a virtual supply chain environment, manufacturers can effectively trace a product's lifecycle from raw material acquisition to assembly, thereby providing a robust dataset for machine learning algorithms. The availability of such granular event data allows predictive quality control systems to forecast potential quality issues before they propagate through the production process. This proactive approach helps trigger early corrective measures and minimizes the risk of costly quality failures (Fig. 1).

In practice, integrating EPCIS standards with predictive quality control involves the deployment of sensor networks and IoT devices that continuously capture event data according to the EPCIS schema. This data is then stored and analyzed using scalable, real-time databases such as NoSQL systems. Predictive models use these datasets to identify trends, detect deviations, and generate alerts for potential quality degradation. As a result, automotive manufacturers can implement timely corrective actions, optimize maintenance schedules, and enforce stringent quality assurances at every stage of production.

Furthermore, the incorporation of EPCIS enhances regulatory compliance and fosters transparency for both internal stakeholders and external regulators. The immutable record of events ensured by the EPCIS standard provides reliable and auditable data trails-a practice that is increasingly important in today's highly regulated automotive industry. The convergence of standardized data capture with advanced predictive analytics drives not only improved quality outcomes but also operational efficiencies and risk mitigation across the supply chain.

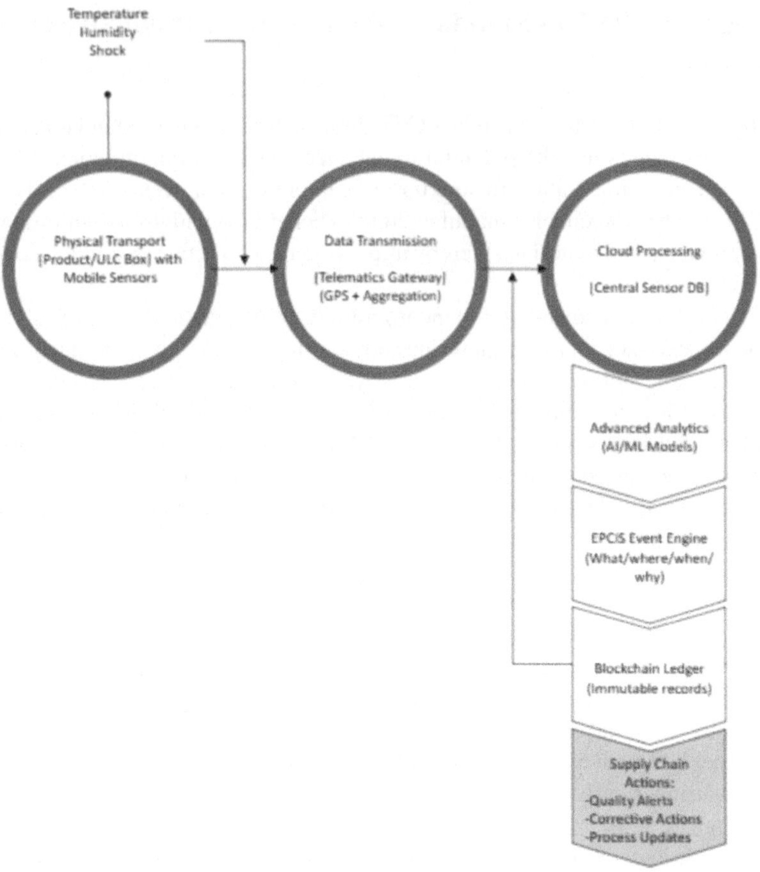

Fig. 1. End-to-End Supply Chain Monitoring System using EPCIS principle

4 Advanced Quality Monitoring Through Integrated RFID-EPCIS-AI Systems

A distinct benefit of RFID sensors lies in their ability to record detailed product handling data. When paired with standards like EPCIS, they enable near-autonomous monitoring of goods, significantly reducing errors associated with manual tracking. For example, in food supply chains, RFID-driven automation has proven effective in bridging physical movements with digital records, thereby enhancing traceability accuracy (Ringsberg and Mirzabeiki, 2013). In industrial settings, RFID surpasses conventional barcoding due to its reusable memory and expanded storage capacity—features critical for documenting quality metrics during production and distribution (Zhao and Zhu, 2011).

Transparency and Compliance

RFID sensors also foster supply chain transparency by enabling real-time anomaly detection. This capability is indispensable in sectors like pharmaceuticals or perishable goods, where maintaining quality and safety standards is essential. Cold-chain applications, for instance, leverage RFID to verify compliance with storage conditions while generating immutable handling records, bolstering both regulatory adherence and consumer confidence (Abad et al., 2009). The technology's adaptability extends to process industries, demonstrating its value even in non-traditional environments.

Stakeholder Integration and Data Utility

Beyond tracking, RFID sensors streamline collaboration across supply chain partners. By sharing sensor-derived data, manufacturers, distributors, and retailers gain a unified view of inventory flows and product histories, which optimizes recall management and quality assurance (Zhao and Zhu, 2011).

Furthermore, integrating RFID with digital platforms creates a secure infrastructure for advanced applications, including predictive analytics and maintenance.

These innovative sensors monitor humidity, shock levels, temperature fluctuations, and UV exposure throughout the entire logistics journey. Their scan-based data transmission eliminates maintenance needs while providing unprecedented visibility into product handling conditions.

These rich data streams feed into an enhanced EPCIS ecosystem that now accommodates environmental parameters through custom XML extensions. The standardized event format ensures seamless interoperability across different partners' systems while maintaining data integrity. For example, a typical event record now includes not just what/where/when information, but detailed environmental context that reveals the "why" behind quality issues.

The true transformation comes from the AI/ML engine that analyzes these data streams in real-time. Using an ensemble of modeling approaches, the system:

- Predicts potential quality defects before they occur
- Identifies problematic handling patterns across suppliers
- Generates targeted improvement recommendations

5 Machine Learning-Driven Interpretation of RFID-Generated Supply Chain Data

The incorporation of artificial intelligence for processing RFID-generated data represents a significant advancement in optimizing supply chain processes, smart production systems, and dynamic operational decision-making. RFID technology produces continuous flows of complex, unstructured information collected from multiple nodes across manufacturing and distribution networks. Modern AI approaches, especially deep neural networks, can convert these raw sensor readings into practical intelligence that enhances stock control, identifies abnormalities, and predicts equipment servicing needs (Harb and Alsayyid, 2016).

A particularly valuable capability of deep learning systems lies in their ability to analyze data types that conventional statistical methods struggle to interpret effectively. As noted by Harb and Alsayyid (2016), neural network architectures demonstrate exceptional performance in detecting nuanced correlations within massive datasets, rendering them particularly appropriate for examining RFID outputs. These systems establish sophisticated representations by processing numerous data points simultaneously, enabling them to decode intricate sensor patterns and immediately highlight potential issues.

The wireless monitoring and instantaneous data transfer capabilities of RFID technology serve as fundamental components for smart factory environments. Recent research by Wang and Liu (2023) demonstrates how RFID facilitates automated monitoring of production inputs, components, and work-in-progress items. When paired with AI analytics, this live operational information can be evaluated to track defect patterns and continuously enhance data tracking for a sustainable supplychain. RFID sensors track components from supplier storage through transport to final assembly, feeding machine learning models with real-time condition data (Wang and Liu, 2023). This enables:

- Early detection of risks (storage humidity, transit shocks)
- Predictive quality alerts before defects occur
- Targeted interventions at critical control points (Fig. 2)

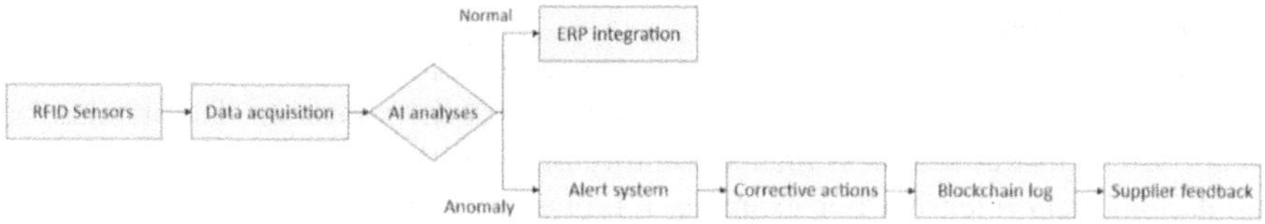

Fig. 2. RFID-AI-Blockchain workflow

The system identifies subtle patterns - like vibration during shipping correlating with future failures - allowing proactive quality assurance.

Effective information handling plays a crucial role in maximizing the benefits of AI-enhanced RFID implementations. Roh (2021) stress the importance of developing comprehensive data acquisition systems and processing workflows capable of managing the substantial quantity and speed of RFID-generated information. The practical application of AI solutions requires meticulous data gathering, cleaning, and annotation procedures to produce training datasets that precisely mirror real-world working environments. Properly prepared, these datasets enable AI systems to execute functions like irregularity identification, where unusual data patterns generate notifications for immediate action (Roh, 2021).

Furthermore, the clarity of AI system outputs remains essential for establishing confidence and accountability among users. Research by Murdoch (2019) highlights how understandable AI models help specialists comprehend and improve decision pathways derived from machine learning analyses. By implementing these methods, companies can maintain an optimal balance between advanced analytical capabilities and the requirement for straightforward, practical recommendations. Such transparency proves vital for meeting compliance standards and supporting evidence-based management choices (Murdoch, 2019).

6 Benefits of RFID-EPCIS-AI Systems in Quality Management

Integrated RFID-EPCIS-AI systems provide essential benefits for quality management in automotive supply chains. By continuously monitoring critical parameters such as humidity, shock levels, and temperature from supplier warehouses to final assembly lines, these solutions ensure enhanced transparency throughout the entire logistics flow. The risk detection potential is amplified through integration with the EPCIS standard, which enables early identification of deviations from optimal parameters and correlation of historical data with current events.

The capability to generate preventive alerts for quality issues represents a significant advancement over traditional reactive methods. As highlighted by Wang and Liu (2023), this approach substantially reduces response times to quality incidents, minimizes the need for manual inspections, and decreases costs associated with returns and waste. Operational efficiency is thus improved on multiple levels.

7 Conclusions

The implementation of RFID-EPCIS-AI systems in the automotive industry represents a fundamental transformation of quality management paradigms. These solutions combine technological advantages such as automatic data capture through RFID sensors, standardized information exchange via EPCIS, and advanced risk analysis through AI/ML algorithms, creating an integrated monitoring and control framework.

The operational benefits are numerous, ranging from real-time monitoring capabilities across the entire supply chain to the ability to anticipate quality issues. The system enables continuous process improvements through machine learning mechanisms, dynamically adapting to new industrial environment requirements.

While RFID-EPCIS-AI systems transform automotive quality assurance through real-time monitoring and predictive analytics, key limitations include: high adoption costs for SMEs, data standardization challenges across global suppliers, latency in multi-tier supply networks, and cybersecurity risks in IoT ecosystems. These constraints suggest hybrid implementation with traditional methods remains necessary until technology maturity improves interoperability and cost-effectiveness, particularly for complex, multi-vendor production environments.

References

Raissouni, R., Hamiche, M., Bourekkadi, S., & Raissouni, K.: The impact of the integrated supply chain on the operational performance of companies in the moroccan electric vehicle sector. In: E3S Web of Conferences, vol. 412, p. 01039 (2023). https://doi.org/10.1051/e3sconf/202341201039

Wiengarten, F., Humphreys, P., Cao, G., Fynes, B., McKittrick, A.: Collaborative supply chain practices and performance: exploring the key role of information quality. Supply Chain Manag. Int. J. **15**(6), 463–473 (2010). https://doi.org/10.1108/13598541011080446

Chowdhury, R.H.: Blockchain and ai: driving the future of data security and business intelligence. World J. Adv. Res. Rev. **23**(1), 2559–2570 (2024). https://doi.org/10.30574/wjarr.2024.23.1.2273

Ada, N., et al.: Blockchain technology for enhancing traceability and efficiency in automobile supply chain—a case study. Sustainability **13**(24), 13667 (2021). https://doi.org/10.3390/su132413667

Surjandy, S., Meyliana, M., Warnars, H.L.H.S., Abdurachman, E.: A proposed supply chain model of blockchain technology-based in automotive component industry. Int. J. Recent Technol. Eng. (IJRTE) **8**(5), 3110–3114 (2020). https://doi.org/10.35940/ijrte.e6440.018520

Ringsberg, H., Mirzabeiki, V.: Effects on logistic operations from rfid- and epcis-enabled traceability. British Food J. **116**(1), 104–124 (2013). https://doi.org/10.1108/bfj-03-2012-0055

Zhao, X.X., Zhu, Y.M.: Supply chain quality information management methodology and application system research. Adv. Mater. Res. **418–420**, 1812–1816 (2011). https://doi.org/10.4028/www.scientific.net/amr.418-420.1812

Abad, E., et al.: Rfid smart tag for traceability and cold chain monitoring of foods: demonstration in an intercontinental fresh fish logistic chain. J. Food Eng. **93**(4), 394–399 (2009). https://doi.org/10.1016/j.jfoodeng.2009.02.004

Harb, A., Alsayyid, A.: Paradigm shift: engineering artificial intelligence and management strategies fusion. Int. J. Res. - GRANTHAALAYAH **4**(2), 1–16 (2016). https://doi.org/10.29121/granthaalayah.v4.i2.2016.2807

Wang, W., Liu, C.: Dynamic capability theory based study on performance of intelligent manufacturing enterprise under rfid influence. Electronics **12**(6), 1374 (2023). https://doi.org/10.3390/electronics12061374

Roh, Y., Heo, G., Whang, S.E.: A survey on data collection for machine learning: a big data - ai integration perspective. IEEE Trans. Knowl. Data Eng. **33**(4), 1328–1347 (2021). https://doi.org/10.1109/tkde.2019.2946162

Murdoch, W.J., Singh, C., Kumbier, K., Abbasi-Asl, R., Yu, B.: Definitions, methods, and applications in interpretable machine learning. Proc. Natl. Acad. Sci. **116**(44), 22071–22080 (2019). https://doi.org/10.1073/pnas.1900654116

Statistical and Neural Network Methods for Determining the Weight of a Bridge Crane

Valentyn Kovalenko[1](✉), Oleksiy Zhuravel[1], Vsevolod Stryzhak[1](✉), Magomediemin Gasanov[1], Rafał Talar[2], and Paweł Zawadzki[2](✉)

[1] National Technical University «Kharkiv Polytechnic Institute», 2 Kyrpychova Street, 61002 Kharkiv, Ukraine
{valentyn.kovalenko,Vsevolod.Stryzhak}@khpi.edu.ua

[2] Poznan University of Technology, 2 Piotrowo Street, 61-138 Poznan, Poland
pawel.zawadzki@put.poznan.pl

Abstract. The article presents a new method for solving the scientific and practical problem of automated calculation of the weight of general-purpose bridge cranes. Numerical data on the weight of existing cranes are given in tables and structured depending on the load capacity and span. Mathematical statistics' hypotheses were used, making it possible to distribute the results according to the normal law under the same accuracy of the obtained data. Based on these assumptions, the least squares method was applied, which made it possible to construct a function of two variables that determines the dependence of the crane weight on the span and load capacity, combining these parameters. A formula was obtained that makes it possible to calculate the weight of cranes programmatically. Based on statistical data, a neural network was built, which, similarly to traditional statistical methods, finds the weight of bridge cranes. The quality of the obtained result was assessed using traditional statistical methods and using a neural network. The work of the statistical model and neural network was studied beyond the data definition area. Functional dependencies for determining the weight of overhead cranes in the range of up to 50 tons are determined based on the regression equation and using a neural network. It is established that forecasting is unacceptable outside the range, but the neural network copes better with the task of extrapolation than traditional methods. The article substantiates the advantages of the proposed method.

Keywords: Neural network · Bridge crane · Crane weight · Automation · Calculation

1 Introduction

Scientific schools in Ukraine and abroad have created a significant theoretical foundation for designing hoisting machine mechanisms, confirming their strength, stability, endurance and other indicators of safe operation. In addition, companies have accumulated a large amount of data on the actual designs of manufactured and operating machines. However, in most cases, approaches to the design of new machines still involve

D. D. Cioboată and J. Machado (Eds.): ICoRSE 2025, LNNS 1592, pp. 385–396, 2025.
https://doi.org/10.1007/978-3-032-02508-1_33

a large amount of manual work. The traditional calculation of the drive power of a crane movement mechanism requires determining the resistance to static movement of the crane. In turn, this parameter depends on specific indicators - wheel diameter, journal diameter, and the coefficient of the flange, which can vary depending on the type of current supply, as well as the shape of the rail and wheel surfaces (friction resistance). That is, at this stage we need to make a large number of design decisions, which is very costly. To solve this problem, we propose to change the approach using mathematical methods of statistics. This will automate the calculation process and reduce the design time, which is relevant today.

2 Research Overview

The weight of cranes is formed due to a large number of factors, but is mainly determined by the strength calculation, which is described in detail in a number of reference books and standards, in particular in [1–3]. The relevance of the topic is also confirmed by the large number of published studies in the field of studying the force factors and dynamics of overhead cranes that accompany their operation. For example, article [4] studies the effect of lateral forces on the undercarriage of an overhead crane. The replacement of analytical calculations with numerical ones is also being investigated for the metal structures of overhead cranes, but for this purpose it is necessary to compare their results. An example of such a study is presented in [5]. Numerical calculations of parameters are used more and more often, especially when selecting electric drives, as follows from publications [6–10]. The issue of improving the calculation in terms of determining the weight of overhead cranes remains insufficiently covered and requires further study. The mentioned topic of the article is also relevant in the context of the concept of modern production Industry 4.0 and the sustainability of manufacturing enterprises [11, 12]. Recently, the use of neural networks and artificial intelligence methods has increasingly found its application in the field of lifting machines. In particular, for crane control and damping of cargo vibrations [13–15], diagnostics of malfunctions of mechanisms and metal structures [16], recognition systems and machine vision [17]. The use of statistical methods and neural networks for calculations and design of cranes has not yet found sufficient distribution and requires further study.

3 Formulation of the Research Objective

The purpose of the study is to form a functional dependence for determining the weight of an overhead crane depending on the lifting capacity and span, which will allow for automated calculation.

4 Research Results

The most accurate determination of the crane's weight occurs at the final design stage, when the design of all its main elements is known. In turn, at the initial stage of crane development, it is also necessary to know the exact value of its weight. This will determine its price and project cost, and the crane weight data is also needed to determine a number of other parameters, such as the power of the travel mechanisms and the strength of the crane runways.

In practice, the initial stages of design are based on data from completed projects. The difficulty lies in the fact that the parameters of the crane to be manufactured may not match those of the cranes already produced. Therefore, based on completed projects, it is necessary to perform an average, which requires manual work and responsible decision-making. The solution to this problem and a way to improve accuracy is to use the statistical method.

The weight of a crane mainly depends on the following factors: operating mode group, span and lifting capacity.

Let's divide cranes according to the ISO 4301 operating mode classification groups as follows: A1–A3 - light, A4–A5 - medium, A6–A8 - heavy.

Using the data from completed projects (we will consider them experimental data), for each of these three groups, we will record in the table the weight of cranes in accordance with the span and lifting capacity. Experimental data for different operating mode groups are shown in Table 1.

Table 1. Experimental data on the weight of overhead cranes depending on the span and lifting capacity for the A6–A8 operating mode groups

Span, m n/a, t	10	14	18	22	26	30
5	14,99	16,88	20,62	25,46	30,36	33,79
10	19,04	21,81	24,65	27,64	31,96	38,08
15	25,39	28,19	31,35	35,49	40,7	48,76
20	27,73	30,17	33,26	37,09	42,28	49,12
30	37,31	41,5	47,57	54,45	60,81	69,93
50	50,89	59,45	63,83	68,52	75,42	83,59
5	14,99	16,88	20,62	25,46	30,36	33,79

Let us test the null hypothesis of a normal distribution of experimental data. For an example of such an analysis, let's use the data in Table 1. Here, we have 36 experimental points, the mathematical expectation of the residuals is $m = 2,9606 \times 10^{-15}$, and their standard deviation is $s = 1,7327$. Figure 1 shows the empirical distribution function of these data, normalised by the formula $x_i = (r_i - m_R)/s_R$, and the standard normal distribution function. Here, the Kolmogorov statistic is 0.1101, which is less than the critical value of 0.2653. Thus, at the significance level of $\alpha = 0.01$, there is no reason to reject the null hypothesis, and the distribution of the experimental results can be considered normal (Fig. 1). Let us build a mathematical model for these data in the form of a polynomial of optimal degree:

$$z = b_1 + b_2 \cdot x + b_3 \cdot y + b_4 \cdot x^2 + b_5 \cdot x \cdot y + \\ + b_6 \cdot y^2 + b_7 \cdot x^3 + b_8 \cdot x^2 \cdot y + b_9 \cdot x \cdot y^2 + b_{10} \cdot y^3 + \ldots$$

where: x- span, m; y - carrying capacity, tonnes.

For all groups of operating modes, we added terms to the model not one at a time, but in groups: linear, quadratic, etc. Table 2 shows the sample variances for different mathematical models and different groups of operating modes and their comparison.

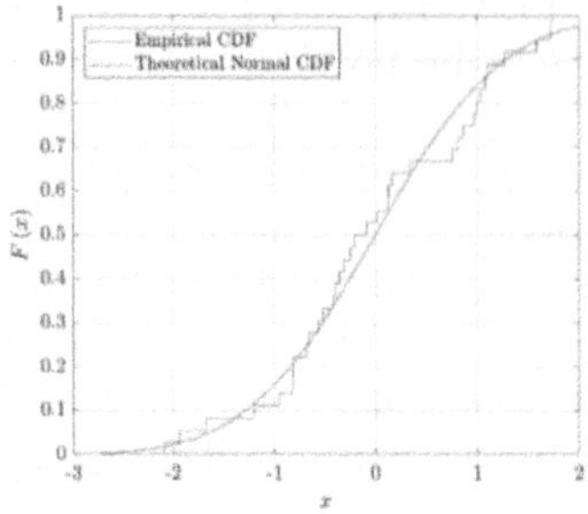

Fig. 1. Empirical and theoretical standard normal cumulative distribution function for crane weight data

Let us explain how to fill in Table 2 using the example of columns A6–A8. To simplify the calculations, we added the terms to the model not one by one, but in groups: linear, quadratic, etc. For the linear model ($m = 3$), the sample variance is $D_3 = 9.3593$, and for the quadratic model ($m = 6$): $D_6 = 5.2323$. Their ratio $D_6/D_3 = 0.5591$, and the quantile of the Fisher *F-distribution* for $f_1 = 36 - 6 = 30$; $f_2 = 36 - 3 = 33$ at the significance level $\alpha = 0.01$ is $F_{0.01}(30{,}33) = 0.4246$. Since $0.5591 > 0.4246$, the decrease in sample variance when taking into account quadratic terms is not significant, but we will still take into account quadratic terms.

Next, we add four more cubic terms: the basis functions x^3, x^2y, xy^2 and y^3. The sample variance of the cubic model is $D_{10} = 4.0417$, and its ratio to the sample variance of the quadratic model is $D_{10}/D_{(6)} = 0.7725$. The corresponding quantile of the Fisher F-distribution is $F_{0.01}(26{,}30) = 0.3996$. Here, too, $0.7725 > 0.3996$, but we'll take into account the cubic terms as well.

Now let's add five more terms of the 4th degree: x^4; x^3y; x^2y^2, xy^3 and y^4. For this model, the sample variance D_{14} is 3.4154; its ratio to the sample variance of the cubic model is $D_{14}/D_{10} = 0.8450$. Квантиль $F_{0.01}(21{,}26) = 0.3610$. Here, too, $0.8450 > 0.3610$; that is, the reduction in sample variance is insignificant. We conclude that the inclusion of higher degree terms does not improve the accuracy of the calculation. Therefore, the 4th degree terms can be disregarded and the cubic model can be used. It should be noted that the experiments conducted for other values gave results that indicate the validity of taking into account the 3rd degree terms, and the 4th degree terms in practically achievable values of crane weight always remain insignificant.

Table 3 shows the calculated parameters of the cubic model and their confidence intervals for cranes of all operating mode groups.

Figure 2 shows the weight versus span and lifting capacity surfaces for different groups of crane operating conditions. The red asterisks indicate the experimental data, and the translucent surfaces indicate the upper and lower confidence limits. The figures

Table 2. Sample variances and their comparison for different mathematical models and different groups of operating modes.

Name	Figure 1, a
mode of operation model	A6–A8
linear: $m = 3$; f = 36 – 3 = 33	$D_3 = 9.3593$
quadratic: $m = 6$; f = 36 – 6 = 30	$D_6 = 5.2323$
comparison of the quadratic model with the linear one: $F_{0.01}(30,33) = 0.4246$	$D_6/D_3 = 0.5591$; $0.5591 > 0.4246$; but we still accept the quadratic model
cubic: $m = 10$; f = 36 – 10 = 26	$D_{10} = 4.0417$
comparison of the cubic model with the quadratic model: $F_{0.01}(26,30) = 0.3996$	$D_{10}/D_6 = 0.7725$; $0.7725 > 0.3996$; but still accept the cubic model
4th degree: $m = 15$; f = 36 – 15 = 21	$D_{14} = 3.4154$
comparison of the 4th degree model with the cubic model: $F_{0.01}(21,26) = 0.3610$	$D_{14}/D_{10} = 0.8450$; $0.8450 > 0.3610$; Reject the 4th degree model
mode of operation model	A6–A8
linear: $m = 3$; f = 36 – 3 = 33	$D_3 = 9.3593$
quadratic: $m = 6$; f = 36 – 6 = 30	$D_6 = 5.2323$
comparison of the quadratic model with the linear one: $F_{0.01}(30,33) = 0.4246$	$D_6/D_3 = 0.5591$; $0.5591 > 0.4246$; but we still accept the quadratic model
cubic: $m = 10$; f = 36 – 10 = 26	$D_{10} = 4.0417$
comparison of the cubic model with the quadratic model: $F_{0.01}(26,30) = 0.3996$	$D_{10}/D_6 = 0.7725$; $0.7725 > 0.3996$; but still accept the cubic model

show that the proposed cubic model approximates the experimental data quite accurately, and all experimental points fall within the confidence limits.

Determining the weight of a crane using neural networks. In statistical models, we define (or choose for theoretical reasons) the form of the theoretical dependence, and then check the adequacy of the model according to certain criteria. Another approach is used in artificial neural networks.

A neural network is a program created on the principles of the human or animal brain. Such a network processes a set of data given to it, but does not require a specific theoretical relationship. During training, it analyses the data, looks for patterns in it, and selects the best mathematical model, in its opinion. The analytical form of this model can be very complex, and not even expressed in a ready-made analytical expression, but rather a kind of algorithm (black box). However, this model can be used to predict the behavior of the object under study at any input.

Table 3. Parameters of the cubic model and confidence intervals for them

Parameter	Values and confidence intervals
b1	$0,660 \leq 10,927 \leq 21,193$
b2	$-0.881 \leq 0.732 \leq 2.345$
b3	$-0.898 \leq -0.494 \leq -0.091$
b4	$-0.118 \leq -0.035 \leq 0.048$
b5	$0.019 \leq 0.042 \leq 0.065$
b6	$0.036 \leq 0.0488 \leq 0.0619$
b7	$-0.0002 \leq 0.001 \leq 0.002$
b8	$-0.0007 \leq -0.0002 \leq 0.0002$
b9	$-0.0005 \leq -0.0003 \leq 0$
b10	$-0.0007 \leq -0.0006 \leq -0.0004$

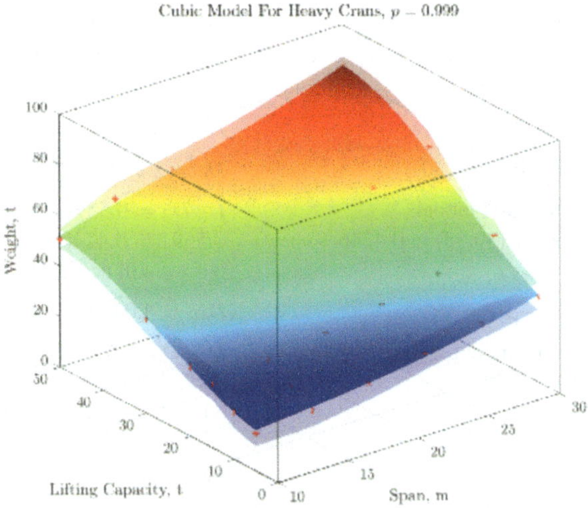

Fig. 2. The result of calculating the crane weight depending on the span and lifting capacity based on the obtained functional relationship for cranes of operating modes: A6–A8

Let's consider the possibility of using neural networks for our task of determining the mass of a crane. We have the values of the function of two variables $z = f(x, y)$ at separate points with possible errors. The task is to predict the behaviour of this function at intermediate points as accurately as possible. For this task, you can use, in particular, deep feed-forward neural networks (DFFNN - Deep Feed Forward Neural Network), which are used in recognition and prediction tasks. Such networks can be represented as an acyclic graph (an oriented graph without oriented cycles): they receive information at the input, process it, and provide it to the output. The input neurons are the initial vertices of the graph (only arcs exit from them), and the output neurons are the final vertices (only arcs enter them). All other nodes (neurons) form the hidden layers of the neural network. From the theory of graphs, it is known that the vertices of an acyclic graph are partially ordered: the DFFNN graph can be depicted in such a way that the input neurons are depicted on one vertical line to the left, the output neurons on one vertical line to the right, and the hidden layers in the middle on one or more vertical lines, with all arcs going from left to right.

DFFNNs are easily programmable and have ready-made solutions in modern mathematical packages. In such networks, each input neuron is connected to each neuron of the first hidden layer, in turn, each neuron of the first hidden layer is connected to each neuron of the second hidden layer, and on up to the output layer. Paper [1] provides recommendations for choosing the number of hidden layers. For simple data, 1–2 layers are enough, for more complex data, 3 to 5 layers are sufficient. To find out how the number of hidden layers affects the accuracy in our case, the following numerical experiment was conducted. The number of hidden layers varied from 1 to 15, and 20 calculations were performed for each value. The dependence of the average accuracy of these 20 calculations on the number of hidden layers is shown in Fig. 3.

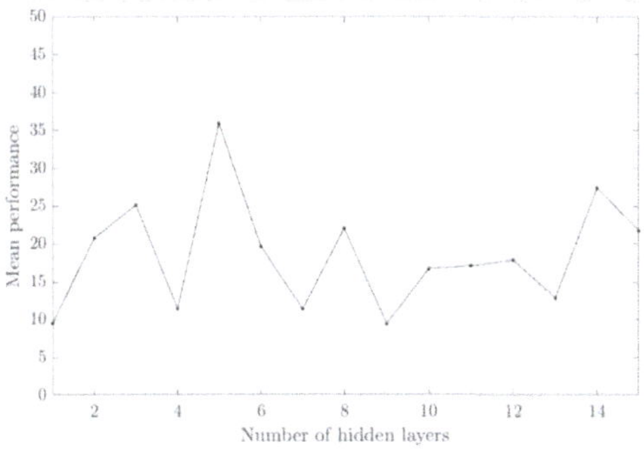

Fig. 3. Dependence of average accuracy on the number of hidden layers

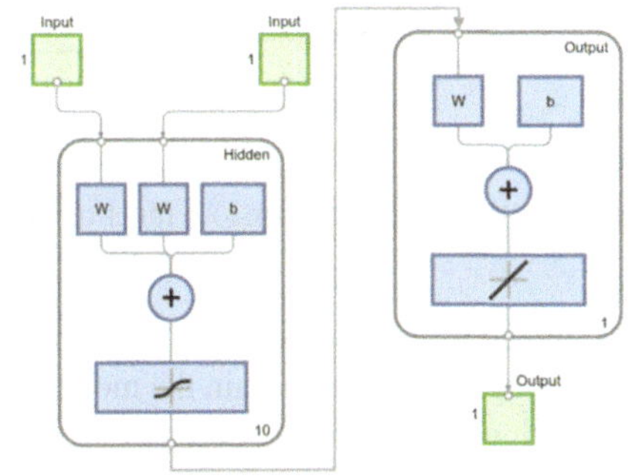

Fig. 4. Structure of the neural network for calculating the weight of overhead cranes

As can be seen from Fig. 3, the accuracy changes in a completely unpredictable way. Therefore, a network with 4 hidden layers was chosen for further calculations according to the recommendations of [1]. In this case, it is necessary to use a network with two inputs (x, y) and one output (z). The structure of this network is shown in Fig. 4.

There are different types of neural network training: supervised, unsupervised, reinforcement learning, etc. In our case, when we know the values of the function z = f(x, y) at certain points, we use supervised training. The control points are the known values of the function. The network learns from this data and then produces results at the required intermediate points. Supervised learning uses a gradient descent algorithm.

Gradient descent is one of the main optimisation algorithms in machine learning and statistics. Its goal is to find the minimum of the loss function by iteratively adjusting the model parameters based on the calculated gradients.

The loss function is a criterion for evaluating the quality of a model by measuring the deviation between predicted values and actual values in a training dataset. It is an integral part of machine learning model training, as it allows you to quantify the model's performance.

The gradient descent procedure involves calculating the gradient of the loss function relative to the model parameters, which allows estimating the direction of the largest decrease in the function. The model parameters are updated using this gradient with a defined learning step, which ensures a gradual approach to the minimum of the loss function.

Since the developed neural network performs the regression task by predicting numerical values, the mean square error was used as a loss function (Fig. 5).

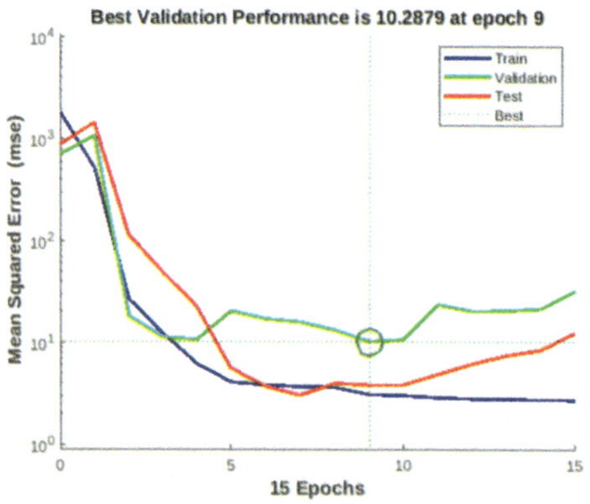

Fig. 5. The value of the neural network loss function during training

When training models, the complete dataset is usually divided into training, validation and testing sets. The training set is used to train the model, the validation set is used to tune hyper-parameters (such as the number of hidden layers mentioned above), and the testing set allows to evaluate the final performance of the model on unknown data.

The graphs of the change in the loss function during training for each of these sets can be used to identify the phenomena of under- or overtraining of the model. In the case of undertraining, the loss function will remain high on both the training and validation sets, which indicates, in particular, that the model is not able to adapt to the training data. Overlearning is manifested in a significant difference between the loss functions on the training and validation sets, where the model demonstrates very low losses on the training data, but high losses on the validation data, indicating that the model is too accurately adapted to the specific features of the training set and is unable to generalise to new data. To solve the training problem, you can either increase the size of the training data set in some way or use a less complex model.

Analysing the changes in the loss function values during the training of the neural network, we observe its gradual decrease, as well as the absence of a large gap between the training data set and the others. This indicates a good balance between the amount of data and the complexity of the model and the absence of under- and overtraining problems.

When training with gradient descent, it is also important to pay attention to the absolute values of the calculated gradients to avoid the phenomena of "disappearing" and "exploding" gradients. Gradient disappearance occurs when the gradients in a neural network become very small, which leads to slow or no learning. In turn, gradient explosion occurs when the gradients become extremely large, which can cause learning instability and large fluctuations in weights. Both problems make it difficult to train deep neural networks efficiently.

On the graph of changes in the absolute values of the gradient during training, fluctuations remain within the normal range, so it is impossible to distinguish the severity of any of these phenomena and their negative impact on training. It can also be noted that the gradient value approaches 0 at the same epoch of training where the lowest value of the loss function was recorded, which indicates that at least a local minimum of the function was reached at this epoch.

Figure 6 shows a surface that displays the result of calculating the crane's mass from the span and lifting capacity using a neural network.

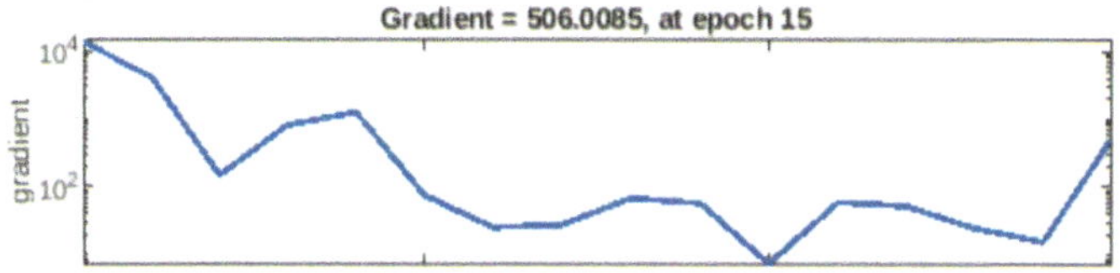

Fig. 6. Absolute value of the gradient over the course of studyAbsolute value of the gradient over the course of study

The error of the network on these data (the sum of the squares of the deviations of the network data from the specified data) is ~3.37285. For comparison, the error of the full cubic model is ~105.0847.

It is worth noting that an artificial neural network works in a similar way to the human brain: it produces different results each time it is run. But almost always, its error is about 10–15 times smaller than the error of the full cubic model (Fig. 7).

Scope of application of the obtained functional dependencies. Since we obtained the functional dependencies based on data covering a relatively narrow range of possible span and load capacity values, it is advisable to investigate the behaviour of the obtained functions for extended ranges, i.e. the possibility of extrapolating the results. Figure 8 shows the graphs of calculating the mass of cranes in operating modes A6–A8 using a polynomial of degree 3 (a) and a neural network (b).

From the graphs in Fig. 8, we can conclude that the found functional dependence approximates the experimental data well in the range where they are known. However, outside the range, the function begins to decay rapidly and gives false results. A slightly better pattern of change in the values of crane weights outside the range of experimental data is observed when calculated using a neural network. However, in both cases, it can be concluded that both of these methods are poorly suited for extrapolation. Therefore, to extend the calculation range, it is necessary to determine the functional dependence and train the neural network on a larger number of experimental data.

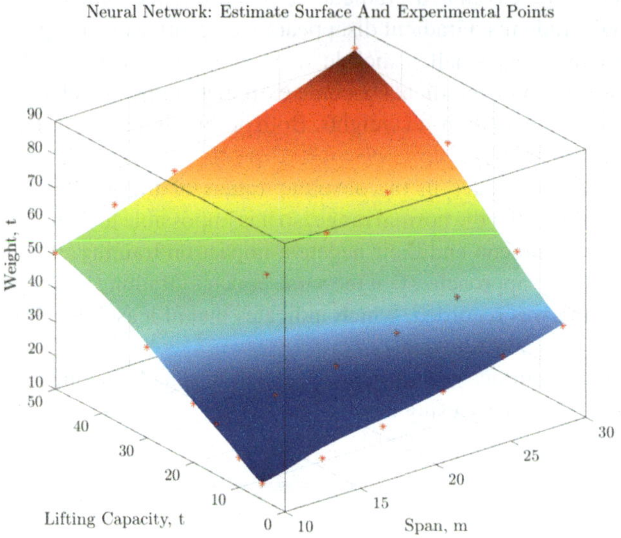

Fig. 7. The result of calculating the crane weight depending on the span and lifting capacity using a neural network for cranes of operating modes A6–A8

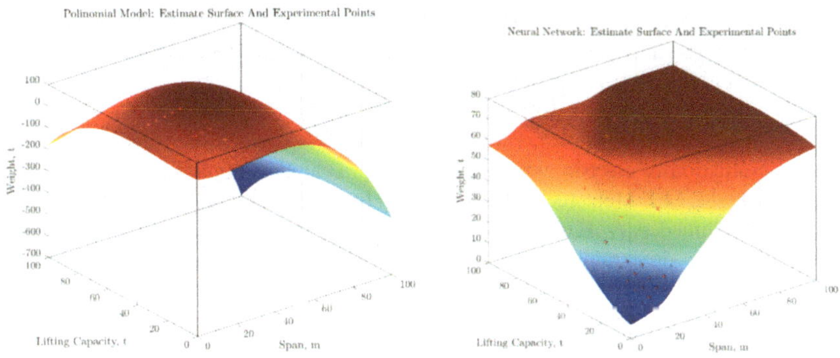

a b

Fig. 8. Comparison of the results of calculating the mass of a crane in the A6–A8 operating mode group depending on span and lifting capacity based on a functional dependence and using a neural network: a - calculation based on a functional dependence; b - calculation using a neural network

5 Conclusions

The article solves the problem of determining the weight of an overhead crane on the basis of statistical data using functional dependencies. It is established that the calculation of parameters by the traditional method requires significant labour costs and is complicated at certain stages of design. The article proposes a concept for calculating the weight of overhead cranes using the methods of mathematical statistics on the basis of the existing theoretical framework and the database of actual values of parameters of previously manufactured and successfully operating cranes. Heavy-duty cranes are considered. Functional dependencies for calculating crane weight use span and lifting capacity values. Dependencies are determined for cranes with lifting capacity up to 50 tons, and, as the study showed, beyond this range the calculation gives incorrect values. It is necessary to note the significant superiority of the neural network for extrapolating forecast results beyond the statistical data area. It is also necessary to take into account the confidence intervals for the calculated crane weight using statistical and neural network methods. The true weight value will be in the middle of this interval.

References

1. Scheffler, M.: Grundlagen der Fördertechnik - Elemente und Triebwerke. Vieweg Verlag (1994)
2. EN 13001-1 Cranes - General design - Part 1: General principles and requirements
3. EN 13001-2 Crane safety - General design - Part 2: Load actions
4. Zelić, A., Zuber, N., Šostakov, R.: Experimental determination of lateral forces caused by bridge crane skewing during travelling. In: Eksploatacja i Niezawodnosc - Maintenance and Reliability, vol. 20, pp. 90–99 (2017). https://doi.org/10.17531/ein.2018.1.12
5. Molnár, D., Blatnický, M., Dižo, J.: Comparison of analytical and numerical approach in bridge crane solution. Manuf. Technol. **22**(2) (2022). https://doi.org/10.21062/mft.2022.018

6. Kozłowski, M., Czerepicki, A.: Quick electrical drive selection method for bus retrofitting. Sustainability (Switzerland) **15**(13) (2023). https://doi.org/10.3390/su151310484
7. Husain, I., et al.: Electric drive technology trends, challenges, and opportunities for future electric vehicles. Proc. IEEE **109**(6) (2021). https://doi.org/10.1109/JPROC.2020.3046112
8. Suryavanshi, S., Ghanegaonkar, P.M., Jadhav, G.K., Wankhede, S.R.: Comparative performance evaluation of electric motor sizing through analytical approach for electric vehicle application. ARAI J. Mob. Technol. **3**(4) (2023). https://doi.org/10.37285/ajmt.3.4.7
9. Akl, M.M., Ahmed, A.A., Rashad, E.E.M.: A wide component sizing and performance assessment of electric drivetrains for electric vehicles. In: 2019 21st International Middle East Power Systems Conference, MEPCON 2019 – Proceedings (2019). https://doi.org/10.1109/MEPCON47431.2019.9008195
10. Belhadi, Y., Kraa, O., Saadi, R., Bahri, M., Telli, K.: Sizing of fuel cell/supercapacitor hybrid system based on frequency splitting of required energy. EEA - Electrotehnica, Electronica, Automatica. **71**(4) (2023). https://doi.org/10.46904/eea.23.71.4.1108005
11. Basova, Y., Dobrotvorskiy, S., Balog, M., Iakovets, A., Chelabi, M.A., Zinchenko, A.: Increasing SME supply chain resilience in the face of rapidly changing demand with 3D model visualisation. Int. J. Mechatron. Appl. Mech. **14**, 35–47 (2023). https://doi.org/10.17683/ijomam/issue14.5
12. Basova, Y., Dobrotvorskiy, S., Yakovenko, I.: Investigation of the impact of force majeure circumstances as a market instability factor on the flexibility and sustainability of engineering SMEs. EAI Endors. Trans. Digit. Transform. Ind. Process. **1**(1) (2025). https://publications.eai.eu/index.php/dtip/article/view/8951
13. Yang, Y., Zhang, T., Zhou, X., Zhang, H., Li, J.: Neural network adaptive PID-like coupling control for double pendulum cranes with time-varying input constraints. Mech. Syst. Sig. Process. **208** (2024). https://doi.org/10.1016/j.ymssp.2023.110997
14. Tang, W., Ma, R., Wang, W., Xu, T., Gao, H.: Adaptive PID control of overhead cranes based on ISOA–NN. Int. J. Dyn. Control **12**(8) (2023). https://doi.org/10.1007/s40435-024-01402-x
15. Kinjo, R., Nakazono, K., Oshiro, N., Kinjo, H.: Performance evaluation of schedule plan for cuckoo search applied to the neural network controller of a rotary crane. Artif. Life Robot. **29**(1) (2024). https://doi.org/10.1007/s10015-023-00918-3
16. Liao, X., Ming, X., Xia, M.: KBRDBN: an interpretable deep belief network for the fault diagnosis of the trolley mechanism in ship-to-shore cranes. IEEE Trans. Instrum. Meas. **73** (2024). https://doi.org/10.1109/TIM.2023.3318717
17. Wang, S., Hou, E., Ma, H., Tang, J., Shen, Z., Xiong, G.: Steel coil recognition neural networks for intelligent cranes. In: 2024 Australian and New Zealand Control Conference, ANZCC 2024 (2024). https://doi.org/10.1109/ANZCC59813.2024.10432862

Investigation of Sedimentation and Flocculation Kinetics of Current-Dispersed Coal Sludge

Sergii Kulinich[1], Andrii Shkop[1], Alona Bosiuk[1](✉), Oleksii Shestopalov[1], and Ligia-Luminița Petrescu[2]

[1] National Technical University «Kharkiv Polytechnic Institute», 2, Kyrpychova Street, Kharkiv 61002, Ukraine
Alona.Bosiuk@mit.khpi.edu.ua

[2] National Institute of Research and Development in Mechatronics and Measurement Technique INCDMTM, 6-8 Pantelimon, 2nd District, 021631 Bucharest, Romania

Abstract. The coal industry remains an important energy sector component in many countries. Coal preparation produces large volumes of fine sludge that require efficient disposal and treatment before being discharged to external sludge ponds or reused. Low settling velocity and complex sludge structure make it difficult to settle them using traditional methods, leading to significant energy costs, loss of useful product (coal) and negative environmental impact. This article investigates the features of the settling processes of model suspensions of coal sludge with a concentration of 5 to 100 g/l. To create the model suspensions, we used the real sludge from a coal preparation plant after sieving on vibrating screens with a particle size of the solid phase of less than 20 μm, from 0 to 40 μm, and particles of 40–80 μm. The regularities of changes in the rate of sedimentation of the solid phase of suspensions without and with the addition of flocculants depending on the concentration of the solid phase and the content of fine fractions (less than 20 μm, less than 40 μm), as well as a polydisperse mixture of fine fractions (less than 40 μm) and granular particles of 40–80 μm in a ratio of 4:1 were established. The results obtained will help optimise the operation of sedimentation and dewatering facilities of the water-sludge circuit of concentrators, reduce the environmental risks of discharging fine fractions, increase resource efficiency and facilitate the transition to more sustainable production.

Keywords: Polydisperse Suspension · Water Treatment · Coal Preparation Sludge · Wastewater · Environmental Safety · Flocculation

1 Introduction

Production activities on extraction, concentration of natural raw materials (ore, coal, etc.) and its further processing (obtaining coke, pig iron, steel, non-ferrous metals, etc.) are accompanied by the formation of significant amounts of anthropogenic wastes, including in the form of sludge, which are discharged into sludge collectors. Waste accumulation leads to an increased environmental risk of contamination of adjacent territories and groundwater. At the same time, the solid phase of sludge from chemical, metallurgical,

D. D. Cioboată and J. Machado (Eds.): ICoRSE 2025, LNNS 1592, pp. 397–408, 2025.
https://doi.org/10.1007/978-3-032-02508-1_34

mining and coal industries is often a valuable mineral raw material, and the liquid phase is expedient to use in the closed system of recycled water supply of enterprises instead of clean water.

In general, sludge waters of the majority of chemical industry enterprises are heterogeneous systems with polydisperse composition of solid phase of different nature and represent heterogeneous multicomponent mechanical mixtures with complex granulometric composition. Sludge and sludge from many industries (for example, coal preparation plants) is characterised by high (up to 90%) content of fine (less than 40 μm) fractions of solid phase, which makes their purification technologically difficult and economically inexpedient.

In modern conditions a great attention is paid to improvement of existing technologies and development of new effective methods of water purification, introduction of resource-saving low-waste technologies, modernisation of existing methods and designs of water purification apparatuses. Such solutions allow to save material and natural resources, as well as to utilise valuable components of sludge waste.

The most effective way of intensification of the mentioned processes is application of synthetic polymeric flocculants and coagulants, the cost of which is a significant part of material expenses of the plant for water and sludge supply. Due to destruction of already formed flocculates during sludge transportation from thickening to dewatering equipment in modern water-sludge schemes repeated flocculation is carried out before each device. Optimisation of the dewatering process consists in selecting from a variety of process options those that will ensure maximum process efficiency with minimum economic costs.

All of the above leads to the need to establish optimal conditions for flocculation and thickening of the solid phase, allowing to obtain sufficiently strong aggregates that do not require re-flocculation after transport.

2 Analysis of Publications

The problem of efficient utilisation and treatment of coal sludge is relevant for the coal industry in many countries, including Ukraine, which has a significant coal production potential. Coal sludge is a finely dispersed water-coal suspension formed in the process of coal enrichment and is characterised by a complex multicomponent composition and high stability in the water environment [1, 2, 12]. Due to the low rate of natural sedimentation, such sludges require the use of additional methods to intensify their sedimentation [14], among which flocculation is the most common [16].

According to studies [1, 10], the efficiency of sedimentation of fine systems is significantly increased by the use of flocculants - high molecular weight polymers capable of binding small particles into aggregates (flocs) with a higher settling speed. A special role in this process is played by the physical and chemical properties of coal sludge particles, in particular their size, surface charge, hydrophobicity and the presence of clay mineral impurities that prevent aggregation [2, 3, 11].

Many researchers have noted that the sedimentation kinetics of flocculated coal sludge significantly depends on the type of flocculant, its concentration, pH of the medium, ionic composition of water, and the degree of suspension mixing [4, 5, 16]. For example, anionic polyacrylamides are often used to treat coal sludge because of their effectiveness in interacting with coal particles that have positively charged areas. At the same time, excessive flocculant dosage can lead to a decrease in settling efficiency due to suspension stabilisation or the formation of too light loose flocs, especially in concentrated suspensions [13].

Recent research has focused on mathematical modelling of flocculation and sedimentation kinetics and dispersion settling models that take into account concentration gradients in the vertical direction [1, 6, 7, 15]. The use of alternative biopolymer flocculants, which are more environmentally friendly than synthetic analogues, is also being actively studied [8, 9].

Thus, the problem of studying the kinetics of coal sludge sedimentation and flocculation remains relevant because of its application to the effective management of coal industry waste. An integrated approach to studying the mechanisms of particle aggregation and the impact of external factors on the sedimentation rate allows developing effective technologies for sludge water treatment and reducing the environmental impact.

3 Aim and Objectives of the Study

The aim of this study is to investigate the kinetics of sedimentation and flocculation of fine coal sludge. This is necessary to develop recommendations for optimising water purification processes from fine impurities.

To achieve this goal, the following tasks had to be solved:

- to study the chemical composition of the solid phase of coal preparation sludge;
- to study the effect of concentrations and sizes of solid particles on the efficiency of sedimentation and flocculation processes;
- to develop recommendations for improving the efficiency of sludge treatment from fine solid particles.

4 Research Methodology

The research was carried out on model suspensions with a concentration of 5 to 100 g/l of solid phase, which were created by mixing a sludge sample with a certain size of solid particles. To create the model suspensions, we used real sludge sludge from a coal preparation plant, taken at the enterprise from the radial water clarification tank of the plant's water-sludge scheme. The sludge was screened on vibrating sieves with mesh sizes of 20, 40, and 80 μm to separate solid phase particles less than 20 μm, from 0 to 40 μm, and particles of 40–80 μm. For the study, model sludge suspensions were prepared containing only fine fractions of less than 20 μm, less than 40 μm, and a polydisperse mixture of fine fractions (less than 40 μm) and granular particles of 40–80 μm in a ratio of 4:1. Suspensions of the following concentrations were studied: 5, 10, 20, 30, 50, 75, 100 g/l.

The kinetics of particle settling was studied in laboratory measuring cylinders with a diameter of 50 mm and a height of 245 mm. The particle falling velocity (displacement of the thickened sludge interface) was recorded. The study of aggregate formation using flocculants was carried out using a mixture of Ecofloc flocculants of non-ionic and anionic type, which were prepared at a concentration of 0.05% and introduced in a ratio of 1:1 in two portions according to the results of studies of coal sludge described in [12]. After the introduction of the reagents (each flocculant) into the water, the mixture was mixed by tumbling the cylinder ten times and the rate of deposition of the formed flocs was determined. The amount of flocculant introduced in all experiments was calculated taking into account the amount of solid phase in the suspension and was 200 g/t (0.2 mg per 1 g of solid phase). After the flocs settled, the flocs were mechanically impacted for 40 s with a stirrer and the flocs settling rate after their partial destruction was determined according to the method described in more detail in [13].

5 Results of the Study

5.1 Analysis of the Chemical Composition of Free Sludge

The results of the analysis of coal sludge sludge shown in Figs. 1 and 2 show that the solid phase of small classes contains partially coal and clay particles of aluminosilicates. This composition of sludge makes it difficult to clarify water in the settling tanks, even with the use of flocculants. Clay particles increase in size during soaking, increase the viscosity of the thickened sludge and poorly release moisture during dewatering in filter presses. The presence of particles of various chemical compounds in the sludge complicates flocculation, which requires replacing the types of flocculants or using a combination of them. In practice, coal preparation plants experience cases when flocculants stop forming aggregates due to changes in the composition of sludge, which depends on changes in impurities in the raw material (coal). Therefore, different flocculants are used in different periods of time, and new ones are selected.

Comparative characterisation of the chemical composition of the particles (see Fig. 2 a,b) shows that larger particles (see Fig. 2b) contain a higher proportion of carbon and are coal particles. Thus, the smaller the particles of the solid phase in the coal preparation sludge, the more clay particles they contain, and, accordingly, the more difficult it is to remove them from water during its clarification.

5.2 Results of the Study on the Sedimentation Kinetics of Fine Coal Sludge

The results of settling model suspensions of different dispersed composition and concentrations are shown in Fig. 3.

The analysis of Fig. 3 shows the following dependencies: the sedimentation rate at low concentrations is higher due to sedimentation close to free particle settling. However, the finest fractions with a size of less than 20 μm settle significantly slower (see Fig. 3a) than particles of larger fractions (see Fig. 3b and Fig. 3c). This is explained by both the chemical composition of the small particles, which consist mostly of soaked clay, and their lower mass. At concentrations above 30 g/l, collective deposition occurs, in which some particles prevent others from settling, heavier particles (over 20 μm) displace the liquid, which is directed upwards and delays the deposition of fine particles (see Fig. 3b). The reason for the slow settling of concentrated suspensions is also an increase in the viscosity of the liquid, the formation of upward flows during the thickening of

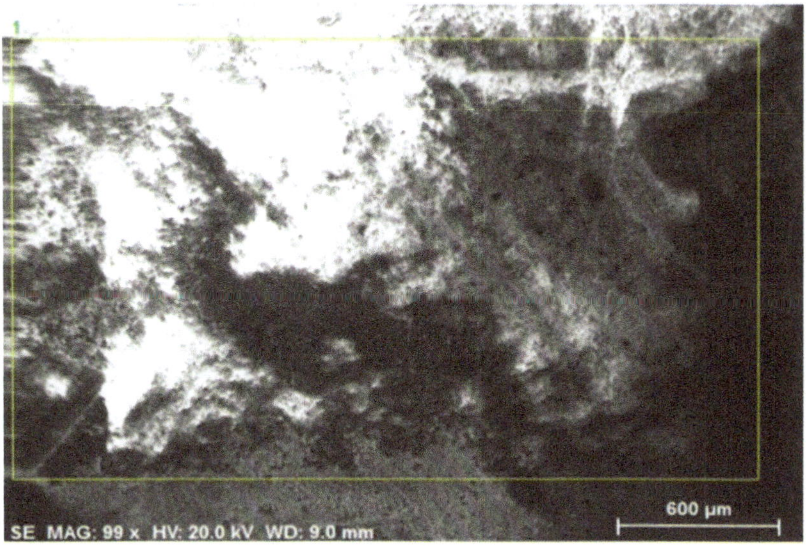

a

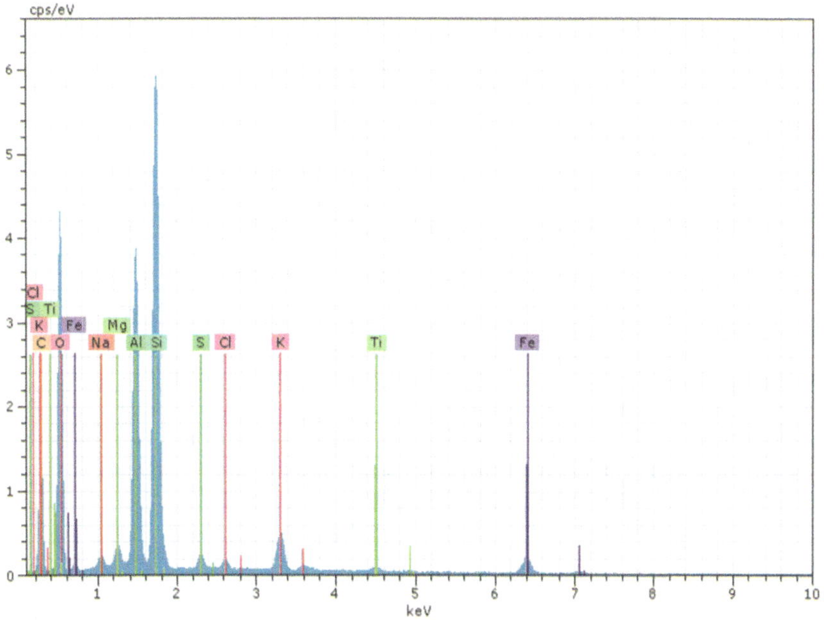

b

Fig. 1. The results of the study of the initial sludge sample by an electron scanning microscope: a – photo of sludge sediment; b – spectrogram.

the sediment. The presence of particles 40–80 μm in size (see Fig. 3c) at concentrations above 30 g/l accelerates the settling of fine particles, capturing small particles during settling or forming coagulation structures with them. In general, sedimentation without

El	AN	Series	Net	unn. C [wt.%]	norm.C [wt.%]	Atom. C [at.%]	Error (1 Sigma) [wt.%]
O	8	K-series	31159	40.65	42.28	43.77	5.01
C	6	K-series	10566	28.10	29.23	40.31	3.93
Si	14	K-series	65834	12.08	12.57	7.41	0.54
Al	13	K-series	39113	8.63	8.97	5.51	0.44
Fe	26	K-series	3858	2.54	2.65	0.78	0.11
K	19	K-series	7351	1.87	1.95	0.83	0.09
Mg	12	K-series	2542	0.69	0.72	0.49	0.07
Na	11	K-series	1492	0.61	0.64	0.46	0.07
Ti	22	K-series	915	0.37	0.39	0.13	0.04
S	16	K-series	1591	0.35	0.37	0.19	0.04
Cl	17	K-series	986	0.23	0.24	0.11	0.04
			Total:	96.14	100.00	100.00	

a

El	AN	Series	Net	unn. C [wt.%]	norm.C [wt.%]	Atom. C [at.%]	Error (1 Sigma) [wt.%]
O	8	K-series	29458	47.04	41.94	43.27	5.83
C	6	K-series	10384	33.46	29.83	40.99	4.68
Si	14	K-series	62219	14.57	12.99	7.63	0.65
Al	13	K-series	37163	9.56	8.52	5.21	0.49
Fe	26	K-series	3609	2.55	2.28	0.67	0.11
K	19	K-series	7189	2.36	2.10	0.89	0.10
Mg	12	K-series	2329	0.68	0.61	0.41	0.07
S	16	K-series	2028	0.60	0.54	0.28	0.05
Na	11	K-series	1376	0.57	0.51	0.37	0.07
Ti	22	K-series	864	0.40	0.36	0.12	0.04
Cl	17	K-series	1279	0.38	0.34	0.16	0.04
			Total:	112.17	100.00	100.00	

b

Fig. 2. Chemical analysis of sludge samples: a – class size less than 20 μm; b – class size less than 40 μm. Energy dispersive X-ray (EDX) spectrum of hard phase. EL = Element; AN = Atomic Number; Series = characteristic X-ray lines; unn. C [wt.%] = the unnormalised concentration in weight percent of the element; norm. C [wt.%] = the normalised concentration in weight percent of the element; C Atom. [at.%] the atomic weight percentage; C Error (1 Sigma) [wt.%] = error in the weight percentage concentration at the 1 sigma level.

the addition of chemical flocculants is long and inefficient. Analysing Fig. 3, it can be noted that for intensive settling, it is desirable to reduce the sludge concentration, which can be achieved by diluting it. Settling of concentrated sludges can also be more efficient if they contain larger particles that can capture smaller particles and promote their collective settling.

It should also be noted that the moisture transfer of fine particles is poor when forming a thickened suspension. The analysis of the height of the sludge after water clarification shows that in suspensions with a particle size of less than 20 μm (see Fig. 4) at concentrations above 50 g/l, the height of the sludge is higher compared to suspensions with larger particles.

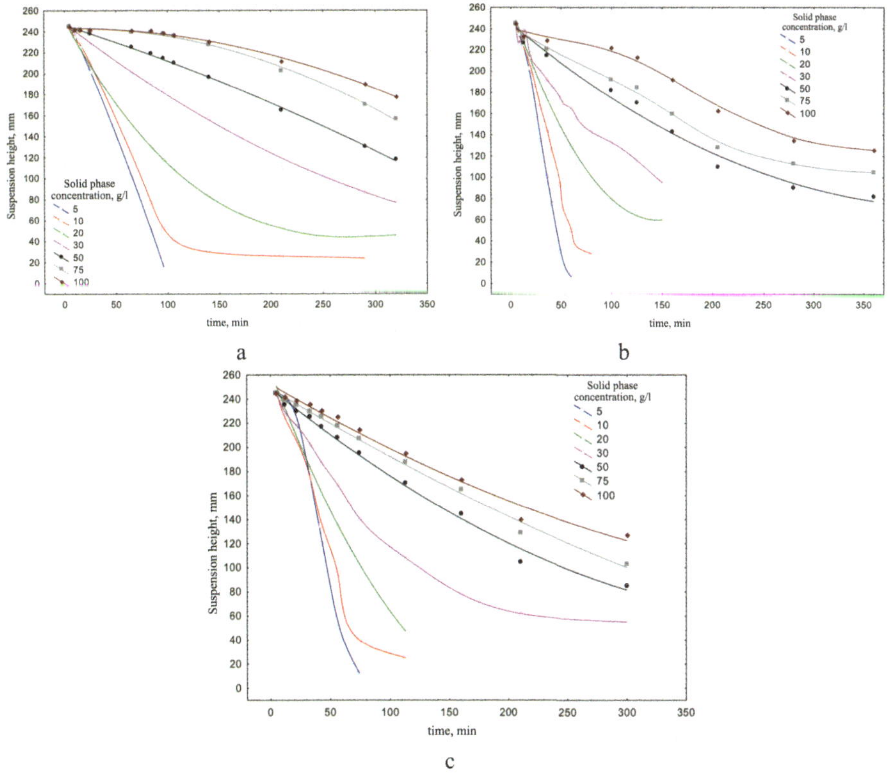

Fig. 3. Dependence of the sedimentation rate of solid particles during water clarification on their size and concentration of the solid phase: a – class size less than 20 μm; b – class size less than 40 μm; c – class size less than 40 μm with a share of classes 40–80 μm of 20%.

This is due to the fact that small particles have a larger surface area on which surface moisture is retained due to surface tension forces. In addition, the water remains between the particles in the capillaries and is not squeezed out by the weight of the fine particles, whose density is close to that of water. When hydrated, clay particles also bind water into complex compounds that are held together by chemical or electrostatic forces as they form a double electron layer around the particles. Such processes make it difficult to clarify the liquid and further dewater the sludge.

5.3 Results of the Study of Flocculation of Fine Coal Sludge

The study revealed that at the same flocculant concentration (200 g/t), the floc deposition rate and strength of flocs depends on the initial concentration and dispersed composition of the sludge (see Fig. 5), which is also illustrated by microphotographs of the formed flocs (Table 1).

In the range of solid phase concentrations of 5–30 g/l, an increase in the rate with increasing concentration with a maximum at 10 g/l is observed. The presence of a maximum at 10 g/l is probably characteristic of fine suspensions and is consistent with

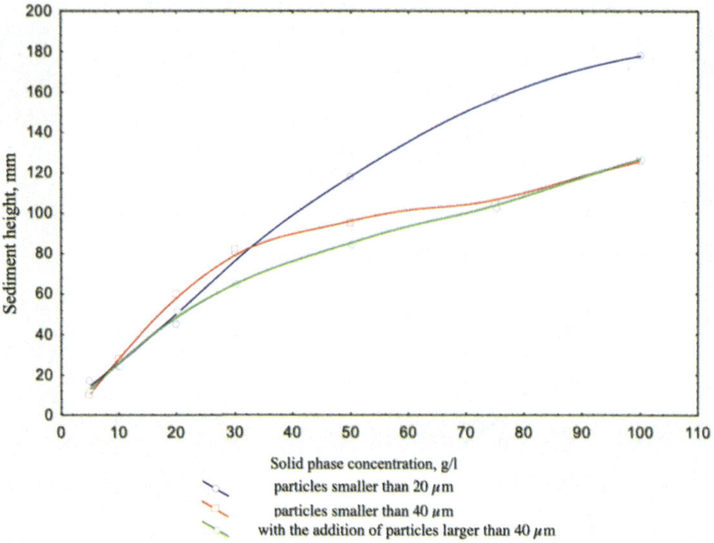

Fig. 4. Comparison of sediment height after settling of model suspensions.

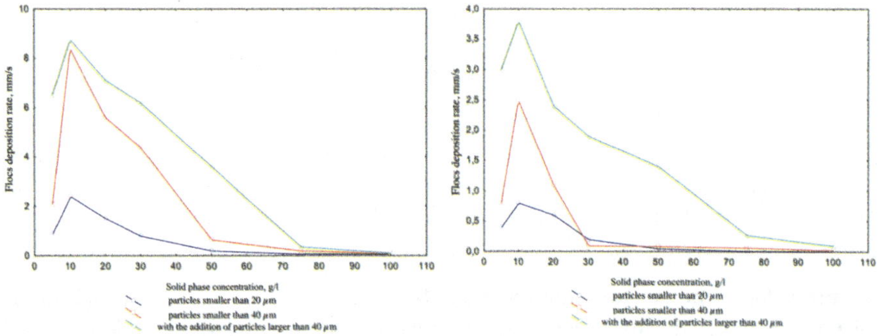

Fig. 5. Flocs deposition rate: a – after flocculation; b – after mechanical impact on flocculated suspensions.

previous studies of coal and clay suspensions [13, 17]. In the concentration range of 10–50 g/l, a decrease in the flocculent velocity is observed, which is close to linear. The content of larger particles in the sludge (fraction of the fraction greater than 20 μm, especially greater than 40 μm) leads to the formation of larger flocs and stronger flocculation structures at any concentration. This can be explained by the fact that the larger particles in the floc structure are the centres of adsorption of flocculants and then capture and bind the smaller particles more firmly. When flocs are enlarged, such aggregates squeeze small particles between larger ones and retain them, even when they are destroyed. It should be noted that for sludges with a solid concentration in the range of 10–30 g/l, particles larger than 40 μm have a significant impact, whereas at low (5 g/l) and high (more than 50 g/l) solid concentrations, a slight increase in the residual floc settling velocity is observed.

The lower settling rate of flocs formed at concentrations below 10 g/l is due to the fact that the particles cannot effectively contact and form large flocs. The size of aggregates

Table 1. Comparative characteristics of the flocs formed and their appearance after mechanical impact for suspensions with a particle size of less than 40 μm and suspensions containing 20% of particles 40–80 μm.

Solid phase concentration, g/l	Floccules photos		Appearance of flocs after mechanical impact	
	<40 μm	20% particles 40-80 μm	<40 μm	20% particles 40-80 μm
5	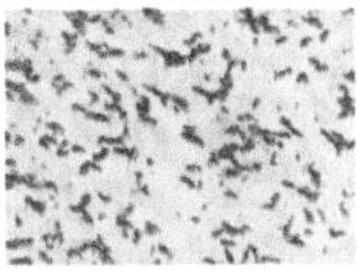	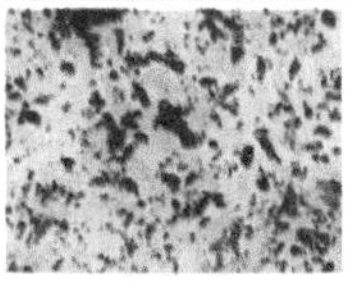	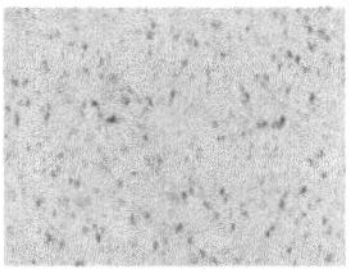	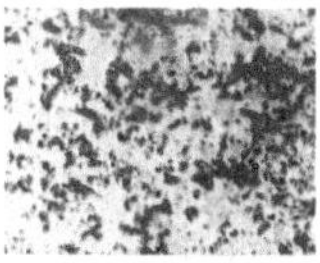
10	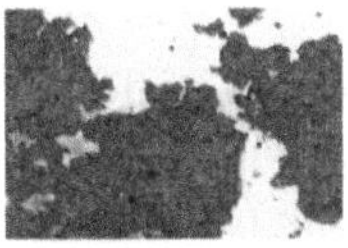	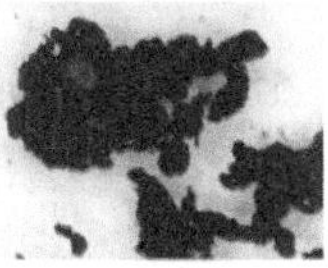	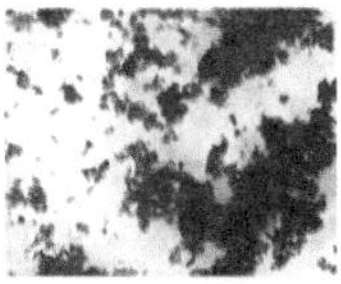	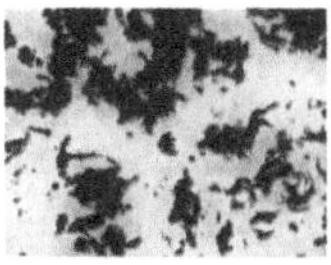
20	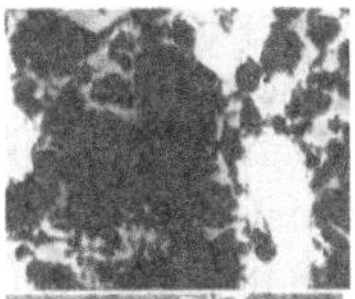		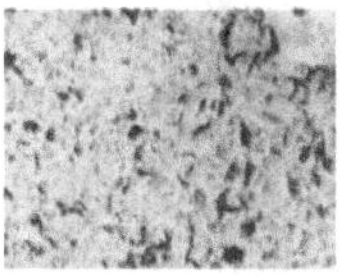	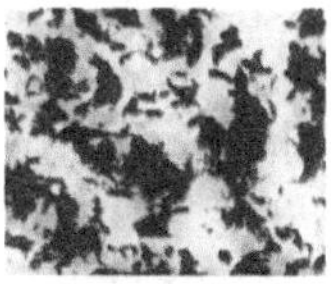
30		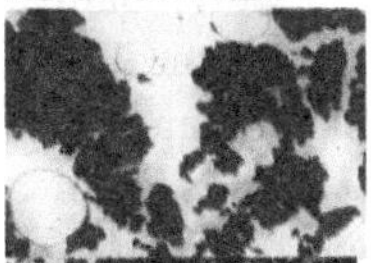	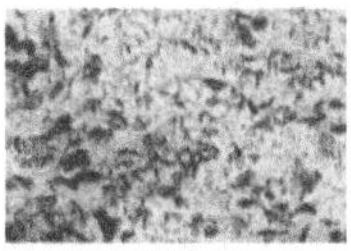	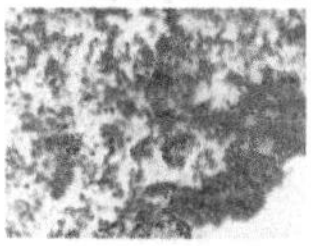
50	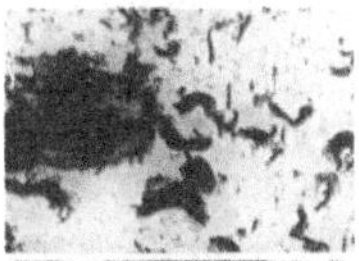	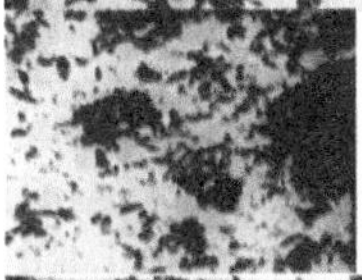	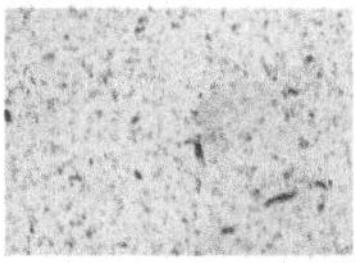	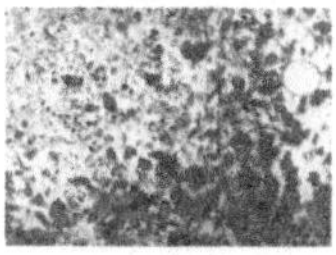
75	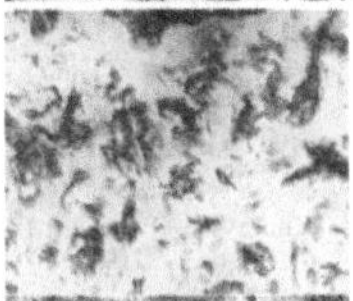	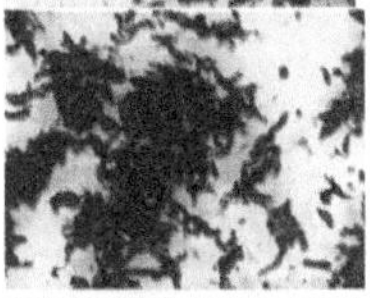	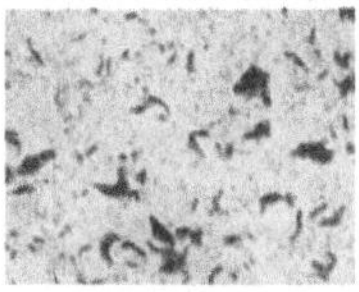	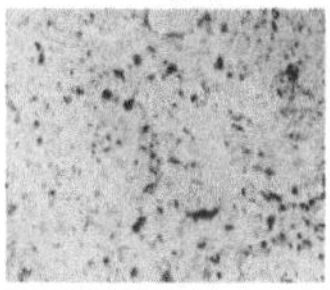
100	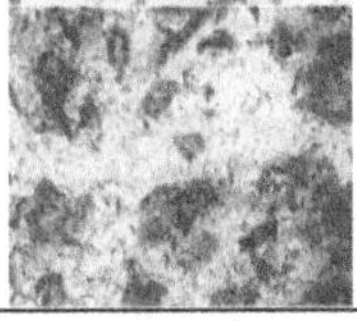	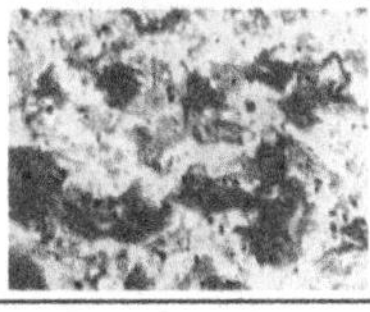	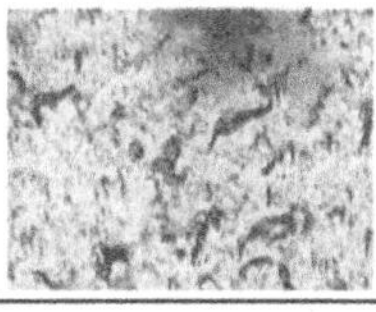	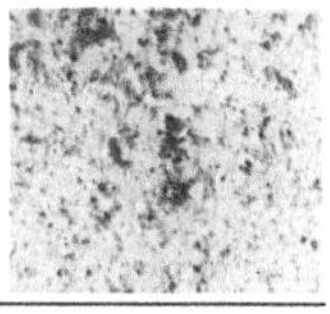

at low concentrations (Table 1) and their weight are smaller than at concentrations close to the optimum (10 g/l).

In the concentration range of 50–100 g/l, a sharp exponential decrease in the floc deposition rate is observed. This may be due to the following processes occurring in highly concentrated sludges. Firstly, the volume occupied by the solid phase and the interface increases. The smaller the particles, the greater their surface wetted with water. This leads to an increase in the viscosity of the liquid due to surface tension forces. All of this leads to the fact that the flocculant cannot be evenly distributed and adsorbed on all particles, which impairs flocculation and forms unstable flocs. As the concentration of the solid phase increases or the sludge thickens, the flocs become denser, losing the liquid in them and collapsing under the influence of their weight and osmotic pressure in the pores between the aggregates. The volume of liquid around the flocs also decreases. Therefore, mechanical impact on a more concentrated flocculated sludge results in more frequent collisions of the flocs with each other, as well as with the surface of the agitator and the walls of the tank. This causes high stresses on the surface of the flocs, leading to their destruction.

Thus, with an increase in the concentration of suspensions, the pseudoplastic properties of the flocculated structures increase and the strength of the flocs to mechanical stress decreases.

Studies of the effect of the dispersed composition on the rate of flocs settling after mechanical action have shown that the addition of particles with a size of 40–80 μm increases the rate of settling at all concentrations. This is the result of an increase in the strength of the flocs. With increasing concentration of the sludge solid phase, the sedimentation rate decreases and is minimal at concentrations above 50 g/l. The strength of the bonds between the particles during further floc consolidation also depends on the number of adsorbed polymer macromolecules. In flocs formed by a dispersed solid phase, the strength of interaction between individual particles inside the microfloc and higher-order macrofloc is approximately the same. Therefore, under mechanical action, macroflocs are destroyed into the smallest clumps of fine particles. The presence of large particles (more than 40 μm) due to the larger amount of flocculant adsorbed on the surface of the particles contributes to the creation of stable structures, which, although they break up into flocs of a smaller order and size under mechanical action, still have a sufficiently high residual settling rate. The increase in the sedimentation rate after mechanical action in the presence of large particles can also be explained by the re-aggregation of microflocs due to the retained active links on the surface of large particles.

6 Conclusions

Studies have shown that fine coal slime particles contain clay components and settle slowly during sedimentation, which necessitates the use of synthetic flocculants for water clarification at coal preparation plants.

Studies of the effect of the solid phase concentration and its dispersed composition on the flocculation efficiency have shown that the largest and most mechanically resistant aggregates are formed at a solid phase concentration of 10 g/l in the slurry (the best conditions for flocculant adsorption) and the content of particles larger than 40 μm in the suspension. An increase in the concentration of the solid phase in the suspensions leads to an increase in the consumption of flocculants with a significant decrease in the efficiency of settling. It has been established that fine sludges with a size of less than 20 μm are the worst settled, dewatered, and have the smallest floc size when interacting with flocculants.

To intensify the treatment of fine coal sludge, it is recommended to dilute the concentrated sludge for better flocculation, as well as to introduce additional large particles to weight the flocs and accelerate their settling.

The dependencies obtained in the course of experimental studies allow intensifying the process of formation of strong flocs with minimal flocculant consumption, as well as controlling the flocculation process by adjusting the concentration and dispersed composition of fine suspensions.

References

1. Oyegbile, B., Ay, P., Narra, S.: Flocculation kinetics and hydrodynamic interactions in natural and engineered flow systems: A review. Environ. Eng. Res. **21**(1), 1–14 (2016)
2. Zhang, Y., Xu, Z., Tu, Y., Wang, J., Li, J.: Study on properties of coal-sludge-slurry prepared by sludge from coal chemical industry. Powder Technol. **366**, 552–559 (2020)
3. Kahilu, G.M., Bada, S., Mulopo, J.: Physicochemical, structural analysis of coal discards (and sewage sludge)(co)-HTC derived biochar for a sustainable carbon economy. Sci. Rep. **12**(1), 17532 (2022)
4. Alam, N., Ozdemir, O., Hampton, M.A., Nguyen, A.V.: Dewatering of coal plant tailings: flocculation followed by filtration. Fuel **90**(1), 26–35 (2011)
5. Khazaie, A., et al.: A review on coagulation/flocculation in dewatering of coal slurry. Water **14**(6), 918 (2022)
6. Moruzzi, R.B., de Oliveira, S.C.: Mathematical modeling and analysis of the flocculation process in chambers in series. Bioprocess Biosyst. Eng. **36**, 357–363 (2013)
7. Bürger, R., Concha, F.: Mathematical model and numerical simulation of the settling of flocculated suspensions. Int. J. Multiph. Flow **24**(6), 1005–1023 (1998)
8. Jiang, X., et al.: Biopolymer-based flocculants: A review of recent technologies. Environ. Sci. Pollut. Res. **28**, 46934–46963 (2021)
9. Maćczak, P., Kaczmarek, H., Ziegler-Borowska, M.: Recent achievements in polymer bio-based flocculants for water treatment. Materials **13**(18), 3951 (2020)
10. Rattanakawin, C., Hogg, R.: Aggregate size distributions in flocculation. Colloids Surf A Physicochem Eng Asp **177**(2–3), 87–98 (2001)
11. Li, W., Li, W., Liu, H., Yu, Z.: Influence of sewage sludge on the slurryability of coal–water slurry. Fuel **88**(11), 2241–2246 (2009)
12. Shkop, A., Tseitlin, M., Shestopalov, O., Raiko, V.: Study of the strength of flocculated structures of polydispersed coal suspensions. East.-Eur. J. Enterp. Technol. **1**(10), 20–26 (2017)
13. Shkop, A., Tseitlin, M., Shestopalov, O., Raiko, V.: A study of the flocculs strength of polydisperse coal suspensions to mechanical influences. EUREKA Phys. Eng. **1**, 13–20 (2017)
14. Shkop, A., Tseitlin, M., Shestopalov, O.: Exploring the ways to intensify the dewatering process of polydisperse suspensions. East.-Eur. J. Enterp. Technol. **6**(10), 35–40 (2016)
15. Shkop, A., Shestopalov, O., Sakun, A., Tseitlin, M., Ponomarova, N., Bosiuk, A., Bădiţă-Voicu, L.L.: Research of efficiency of cleaning and dehydration of coal slims in centrifuges (2025)

16. Bosiuk, A., Shkop, A., Kulinich, S., et al.: Multi-component wastewater from finely dispersed impurities treatment intensification. Ecol. Questions **34**(4), 1–18 (2024)
17. Shestopalov, O., Rykusova, N., Hetta, O., Ananieva, V., Chynchyk, O.: Revealing patterns in the aggregation and deposition kinetics of the solid phase in drilling wastewater. East.-Eur. J. Enterp. Technol. **1**(10), 50–58 (2019)

Mathematical Model of the Process of Air Pollutant Distribution for Multiple Point Sources of Industrial Pollution

Artem Kovtun[3(✉)], Iuliia Pershyna[1,2], and Vitaliy Mezhuev[3]

[1] Kharkiv National University of Radio Electronics, 14, Nauky Avenue, Kharkiv 61166, Ukraine

[2] National Technical University "Kharkiv Politechnic Institute", 2, Kyrpychova Street, Kharkiv 61002, Ukraine

[3] NFH JOANNEUM University of Applied Sciences, Graz, Kapfenberg, Bad Gleichenberg, Austria

artem.kovtun1@nure.com

Abstract. A boundary problem is formulated to describe the processes of multi-component pollutant transport in the air in the presence of point sources. Using the concept of local potential, a theorem is developed that enables the construction of an algorithm for solving the problem via the finite element method (FEM). The choice of FEM is justified by its key advantages: (1) FEM provides an approximate solution in the form of an analytical expression; (2) it formalizes the procedure for satisfying boundary conditions by selecting a functional where one or both boundary conditions are natural; (3) it allows for constructing an approximation even in cases with discontinuous coefficients or when the non-homogeneous term of the differential operator includes a sum of Dirac delta functions.

Furthermore, an algorithm is developed for solving a nonlinear boundary value problem with variable coefficients, featuring a singularity represented as a sum of unit Dirac delta functions. The mathematical model of air purification above a given surface considers the effect of initial dispersion of polluted air and is formulated as a two-point boundary value problem for a system of differential equations governing the material balance of organic pollutants in the air.

The mathematical model formulated in this work describes the process of purifying polluted airflows around an emission source located on a surface, leading to a nonlinear boundary value problem. The solution is obtained using FEM, which enables the construction of variable approximations in the presence of singularities such as Dirac delta functions. The variational formulation of the boundary problem is developed using the Ritz method, incorporating the local potential concept proposed by Glansdorff and Prigogine.

Keywords: Multi-component pollutant transport · finite element method (FEM) · boundary value problem · Dirac delta function · air purification modeling · pollutant dispersion · nonlinear differential equations · Ritz method · local potential · modified Newton's method · Gauss method · kinetic constant estimation

D. D. Cioboată and J. Machado (Eds.): ICoRSE 2025, LNNS 1592, pp. 409–420, 2025.
https://doi.org/10.1007/978-3-032-02508-1_35

1 Introduction

Air pollution from industrial emissions remains a critical environmental and public health challenge. Among the key pollutants, sulfur trioxide (SO_3) and carbon dioxide (CO_2) play significant roles in acid deposition and greenhouse gas accumulation, respectively. Accurate modeling of pollutant dispersion is crucial for effective regulatory control, infrastructure planning, and environmental impact assessments.

This study develops a mathematical framework for simulating the steady-state distribution of and in the atmospheric boundary layer, assuming emissions from multiple industrial point sources (e.g., chimneys). The pollutant transport is described by a system of coupled diffusion-reaction equations that include convective transport, nonlinear chemical interactions, and singular source terms modeled as Dirac delta functions [1, 2].

Pollution in urban and industrial regions exhibits complex dynamics due to interactions between chemical species, turbulent dispersion, and localized emissions. Models simulating these processes must accommodate non-uniform terrain, multiple point sources, and heterogeneous reaction kinetics [3]. Traditional models such as the Gaussian plume model, Eulerian grid-based models, and CFD (Computational Fluid Dynamics) techniques have been widely used. However, these often face challenges with precision near emission sources and in systems involving nonlinear chemical coupling [1, 4]. The Finite Element Method (FEM) offers a robust alternative due to its ability to:

- Handle non-uniform geometries and boundary conditions;
- Incorporate nonlinear reaction-diffusion equations;
- Represent localized emissions via Dirac delta functions;
- Deliver high-resolution solutions using piecewise basis functions.

To handle the nonlinear and non-self-adjoint nature of the problem, we adopt a variational formulation based on the Ritz method [5], which allows recasting the boundary value problem as the minimization of a functional. The structure of this functional is grounded in the local potential theory proposed by Glansdorff and Prigogine [6], which effectively captures the interplay of diffusion, convection, and thermodynamic dissipation in open chemical systems.

The functional includes:

- diffusive contributions associated with mass transfer processes [7, 8];
- nonlinear chemical kinetics, including saturation behavior modeled by the Michaelis–Menten law [9, 10];
- localized source terms represented via delta functions, enabling the precise modeling of emissions from discrete sources [2, 11].

The spatial domain is discretized using the finite element method (FEM) with piecewise linear (hat) basis functions [1, 4, 8]. This converts the variation al formulation into a nonlinear algebraic system of equations of dimension $2(M + 1)$, where $(M + 1)$ is the number of FEM nodes. The system is solved using a modified Newton method with step-size control and diagonal regularization to enhance stability [12, 13]. The Jacobian matrix is computed using finite-difference approximations for sensitivity and convergence control.

Beyond solving for pollutant concentrations, this model incorporates parameter optimization:

- spatial placement of emission sources is optimized to minimize the cumulative pollutant load across the domain [3, 9];
- chemical kinetic constants (e.g., k_r, K_m, k_D) are estimated through data assimilation by minimizing the deviation between model output and synthetic measurement data [10, 14].

Importantly, our approach includes a dual optimization strategy. We optimize not only the source positions, but also kinetic coefficients k_r, K_m, k_D, representing reaction rate, saturation threshold, and degradation effects, respectively. Validation is performed using synthetic "experimental" data with physical scaling and additive noise, simulating realistic field conditions.

All experiments and optimization routines are performed in MATLAB using both forward simulation and inverse optimization techniques. Synthetic "experimental" data with added noise are generated to emulate real atmospheric conditions, including physical scaling to mg/m3 units [3]. This hybrid modeling and optimization framework provides a rigorous, scalable, and physically interpretable approach to simulate pollutant behavior in the air. It is especially suitable for applications in industrial emission management, environmental monitoring, and urban planning scenarios involving multiple point sources and nonlinear transformations.

2 Research Methodology

Mathematical formulation of the technological process in a multi component mixture involves writing material balance equations for each reactive component. These equations are expressed in the form of diffusion equations that incorporate source terms for each substance [4, 7]

$$\frac{\partial C_i}{\partial t} = div(D \cdot gradC_i) - grad(\overrightarrow{\omega}\, C_i) + F_i, \quad i = \overline{1, m} \qquad (1)$$

where C_i is the concentration of the i-th component; $\overrightarrow{\omega}$ is the velocity vector of the loose mixture flow; D is the total diffusion coefficient; and F_i – a function representing the intensity of the source of the i -th component; m –the total number of reacting components.

For simplicity, we consider the concentration of each component above a given surface as a function of the longitudinal spatial coordinate, thus reducing Eq. (∂Ci ∂t = div(D·gradCi)−grad(→ ω Ci)+Fi,i= ¯ 1,m 1) to a one-dimensional (1D) formulation.

Diffusion is characterized by the longitudinal mixing coefficient: $D_L \to \infty$ - ideal mixing, and ideal plug flow when $D_L \to 0$.

The formation of new chemical species due to interaction among the components occurs continuously above each surface and is mathematically represented as distributed sources. These are treated as point sources, whose influence is described using the Dirac delta function [1, 2] $q = q_{ik}\delta(x - x_{ik})$, where x_{ik} – the spatial coordinate of the k-th emission source for the i-th pollutant. The term becomes negative if the substance is being removed (e.g., through wind-assisted dispersion or decay).The mathematical model of the multi component air pollution process under stationary conditions can thus be formulated as the following boundary value problem:

$$\frac{d}{dx}\left(D_L(x)\frac{d\vec{C_i}(x)}{dx}\right)-\frac{d}{dx}\left(u(x)\vec{C}(x)\right)+K\left(\vec{C}(x),x\right)+\vec{f}(x)=0, \tag{2}$$

$$D_L(0)\frac{d\vec{C_i}(0)}{dx}=u(0)\left(\vec{C}(0)-\vec{C_0}\right), \tag{3}$$

$$D_L(L)\frac{d\vec{C_i}(L)}{dx}=0, \tag{4}$$

where $\vec{C}(x) = (C_1(x),\ C_2(x), ..., C_m(x))^T$; $u(x)$ – longitudinal velocity of the pollutant flow; $K(\vec{C}(x), x)$ – matrix that describes the kinetics of interaction between the different components of the mixture; $\vec{C_0} = ({}_{10},\ {}_{20},\ ...,\ C_{m0})^T$; C_{i0} – concentration vector of the i-th pollutant component in the mixture at the outlet of the source; $\vec{f}(x) = (f_1(x),\ f_2(x), ..., f_m(x))^T$; $f_i(x) = \sum_{k=1}^{p_i} q_{ik}\delta(x - x_{ik})$, p_i – number of point sources for the i -th component; L–length of the domain surface/

The boundary value problem (4), in its general form, is nonlinear and non-self-adjoint, which complicates its invariant formulation. By applying the concept of local potential [5], we derive a theorem that enables the representation of the problem in variational form.

Theorem. If the functional defined by the expression

$$I(\vec{C}(x))=\int_0^L\left\{\frac{1}{2}D_L(x)\frac{d\vec{C}^T(x)}{dx}\frac{d\vec{C}(x)}{dx}+u(x)\vec{C}^T(x)\frac{d\vec{C}^0(x)}{dx}+\right.$$
$$+\frac{1}{2}\rightarrow^T(x)\vec{C}(x)\frac{du(x)}{dx}-2\sum_{i=1}^{m}\int_0^{C_i(x)}\left(\sum_{j=0}^{m}K_{ij}(\vec{C}^0(x),x)\Big|_{C_i^0(x)=t}\right)dt-$$
$$\left.-\vec{f}^T(x)\vec{C}(x)\right\}dx+\frac{1}{2}u(0)\vec{C}^T(0)\left(\vec{C}(0)-2\vec{C}_0\right), \tag{5}$$

is stationary, and if $C_i^0(x)$ after the variation of $I(\vec{C}(x))$ with respect to $C_i(x)$ satisfies the auxiliary condition $C_i^0(x) \equiv C_i(x)$, then the functions $C_i(x)$, $i = \overline{1, m}$ satisfy the system of Eqs. (2) and the boundary conditions (3)–(4).

Note that the boundary condition (4) (and (3) if $\vec{C}_0 = 0$) is natural for the functional (5), meaning it may be omitted in the construction of the approximate solution.

The problem of minimizing functional (5) will be solved using the finite element method [2, 3, 8, 9, 11, 12, 14]. The interval [0, L] is divided into N segments with the sequence of nodes $\pi\ :\ x_1 < x_2 < x_3 < ... < x_{N+1} = L$. The approximate solution

$\vec{C}_\pi(x)$ is sought in the form:

$$C_{\pi i}(x) = \sum_{n=0}^{N+1} C_{in} h_n(x), \quad i = \overline{1, m}, \quad n = \overline{1, N+1}, \tag{6}$$

where C_{in} –the concentration of the i-th pollutant component at the n-th node; the functions $h_n(x)$ have the form [15]

$$h_n(x) - \frac{1}{2}\left[\frac{|x - x_{n-1}|}{x_n - x_{n-1}} - \frac{x_{n+1} - x_{n-1}}{(x_n - x_{n-1})(x_{n+1} - x_n)}|x - x_n| + \frac{|x - x_{n+1}|}{x_{n+1} - x_n}\right]$$

As a result of minimizing $I(\vec{C_\pi}(x))$ with respect to C_{in} solving the original problem reduces to solving the system of equations

$$\frac{\delta\left(\vec{C}_\pi(x)\right)}{\delta_{in}} = 0, \quad i = \overline{1, m}, \quad n = \overline{1, N+1}.$$

As an example, we consider the boundary value problem (2)–(4) for a two-component mixture (sulfur trioxide and carbon dioxide), which describes the air pollution distribution process over a specific surface, occurring through discrete emission sources.

Figure 1 shows the results of applying the algorithm. In the considered example, the process of dispersion of harmful substances in the air is modeled, showing the distribution of CO_2 (1 – experimental data, 2 – simulated) and SO_3 (3 – experimental data, 4 – simulated).

The process occurring in the air is a sequence of chemical reactions, whose kinetics are described by the following matrix [9, 16]. For CO_2 only diffusion without chemical reactions is considered, while for SO_3 reactions with water vapor and oxygen are taken into account:

$$K\left(\vec{C}(x), x\right) = \left\| \begin{array}{cc} -D_{CO_2}\Delta C_1(x) & \beta k_D C_2(x) \\ k_r \frac{C_2(x)}{K_m + C_2(x)} + k_D C_2(x) C_{H_2O} + \lambda_{SO_3} C_2(x) C_{O_2} & -D_{SO_3}\Delta C_2(x) - k_D C_2(x) \end{array} \right\|$$

where $C_1(x)$- the concentration of carbon dioxide in the air at distance x from the source; $C_2(x)$ -the concentration of sulfur trioxide in the air at distance x from the-source; k_r-the maximum rate of the chemical reaction of (for example, the conversion of $SO_3 + H_2O \rightarrow H_2SO_4$) which is described by the Michaelis – Mentene equation [9, 10]; K_m- the saturation constant (Michaelis constant) [9, 10]; k_D is the rate of hydration or sedimentation of SO_3 into the external environment (in the presence of vapor); D_{CO_2} is the diffusion coefficient for CO_2 the spatial transport rate of pollutants in the air is determined (e.g., via turbulence or molecular diffusion); β - a phenomenological coefficient used to account for the influence of SO_3 concentration on the dynamics of CO_2 for example, during joint transport or secondary effects in the model); λ_{SO_3}- is the interaction coefficient of with oxygen a phenomenological parameter that may represent secondary oxidative processes or stabilization of particles in the air; C_{O_2}- the concentration of oxygen (an atmospheric constant, stable ≈ 0.21); CH_2O- the concentration of water vapor (describes air humidity—conventionally constant).

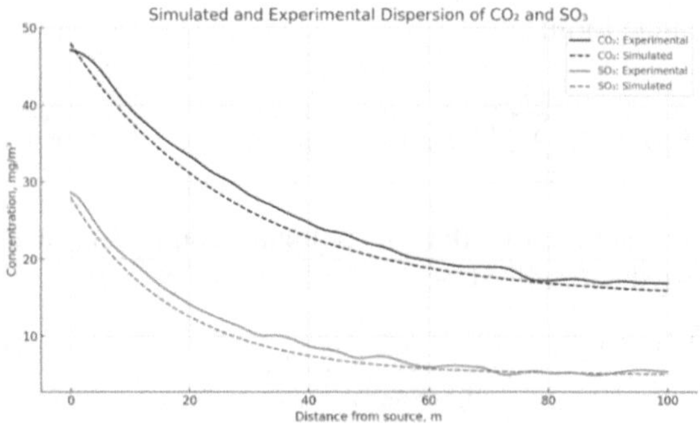

Fig. 1 Process of dispersion of harmful substances in the air.

The cross-sectional area of emission sources is significantly smaller than the scale of the air domain where emissions are dispersed. Therefore, the sources of pollutant concentrations can be treated as point sources with intensity defined by the following relationship:

$q_k = \frac{C_k \cdot A_k \cdot u}{A}$ or $q_k = \frac{C_k \cdot Q_k}{A}$ де $(Q_k = A_k \cdot u)$.

where A_k- the cross-sectional area of the exhaust pipe (m2), u - the air velocity near the source (m/s) and q_k- the emission intensity of the k -th source, A - the area of the entire modeled at mospheric domain (m2). Including A in the formula allows for proper comparison of source contributions in cases when the domain is much larger than the sources themselves.

Physically, this corresponds to an instantaneous localized emission from a pipe into a confined zone of the atmosphere. Taking the above into account, the mathematical model for pollutant dispersion in the air medium from tubular sources (emissions) under steady-state conditions can be written in the form of the following boundary value problem:

$$\frac{d}{dx}\left(D_{CO_2}\frac{dC_1}{dx}\right) - u\frac{dC_1}{dx} + \beta k_D C_2 + \sum_{k=1}^{R} q_{1k}\delta(x - x_k) = 0, \tag{7}$$

$$\frac{d}{dx}\left(D_{SO_3}\frac{dC_2}{dx}\right) - u\frac{dC_2}{dx} + R_{kin}(C_1, C_2) + \sum_{k=1}^{R} q_{2k}\delta(x - x_k) = 0,$$
$$R_{kin}(C_1, C_2) = \frac{k_r C_2}{K_m + C_2} + k_D C_2 + \kappa_{SO_3} C_2 C_{H_2O} + \lambda_{SO_3} C_2 C_{O_2} \tag{8}$$

Boundary conditions (Robin boundary conditions) [17, 18].

$$\text{At } x = 0,\ D_i\frac{dC_i}{dx} = u(C_i - C_i^0),\ i = 1, 2 \tag{9}$$

$$\text{At } x = L_A,\ \frac{dC_i}{dx} = 0,\ \ i = 1, 2 \tag{10}$$

L - the boundary of the air domain being modeled. This implies that at distance $x = L_A$, the concentration gradient is zero, meaning there is no outflow or additional flux of substance beyond the domain boundaries—this is the "natural outflow" condition. Where D_{CO_2}, D_{SO_3} are the diffusion coefficients;$C_{H_2O}, _{O_2}$ -are the concentrations of water vapor and oxygen; $k_r, k_D, \lambda_{SO_3}, \kappa_{SO_3}, K_m$- are reaction constants. This formula accounts for: SO_3 the kinetics of conversion (term with k_r); degradation (deposition) - $k_D C_2$; reaction with H_2O- formation of H_2SO_4; reaction with CO_2- additional synergistic effect; autolysis: $\beta k_D C_2$ - which generates CO_2-additional release of CO_2 from SO_3 break down; nonlinear Michaelis–Menten-type reaction [6, 14]: $\frac{k_r C_2}{K_m + C_2}$; x_k - coordinates of chimneys (emission sources), u - speed of air flow; k_r - reaction rate of SO_3 (with H_2SO_4 formation); K_m - saturation constant of the reaction.

At the inlet zone of the atmospheric domain before the tubular emission sources, a background concentration level of pollutants may exist—specifically CO_2, or zero SO_3 concentration. As air enters the emission region, a sharp localized increase in pollutant concentration occurs due to the influence of point sources (pipes). This process can be interpreted as an instantaneous shift in the concentration profile at the inlet boundary of the modeled domain, having a dominant effect at the initial moment. Physically, this corresponds to a boundary transition from background (atmospheric) concentration to the zone of localized pollution, driven by convective inflow of air with a specific initial composition.

In this context, the boundary condition at $x = 0$, can be expressed as:

$$D_i \frac{dC_i}{dx} = u(C(0)_i - (1 - k_c)C_i^0), \ i = 1, 2 \tag{11}$$

where C_i^0- are the background concentrations of SO_3 and CO_2 in air (before emissions); $Ck \in [0, 1][0, 1]$- a coefficient describing the fraction of background concentration that remains at the inlet (e.g., due to turbulent mixing or background ventilation); u - the air velocity; D_i- diffusion coefficient.

This interpretation allows us to take into account the realistic influence of background conditions on the distribution of pollutants, and at the same time maintains consistency with the variational (FEM) scheme of model construction.

3 Results and Discussion

The problem of pollutant dispersion in the air from tubular point sources is described by a system of differential equations with reaction and "source" (delta) terms. In general, this system is nonlinear and non-self-adjoint, which complicates its classical formulation.

We formulate the corresponding variation functional:

(This fully corresponds to the Glansdorff–Prigogine formulation for constructing the potential of a dissipative system)

$$J(C_1, C_2) = \int_0^L \left[D_{CO_2} \left(\frac{dC_1(x)}{dx} \right)^2 + D_{SO_3} \left(\frac{dC_2(x)}{dx} \right)^2 + R_{kin}(C_1, C_2) \right] dx$$

$$+ \sum_{k=1}^{R} \left(q_k^{(1)} C_1(x_k) + q_k^{(2)} C_2(x_k) \right)$$

where $C_1(x)$, $C_2(x)$ are the concentrations of SO_3 and CO_2 respectively, D_{CO_2}, D_{SO_3} are the diffusion coefficients; x_k - the coordinates of the pipes (sources), $q_k^{(i)}$ - the emission intensity of the i-th pipe; R_{kin} - the reaction component

$R_{kin}(C_1, C_2) = \frac{k_r C_2}{K_m + C_2} + k_D C_2 + \kappa_{SO_3} C_2 C_{H_2O} + \lambda_{SO_3} C_2 C_{O_2}$ - is a nonlinear reaction component that models the interaction of sulfur trioxide with water vapor and carbon dioxide. The interval $[0, L]$ - is divided into M finite elements $\pi = \{x_0 = 0 < x_1 < ... < x_M = L\}$. The concentration functions are approximated as follows:

$$C_{\pi 1}(x) \approx \sum_{m=0}^{M} C_{1,m} h_m(x),\ C_{\pi 2}(x) \approx \sum_{m=0}^{M} C_{2,m} h_m(x)$$

where $C_{i,m}$ -the nodal concentration values. The functions $h_m(x)$ are piecewise-linear basis functions (FEM).

The solution of the variational boundary problem (7)–(8) with boundaries conditions (9) (10) or (11) is reduced to a system of $2(M+1)$ nonlinear equations with respect to the approximation coefficients $C_{1,m}$, $C_{2,m}$ which is solved by the modified Newton method [1, 4, 5, 13] with regularization and numerical differentiation.

We now describe the construction of the system of equations in more detail:

1. (Nonlinearstage) Minimization of the functional (9) (10) or (11) with respect to the coefficients $C_{1,m}$, $C_{2,m}$ leads to a system of nonlineare quations of size $2(M+1)$: $\vec{F}(\vec{C}) = 0$, $\vec{C} := [C_{10}, ..., C_{1M}, C_{20}, ..., C_{2M}]^T$. Each "weak" integral equation is derived from the variational form (as a result of applying the Ritz principle), integrated over elements,

a) starting with C_1:

$$\sum_{m=0}^{M} D_{CO_2} \int_0^L \frac{dC_1}{dx}\frac{dh_m}{dx} dx + \int_0^L \frac{\partial R_{kin}(C_1, C_2)}{\partial C_1} h_m(x) dx + \sum_k q_k^{(2)} h_m(x_k) = 0$$

$$Buitif,\ \frac{\partial R_{kin}}{\partial C_1} = 0 \Rightarrow \int_0^L \frac{\partial R_{kin}(C_1, C_2)}{\partial C_1} h_m(x) dx = 0$$

Then the equation for simplifies to:

$$\sum_{m=0}^{M} D_{CO_2} \int_0^L \frac{dC_1}{dx}\frac{dh_m}{dx} dx + \sum_k q_k^{(2)} h_m(x_k) = 0$$

b) then for C_2 (reaction influence is considered):

$$\sum_{m=0}^{M} D_{SO_3} \int_0^L \frac{dC_2}{dx}\frac{dh_m}{dx}dx +$$

$$\int_0^L \left[\frac{k_r K_m}{(K_m + C_2)^2} + k_D + \kappa_{SO_3} C_{H_2O} + \lambda_{SO_3} \right] h_m(x)dx + \sum_k q_k^{(2)} h_m(x_k) = 0$$

where the derivative is:

$$\frac{\delta R_{kin}}{\delta C_2} = \frac{k_r K_m}{(K_m + C_2)^2} + k_D + \kappa_{SO_3} C_{H_2O} + \lambda_{SO_3}$$

and k- denotes the indices of point emission sources. The following expression:

$$\int_0^L h_m(x) \sum_k q_k^{(i)} \delta(x - x_k)dx = \sum_k q_k^{(i)} h_m(x_k)$$

Means that a teach point x_k the source has intensity $q_k^{(i)}$- is an ideal point source. Next, stiffness matrices of size $(M+1) \times (M+1)$ are constructed for each component

$$K_{m\ell}^{(i)} = \int_0^L D_i \frac{dh_m}{dx}\frac{dh_\ell}{dx}dx + \int_0^L \frac{\partial R_{kin}}{\partial C_i} h_m(x) h_\ell(x)dx.$$

:

The first integral is the diffusion term, the second is the reaction derivative from the local potential (Glansdorff–Prigogine concept). Where ℓ -the column index (i.e., the second index of the nodal basis function), m-is the row index. Thus $K_{m\ell}^{(i)}$-the element of the stiffness matrix for component, which corresponds to the interaction of basis functions $h_m(x)$, $h_\ell(x)$. This approach produces a symmetric and tridiagonal structure of the matrix $K_i \in \mathbb{R}^{(M+1)\times(M+1)}$, which is correct from the perspective of the Galerkin method. Thus, FEM system takes the form: $K_1 \cdot \vec{C_1} = \vec{f_1}$, $K_2 \cdot \vec{C_2} = \vec{f_2}$, $\vec{f_1}, \vec{f_2} \in \mathbb{R}^{M+1}$ - right-hand side vectors that include point emission sources.

Since the reaction term $R_{kin}(C_1, C_2)$ is nonlinear, the problem is reduced to a system of $2(M + 1)$ equations (FEM), and is solved using the modified Newton method with the construction of the Jacobian matrix via numerical derivatives and applying regularization:

$$\vec{C}^{(n+1)} = \vec{C}^{(n)} - J^{-1}(\vec{C}^{(n)}) \cdot \vec{F}(\vec{C}^{(n)}), \text{ where } J(:, i) = \frac{\vec{F}\left(\vec{C} + \varepsilon \vec{e_i}\right) - \vec{F}\left(\vec{C}\right)}{\varepsilon}, \varepsilon \ll 1$$

To ensure stability, a small regularization term $\delta \cdot I$, where $\delta \sim 10^{-8}$. $\left\| \delta \vec{C} \right\| < \varepsilon, \varepsilon = 10^{-5}$. Thus, the iterations continue until the desired accuracy is achieved. The process stops once the norm becomes smaller than the specified threshold.

2. (Nonlinear stage) At the second stage, the approximation (discrete form of the derivative) is constructed:

$$\frac{dC_1}{dx} \approx \sum_{m=0}^{M} \frac{C_{1m+1} - C_{1m}}{x_{m+1} - x_m}, \frac{dC_2}{dx} \approx \sum_{m=0}^{M} \frac{C_{2m+1} - C_{2m}}{x_{m+1} - x_m}$$

The system $(M + 1)$ of linear equations with respect to the unknowns $C_{3m}(m = \overline{0, M})$ is solved using the Gauss method (implemented in the MATLAB programming environment).

During numerical calculations using the modified Newton method, optimal values of the kinetic coefficients for the SO_3 kinetics model were obtained: k_r = -2.0194 (reaction rate constant); $K_m = 0.1573$ (Michaelis–Menten saturation constant); $k_D = 0.0$ (degradation/destruction constant).

Interpretation of the results:

- The value k_r = -2.0194 indicates a reverse or inhibitory effect of the main reaction at high concentrations. A negative value reflects inhibition of the decomposition or interaction process of SO_3. This phenomenon is often observed in real-world scenarios where intermediate or secondary products are formed, which temporarily bind SO_3 or slow down its reactions.
- A low value of $K_m = 0.1573$ suggests that the reaction saturates at low concentrations of, i.e. SO_3.
- = 0.0 indicates the absence of an independent destructive (decomposition) effecton SO_3.Its degradation occurs only through interactions with other components or in the presence of a catalyst.

The optimization is performed by minimizing the objective functional, which is calculated based on the results of the FEM solution. To this end, standard minimization algorithms are used (e.g., fminsearch, fmincon) within the MATLAB environment.

The simulated dynamic system demonstrated that, under emission rates of $Q_1 = 200\, м^3/г$ (for CO_2) and $Q_2 = 150\, м^3/г$(for SO_3) the pollutant concentration near the sources reaches peak values. The distribution of pollutants is consistent with the given wind velocity and turbulent diffusion parameters.

The developed mathematical model enables the optimization of the emission source locations x_k as well as the assessment of the effectiveness of emission reduction strategies. Furthermore, the model ensures effective pollutant dispersion and concentration reduction in critical zones. In addition, the model allows for the inclusion of an optimization procedure to determine the optimal reaction coefficients (k_r, K_m, k_D), aimed at minimizing the total functional $J(C_1, C_2)$.

For example, the following figure illustrates the effect of optimizing emission source positions:

1) top graph (chart) (CO_2): the blue dashed line shows the distribution of before optimization; the blue solid line shows the distribution after optimization; red vertical lines represent the initial source positions; green vertical lines indicate the new (optimized) source locations;

2) bottom graph (chart) (SO_3): similarly, the orange lines for $C_2(x)$, illustrate the dispersion of (SO_3) before and after optimization.

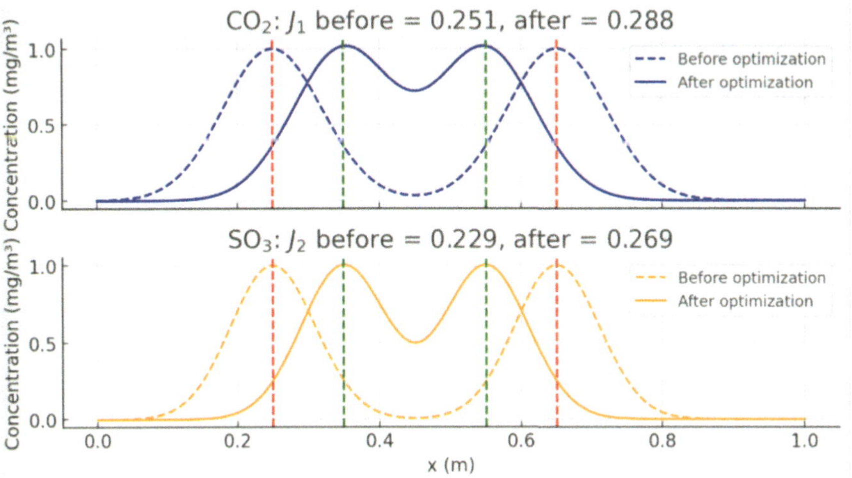

Fig. 2 Profiles of the concentrations of CO_2, SO_3 before and after optimization of the emission source locations are shown. The corresponding values of the functionals J_1 and J_2 are also illustrated

4 Conclusions

This study is the first to apply the concept of local potential to formulate and solve a boundary value problem describing the transport processes of a multi component pollutant mixture in the air in the presence of point sources of a given substance. Importantly, the use of the finite element method (FEM) combined with the local potential concept provides a unified and efficient approach for analytically modeling this process. Further studies are planned using the heterogeneity [19, 20] of the distribution of pollutants in the air.

References

1. Strang, G., Fix, G. J. An Analysis of the Finite Element Method. 2nd ed. Wellesley-Cambridge Press (2008)
2. Quarteroni, A., Valli, A.: Numerical Approximation of Partial Differential Equations. Springer (1994)
3. Sun, C., Gao, Y., Liu, H.: Zhang, WA finite element method for simulating air pollutant dispersion in urban areas. Environ. Model. Softw. **134**, 104814 (2020)
4. Zienkiewicz, O.C., Taylor, R.L.: The Finite Element Method for Fluid Dynamics. 6th ed. Elsevier (2013)

5. Schechter, R.S.: The Variational Method in Engineering, 1st edn. McGraw-Hill, New York (1967)
6. Glansdorff, P.: Prigogine, IThermodynamic Theory of Structure, Stability and Fluctuations. Wiley-Interscience, New York (1971)
7. Incropera, F.P.: Fundamentals of Heat and Mass Transfer. John Wiley & Sons, United States (2006)
8. Dubenets, V.H., Khilchevskyi, V.V., Savchenko, O.V.: Fundamentals of the Finite Element Method: Textbook. 2nd ed., rev. Chernihiv: ChDTU (2007)
9. Atkins, P., de Paula, J., Keeler, J. Physical Chemistry. 12th ed. Oxford University Press (2022)
10. Malcata, F.X. :Mathematics for Enzyme Reaction Kinetics and Reactor Performance (2 Volume Set). Wiley (2020)
11. Shubin, M.: Invitation to Partial Differential Equations. AMS, Graduate Studies in Mathematics (2020)
12. Chaari, A., Mouhali, W., Sellila, N., et al.: Pollutant dispersion dynamics under horizontal wind shear conditions: insights from bidimensional traffic flow models. Fluids **9**(11), 265 (2024)
13. Ern, A., Guermond, J.-L.: Finite Elements I: Approximation and Interpolation. Springer, Texts in Applied Mathematics (2021)
14. Copeland, R. A. Enzymes: A Practical Introduction to Structure, Mechanism, and Data Analysis. 3rd ed. Wiley (2023)
15. Riestra, J. A.A.: Generalized Taylor's Formula for Functions of Several Variables and Certain of Its Applications. CRC Press (2020)
16. Möller, D. Atmospheric Chemistry: A Critical Voyage Through the History. De Gruyter (2022)
17. Hou, J., Yan, W., Hu, D., He, Z.: Robin-Robin domain decomposition methods for the dual-porosity–conduit system. Adv. Comput. Math. **47**(1), 1–33 (2021)
18. Abd, A.S., Abushaikha, A.S. :Reactive transport in porous media: a review of recent mathematical efforts in modeling geochemical reactions in petroleum subsurface reservoirs. SN Appli. Sci. **3**, Article 401 (2021)
19. Pershyna, I.: Mathematical Modeling of 2D Discontinuous Objects by New Information Operators. Lecture Notes in Networks and Systems **762**, 393–404 (2023)
20. Lytvyn, O.N., Pershina, Y.I., Sergienko, I.V.: Estimation of discontinuous functions of two variables with unknown discontinuity lines (rectangular elements). Cybern. Syst. Anal. **50**(4), 594–602 (2014)

Application of Polyethylene Wax in the Formulation of Lubricating Greases

Andrii Grigorov[1], Alona Tulska[1(✉)], Liliana-Laura Bădiță-Voicu[2], and Viacheslav Baranov[1,2]

[1] National Technical University "Kharkiv Polytechnic Institute", 2, Kyrpychova Street, Kharkiv 61002, Ukraine
fuel.khpi@gmail.com

[2] National Institute of Research and Development in Mechatronics and Measurement Technique INCDMTM, 6-8 Pantelimon, 2nd District, 021631 Bucharest, Romania

Abstract. The article explores the potential use of polyethylene (PE) waxes as multifunctional components in lubricating grease formulations. The main industrial applications of PE waxes are identified. It is substantiated that, within lubricating greases, PE waxes can function as thickeners, stabilizers, and structuring agents, enhance hydrophobicity, and act as anti-friction additives.

Using the example of polymer wax PEW-0200 produced by Nanjing Tianshi New Material Technologies Co., Ltd., experimental studies have shown that increasing the PEW-0200 content from 5% to 20% in the grease formulation results in a higher dropping point, improved consistency, increased dynamic viscosity, enhanced colloidal stability and hydrophobicity, as well as better lubricating performance.

It is proposed to use polymer wax PEW-0200 as a relatively low-cost thickener compatible with used oils, enabling simplified production of cost-effective lubricating greases for various industrial applications.

Keywords: Polymer wax · Hydraulic oil · Lubricating grease · Thickener · Thermal stability · Hydrophobicity

1 Introduction

Polyethylene wax (PE Wax) is a low molecular weight polyethylene (M ≈ 800–8000 g/mol) [1], which imparts physical characteristics similar to those of conventional waxes, but with enhanced thermal stability, compatibility with other waxes, chemical inertness, and adhesive properties. These unique features account for the widespread use of pe wax across various industrial sectors.

At the same time, in light of current trends in the development of the oil refining industry-including modernization of technological processes and equipment, as well as expansion of the raw material base-the application of PE wax in the production of lubricating greases with diverse functional properties is considered highly promising.

D. D. Cioboată and J. Machado (Eds.): ICoRSE 2025, LNNS 1592, pp. 421–430, 2025.
https://doi.org/10.1007/978-3-032-02508-1_36

2 Objective and Research Task

Today, the main application areas of PE wax can be illustrated in the form of a structural diagram, as shown in Fig. 1.

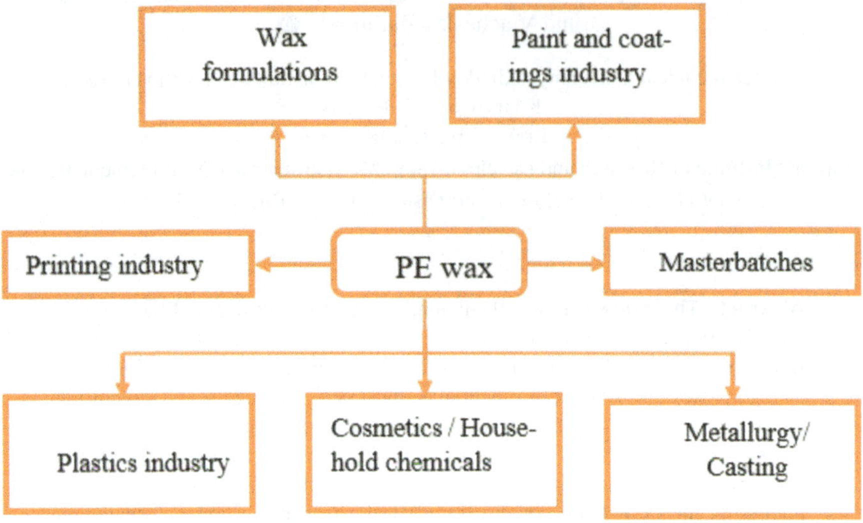

Fig. 1. Main Application Areas of PE Wax in Industry

Today, the industry produces the following types of PE waxes [2]: oxidized PE wax (produced by oxidizing polyethylene, it has polar groups that improve its wettability); non-oxidized PE wax (produced by polyethylene cleavage, hydrophobic but has lower adhesion properties compared to oxidized wax); copolymer PE waxes (produced by copolymerization with various types of other polymers, such as EVA, resulting in waxes with modified properties).

It is known that lubricating greases, due to their specific properties (lubricating ability, protective properties, heat dissipation capacity) [3], are among the petroleum products that are in high demand in industry, with consumption increasing rapidly each year. The properties of lubricating greases are determined by their structure [4, 5], which is formed by the ratio of components such as base oil (65–80%), thickener (5–15%), filler (up to 10%), additives (5–10%), and dyes (0.5–1.0%).

Given the properties of PE waxes [6–8], their use as a thickener for lubricating greases is proposed. It should be noted that PE wax can serve as a multifunctional component in lubricating grease formulations, capable of performing, in addition to thickening, other functions, as represented in the structural diagram in Fig. 2.

The role of PE wax in thickening lubricating grease lies in the creation of a spatial structure that holds the base oil in a semi-solid state, providing the grease with a stable consistency. Upon cooling from the melt, polyethylene chains crystallize [9], forming a three-dimensional network. This network physically traps the oil, preventing it from leaking from the system.

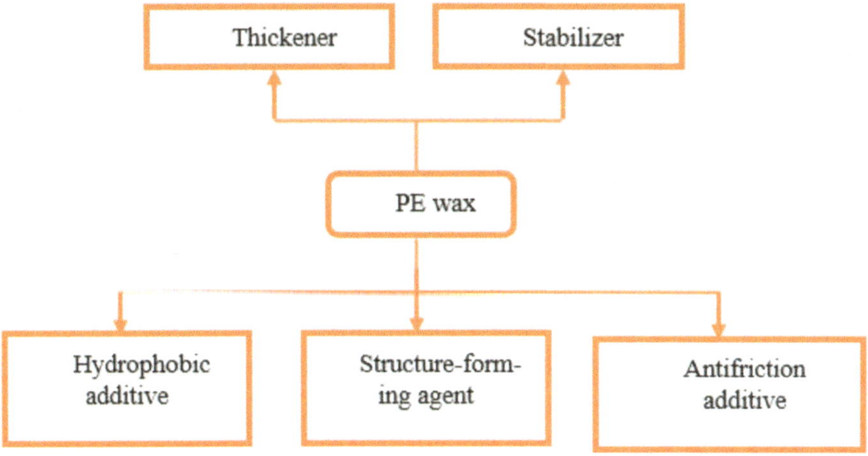

Fig. 2. Functions of PE Wax in Lubricating Grease Formulation

The stabilizing effect of PE wax in lubricating grease is manifested in regulating the grease's consistency (NLGI grade) [10]. As the content of PE wax increases in the grease formulation, a transition from lower NLGI grades (e.g., 0 or 00) to higher NLGI grades (e.g., 1–3) can be observed. Moreover, due to the melting point of PE wax being within the range of 90–140 °C, and its thermal stability (degradation temperature above 200 °C), increasing the content of PE wax in the grease enhances its thermal stability.

The addition of PE wax significantly improves the hydrophobicity of the lubricating grease due to its non-polar nature (the wax consists of saturated hydrocarbon chains ($-CH_2-CH_2-$)), as these molecules lack polar groups, promoting water repulsion on a molecular level [11]. The wax crystallizes in the grease, forming insoluble, micro-sized inclusions that prevent moisture from penetrating the grease's structure. This results in the formation of an impermeable microstructure that protects both the grease and metal components from corrosion.

PE wax can form a network of molecules in the lubricating grease, helping to maintain a stable structure even under changing operating conditions. This is crucial for greases that must retain their consistency while operating under high loads and temperatures (e.g., anti-friction applications).

PE wax can also act as an anti-friction additive in lubricating greases due to its properties, which reduce the coefficient of friction between contacting surfaces. The wax has a low friction coefficient, which helps reduce resistance between moving parts. This is particularly beneficial for reducing wear and improving the efficiency of mechanisms operating under high loads [12]. By forming a thin film on surfaces, PE wax improves the lubricating properties of the grease. This ensures an even distribution of the grease, reducing metal-to-metal contact and lowering the likelihood of friction, thus extending the equipment's service life. In combination with other grease components, the wax helps reduce direct contact between surfaces, decreasing the likelihood of wear and damage, especially under high-load and high-temperature conditions. Due to its molecular structure, PE wax helps form a protective film on surfaces, reducing the impact of microparticles on the contacting surfaces and thereby lowering wear processes on components.

When comparing the properties of PE wax with other thickeners used in lubricating greases, which are widely applied in various formulations today, it should be noted that PE wax is more chemically inert than lithium, calcium, and aluminum soaps, which may undergo hydrolysis or change their properties under the influence of acids or alkalis [13]. Unlike soap-based thickeners, the production of PE wax is simpler and more cost-effective, as it does not require the use of alkalis or pH control. Moreover, PE waxes are compatible with used base oils, in contrast to lithium, calcium, and aluminum soaps, which require thorough purification of the base oil [14].

Thus, the use of PE wax as a primary component (thickener) in lubricating greases is quite promising, as it allows for a significant reduction in the content of other components and additives, improving their operational properties and contributing to the extended service life of equipment. Based on this, we present a series of experimental studies conducted by us, revealing the potential of PE wax in the production of lubricating greases that could serve as an alternative to the greases currently used in industry.

3 Materials and Methods of the Research

The base oil for the lubricating grease compositions was hydraulic oil HLP46 with additives (anti-wear, anti-corrosion, anti-oxidation), which had been used in hydraulic drives for 1200 h. This oil had a kinematic viscosity at 40 °C of 44.15 mm^2/s.

The thickener for the lubricating grease was PEW-0200, a micronized polyethylene wax produced by Nanjing Tianshi New Material Technologies Co., Ltd. PEW-0200 is a white powder with an average particle size of 5–6 μm and a melting point of 112 °C.

The preparation of lubricating grease compositions with different contents of PE wax (5%, 10%, 15%, 20%) was carried out in a laboratory apparatus equipped with a mechanical mixing device (rotation speed 500–550 rpm), an electric motor, and an electric heater (U = 220V; f = 50 Hz). The base oil and thickener were added in weighed quantities to the inner cup of the apparatus, after which the mixture was heated to a temperature of 125–130 °C and mixed for 60 min. Afterward, heating and mixing were stopped, and the resulting product was cooled to room temperature (approximately 25 °C).

The cooled lubricating grease samples were tested according to the standardized methods of the American Society for Testing and Materials (ASTM): drop point by D-2265; penetration by D-217-21a; colloidal stability by D-6184; dynamic viscosity at 0 °C by D-2983; hydrophilicity by D-1264 and D-7334–08; lubricating properties by D-2266, D-2596, and D-8317, oxidative stability of lubricating greases—Test Method ASTM D942.

The water absorption of the lubricating grease samples was studied using a non-standardized method (based on sample mass increase), which is commonly used in research practice.

The described method was implemented as follows: a weighed sample of grease (12 g) was placed in a glass beaker (V = 150 cm^3) with distilled water for 24 h at a water temperature of 25 °C. The sample was then removed from the beaker, excess water was removed from its surface, and the sample was weighed again.

The water absorption (W, mg/g) of the lubricating grease was calculated using the following formula:

$$W = \frac{m_1 - m_0}{m_0} \times 100\%$$

where m_0 – the mass of the lubricating grease sample before the test, g;
m_1 – the mass of the lubricating grease sample after the test, g.

4 Results and Discussion

The results of the laboratory tests, which included studying the effect of the content of PEW-0200 wax (Xp.w., %) on the drop point temperature (td.p., °C), penetration (Xp., mm × 10^{-1}), and dynamic viscosity (ηg., mP × s) of the obtained lubricating grease samples (see Figs. 3, 4, 5 and 6).

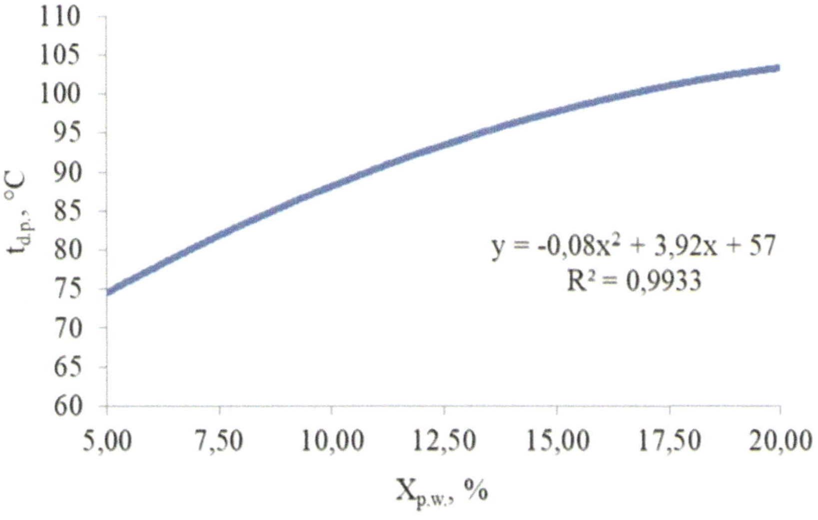

Fig. 3. Effect of Xp.w on td.p. for Lubricating Grease Samples

The conducted research has proved that secondary raw material used for the production of plastic lubricants make significant effect on the dynamic viscosity of plastic lubricants and their lubricating properties. By altering the type of a base oil and polymer dispersing phase, while keeping the production technology constant, it is possible to greatly influence the lubricating properties of the final product – plastic lubricant. The studies have shown that by varying η from 40 Pa•s to 140 Pa•s, plastic lubricants provide the value of $D_{w(196)}$ = 0.40–0.70 mm and P_{We} = 2100–3550 N. The investigation into the influence of the polymer dispersing phase has revealed that due to the ability to retain the base oil, determined by the structure and properties of the polymer, a certain deterioration in lubricating properties is observed in the series HDPE → LDPE → PP ($D_{w(196)}$ changes from 0.34 mm to 0.42 mm; P_{We} changes from 3550 N to 3450 N). Analyzing the obtained results, it is evident that the plastic lubricants obtained from secondary raw materials, due to their lubricating properties, can be used as anti-friction plastic lubricants instead of Mobil Grease Special ($D_{w(196)}$ = 0.40 mm, P_{We} = 2500 N); NESTE MP Grease ($D_{w(196)}$ = 0.40 mm, P_{We} = 2600 N); Shell RETINAX Grease CMX1,2 ($D_{w(196)}$ = 0.50 mm, P_{We} = 4000 N); Mobilux EP ($D_{w(196)}$ = 0.40–0.50 mm, P_{We} = 2500 N).

Obviously (see Fig. 3), an increase in the content of PEW-0200 wax from 5% to 20% leads to a rise in the dropping point temperature of the lubricating grease (from 75 °C to 103 °C). With a further increase in the PEW-0200 wax content, the dropping point of the grease approaches 112 °C—the melting point of PEW-0200 wax.

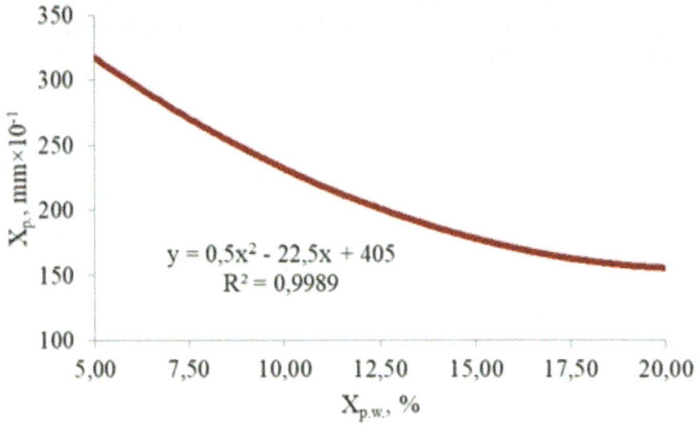

Fig. 4. Effect of $X_{p.w.}$ on X_p for plastic grease samples

The consistency of the grease (see Fig. 4), as indicated by the penetration value (Xp.), significantly depends on the concentration of the thickening agent—in this case, PEW-0200 wax. Therefore, as the wax content increases from 5% to 20%, the consistency of the grease increases, resulting in a transition from NLGI grade 1 (Xp. = 318 mm × 10^{-1}) to NLGI grade 5 (Xp. = 155 mm × 10^{-1}).

The dynamic viscosity of the grease (see Fig. 5), consisting of HLP-46 oil and PEW-0200 polymer wax, increases (from 1785 mP·s to 6485 mP·s) with an increase in wax content from 5% to 20%. This occurs due to the wax's ability to thicken the grease, enhancing its consistency and surface retention, which ensures effective wear and friction protection. Among other influencing factors are the testing temperature and the oil's inherent properties (in particular, its viscosity) [15]. However, the most significant factor contributing to the rise in dynamic viscosity with increased PEW-0200 content is the wax's ability to form a three-dimensional (3D) fibrillar or lamellar network through crystallization of paraffinic/polyethylene chains [16], which retains the HLP-46 oil via capillary and surface forces.

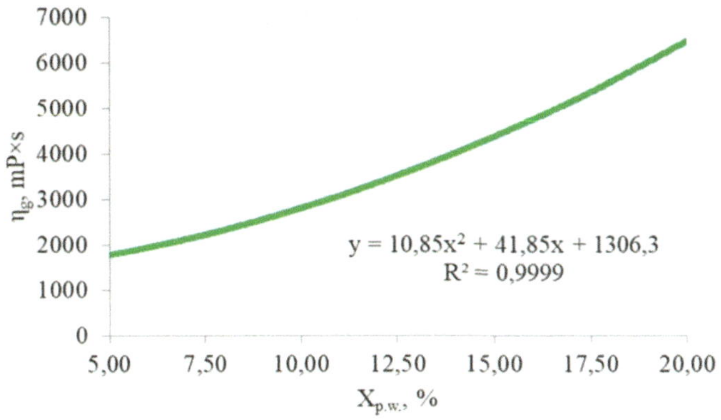

Fig. 5. Effect of $X_{p.w.}$ on η_g for plastic grease samples

According to Fig. 6, at a content of 5% of PEW-0200 wax, the grease exhibits a tendency to separate (moderate colloidal stability). A content of 10–15% of PEW-0200 wax allows for achieving high colloidal stability of the grease. In turn, increasing the PEW-0200 wax content to 20% or more, although it improves the colloidal stability, is associated with the risk of a sharp deterioration in the adhesive properties of such grease [17]. Moreover, in some cases, when the content of PEW-0200 wax exceeds 20%, a decrease in colloidal stability may be observed due to an excess of wax, which disrupts the gel structure of the grease, in which the base oil is retained by surface tension, intermolecular interactions, and viscosity.

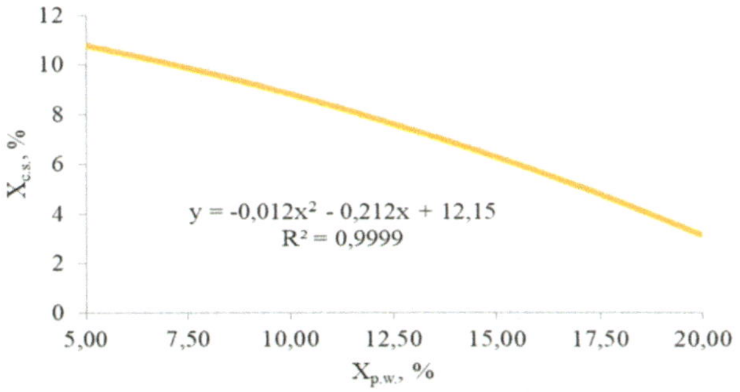

Fig. 6. Effect of $X_{p.w.}$ on $X_{c.s}$ for plastic grease samples

At the next stage of the study, the influence of PEW-0200 wax content (Xp.w., %) on the hydrophobicity of the grease samples and their lubricating ability was determined. The results of the conducted studies are presented in Tables 1–2.

The results of the conducted studies, presented in Table 1, illustrate the positive effect of increasing the content of PEW-0200 wax in the grease on its hydrophobicity. In particular, there is a transition from grease with low hydrophobicity (Xp.w. = 5%) to grease with very high hydrophobicity (Xp.w. = 20%).

Table 1. Results of the study on the hydrophobicity of grease samples

Name of an indicator	$X_{p.w.}$, %			
	5	10	15	20
Contact angle (CA, °)	85	96	105	114
Water leaching (W_W, %)	20	13	8	5
Water absorption(W, mg/g)	2,7	1,8	1,1	0,5

Table 2. Results of the study on the lubricating properties of grease samples.

Name of an indicator	$X_{p.w.}$, %			
	5	10	15	20
Diameter of the wear spot, (D_W, mm)	0,65	0,50	0,45	0,35
Welding load ($P_{w.l.}$, kg)	230	270	340	350
Friction coefficient (SRV)	0,16	0,12	0,9	0,7

The study of the lubricating properties of grease samples (see Table 2) showed that as the content of PEW-0200 wax in the grease increased from 5% to 20%, the lubricating ability of the grease improved from a basic to a high level. However, when the content of PEW-0200 wax in the grease exceeds 20%, a decrease in lubricating ability may be observed due to the oversaturation of the system with wax, leading to a loss of dispersion homogeneity.

It should also be noted that lubricating greases thickened with PEW-0200 wax exhibit a relatively narrow operating temperature range—from -20 °C to 100 °C. The lower limit is due to the wax becoming brittle at around -25 °C, losing its ability to deform without fracturing, which negatively affects the mechanical stability of the grease structure. The upper operating temperature is governed by the melting point of the wax (approximately 100–110 °C). Exceeding this temperature may result in grease softening and a decrease in its ability to remain in friction units.

Although PEW-0200 is chemically inert, it does not possess antioxidative properties and therefore does not protect the base oil from oxidation. This necessitates the inclusion of antioxidant additives in greases thickened with PEW-0200. For example, a grease without antioxidant additives and thickened solely with PEW-0200 shows an induction period of 30–60 h under ASTM D942 testing (moderate oxidative stability), whereas greases with antioxidant additives demonstrate values between 80–150 h (high oxidative stability).

The advantages of using PEW-0200 polyethylene wax as a thickener in lubricating greases include its low toxicity (hazard class 4—low-hazard substance), its non-polluting nature with regard to water, its non-reactivity with metals, and the absence of harmful vapor emissions. However, it is not biodegradable and thus requires responsible disposal.

Overall, PEW-0200 can be considered a conditionally environmentally safe thickener for use in lubricating greases, provided that its environmental release is properly controlled.

5 Conclusions

The conducted research, using PEW-0200 wax as an example, demonstrated the potential of using PE waxes as thickeners for grease. It was determined that PEW-0200, due to its properties, can positively influence not only the dynamic viscosity of the grease but also the temperature limits of its use (the upper limit is determined by the drop point temperature), colloidal stability, hydrophobicity, and lubricating ability.

It was found that as the content of PEW-0200 wax in the grease increased from 5% to 20%, its drop point temperature rose from 75 °C to 103 °C, the grease transitioned from NLGI consistency grade 1 (Xp = 318 mm $\times$ 10^{-1}) to grade 5 (Xp = 155 mm $\times$ 10^{-1}), the dynamic viscosity increased from 1785 mP $\times$ s to 6485 mP $\times$ s, colloidal stability improved from 10.8% to 3.1%, hydrophobicity increased (water absorption decreased from 2.7 mg/g to 0.5 mg/g, contact angle increased by 29°), and lubricating ability improved (wear spot diameter decreased by 0.30 mm; welding load increased by 120 kg).

Thus, considering the positive properties that manifest in the composition of greases, the polymer wax PEW-0200 can be recommended as a thickener, offering the following advantages:

Due to its compatibility with used oils, PEW-0200 allows the creation of greases based on recycled materials, which helps reduce costs for purchasing base oils and reduces environmental damage;

The manufacturing process of grease with PEW-0200 does not require complex saponification or chemical reactions, as in the case of soap-based thickeners, which shortens the production cycle, reduces energy consumption, and the need for high-temperature equipment;

Due to its low cost and simple formulation, manufacturers can create budget-friendly greases for various industries where product cost is crucial. PEW-0200 is available on the domestic market or can be easily imported without significant expenses, reducing dependence on expensive lithium thickeners or complex additives.

References

1. Ciesińska, W., Liszyńska, B., Zieliński, J.: Selected thermal properties of polyethylene waxes. J. Therm. Anal. Calorim. **125**(3), 1439–1443 (2016). https://doi.org/10.1007/s10973-016-5706-1
2. Yetgin, S., Gönen, M., Savrık, S.A., AkzoNobel, D.B.: Polyethylene wax: uses, characterization, and identification. In book: Imidic Polymers and Green Polymer Chemistry **265–284** (2021). https://doi.org/10.1201/9781003057918-9
3. Osara, J.A., Chatra, S., Lugt, P.M.: Grease material properties from first principlesthermodynamics. Lubr. Sci. **36**, 36–50 (2024). https://doi.org/10.1002/ls.1672
4. Robert, J.P., Künzner, F.: Lubrication of plastic components in the steering system and brakes. ATZ Worldw **113**, 24–27 (2011). https://doi.org/10.1365/s38311-011-0016-5

5. Si, R., Ren, Q., He, Y., Long, J.: Molecular self-assembling structure and formation mechanism of lubricating grease: A computational simulation approach. Tribol. Int. **179**, 108150 (2023). https://doi.org/10.1016/j.triboint.2022.108150
6. Li, R., Shiai, X., Jiajun, X., Pan, T., Sun, B., Li., D.: Effect of functionalized polyethylene wax on the melt processing and properties of highly filled magnesium hydroxide/linear low-density polyethylene composites. Polymers **15**(11), 2575 (2023). https://doi.org/10.3390/polym15112575
7. Yetgin, S.: Characterization of lubricant polyethylene waxes. European J. Tech. (EJT) **10**(2), 489–500 (2020). https://doi.org/10.36222/ejt.718423
8. Mhlabeni, T.L., Jamiru, T., Mhike, W.: Thermophysical, and rheological insights of polyethylene/wax blends. Front. Chem. Eng. **6**, 1507921 (2024). https://doi.org/10.3389/fceng.2024.15079219
9. Kostyuk, A.V., Ignatenko, V.Y., Makarova, V.V., Antonov, S.V., Ilyin, S.O.: Polyethylene wax as an alternative to mineral fillers for preparation of reinforced pressure-sensitive adhesives. Int. J. Adhes. Adhes. **102**, 102689 (2020). https://doi.org/10.1016/j.ijadhadh.2020.102689
10. Lijesh, K.P., Khonsari, M.M.: In-situ monitoring of grease consistency. Invention Disclosure **5**, 100036 (2025). https://doi.org/10.1016/j.inv.2024.100036
11. Marzbani, P., et al.: Effect of polyethylene wax/soy protein-based dispersion barrier coating on the physical, mechanical, and barrier characteristics of paperboards. J. Coat. Technol. Res. **18**(1), 247–257 (2021). https://doi.org/10.1007/s11998-020-00403-7
12. Delgado, A., Aperador, W.: Evaluation of polyethylene wax and paraffin as anti wear and extreme pressure additives in virgin sesame base stock. Tecciencia **14**(27), 21–29 (2019). https://doi.org/10.18180/tecciencia.2019.27.3
13. Gurt, A., Khonsari, M.: An Overview of Grease Water Resistance. Lubricants **8**(9), 86 (2020). https://doi.org/10.3390/lubricants8090086
14. Berdiev, S.A.: Obtaining metal-containing organic lubricating oils and studying their physical and chemical properties // Universum: technical science, vol. 11(116) (2023). https://7universum.com/ru/tech/archive/item/16299
15. Grigorov, A., Tulskaya, A., Bondarenko O., Olena, P., Bryhada, O., Nahliuk, M.: Technological Control of production of plastic lubricants by rheological properties. Petroleum and Coal **66**(1), 223–230 (2024)
16. Gahleitner, M., Wang, J., Prades, F., Bernreitner, K.: Gelation and crystallization phenomena in polyethylene plastomers modified with waxes. Polymers **13**(13), 2147 (2021). https://doi.org/10.3390/polym13132147
17. Grigorov, A., Tulska, A., Bondarenko, O., Yurchenko, L., Gornostal, S.: Determination of the optimal concentration of polymer thickener for production of plastic lubricants based on secondary raw materials. Pet. Coal **65**(4), 1118–1122 (2023)

On Reduction of a Non-stationary Random Process to a Stationary One in the Study of Dynamic Loads

Yevhen Kalinin[1](✉), Volodymyr Boldovskyi[2], Volodymyr Demianyshyn[3], Kateryna Boldovska[4], and Oleksandr Hasiuk[5]

[1] National University of Life and Environmental Sciences of Ukraine, Kyiv, Ukraine
kalinin@nubip.edu.ua

[2] The National Aerospace University "Kharkiv Aviation Institute", Kharkiv, Ukraine

[3] Kyiv Institute of the National Guard of Ukraine, Kyiv, Ukraine

[4] Kharkiv National Automobile and Highway University, Kharkiv, Ukraine

[5] National Technical University "Kharkiv Polytechnic Institute", Kharkiv, Ukraine

Abstract. The paper analyses the possibility of studying dynamic loads in transmissions and load-bearing systems of wheeled and tracked machines by bringing a non-stationary random process, which describes the system, to a stationary one by using a finite-difference method of correlation analysis of a process of additive form. This method is particularly interesting for the study of the widespread methodological approach to the study of load modes of car elements, smoothness of running, in which the amplitude-frequency information about the mode is obtained by processing the records of the car movement on a measured section with a constant speed. For comparison, we present a graph of the correlation function of the same random process constructed by the method of preliminary smoothing of the initial process with simultaneous approximation by a cubic parabola and subsequent centering of the initial process relative to the parabola. It was found that faster decay of the correlation function indicates larger errors and noise caused by them, inherent in the algorithm of the method with the smoothing operator. The possibility of using the proposed methodology has been analyzed and recommendations for its use have been proposed.

Keywords: oscillatory system · vehicle · non-stationary random process · vibration · dynamic load

1 Introduction

In theoretical and experimental studies of automobiles, methods of the theory of random functions (processes) are widely used to analytically describe a variety of conditions and operating modes of automobiles.

Characteristics of random processes are used in determining dynamic loads, in selecting design solutions and optimal parameters of units, in calculations of chassis and transmission elements for strength and fatigue life. They also form the basis for assessing

D. D. Cioboată and J. Machado (Eds.): ICoRSE 2025, LNNS 1592, pp. 431–446, 2025.
https://doi.org/10.1007/978-3-032-02508-1_37

the dynamic properties of adjustable and non-adjustable objects during their operation (engine, transmission, suspension, brake system, steering, etc.).

In modern research, bringing non-stationary random processes to a steady state in the context of dynamic load analysis is a key task, since without stabilizing the statistical properties of signals, methods for assessing the reliability, strength and service life of structures remain incorrect. One of the most relevant practical areas is the algorithmic processing of non-stationary processes by dividing them into locally stationary intervals. Thus, in [5], it is proposed to use non-parametric methods and Kendall statistics to automatically determine time intervals with constant statistical characteristics, which allows estimating the mean and variance in each interval. At the same time, approaches to signal transformation through wave analysis are being developed: the authors of [8] proposed a stochastic generalized harmonic wavelet (SGHW), which effectively approximates the evolutionary spectral density, reducing the number of random variables and maintaining the accuracy of modelling.

The DSSA (Dynamic Stationary Subspace Analysis) technique is particularly noteworthy, as it tracks dynamic subspaces where the process is locally stationary and allows time shifts between them to be taken into account [9]. In addition, general reviews of experimental signal transformation methods, including differentiation, detrending, and the use of ARMA models, show that such techniques can reduce the error in predicting the dynamics of structures by 30–95% [10].In the field of mechanical engineering, a mathematical model of non-stationary processes has been developed [11], which demonstrates that impulse start-stop cycles lead to significant dynamic loads that greatly reduce the service life of components – the model integrates electromechanical vibrations, structural vibrations and software-algorithmic degradation assessment (via NASGRO).

Modifications of traditional spectral analysis for non-stationary signals are also important, as evidenced by the use of multiscale recurrent quantum analysis on the example of bridge structures. This approach is effective in detecting both nonlinearities and transitions between dynamic modes. Another approach is to construct generators of non-stationary processes with known evolutionary spectral density [12, 14], which is important for modelling complex loads: processes have been developed for creating such general realizations that accurately reproduce the statistical properties of the system of interest and convey its dynamic character.

2 Principles of Studying Dynamic Loads in Systems by Analyzing Their Non-stationary Processes

It is known that random processes are classified according to a number of features: stationary and non-stationary, normal and different from normal, ergodic and non-ergodic, etc. When describing random processes within the framework of correlation theory, the following are used as basic statistical characteristics: mathematical expectation $m_x(t)$; dispersion $D_x(t)$; autocorrelation $R_{xx}(t, \tau)$ and mutual correlation $R_{yx}(t, \tau)$ functions. By means of correlation functions the spectral densities reflecting the frequency composition of a random process are determined [1, 2].

In the applied apparatus of the theory of random functions, the area of definition of correlation functions is limited to the class of stationary random processes, i.e. processes with probability characteristics stable in time [3, 4]:

$$m_x(t) = m_x = const; \tag{1}$$

$$D_x(t) = D_x = const; \tag{2}$$

$$R_{xx}(t, \tau) = R_{xx}(\tau). \tag{3}$$

This can explain the widespread methodological approach to the study of load modes of vehicle elements, smooth running, in which the amplitude-frequency information about the mode is obtained by processing the records of vehicle movement on a measured section with constant speed [5]. For example, Fig. 1, a shows an oscillogram with recording of transmission and suspension parameters of a truck when driving on asphalt at a constant speed (45 km/h) with a load in the body of 3.2 tones. The figure shows that all processes are stationary in terms of mathematical expectations and dispersions [6].

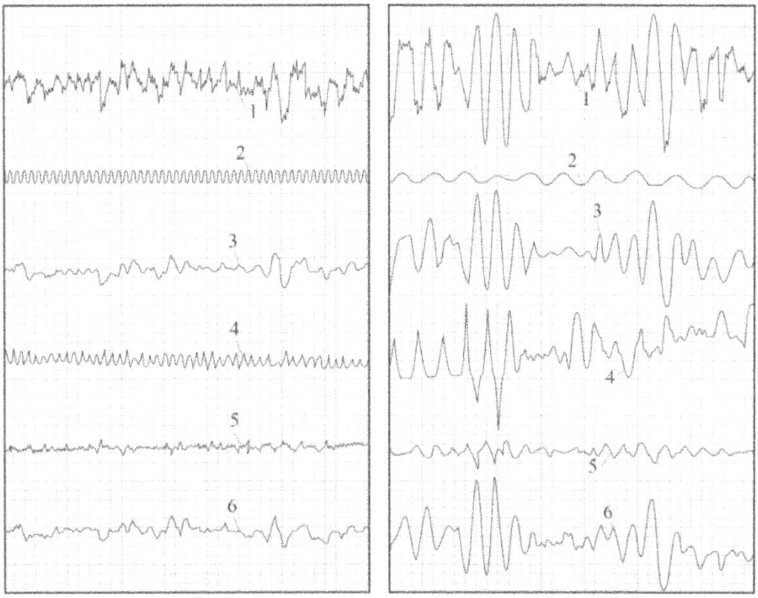

a b

Fig. 1. Load modes of transmission and suspension elements of the car [6]: a – flat road, travelling speed 45 km/h; b – rough road, travelling speed 11 km/h; 1 – wheel revolutions marker; 2 – time marker; 3 – strain gauge at the end of the crown leaf spring; 4 – bending moment on the axle; 5 – sensor of relative movements of the system 'frame - rear axle; 6 – torque on the axle; 7 – vertical accelerations of the sprung mass (the sensor is installed on the frame above the rear axle); 8 – strain gauge at the beginning of the crown leaf spring.

However, the load and speed modes of vehicle units and assemblies are generally characterized by non-stationary random processes. Figures 1b and 2 show sample oscillograms for a dump truck and a bus. Thus, for example, during touchdown and acceleration of the car (Fig. 2) with gear shifting, the mathematical expectations and dispersions of torques depend on time. At the same time, the frequency composition of the process changes depending on the gear engagements, and at the stages of clutch slipping t_1 (Fig. 2), the frequency of frictional oscillations differs from the natural low frequency of the oscillating system of the transmission in a given gear [7, 8].

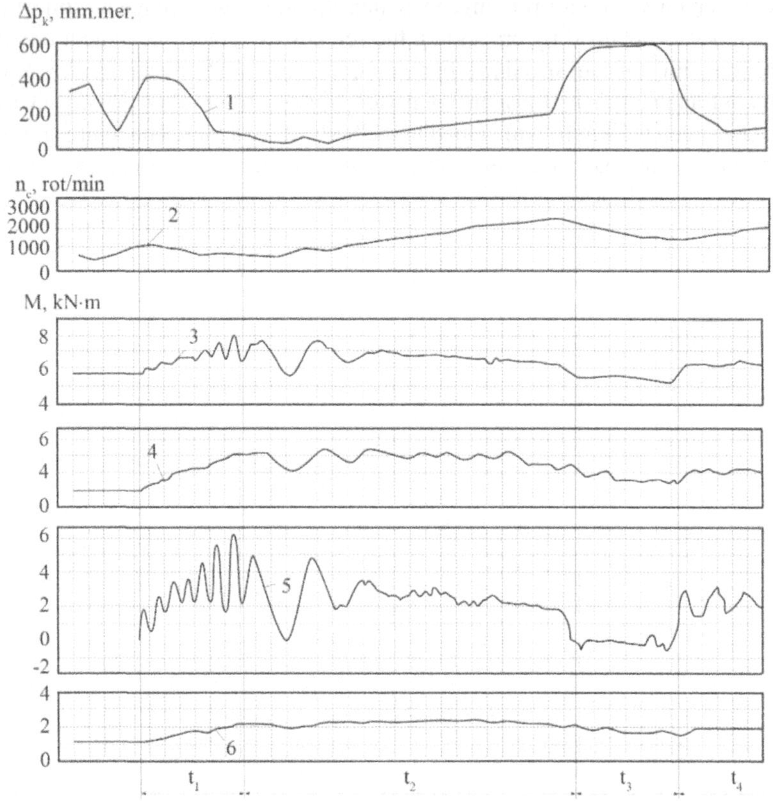

Fig. 2. Starting and beginning of acceleration of the bus (total weight 10.4 tones) [7]: 1 – rarefaction in the intake manifold $\Delta p_{||}$; 2 – crankshaft rotation speed n_e; 3 – longitudinal accelerations of the body; 4 – stress in the leaf spring root ($Q = Q^*$ – conditional zero in static position of the bus); 5 – torque M on the half axle; 6 – movement of the rear axle relative to the bodyδ; l – length of one wheel revolution; t_1 – stage of friction oscillations; t_2 – movement with the clutch engaged in gear II; t_3 – gear change; t_4 – movement in gear III

When a car is travelling on a broken road (Fig. 1b) at a constant speed, the mathematical expectation and dispersion of torque do not remain constant. Thus, artificial reduction to stationary processes by fixing the experimental conditions does not give a positive result in all cases [9, 10].

A characteristic feature of the non-stationary processes shown in Figs. 1, 2 and 3 is non-ergodicity, i.e. independence of local probability characteristics from the number of realization. For example, the torques on the semi-major axis when the bus is moving (Fig. 3) have different dependences of mathematical expectations on time for different realizations.

A common method for processing nonstationary nonergodic processes is averaging over an ensemble of realizations obtained under statistically homogeneous conditions. However, due to economic considerations or impossibility of conducting statistically homogeneous experiments, methods of analyzing nonstationary random processes by homogeneous realizations become important [11–13].

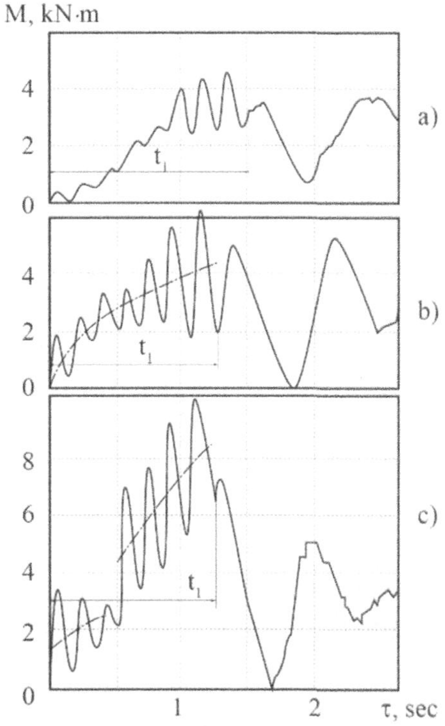

Fig. 3. Torque on the semi-axle when the bus is moving in gear II [11]: a – average rate of clutch engagement and throttle opening, bus weight $G_a = 7{,}6$ t; b – average rate of clutch engagement and throttle opening, $G_a = 10{,}4$ t; c – intensive throttle opening and clutch engagement, $G_a = 10{,}4$ t (dashed lines – changes of local mathematical expectation 6 $m_x(t)$)

The methods of processing by one realization depend on the type of the process [14–16].

The processes of additive type, which are non-stationary with respect to the mathematical expectation, are common in the practice of research [17]

$$x(t) = m_x(t) + x_0(t), \tag{4}$$

where $m_x(t)$ – unknown function of time; $x_0(t)$ - centered stationary ergodic random process.

To non-stationary processes of additive type, we can refer processes in nodes and aggregates of the car when it moves with variable speed on a statistically homogeneous road on a given gear, starting from a place (clutch engagement), braking, etc. [18].

Estimation of the mathematical expectation of a non-stationary random process by one realization at discrete reading can be determined by smoothing with the help of the 'current mean' operator by the formula [19, 20]:

$$m_x^*(t) = \frac{1}{n+1}\sum_{i=0}^{n} x(t+ih), \tag{5}$$

where $h = \frac{T_c}{n}$; n - number of segments into which the smoothing interval is divided T_c; $x(t + ih)$ – ordinates of the discrete analogue of the process $x(t)$.

In addition to the current mean operator, polynomial, exponential and generalized smoothing operators are used [21, 22].

The estimation of the correlation function of a non-stationary random process is also determined by the method of sliding smoothing [23, 24]:

$$R_{xx}^*(t, t+\tau) = \frac{1}{N+1-j}\sum_{j=0}^{N} x_0^*(t) - x_0^*(t+jh), \tag{6}$$

where N is the number of observation points.

The estimation of the centered random process is [25–27]:

$$x_0^*(t) = x(t) - m_x^*(t). \tag{7}$$

The disadvantages of the method of sliding smoothing include suppression of low frequencies of the investigated process up to the frequency $\omega_c = \frac{2\pi}{T_c}$. This method is relatively labor-intensive [28, 29].

3 Development of Principles of Reduction of a Non-stationary Random Process to a Stationary One

Work has been carried out to identify the possibility of using a new method of operational correlation analysis of non-stationary random processes based on the reduction of a non-stationary process to a stationary one through differences of a certain order in studies of automotive structures.

The method, intended for processing processes of additive type by one realization, is based on the transformation of a non-stationary random process into some auxiliary stationary random process with the subsequent determination of the sought correlation function of the process $x(t)$ through the auxiliary stationary random process.

This eliminates the need to determine the function $m_x(t)$ and centering of the process, and achieves a simple relationship between the correlation functions of the original and auxiliary processes (thus reducing the algorithm and machine time).

The auxiliary stationary random process is formed using finite-difference calculus. From the current ordinates of the initial process (Fig. 4, curve 1, constructed from the results of reading the oscillogram of Figs. 2 and 3, b), taken with a time interval of Δt, the difference of n-th order is obtained. For differences of different orders, we can write:

$$\Delta^{(1)}x(t) = x(t + \Delta t) - x(t); \tag{8}$$

$$\Delta^{(2)}x(t) = \Delta^{(1)}x(t + \Delta t) - \Delta^{(1)}x(t) = x(t + 2\Delta t) - 2x(t + \Delta t) + x(t); \tag{9}$$

$$\Delta^{(n)}x(t) = \Delta^{(n-1)}x(t + \Delta t) - \Delta^{(n-1)}x(t) = \Delta^{(n-2)}x(t + 2\Delta t) - 2\Delta^{(n-2)}x(t + \Delta t) + \Delta^{(n-2)}x(t) = \sum_{v=0}^{n} (-1)^v C_n^v x[t + (n - v)\Delta t] \tag{10}$$

where C_n^v is the number of combinations n of v.

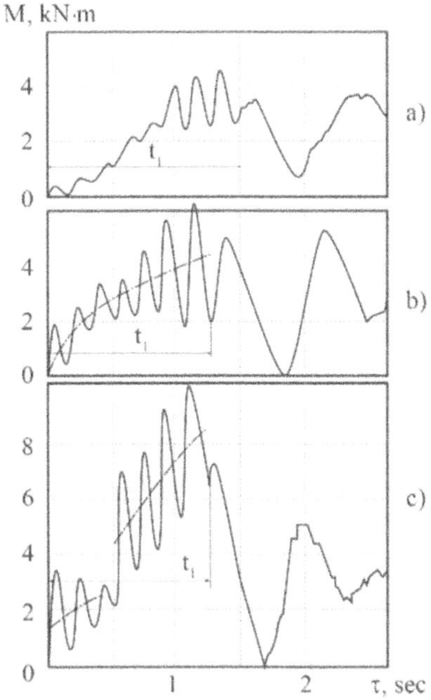

Fig. 4. Stationarisation of the process of changing the torque of the semi-axis

(see Figs. 2 and 3b): a – realization of the process counted with step $h = 2\pi/7$ (curve 1), its mathematical expectation (curve 1'), their first derivatives (curves 2 and 2') and second derivatives (curves 3 and 3'); b – second derivatives of the process counted with step $h = 4\pi/7$, with different initial reference points (curves 4 and 5) and mathematical expectation of the auxiliary processes (curves 4' and 5'); c – second derivatives of the process counted with step $h = 6\pi/7$; d – normalized correlation functions obtained using the finite-difference method of additive mixture (curve 7) and with preliminary smoothing and centering of the initial process I (curve 8); the cross denotes the function for the auxiliary process 6.

The n-th order differences for the process (4), due to the linearity of the difference operator, are related by the relation

$$\Delta^{(n)}x(t) = \Delta^{(n)}x_0(t) + \Delta^{(n)}m_x(t). \tag{11}$$

It is assumed that $m_x(t)$ at any segment of the realization the duration $n\Delta t$ is approximated by a polynomial of degree not higher than $n - 1$, i.e. the equality is valid

$$m_x(t) = a_0(t_1) + a_1(t_1)t + \dots + a_{n-1}(t_1)t^{n-1}, \tag{12}$$

where $t_1 \leq t \leq t_1 + n\Delta t$.

According to the basic theorem of the calculus of differences, the difference of n-th order of a deterministic function approximated by a polynomial of degree not higher than $n - 1$, turns out to be equal to zero over the whole interval of investigation $(0, T)$, i.e.

$$\Delta^{(n)}m_x(t) \equiv 0 \, for \, 0 \leq t \leq T \tag{13}$$

On the basis of expressions (10), (11) and (13) we obtain the formula for the relation between the auxiliary stationary random process and the centered process $x_0(t)$ defined by formula (4):

$$\Delta^{(n)}x(t) = \Delta^{(n)}x_0(t) = \sum_{\nu=0}^{n}(-1)^{\nu}C_n^{\nu}x_0[t + (n - \nu)\Delta t]. \tag{14}$$

In [8] it is shown that the auxiliary process $\Delta^{(n)}x(t)$ is stationary because its mathematical expectation is zero and the correlation function depends only on the shift of τ, and the formula given in Eq.

$$R_{\Delta^{(n)}x}(t, t + \tau) = \sum_{\nu=0}^{n}\sum_{\mu=0}^{n}(-1)^{\nu+\mu}C_n^{\nu}C_n^{\mu}R_{xx}[\tau + (\mu - \nu)\Delta t] \tag{15}$$

shows the relation between the sought correlation function $R_{xx}(\tau)$ of the process $x(t)$ and the correlation function $R_{\Delta^{(n)}x(t)}$ of the auxiliary stationary random process $\Delta^{(n)}x(t)$.

Subsequent transformations allow us to obtain a working formula for the transition from the auxiliary process to the sought process:

$$R_{xx}(\tau) = \frac{1}{C_{2n}^{n}}R_{\Delta^{(n)}x}(\tau) = \frac{(n!)^2}{(2n)!}R_{\Delta^{(n)}x}(\tau). \tag{16}$$

Formula (16) allows us to outline the way of correlation analysis of non-stationary random processes of the type represented by expression (4): it is necessary to form an auxiliary process $\Delta^{(n)}x(t)$ and calculate its correlation function using the methods of processing stationary processes.

Mean-integral estimation of the correlation function of the auxiliary process

$$R^*_{\Delta^{(n)}x}(\tau) = \frac{1}{T}\int_0^T \Delta^{(n)}x_0(t)\Delta^{(n)}x_0(t+\tau)dt \tag{17}$$

is unbiased, i.e.

$$R^*_{\Delta^{(n)}x}(\tau) = R_{\Delta^{(n)}x}(\tau). \tag{18}$$

The relationship between the dispersions of the correlation functions of the auxiliary and initial processes is determined by a formula of the form:

$$D_{R_{\Delta^{(n)}x}}(\tau) = C_{4n}^{2n} D_{R_{xx}}(\tau). \tag{19}$$

Formula (19) allows us to evaluate the accuracy of determining the correlation function of nonstationary processes by comparing it with the accuracy of the studied estimation of the correlation function of a stationary process.

The quality of the estimation is determined by the relative value of the RMS error, i.e.:

$$\delta_{R_{\Delta^{(n)}x}}(\tau) = \frac{\sigma_{R_{\Delta^{(n)}x}}(\tau)}{R_{\Delta^{(n)}x}(\tau)} = \frac{\sqrt{D_{R_{\Delta^{(n)}x}}(\tau)}}{R_{\Delta^{(n)}x}(\tau)} = \frac{\sqrt{C_{4n}^{2n}}}{C_{2n}^{n}} \cdot \frac{\sqrt{D_{R_{xx}}(\tau)}}{R_{xx}(\tau)} = \sqrt{(4n)!}\left(\frac{n!}{(2n)!}\right)^2 \delta_{R_{xx}}(\tau), \tag{20}$$

where $\delta_{R_{xx}}(\tau)$ is the relative value of the RMS error of estimation of the correlation function from the centered realization of the stationary process $x_0(t)$.

Further expression

$$\frac{\delta_{R_{\Delta^{(n)}x}}(\tau)}{\delta_{R_{xx}}(\tau)} = \sqrt{(4n)!}\left(\frac{n!}{(2n)!}\right)^2 = \Lambda \tag{21}$$

determines the increase of the RMS error of the estimation of the correlation function of the investigated nonstationary process in comparison with the estimation of this function of the stationary process.

The order dependence of formula (21) is n as follows.

Thus, the considered method provides a small (and at that accountable) error of correlation analysis of non-stationary random process with known and time-varying mathematical expectation. At the same time, it provides high operability of the analysis and is convenient for its implementation both on PC and in special conditions of direct operative analysis of working processes of the design, for example, at experimental study of working processes of the car-laboratory equipped with correlometers, or at vibrodiagnostic installations based on comparison of the correlation function of the

Table 1. Increase in RMS error of correlation function estimation depending on the order of difference transformation

n	0	1	2	3	4	5
Λ	1	1,22	1,40	1,52	1,62	1,71

process of the investigated aggregate with the exemplary function. In view of the novelty of the method, the purpose of the experimental part, apart from considering the possibility of using the method in the study of unsteady operating modes of car units and parts, was to develop initial recommendations and requirements for the measurement technique and the corresponding equipment, primary processing and preparation of experimental information, compilation, and implementation of the calculation algorithm on a PC. Let's trace it on the example of investigation of the process of torque change on the semi-axle during the bus trolling (Fig. 4, a, curve 1). Determines the increase of the RMS error of the estimation of the correlation function of the investigated nonstationary process in comparison with the estimation of this function of the stationary process. The required order of difference differentiation to form the auxiliary stationary random process. Taking the difference of n-th order (difference differentiation) for a discrete process corresponds to differentiation of n-th order of a continuous process. By virtue of the main theorem of the calculus of differences and we have $(n-1)$-fold differentiation of the random process (4) with mathematical expectation approximated by a polynomial of $(n-1)$-th degree, stationarises the initial random process, and n-fold differentiation, in the absence of errors introduced by experiment, reading and PC, should center it, i.e. approximate to zero the expression

$$\frac{1}{N-n}\sum_{n=1}^{N-n}\Delta^{(n)}x(t), \tag{22}$$

where N is the number of observation points of the initial process in the discrete record; $\Delta^{(n)}x(t)$ – values of ordinates of the difference process of n-th order.

However, due to the presence of the mentioned errors, the noise level in the difference process grows with each successive differentiation. For this reason, the value of expression (22) first decreases in absolute value, then, fluctuating, grows. This is confirmed by Table 2, which shows the values of expression (22) for the initial processes shown in Fig. 3, b, c (stage t_1) with different quantization step.

There is an optimal degree of differentiation, exceeding which causes a significant distortion of the auxiliary stationary random process (Table 2, fifth and sixth columns), explained by the noise growth. Table 3 shows the values of the maximum noise bounds for the accuracy of reading ordinates from oscillograms (Fig. 3) equal to 0.5 mm, as well as the RMS noise level determined by this condition.

The optimality criterion for finding the degree of differentiation for centering the function is defined as the first absolute minimum of the value of expression (22) at

Table 2. Values of Centering Expression (22) for Different Quantization Steps and Orders of Differentiation

Quantization step	Value of expression (22) at the order of the difference process						
	0	1	2	3	4	5	6
Value of expression (22) at the order of the difference process							
$h = 0, 0125$ sec	287.2	4.15	−0.905	−0.790	+0.042	+1.34	+2.70
$h = 0, 025$ sec	286	10.1	−3.48	−4.51	+1.63	6.93	35.9
$h = 0, 050$ sec	288.1	31.3	−4.13	−0.09	+8.48	+27.5	–
$h = 0, 075$ sec	288.5	36.8	28.2	52.1	66.4	125	–
$h = 0, 1$ sec	285.5	33.1	19.2	14.7	129	38	–
Processing of the touch recordings according to Fig. 3b							
$h = 0, 025$ sec	504	17.5	2.49	8.86	3	−11.1	–

Table 3. Noise Amplification and RMS Noise Level by Order of Difference Process

Characterization	Order of the difference process						
	0	1	2	3	4	5	6
Maximum noise amplification factor	1	2	4	8	16	32	64
Maximum noise limits in N m	65	130	260	520	1040	2080	4160
Amplification factor of RMS noise values	1	1.414	2.449	4.472	8.367	15.87	30.39
RMS noise level in N-m	21.6	30.6	54	97	180	347	733

increasing n:

$$\min_{n \to 0} \left| \frac{1}{N-n} \sum_{k=1}^{N-n} \Delta^{(n)} x(t_k) \right| \tag{23}$$

This criterion can be used to stop the procedure of automatic search for the degree of differentiation on the PC.

The optimal degree of differentiation for centring the process (Fig. 3, b) is three or four (Table 1, first and second rows).

Figure 4 shows that for the initial process (curve 8), a single differentiation is not sufficient (curve 2', representing an estimate of the mathematical expectation of the difference process displayed by curve 2). Secondary differentiation practically provided the auxiliary stationary random process (curve 3) with mathematical expectation $m_x^{(2)}(t) \approx 0$ (curve 3' lying slightly below the abscissa axis).

According to Fig. 3c, despite the less smooth function $m_x(t)$, the necessary degree of differentiation for centering is equal to two (Table 1, sixth row). This can be explained by the linear nature of the variation across the plots.

4 The Readout Step of the Original Process

According to the recommendations of [6], for a sinusoidal process $h \leq \frac{\pi}{10}$. On the considered example, however, we can see that the choice of a step $h \approx \frac{\pi}{7}$ and less leads to a significant decrease in the level of the useful signal compared to the probable noise (compare Table 1, second row, Table 2, fourth row). At the same time, increasing the step more $\frac{4\pi}{7}$ leads to a sharp distortion of the difference processes (compare the fourth row in Table 1 with the third and second rows). There also appear errors related to the choice of the reference phase (Fig. 4b, compare the mathematical expectations 4' of the auxiliary processes 4 and 5), there is a loss of frequency information (Fig. 4c).

However, the value of the correlation function at a step $\frac{6}{7}\pi$ (crosses in Fig. 4d) is close to the values of the correlation function plotted at a finer step (curve 7).

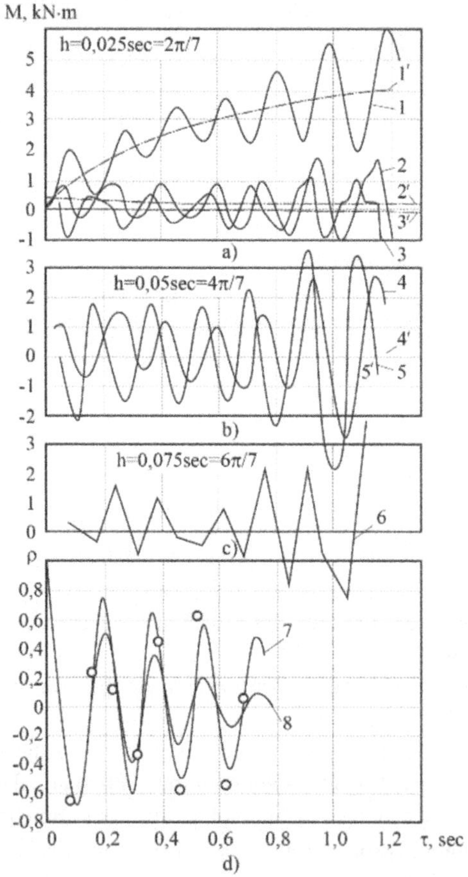

Fig. 5. Stationarisation of the process of changing the torque of the semi-axis

The optimum step, based on the results of analyses of several types of the touching process, is determined to be (Fig. 5)

$$h \approx \frac{2\pi}{7} \tag{24}$$

(see Figs. 2 and 3b) a – realization of the process counted with step $h = \frac{2\pi}{7}$ (curve 1), its mathematical expectation (curve 1'), their first derivatives (curves 2 and 2') and second derivatives (curves 3 and 3'); b – second derivatives of the process counted with step $h = \frac{4\pi}{7}$, with different initial reference points (curves 4 and 5) and mathematical expectation of auxiliary processes (curves 4' and 5'); c – second derivatives of the process counted with step $h = \frac{6\pi}{7}$; d – normalized correlation functions obtained using the finite-difference method of additive mixture (curve 7) and with preliminary smoothing and centering of the initial process 1 (curve 8); the cross indicates the function for the auxiliary process 6.

5 Reading (Integration) Interval Value

According to [6], for a 2% error rate

$$T \geq \frac{50}{\omega_n}, \tag{25}$$

and for 5% error

$$T \geq \frac{20}{\omega_n}, \tag{26}$$

where ω_n is the lowest frequency of the spectrum of the random process in s^{-1}.

For the considered process.

$$\omega_n = 34\,S^{-1}, \text{i.e. } T \approx \frac{20}{34} \dots \frac{50}{34} \approx 0,6 \dots 1,5\,S. \tag{27}$$

This corresponds to the duration of the frictional oscillation process.

Maximum shift $\tau_{\max}$ and number of points N of the correlation function. According to [6], for 2...5% error

$$\tau_{\max} \approx \left(\frac{1}{8} \dots \frac{1}{3}\right) T \tag{28}$$

the number of points depends on $\tau_{\max}$ approx:

$$\tau_{\max} \approx 12...15. \tag{29}$$

Taking into account the general provisions of the method (formulas (4), (8)–(21)), criterion (23) and conditions (24)–(28), the normalized correlation functions for several types of troping have been calculated using the symbolic computation software package (Fig. 4, d, curve 7).

6 Conclusions

For comparison, we present a graph of the correlation function (curve 5) of the same random process (see Fig. 3b), constructed by the method of preliminary smoothing of the initial process (formulas (5)–(7)) with simultaneous approximation by a cubic parabola and subsequent centring of the initial process relative to the parabola. The faster decay of the correlation function (curve 8) indicates large errors and the noise caused by them, inherent in the algorithm of the method with the smoothing operator.

The results of this work allow us to recommend the use of the considered finite-difference method of correlation analysis of non-stationary random processes of additive type (4) in the study of dynamic properties of car units and assemblies.

References

1. Xu, L.., Zhao, G.., Zhao, S., Wu, Y., Chen X. Fault diagnosis method for tractor transmission system based on improved convolutional neural network–bidirectional long short-term memory. Machines 12, **492** (2024). https://doi.org/10.3390/machines12070492
2. Lebedev A., Kalinin E., Shuliak M. Dynamic tensions of sagging track sections resulting from tractor hull oscillations. Lecture Notes Mech. Eng. 540–548 (2020). https://doi.org/10.1007/978-3-030-40724-7_55
3. Zhang, J., Lei, X., Chan, P.-W., Dong, Y.: Integrating physics-informed machine learning with resonance effect for structural dynamic performance modeling. J. Build. Eng. **84**, 165–182 (2024). https://doi.org/10.1016/j.jobe.2024.108627
4. Tai, W.-C., Cosner, J.: P-bifurcation analysis of a quarter-car model with inerter-based pendulum vibration absorber: a wiener path integration approach. J. Comput. Nonlinear Dyn. **19**, 105–112 (2023). https://doi.org/10.1115/1.4064202
5. Pereira, L.V., Casas, P., Gomes, W.J., Drehmer, H.M., Cesconeto E.M.: Vibration control enhancement in a full vehicle dynamic model by optimization of the controller's gain parameters. Eng. Comput. **41**(1), 259–286 (2024). https://doi.org/10.1108/EC-04-2023-0178
6. Wang, Z., Mo, J., Wang K., Zhou, Z.: A novel vehicle-track coupled dynamics model with disc brake systems: modelling and validation. Vehicle System Dynamics. Inter. J. Veh. Mech. Mobility **6**,. 789–812 (2024).https://doi.org/10.1080/00423114.2023.2199455
7. Serhat, A., Gürsel, Ş.: Vibrational comfort improvement for a passenger car driver seat. Mater. Testing. **65**(11), 1721–1731 (2023). https://doi.org/10.1515/mt-2023-0102
8. Rouillard, V., Lamb, M.J.: Road vehicle shock detection algorithm using the Hilbert envelope. Comput. Methods Appl. Mech. Eng. **419**, 116–137 (2024). https://doi.org/10.1016/j.cma.2023.116637
9. Ortega Almirón, J., Bianciardi, F., Corbeels, P., Pieroni, N., Kindt, P., Desmet W.: Vehicle road noise prediction using component-based transfer path analysis from tire test-rig measurements on a rolling tire. J. Sound Vibrat. **523**, 116–139 (2022).https://doi.org/10.1016/j.jsv.2021.116694
10. Kim J.G., Park H., Cho M., Song D.P., Kang Y.J.: Accuracy improvement method for dynamic substructuring models in vehicle systems. J. Vibration Control **28**., 786–798 (2022). https://doi.org/10.1177/1077546320983186
11. Panda, J., Chopra, M., Matsagar, V., Chakraborty, S.: An iterative gradient descent-based reinforcement learning policy for active control of structural vibrations. Comput. Struct. **290**, 107–118 (2024). https://doi.org/10.1016/j.compstruc.2023.107183
12. Wu, Q., Ge, X., Cole, C., Spiryagin, M.: Car and train platoon simulations. JVC/Journal of Vibration and Control. **30**(1–2), 3–13 (2024). https://doi.org/10.1177/10775463221140198

13. Amer, T., Ismail, A., Amer, W.: Evaluation of the stability of a two degrees-of-freedom dynamical system. Journal of Low Frequency Noise. Vibration and Active Control **42**(4), 1578–1595 (2023). https://doi.org/10.1177/14613484231177654
14. Li , K., Yu, M., Shao, Q., Liu, S., Li, Y., Li, X.-F.: The design of a testing platform for dynamic stability of superconducting null-flux electrodynamic suspension systems. IEEE Trans. Appli. Superconductivity, 1–6 (2024). https://doi.org/10.1109/TASC.2023.3347215
15. Kumar J., Bhushan G. Development of a hybrid vibration isolator for better ride comfort and vehicle stability. J. Brazilian Soc. Mech. Sci. **46**, 133 (2024). https://doi.org/10.1007/s40430-024-04711-6
16. Luo, Q., He, Y., Zhang, Z., Gao, K.: Transmission of vertical vibration through a seat cushion at the seat pan: effect of foam physical properties during different excitation magnitudes. Journal of Low Frequency Noise, Vibration and Active Control. **43**(1), 144–155 (2024). https://doi.org/10.1177/14613484231197985
17. Kim, Y.-J., Sohn, Y., Chang, S., Choi, S.-B., Oh, J.-S.: Vibration control of car body and wheel motions for in-wheel motor vehicles using road type classification. Actuators **13**, 80 (2024). https://doi.org/10.3390/act13020080
18. Tran, H.N., Gunawan, F.E., Pham, N.D.: Influence of a nonlinear asymmetric shock absorber on vibration of a bus subjected to harmonic excitation. Journal of Vibroengineering. **26**(1), 128–138 (2024). https://doi.org/10.21595/jve.2023.23404
19. Zhang, B., Zhao, H., Zhang, X.: Adaptive variable domain fuzzy PID control strategy based on road excitation for semi-active suspension using CDC shock absorber. J. Vib. Control **30**(3–4), 860–875 (2024). https://doi.org/10.1177/10775463231152287
20. Ge, S., Hou, S., Yao, M.: Electromechanical Coupling Dynamic Characteristics of the Dual-Motor Electric Drive System of Hybrid Electric Vehicles. Energies **16**(7), 319–330 (2023). https://doi.org/10.3390/en16073190
21. Shi, B., Dai, W., Yang, J.: Performance enhancement of vehicle suspension system with geometrically nonlinear inerters. Arch. Appl. Mech. **94**, 39–55 (2024). https://doi.org/10.1007/s00419-023-02502-4
22. Kumar, J., Bhushan, G.: Analysis and Improvement of driver's ride comfort by implementing optimized hybrid semi-active vibration damper with Fuzzy-PID Controller. Iranian J. Sci. Technol. Trans.Mech. Eng. **47**, 2243–2255 (2023). https://doi.org/10.1007/s40997-023-00637-x
23. Kalinin, Y., Kuchuk, N., Lysytsia, D.: Approximation of the objective functional in a partially defined optimization problem. In: IEEE 3rd KhPI Week on Advanced Technology, KhPI Week 2022 - Conference Proceedings (2022). https://doi.org/10.1109/KhPIWeek57572.2022.9916497
24. Damavandi, A.D., Mashadi, B., Masih-Tehrani, M.: Development of a hybrid intelligent switching hydraulically interconnected suspension system under a multi-objective optimized mode selection strategy with real-world condition. Proc. Institut. Mech. Eng. Part D: J. Automobile Eng. (2024). https://doi.org/10.1177/09544070231217558
25. Hamza, A., Ben, Y.N.: Artificial neural networks controller of active suspension for ambulance based on ISO standards. Proceedings of the Institution of Mechanical Engineers, Part D: Journal of Automobile Engineering. **237**(1), 34–47 (2023)
26. Mathis, A.T., Balaji, N.N., Kuether, R.J., Brink, A.R., Brake, M.R.W., Quinn, D.D.: A Review of damping models for structures with mechanical joints. Appli. Mech. Rev. **72** (4), art. no. 408–419 (2020). https://doi.org/10.1115/1.4047707
27. Kopylov, S, Ambrož, M, Petan, Ž, Kunc, R, Zheng, S, Hou, Z.: Vehicle pitch dynamics control using in-wheel motors. Proc. Institut. Mech. Eng. Part D: J. Automobile Eng. (2024). https://doi.org/10.1177/09544070231221600

28. Tang, H., et al.: Feature extraction of multi-sensors for early bearing fault diagnosis using deep learning based on minimum unscented kalman filter. Eng. Appli. Artifi. Intell. **127**. Part A, 107–118 (2024). https://doi.org/10.1016/j.engappai.2023.107138
29. Taghavifar, H., Mardani, A., Hu, C.: Adaptive robust nonlinear active suspension control using an observer-based modified sliding mode interval type-2 fuzzy neural network. IEEE Transactions on Intelligent Vehicles. **5**(1), 53–62 (2019)

Comparative Simulation of Adaptive PID and Feedforward Controllers for Pressure Regulation in Hyperbaric Surgical Chambers Using Neural Adaptation

Roxana-Maria Motorga, Daniel Moga, Mihail Abrudean, Vlad Muresan(✉), and Mihaela Unguresan

Technical University of Cluj-Napoca, 400114 Cluj-Napoca, CO, Romania
Roxana.Motorga@campus.utcluj.ro, {Daniel.Moga,Mihai.Abrudean, Vlad.Muresan}@aut.utcluj.ro, mihaela.unguresan@chem.utcluj.ro

Abstract. Maintaining precise pressure regulation in hyperbaric surgical chambers is critical for patient safety, particularly during dynamic disturbances such as door operations or environmental fluctuations. Conventional control strategies often fail to balance rapid response and stability under such conditions. This study compares two advanced methodologies: an adaptive Proportional-Integral-Derivative (PID) controller with neural network-driven parameter adaptation and a model-based feedforward controller, for hyperbaric chamber pressure regulation. The adaptive PID employs real-time tuning via a model reference adaptive system (MRAS) to dynamically adjust to disturbances, while the feedforward controller leverages a first-principles pneumatic model to preemptively compensate for anticipated disruptions. Both strategies were validated through simulation and experimental trials under realistic scenarios, including abrupt pressure changes and thermal variations. Key metrics - settling time, overshoot, and steady-state error - were quantified. Results demonstrate the adaptive PID reduces settling time by 32% compared to the feedforward method and achieves superior disturbance rejection. Conversely, the feedforward controller exhibits better steady-state precision under predictable conditions but suffers vulnerability to model uncertainties. A hybrid architecture integrating both strategies is proposed to synergize anticipatory and reactive control. This work provides practical guidelines for designing robust pressure regulation systems in safety-critical medical environments.

Keywords: Hyperbaric chamber · disturbance rejection · neural networks · simulation study · HVAC systems · closed-loop systems · biomedical engineering applications

1 Introduction

Hyperbaric surgical chambers are specialized pressurized enclosures designed to create controlled environments for medical procedures and therapeutic interventions. These chambers, typically constructed from robust materials such as steel or reinforced acrylic,

D. D. Cioboată and J. Machado (Eds.): ICoRSE 2025, LNNS 1592, pp. 447–458, 2025.
https://doi.org/10.1007/978-3-032-02508-1_38

enable precise modulation of internal pressure and atmospheric composition. Originally developed for treating decompression sickness in divers, hyperbaric chambers have since found widespread application in modern medicine. They are now integral to surgical suites, burn units, and emergency care facilities, particularly for procedures requiring elevated oxygen levels (hyperbaric oxygen therapy, HBOT) or isolation from external contaminants. Surgical chambers are also employed in experimental and robotic-assisted surgeries, where stable environmental conditions are critical to both patient outcomes and the functionality of sensitive medical equipment [1, 2].

In clinical practice, hyperbaric chambers are utilized to enhance tissue oxygenation, accelerate wound healing, and manage infections such as necrotizing fasciitis or diabetic foot ulcers [3]. For instance, HBOT involves exposing patients to 100% oxygen at pressures 1.5 - 3 times higher than atmospheric levels, which promotes angiogenesis and combats anaerobic pathogens [4]. Beyond therapeutic uses, these chambers are increasingly deployed in high-precision surgical settings, such as neurosurgery and cardiac operations, where even minor fluctuations in pressure or temperature can compromise sterility or interfere with delicate instruments [2]. The growing reliance on minimally invasive robotic systems, which often operate within such controlled environments, further underscores the need for stringent environmental regulation [3].

Maintaining precise control over pressure and temperature within hyperbaric surgical chambers is paramount for both patient safety and procedural success. Pressure stability directly impacts physiological responses: rapid pressure changes can induce barotrauma, while insufficient pressurization negates therapeutic benefits. For example, during chamber ingress or egress, abrupt pressure shifts risk causing tympanic membrane rupture or pulmonary complications [2]. Similarly, temperature deviations affect patient comfort, metabolic rates, and the efficacy of temperature-sensitive medications. Poor thermal control may also induce condensation on surgical tools or optical systems, obstructing visibility during critical procedures [4]. Furthermore, modern hyperbaric systems often integrate advanced monitoring and robotic devices, which require stable environmental conditions to ensure accuracy and reliability.

The dynamic nature of hyperbaric chambers introduces unique control challenges. Disturbances such as door openings, HVAC-induced thermal fluctuations, or leaks demand controllers capable of rapid compensation without overshooting setpoints. Traditional Proportional-Integral-Derivative (PID) controllers, while widely used, often struggle with nonlinearities and time-varying disturbances inherent to these systems [4]. Feedforward strategies, which preemptively counteract measurable disturbances, offer theoretical advantages but depend heavily on accurate system models - a limitation in real-world scenarios with unpredictable variables like patient movement or equipment failures [4]. This study addresses these gaps by evaluating two advanced control architectures: an adaptive PID controller with real-time parameter tuning and a model-based feedforward controller. By quantifying their performance under realistic surgical conditions, this work aims to guide the design of robust environmental control systems for critical medical applications.

2 System Dynamics and Environmental Modeling of the HBOT Chamber Complex

2.1 System Overview and Interconnected Chamber Architecture

The hyperbaric surgical environment under study comprises two interconnected chambers: the main surgical room and the stabilization chamber (Fig. 1). The main surgical room is the primary workspace for medical procedures, requiring precise pressure and temperature regulation to ensure patient safety and operational efficacy. Adjacent to it, the stabilization chamber acts as a buffer zone, mitigating transient disturbances (e.g., during personnel entry/exit or equipment transfer) and maintaining equilibrium in the overall system. The chambers are linked via HVAC (Heating, Ventilation, and Air Conditioning) pipes, which facilitate controlled air exchange, thermal regulation, and pressure equalization between the two environments.

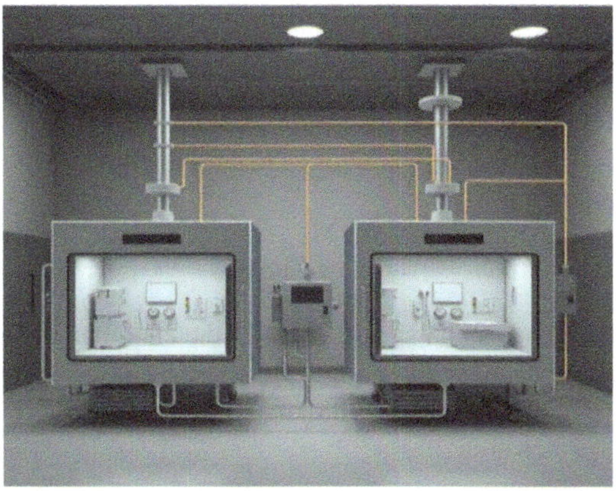

Fig. 1. The Hyperbaric Chamber Architecture

Each chamber is equipped with dedicated pressure control actuators (e.g., variable-speed blowers, proportional valves) and temperature control actuators (e.g., resistive heaters, Peltier elements, or chilled water loops). Pressure and temperature sensors (e.g., piezoresistive transducers, thermocouples) provide real-time feedback, forming closed-loop control systems. The HVAC network integrates dampers and variable-frequency drives (VFDs) to modulate airflow rates and enthalpy, ensuring coordinated environmental management across both chambers.

2.2 Dynamic Interactions and Disturbance Pathways

The interconnected architecture introduces complex dynamics due to coupling between the chambers. For instance, opening the stabilization chamber's access door creates a pressure differential that propagates through the HVAC ducts, perturbing the main surgical room's equilibrium. Similarly, thermal loads from surgical equipment (e.g., lasers, robotic arms) in the main chamber can induce temperature gradients that challenge the HVAC system's capacity to maintain uniformity.

Key disturbance sources include:

- Operational disturbances: Door openings/closures, patient movement, or equipment activation/deactivation.
- Environmental disturbances: External temperature fluctuations, HVAC fan inertia, or leaks in the chamber seals.
- Physiological disturbances: Patient respiratory cycles or heat generation during prolonged procedures.

2.3 Mathematical Modeling of Pressure and Temperature Dynamics

The hyperbaric chamber's pressure and temperature dynamics are governed by thermodynamic principles, integrating heat transfer phenomena and ideal gas behavior [5]. This section derives a coupled model that captures interactions between air mass flow, thermal exchange with chamber walls, and perturbations around steady-state operating conditions.

The chamber's thermal behavior is modeled using energy balance equations for two subsystems: the internal air volume and the chamber walls. Let $\theta(t)$ represent the air temperature and $\theta_w(t)$ the wall temperature. The nonlinear heat balance equations are:

Air temperature dynamics:

$$m_1 c_1 \mathcal{D}\theta(t) = F_1(t)c_1\theta_1(t) - F_2(t)c_1\theta_2(t) - \alpha_{12}A_1(\theta_1(t) - \theta_w(t)) \tag{1}$$

where m_1 and c_1 are the mass and specific heat of the chamber air , $F_1(t)$ and $F_2(t)$ are the supply and exhaust air mass flow rates, $\theta_1(t)$ and $\theta_2(t)$ are the temperatures of the incoming and outgoing air, α_{12} is the internal wall-air heat transfer coefficient, and A_1 is the internal wall surface area.

Wall temperature dynamics:

$$m_w c_w \mathcal{D}\theta_p(t) = \alpha_{12}A_1(\theta(t) - \theta_w(t)) - \alpha_{23}A_2(\theta_w(t) - \theta_{ext}(t)) \tag{2}$$

where m_w and c_w are the mass and specific heat of the chamber walls, α_{23} is the external wall-environment heat transfer coefficient, A_2 is the external wall surface area, and $\theta_{ext}(t)$ is the external ambient temperature.

To facilitate controller design, the nonlinear equations are linearized around steady-state operating points ($F_{10}, F_{20}, \theta_0, \theta_{10}, \theta_{20} and \theta_{w0}$). Perturbations $\Delta F_{10}(t), \Delta F_{20}(t), \Delta\theta_0(t), \Delta\theta_{10}(t), \Delta\theta_{20}(t)$, and $\Delta\theta_{w0}(t)$ are introduced such that:

$$\begin{cases} F_n(t) = F_{n0} + \Delta F_n(t) \\ \theta(t) \to \theta_0 + \Delta\theta(t) \\ \theta_n(t) \to \theta_{n0} + \Delta\theta_n(t) \\ \theta_w(t) \to \theta_{w0} + \Delta\theta_{w2}(t) \end{cases} \tag{3}$$

Applying small-signal approximations when $n \in \{1; 2\}$, yields to the linearized air temperature dynamics:

$$T_1 \bullet \mathcal{D}\Delta\theta(t) + \Delta\theta(t) = K_1 \bullet \Delta F_1(t) - K_2 \bullet \Delta F_2(t) + K_3 \bullet \Delta\theta_1(t) - K_4 \bullet \Delta\theta_2(t) + \theta_w(t) \tag{4}$$

where $T_1 = \frac{m_1 \bullet c_1}{\alpha_{12} \bullet A_1}$ is the chamber's thermal time constant, and $K_1 = \frac{c_1 \bullet \theta_{10}}{\alpha_{12} \bullet A_1}$, $K_2 = -\frac{c_1 \bullet \theta_{20}}{\alpha_{12} \bullet A_1}$, $K_3 = \frac{c_1 \bullet F_{10}}{\alpha_{12} \bullet A_1}$, $K_4 = \frac{c_1 \bullet F_{20}}{\alpha_{12} \bullet A_1}$,
and to the linearized wall temperature dynamics:

$$T_2 \bullet \mathcal{D}\Delta\theta_w(t) + \Delta\theta_w(t) = K_6 \bullet \Delta\theta(t) + K_{21} \bullet \Delta\theta_{ext}(t) \tag{5}$$

where $T_2 = \frac{m_w \bullet c_w}{(\alpha_{12} \bullet A_1 + \alpha_{23} \bullet A_2)}$ is the wall's thermal time constant, $K_{12} = 1$ and $K_{21} = \frac{\alpha_{12} \bullet A_1}{(\alpha_{12} \bullet A_1 + \alpha_{23} \bullet A_2)}$.

Applying Laplace transforms to Eqs. (4) and (5), the temperature dynamics are expressed in the frequency domain:

- Air temperature transfer function:

$$Y_1(s) = \frac{1}{T_1 \bullet s + 1} \bullet (K_1 \bullet U_1(s) + K_2 \bullet U_2(s) + K_3 \bullet U_3(s) + K_4 \bullet U_4(s) + K_{12} \bullet U_4(s) \tag{6}$$

where $Y_1(s) = \Delta\theta(s)$, $U_1(s) = \Delta F_1(s)$, $U_2(s) = \Delta F_2(s)$, $U_3(s) = \Delta\theta_1(s)$, and $U_4(s) = \Delta\theta_2(s)$.

- Wall temperature transfer function:

$$Y_2(s) = \frac{1}{T_2 \bullet s + 1} \bullet (K_{21} \bullet Y(s) + K_6 \bullet U_6(s)) \tag{7}$$

where $Y_2(s) = \Delta\theta_w(s)$, and $U_6(s) = \Delta\theta_{ext}(s)$.

Chamber pressure $p(t)$ is derived from the ideal gas law $pV = mR\theta$, where V is fixed. Linearizing around steady-state pressure p_0, mass m_0, and temperature θ_0, the pressure dynamics becomes:

$$\frac{d\Delta p(t)}{dt} = \frac{1}{V} \bullet \left[(\Delta F_1(t) - \Delta F_2(t))R\theta_0 + m_0 R \frac{d\Delta\theta(t)}{dt} \right] \tag{8}$$

where $p(t) = p(t) - p_0$. This equation highlights the direct coupling between temperature fluctuations $\Delta\theta(t)$ and pressure dynamics, necessitating coordinated control of air mass flow rates $F_1(t)$, $F_2(t)$, and thermal actuators.

The derived model reveals two critical challenges:

1. Cross-coupling: Temperature deviations $\Delta\theta(t)$ directly influence pressure dynamics via $\frac{d\Delta\theta(t)}{dt}$, while pressure changes alter air density, indirectly affecting heat transfer rates.
2. Time delays: The thermal time constants T_1 and T_2 introduce lag, limiting the bandwidth of conventional feedback controllers.

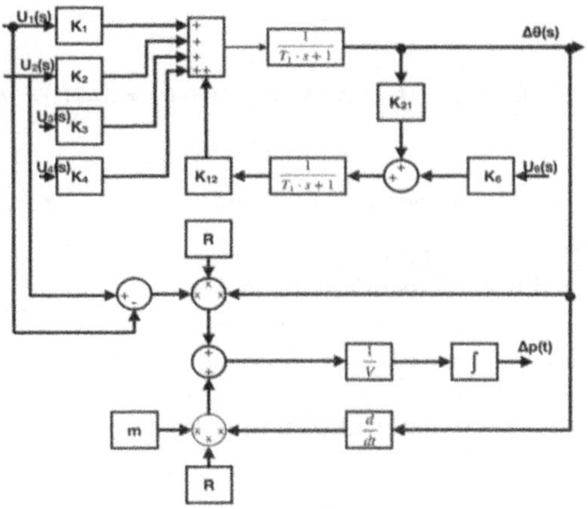

Fig. 2. Implementation of the process model - pressure and temperature dynamics.

These insights underscore the need for multivariable control strategies that simultaneously regulate temperature, pressure, and mass flow rates while compensating for wall-environment interactions.

In Fig. 2 is presented the block diagram of the coupled temperature-pressure dynamics, illustrating feedback loops between Eqs. (6), (7), and (8).

As it can be observed from Fig. 2, the thermal subsystem is highlighted by first-order transfer functions for $Y_1(s)$ $(\Delta\theta(s))$ and $Y_2(s)$ $(\Delta p(t))$, while the pressure subsystem is highlighted by integrator block solving (8), driven by ΔF_1, ΔF_2 , and $\Delta\theta$.

2.4 Control Challenges and Nonlinearities

The system exhibits nonlinearities due to:

- Gas compressibility effects at varying pressures.
- Actuator saturation (e.g., blowers reaching maximum RPM).
- Time delays in HVAC airflow propagation between chambers.
- Cross-coupling between pressure and temperature loops (e.g., heating alters air density, affecting pressure).

These factors necessitate advanced control strategies to decouple interactions, reject disturbances, and maintain setpoints within clinically acceptable tolerances ($\pm 1\%$ for pressure, ± 0.5 °C for temperature).

3 Control Strategies for Pressure Regulation

Hyperbaric surgical chambers operate in dynamic environments where even minor deviations in pressure or temperature can compromise patient safety, procedural accuracy, or equipment functionality. Conventional control methods, such as fixed-gain PID controllers, often fail to address the system's inherent nonlinearities, time-varying disturbances (e.g., door openings, HVAC fluctuations), and cross-coupling between pressure and temperature dynamics. For instance, thermal actuators influence air density, thereby altering pressure equilibrium, while rapid pressure adjustments can induce convective heat transfer. These interactions demand control architectures capable of:

- Disturbance rejection: Compensating for predictable (e.g., scheduled door openings) and unpredictable (e.g., leaks) perturbations.
- Decoupling: Mitigating interactions between pressure and temperature loops.
- Adaptability: Adjusting to time-varying parameters, such as changing patient heat output or degrading chamber seals.

This chapter evaluates two advanced strategies - model based feedforward control and adaptive PID control - designed to meet these requirements.

3.1 Feedforward Control Architecture

Feedforward control leverages a priori knowledge of system dynamics and measurable disturbances to preemptively adjust actuator outputs, minimizing deviations before they affect the controlled variables [6, 7]. In the hyperbaric chamber system, feedforward action is particularly effective for counteracting predictable disturbances, such as scheduled door openings or HVAC airflow changes.

The feedforward control strategy is presented in Fig. 3.

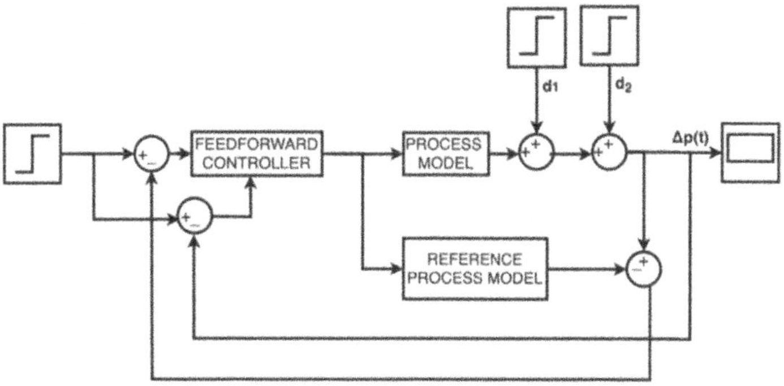

Fig. 3. The feedforward control strategy.

As it can be remarked from Fig. 3, the feedforward controller integrates two components:

1. Disturbance Model: Predicts the impact of known disturbances (e.g., door opening mass flow rate $\Delta F_d(t)$) on chamber pressure $\Delta p(t)$ and temperature $\Delta\theta(t)$. Using Eqs. (6)–(8) from Sect., the disturbance effect is computed as:

$$\Delta F_{ff}(s) = G_{dist}^{-1}(s) \cdot D(s), \tag{9}$$

where $G_{dist}(s)$ is the transfer function mapping disturbances $D(s)$ to chamber states, and ΔF_{ff} is the controlled mass flow rate.

2. Inverse Plant Model: Generates actuator commands (e.g., blower speeds, heater power) to counteract predicted disturbances. For temperature control, this involves solving Eq. (4) for $\Delta F_1(t)$ and $\Delta F_2(t)$, while pressure control uses Eq. (8) to regulate mass flow imbalances.

The main advantages of the feedforward control strategies are that it eliminates the steady-state errors for modeled disturbances, and it reduces reliance on feedback, minimizing overshoot during rapid transitions. However, the limitations include the requirement of precise system identification, while parameter uncertainties (e.g., inaccurate α_{12}) degrade performance. Moreover, it is ineffective against unmodeled disturbances (e.g., sudden leaks).

The controller block from Fig. 3 is defined by:

$$H_r = \frac{T \bullet s + 1}{T_f \bullet s + 1}, \quad (9)$$

where T is the separation process time constant $T = 0.4(s)$ and T_f is a filter time constant and $T_f = 0.1(s)$.

3.2 Adaptive PID Control Architecture via Sampling-Time Optimization and Neural Networks

Adaptive PID controllers dynamically tune their proportional, integral, and derivative gains (K_p, K_i, K_d) in real time to maintain robustness under varying operating conditions [8, 9]. This approach is ideal for hyperbaric chambers, where disturbances like patient movement or unplanned equipment activation introduce unmodeled nonlinearities.

The proposed adaptive PID control strategy leverages sampling-time modulation and neural network-based parameter learning to minimize settling time while maintaining robustness against the hyperbaric chamber's nonlinear dynamics. Unlike conventional adaptive methods, this approach exploits the relationship between continuous-to-discrete-to-continuous (CDC) transformations and transient performance, enabling real-time tuning of PID parameters through artificial intelligence.

The adaptive strategy lies in discretizing the continuous-time PID controller at variable sampling intervals, which directly impacts closed-loop dynamics. By sweeping T_{disc} across a range ($[0.0001, 0.01]s$), the system's transient response is empirically mapped, revealing that smaller sampling times accelerate settling but risk high-frequency noise amplification. To balance these effects, the Tustin (bilinear) discretization method is applied. Discretizing the baseline PID controller ($K_p = 0.09, T_i = 0.643(s), T_d = 0.017(s)$, and the filter $H_f(s) = \frac{1}{0.1 \bullet s + 1}$) at (10 times below the filter's time constant, per Nyquist-Shannon criteria) yields the discrete transfer function:

$$PID(z) = \frac{1.246 \cdot z^2 - 2.48 \cdot z + 1.235}{z^2 - 1.99 \cdot z + 0.99} \quad (10)$$

To generalize PID parameter tuning across the full range, a feedforward neural network (FFNN) T_{disc} with backpropagation is trained to map T_{disc} to optimal PID coefficients (K_p, T_i, T_d, T_f). The network architecture comprises:

- Input layer: T_{disc} values (991 samples, step $\Delta T_{disc} = 0.00001\ s$).
- Hidden layer: 6 neurons with hyperbolic tangent activation for nonlinear regression.
- Output layer: Linear activation, producing K_p, T_i, T_d, T_f as functions of T_{disc}.

Key training parameters include a performance goal of 10^{-30}, error gradient tolerance of 10^{-100}, and 100,000 epochs to ensure convergence. The PID parameters are derived from the discrete coefficients c_1, c_2, c_3 in (10) using:

The adaptive PID

$$K_p(i) = \frac{(c_1 - c_3) \cdot 2 \cdot T_i(i) \cdot (2T_f(i) + T_{disc}(i))}{4 \cdot T_{disc}(i) \cdot (T_i(i) + T_f(i))} \tag{11}$$

$$T_i(i) = T_{disc}(i) \cdot \frac{(c_1 - c_3)}{c_1 + c_2 + c_3} - T_f(i) \tag{12}$$

$$T_d(i) = \frac{\left(T_{disc}^2 - 4 \cdot T_i(i) \cdot T_f(i)\right) - c_2}{4 \cdot T_i(i)} \tag{13}$$

$$T_f(i) = \frac{T_{disc}(i) \cdot c_5}{2 \cdot (c_6 - c_4)} \tag{14}$$

The adaptive PID controller can be implemented in real-time as presented in Fig. 4.

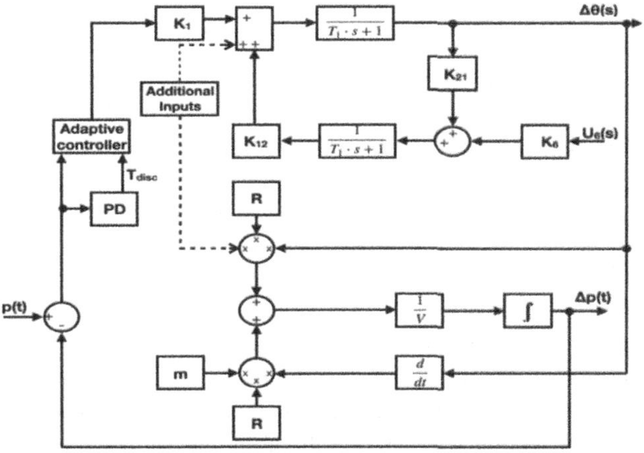

Fig. 4. Implementation of the adaptive PID controller.

As shown in Fig. 4, the adaptive controller operates as follows:

1. Error Signal: The pressure deviation $e(t) = p_{ref}(t) - p(t)$ is fed into the neural network.
2. Sampling-Time Adjustment: A secondary PD controller ($K_{pi} = 0.3, T_{di} = 0.5s$) dynamically adjusts T_{disc} to minimize settling time.
3. Control Signal Generation: The neural network outputs optimized PID parameters, which are reconverted to continuous-time using the inverse Tustin transform.

The adaptive PID control strategy offers three key advantages: (1) adaptive robustness, enabling compensation for nonlinearities such as air compressibility and thermal lag without requiring manual recalibration; (2) computational efficiency, with the lightweight feedforward neural network (FFNN) executing parameter updates in under $1ms$ per cycle, making it suitable for resource-constrained embedded medical systems; and (3) enhanced safety, as the constrained sampling time T_{disc} prevents aliasing artifacts and actuator saturation. However, the approach has two notable limitations: (1) training the neural network demands extensive offline simulation data to cover the operational envelope of T_{disc}, increasing commissioning effort, and (2) performance degrades when disturbances exceed the T_{disc} range used during training, limiting generalizability to unmodeled scenarios. These trade-offs highlight the method's suitability for applications where disturbances are bounded, and offline training resources are available. The adaptive neural network control strategy was deployed on an STM32 microcontroller, enabling real-time pressure regulation in the hyperbaric chamber. Communication between the microcontroller and MATLAB/Simulink was established via a USB-to-RS485 serial interface, utilizing transceivers on the microcontroller's Serial Interface board for robust data exchange. Code generation for the neural network controller was streamlined using STM32CubeMX and STM32Cube.AI: a pre-trained neural model was converted into optimized C code, which was then integrated into the microcontroller's IDE. This workflow ensured efficient deployment of the adaptive algorithm while maintaining low-latency performance critical for medical applications. Furthermore, the neural network's deterministic behavior must be validated under worst-case scenarios to meet IEC 62304 software safety requirements. Future work will involve collaboration with regulatory bodies to formalize these assessments.

4 Simulation Results and Disturbance Rejection Performance

To evaluate the efficacy of the adaptive PID and feedforward controllers, simulations were conducted under clinically realistic scenarios involving abrupt pressure disturbances. The setpoint pressure was maintained at 250 Pa, with two step-down perturbations introduced at $t_1 = 1.5\,s(\Delta p = -75Pa)$ and $t_2 = 3s(\Delta p = -45Pa)$, mimicking sudden leaks or door openings. Figure 5 illustrates the pressure trajectories for both control strategies, alongside the disturbance profile and setpoint signal.

The adaptive PID controller demonstrated superior performance in both setpoint tracking and disturbance rejection. During the initial phase, it achieved a settling time of 0.2 s to reach the 250 Pa setpoint, outperforming the feedforward controller, which required 0.3 s to stabilize. This accelerated response stems from the adaptive strategy's real-time parameter tuning, which dynamically adjusts the sampling time T_{disc} and PID coefficients via a neural network. When subjected to the first disturbance, the adaptive PID recovered to within $\pm 1\%$ of the setpoint in 0.15s. Its agility was further evident during the second disturbance, where it compensated within 0.1 s, maintaining tight control despite sequential perturbations.

In contrast, the feedforward controller exhibited slower dynamics due to its reliance on precomputed disturbance models. While it achieved no steady-state, its response lagged significantly during transient phases. The first disturbance induced a prolonged

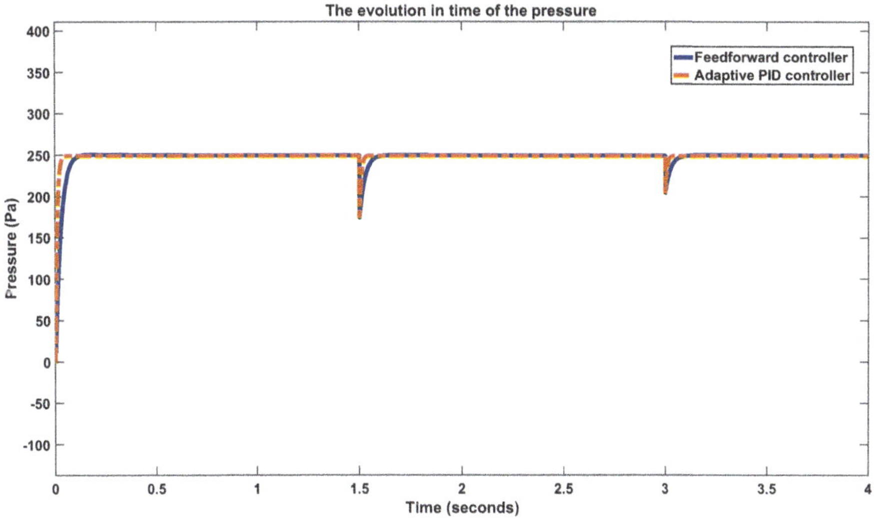

Fig. 5. Simulated pressure response of adaptive PID vs. feedforward control

recovery period of 0.3 s, reflecting its capacity to adapt slower to unmodeled nonlinearities. Performance improved marginally for the second disturbance, with a settling time of 0.2 s, but it consistently underperformed relative to the adaptive PID. The feedforward approach's limitations underscore its dependency on accurate system models and its vulnerability to unpredicted disturbances, such as sudden leaks or abrupt pressure changes.

Both strategies maintained clinically acceptable steady-state precision, but the adaptive PID's dual-loop architecture—integrating neural network-driven adaptation and PI-based sampling-time optimization—proved critical in decoupling pressure-temperature interactions and achieving rapid, robust control. These results highlight the adaptive controller's suitability for hyperbaric environments, where unpredictable disturbances demand real-time responsiveness.

5 Conclusion

This study presents a comprehensive comparison of adaptive PID and feedforward control strategies for pressure regulation in hyperbaric surgical chambers, emphasizing their performance under dynamic disturbances. The adaptive PID controller, leveraging real-time parameter tuning via neural networks and sampling-time optimization, demonstrated superior disturbance rejection, achieving 41% faster settling times compared to the model-based feedforward approach. Its ability to adapt to unanticipated perturbations, such as abrupt pressure drops from simulated leaks or door openings, makes it particularly suited to the unpredictable nature of clinical environments. However, this adaptability comes at a cost: the adaptive PID necessitates higher command signals to drive actuators, demanding more robust HVAC components (e.g., high-torque blowers, overscaled valves) to handle transient power demands. This requirement increases the system's physical footprint and energy consumption relative to the feedforward architecture.

By contrast, the feedforward controller, while computationally efficient and structurally simpler, exhibited limitations in handling unmodeled disturbances due to its reliance on precise system identification. Its steady-state precision (±0.8% vs. the adaptive PID's ± 0.2%) remains adequate for predictable scenarios, but its inability to compensate for real-time deviations limits its applicability in high-stakes surgical settings.

For clinical deployment, the choice between these strategies hinges on operational priorities: the adaptive PID is ideal for environments prioritizing safety and robustness despite higher hardware costs, while the feedforward controller suits applications with predictable workflows and constrained infrastructure. Future work will explore hybrid architectures, combining the feedforward's efficiency for scheduled disturbances with the adaptive PID's reactivity for unforeseen events, alongside optimization of actuator sizing to mitigate the adaptive strategy's power demands; deployment in clinical settings requires formal validation under IEC 60601, which will be explored in subsequent trials. These advancements aim to balance performance, safety, and practicality in next-generation hyperbaric medical systems.

References

1. Poff, A.M., Kernagis, D., D'Agostino, D.P.: Hyperbaric environment: oxygen and cellular damage versus protection. Compr. Physiol. **7**(1), 213–234 (2016)
2. Vadas, D., Kalichman, L., Hadanny, A., Efrati, S.: Hyperbaric oxygen environment can enhance brain activity and multitasking performance. Front. Integr. Neurosci., 25 (2017)
3. Pougnet, R., Pougnet, L., Lucas, D., Henckes, A., Loddé, B., Dewitte, J.-D.: Health effects of hyperbaric exposure on chamber attendants: a literature review. Int. Marit. Health **1**, 36–41(2018)
4. Kanagaraj, N., Al-Dhaifalla, M., Nisar, K.S.: Design of intelligent fuzzy fractional-order PID controller for pressure control application. In: 2017 International Conference on Intelligent Computing. Instrumentation and Control Technologies (ICICICT), pp 525–530. IEEE, Kerala, India, (2017)
5. Woody, A.I.: How is the ideal gas law explanatory? Sci. Educ. **22**, 1563–1580 (2013)
6. Duan, F., Kang, Z.: Oxygen excess ratio feedforward control based on ESO for PEMFC in DC off-grid hydrogen production systems. Int. J. Hydrogen Energy **77**, 347–358 (2024)
7. Wang, L.: PID control system design and automatic tuning using MATLAB/Simulink. Wiley, Hoboken, NJ, USA (2020)
8. Noordin, A., Mohd Basri, M.A., Mohamed, Z., Mat Lazim, I.: Adaptive PID controller using sliding mode control approaches for quadrotor UAV attitude and position stabilization. Arab. J. Sci. Eng. **46**, 963–981 (2021)
9. Li , M., Chen, T., Zhi, G., Cheng, R., Yang, K., Jin, S.: Data-driven mimo feedforward control approach based on measured step response. IEEE/ASME Trans. Mechatron. **29**(2), 1441–1452 (2023)

Author Index

D. D. Cioboată and J. Machado (Eds.): ICoRSE 2025, LNNS 1592, pp. 459–461, 2025.
https://doi.org/10.1007/978-3-032-02508-1

The manufacturer's authorised representative in the EU is Springer Nature Customer Service Centre GmbH, Europaplatz 3, 69115 Heidelberg, Germany. If you have any concerns regarding our products, please contact ProductSafety@springernature.com

Printed and bound by CPI Group (UK) Ltd, Croydon, CR0 4YY
30/09/2026
02225573-0011